Erlangen Earth Conference Series

Series Editor: André Freiwald

Max Wisshak · Leif Tapanila (Eds.)

Current Developments in Bioerosion

Springer

Editors

Dr. Max Wisshak
Universität Erlangen-Nürnberg
Inst. Paläontologie
Loewenichstr. 28
91054 Erlangen
Germany
wisshak@pal.uni-erlangen.de

Dr. Leif Tapanila
Idaho State University
Dept. Geosciences
921 S. 8th Ave. Stop 8072
Pocatello ID 83209-8072
USA
tapaleif@isu.edu

ISBN: 978-3-642-09633-4 e-ISBN: 978-3-540-77598-0

DOI 10.1007/978-3-540-77598-0

Cover design: WMXDesign GmbH, Heidelberg

Printed on acid-free paper

9 8 7 6 5 4 3 2 1

springer.com

Foreword

A. Conrad Neumann[1]

[1] Marine Sciences Department, University of North Carolina, Chapel Hill, NC 27599-3300, USA, (aneumann@email.unc.edu)

Bioerosion: one man's beginning

Just as organisms attack solid substrates, the subject of bioerosion itself cuts across many disciplines. There has been a long reign of biology and geology being practiced together as 'Natural History'. Humboldt, Darwin, John Murray of Challenger Expedition fame, and Alexander Agassiz are only a few of the early workers that did geology and biology as if they were merely different pigments on the same palette. Anyone studying carbonates, the 'rocks that lived', cannot avoid dealing with the intermarriage of biology, geology, and chemistry. Such it is with bioerosion. In Bermuda in the mid-60's when I began work on the biological erosion of limestone coasts, I had no idea I was pioneering. Like many young researchers working alone, I feared that somewhere all this might have been done before. Of course, my fears were well founded. There is nothing new under the sun – nor anything new under the sea for that matter. Every new idea has an old idea to thank. For example, as early as 1883, Nasonov presented beautiful illustrations of *Cliona* borings and the resultant 'ice cream scoop' chips. There is a host of organisms long recognized that erode solid substrates. In marine and fresh water, in carbonates and silicates, in hardrock and soft, biological forces are doing geologic work and writing a geologic history of erosion that goes back to the Archean for microborers and the earliest Phanerozoic for macroborers (Taylor and Wilson 2003). My road to bioerosion was a winding one. In 1959 I was in the South Atlantic on the research vessel 'Atlantis I' of the Woods Hole Oceanographic Institution doing a hydrographic survey of water masses for the International Geophysical Year. I applied to graduate school via radiogram while still at sea. It so transpired that I was to go to Lehigh University to study carbonates with Keith Chave. I was sent to Bermuda to do my Ph.D. on 'something to do with environments of carbonate deposition'. On the advice of Heinz Lowenstam, I ended up in Harrington Sound.

Bermuda and bioerosion

Harrington Sound is a 3.2 km wide, 30 m deep interior lagoon with a tidal range of only 10 cm. It is floored by carbonate sediments in three zones: sand and seagrass to 10 m, an *Oculina* coral zone from 10 to 18 m, and a barren, muddy, dysoxic zone below a sharp seasonal thermocline at 18 m (Neumann 1965). The coastal cliffs of carbonate eolianite are deeply undercut by as much as 4 m. V-shaped notches begin abruptly at lowest low tide level. Sampling underwater by hammer and chisel with my head constantly hitting the often flat roof of the notch wasn't easy, but every sample of the carbonate rock below low tide level told the same story of attack by biological agents. If bioerosion is ever appointed a 'type locality' it ought to be Harrington Sound, Bermuda, with its scenic cliffs undercut and often collapsed due to sub-tidal bioerosion (Fig. 1A). Recognizing the biological action of erosion was unavoidable. In that first summer in Bermuda I saw its effect everywhere. Putting a word to it by sticking 'bio-' onto 'erosion' is a no-brainer. I am not a pioneer in bioerosion simply because I named it. I merely put a term to what had been obvious to many for a long time. What I did manage to do was to survey the undercuts and notches in Bermuda and conduct some crude rate experiments on one common eroder, *Cliona lampa* (Neumann 1966). The bright vermillion surface expression of this sponge should make it a poster child of modern bioerosion. The process of bioerosion is not confined to carbonate substrates as I may have implied. Other workers have shown that non-carbonate substrates can be subject to erosion by biological agents as well. Nor is it confined to marine conditions. Aside from borers, bioerosive surface reduction can also be achieved by scraping the surface (bioabrasion) by organisms that graze on biofilms, or by the pitting of the surface by chemical attack (biocorrosion).

Fig. 1 A The collapse of an eolianite cliff in Harrington Sound, Bermuda, as a result of subtidal bioerosion. **B** Photograph within a notch in Harrington Sound, Bermuda, taken at lowest low tide. The dark discoloration of the notch surface is from a thick (unbrowsed) mat of cyanobacteria

There were some minor adventures in those early days in Harrington Sound. One day as I was snorkeling in the Sound trying to get a handle on the research I might do there. I noticed at one locality near the cliff the water had become cloudy from small white flocculent masses suspended in the water column. I had heard the arguments that fine calcium carbonate could be precipitated spontaneously from saturated seawater while others maintained that a biological intermediary was needed. Here I was witnessing the answer! I swam closer to this white suspended matter, noticed a wavy 'schlieren' effect in the water, and at about that same time I saw the black circular outline of the sewer pipe I was about to swim into. My 'flocculent masses' were toilet paper! This was an ignominious beginning to my field work in recent carbonates (Ever since, I have been suspicious of 'whitings', but let's not stir up that subject). In the 1960's the notch terminus had often revealed bare, recently browsed patches. In some places radial scratches of echinoids were clearly visible

Fig. 2 Drawing of a notched undercut in Harrington Sound, Bermuda and photo of a sawed section of intertidal rock showing the terminus of a notch and also bioerosion below the notch

on the smooth surface of encrusting coralline algae. These observations suggest that the notch itself is more a feature of bioabrasion by browsers, whereas the overall undercutting below the low tide level is the combined effect of many types of borers and excavators. The sawed cross section of an exposed notch revealed in a collapse block (Fig. 2) shows the truncated bore holes of the date mussel, *Lithophaga*, as further indication of bioabrasion in the notch. The once common *Lytechinus* is the suspected notch former. Although seldom seen actively browsing the notch during the day, it is suspected that they work at night (easy to check if it had dawned on me at the time!) I have since witnessed this evening migration of urchins on a concrete wall of a Bahaman marina. In Harrington Sound in the 90's, a thick, felt-like, purple mat coated the notches (Fig. 1B) and the *Lytechinus* urchins so common in the seagrass were not to be seen. It seems the toilets have taken their toll.

Bioerosion deep and dark

Since Bermuda, my studies have taken different twists and turns, but bioerosion has followed me and has showed up in some interesting and diverse situations. One such place is in deep water off Little Bahama Bank in the Straits of Florida, seen from the submersible Alvin. There, at subphotic depths of around 600 m, we saw and studied elongate mounds that we called 'lithoherms' because of their reef-like appearance and their composition of crudely stratified, cemented, carbonate peri-platform ooze. Those under construction reveal an accreting upcurrent thicket of the branching coral *Lophelia*. Most are undergoing active bioerosion, however, which is fed by strong northward bottom currents of the Gulf Stream. Shown here is a view of the irregular terrain on the down-current (leeward) end of a mound where mainly sponge bioerosion appears to be most intense (Fig. 3A). Thin sections show

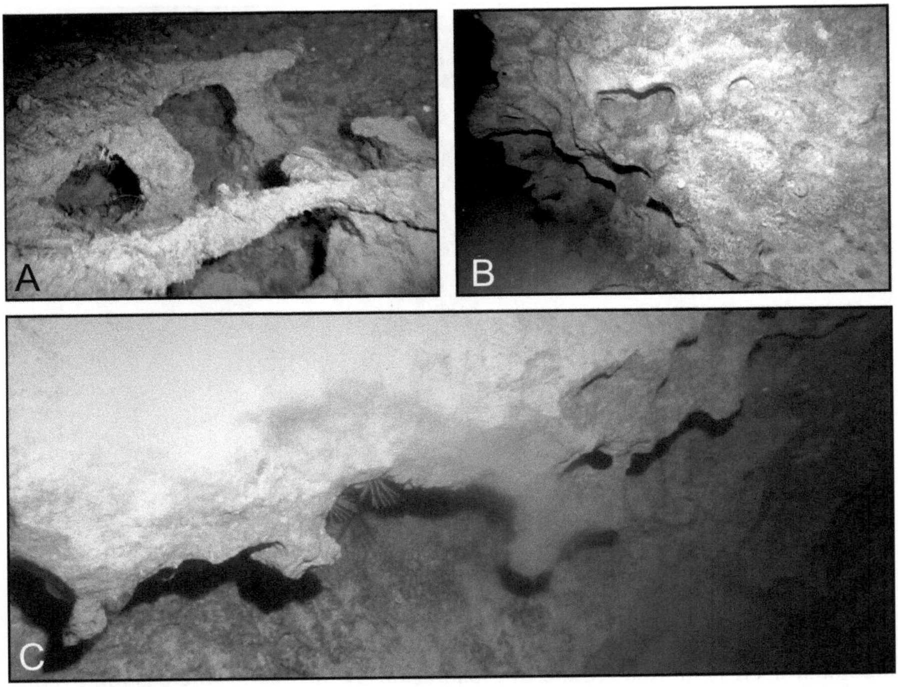

Fig. 3 A Irregular topography of the bioeroded, down-current (northward) end of a lithoherm in the Straits of Florida. The rock is a well-cemented periplatform calcareous ooze. Depth 614 m; Alvin dive 1334; photo C. Pilskaln. **B** Bioeroded crescentic embayments at the base of a steep-sided irregular high in the Northwest Providence Channel, Bahamas. Light colored patches within the embayments are where the exposed yellow patches of boring clionaid sponges can be seen. The embayments are in the order of 20-50 cm across. Depth ~350 m; Alvin dive 844. **C** Overhanging cap ('eaves') of cemented periplatform ooze near the top of a steep-sided irregular high in Northwest Providence Channel, Bahamas. The cap is a steeply sloping mound of pelagic drape that becomes cemented just below the surface. Echinoids and crinoids are observed in the space provided by the overhang. Depth 286 m; Alvin dive 844

a multiple sequence of boring, infilling and cementation (Neumann et al. 1977; Wilber and Neumann 1993). It is no secret that bioeroded hardgrounds can be evidence of ancient bottom currents. Here, aided by accretion of pelagic carbonate and early cementation, mounds resembling reefs form and subsequently undergo attack by vigorous bioerosion. Around the corner in Northwest Providence Channel at depths as great as 1000 m, more work of bioerosion can be seen on the steep escarpments of the margins, canyon walls and at the bases of irregular, isolated build-ups. The steep to vertical walls exhibit distinctive 'scooped out' features, or crescentic embayments, that are visually estimated to be up to 20-50 cm across and 10 cm deep. They are indented more deeply at the top of the embayed area, thus providing a shelter for borers within from the rain of sediment from above. The yellow-orange tissue of clionaid boring sponges can be seen in these sheltered nooks. The bioerosion direction of 'in and up' maintains the crescentic embayments and must result in wall steepening and retreat (Fig. 3B). At the top of the wall in the area of the irregular 50-100 m isolated highs, bioerosion and early cementation create a feature similar to the overhanging eaves of a Swiss chalet (Fig. 3C). From high resolution seismic data, the crests of the isolated highs are seen to be steep domes mantled with a thinly bedded cap of pelagic drape. The sediment on the domed crests becomes progressively cemented down-slope. Where the vertical wall meets the down-slope edge of the cemented drape, the overhang occurs. Organisms can be seen inhabiting the sheltered space under the eaves. Seismic evidence suggests that these steep-sided domes nucleate on pre-existing erosional margin highs and that pelagic drape and early cementation amplify the original relief. Bioerosion appears to give them their steep-sided relief. These features are being dubbed 'drape mounds' and await more sampling in order to warrant publication. They are of interest because, like lithoherms, they represent reef-like mounding of carbonates by accretion and cementation without the role of framework organisms. These observations support bioerosion as a significant process in the formation and maintenance of steep scarps in deep carbonate terrains, including the steepening of pelagic mounds and the headward cutting of canyons. Since those early clumsily wonderful years in Bermuda and the Bahamas, the field of bioerosion in the modern and ancient has spread wide and deep into many new avenues of research that have resulted in many new successes, as this compendium volume attests.

References

Nasanov NK (1883) Zur Biologie and Anatomie der Clione. Z Wiss Zool 39:295-308

Neumann AC (1965) Processes of recent carbonate sedimentation in Harrington Sound, Bermuda. Bull Mar Sci 15:987-1035

Neumann AC (1966) Observations on coastal erosion in Bermuda and measurements of the boring rate of the sponge, *Cliona lampa*. Limnol Oceanogr 11:92-108

Neumann AC, Kofoed JW, Keller GH (1977) Lithoherms in the Straits of Florida. Geology 5:4-10

Taylor PD, Wilson MA (2003) Palaeoecology and evolution of marine hard substrate communities. Earth-Sci Rev 62:1-103

Wilber RJ, Neumann AC (1993) Effects of submarine cementation on microfabrics and physical properties of carbonate slope deposits, Northern Bahamas. In: Rezak R, Lavoie DL (eds) Carbonate Microfabrics. Springer, New York, pp 79-94

Editors preface

A little more than forty years has past since the concept of bioerosion was formally recognized as the biological erosion of hard materials. In that time, it has become apparent from the literature that bioerosional processes affect a wide range of biological and geological systems that cross many disciplines among the sciences. With each discipline comes a different viewpoint on the role and history of bioerosion, and so this book is dedicated to unifying these perspectives and crossing traditional disciplinary boundaries.

The seeds of the present volume were planted at the 5th International Bioerosion Workshop, convened by one of the editors (M.W.) in autumn 2006 in the historic Franconian city of Erlangen (Germany). At this meeting, biologists, palaeontologists and geologists from close to a dozen countries and nationalities, gathered to share their knowledge and participate in fruitful discussions. The diverse spectrum of topics on the agenda ranged from the Early Palaeozoic to the Recent, from tiny microendolithic bacteria to giant macroborers, from field observations to sophisticated CT analyses, and from the Tropics to the Arctic. Originally anticipated as a classical proceedings volume, the majority of the articles compiled here are authored by participants of the workshop, but we have also solicited contributions to widen the range of subfields covered in this volume by presenting novel topics not traditionally associated with the bioerosion research community, such as the article filed by Nicola McLoughlin et al. on microbioerosion in volcanic glass and the contribution on the microbial degradation of bone by Miranda Jans. Other solicited articles provide reviews on classical fields of bioerosion, including Pat Hutchings' contribution on boring polychaetes and Aline Tribollet's article on the role of boring microbiota in coral reefs; new research on the taphonomy of drilling predation is presented by Patricia Kelley. The book is opened by a foreword kindly provided by Conrad Neumann – who was a key player in establishing this field of science by first defining the term 'bioerosion' in the mid 1960s, and is rounded off by an introduction to the widely appreciated online bibliography on marine bioerosion maintained by Mark Wilson. We are indebted to all contributors and are pleased to forward a diverse set of two dozen articles, ranging from specific case studies to extensive reviews, on many aspects of the growing discipline of bioerosion research. Thereby, several new bio- and ichnotaxa are introduced to the literature:

Aka akis sp. n. Bromley and Schönberg
Aurimorpha varia igen. n., isp. n. Wisshak, Seuß and Nützel
Entobia micra isp. n. Wisshak
Entobia nana isp. n. Wisshak
Kardopomorphos polydioryx igen. n., isp. n. Beuck, López Correa and Freiwald
Pyrodendrina cupra igen. n., isp. n. Tapanila
Rhopalia clavigera isp. n. Golubic and Radtke
Trypanites mobilis isp. n. Neumann, Wisshak and Bromley

This volume profited from the peer-review process, during which an international and interdisciplinary board of referees contributed valuable comments and suggestions. On behalf of the editors and authors, their commitment and expertise is herein gratefully acknowledged:

Markus Bertling (Münster, DE)
Lydia Beuck (Erlangen, DE)
Arnaud Botquelen (Le Conquet, FR)
Carlton E. Brett (Cincinnati, US)
Richard G. Bromley (Copenhagen, DK)
Luis Buatois (Saskatoon, CA)
Barbara Calcinai (Ancona, IT)
Gerhard C. Cadée (Texel, NL)
José L. Carballo (Mazatlán, MX)
Elizabeth Chacón (Linares, MX)
Tomas Cedhagen (Aarhus, DK)
Marcos Gektidis (Frankfurt, DE)
Jorge F. Genise (Trelew, AR)
Jordi M. de Gibert (Barcelona, ES)
Ingrid Glaub (Frankfurt, DE)
Stjepko Golubic (Boston, US)
Stephen T. Hasiotis (Lawrence, US)
Leanne J. Hepburn (Colchester, UK)
Markes E. Johnson (Williamstown, US)
Simon R. A. Kelly (Cambridge, UK)
William E. Kiene (Pago Pago, US)
Karl Kleemann (Vienna, Austria)
Lindsey R. Leighton (San Diago, US)
David I. MacKinnon (Christchurch, NZ)
Daniel Martin (Blanes, ES)
Radek Mikuláš (Prague, CZ)
Axel Munnecke (Erlangen, DE)
Claus Nielsen (Copenhagen, DK)
Shirley A. Pomponi (Fort Pierce, US)
Gudrun Radtke (Wiesbaden, DE)
Joachim Reitner (Göttingen, DE)
Michael J. Risk (Ontario, CA)
Eric M. Roberts (Johannesburg, ZA)
Klaus Rützler (Washington, US)
Michael Savarese (Ft. Myers, US)
Charles E. Savrda (Auburn, US)
Felix Schlagintweit (Munich, DE)
Gerhard Schmiedl (Hamburg, DE)
Christine H. L. Schönberg (Oldenburg, DE)
Joachim Schönfeld (Kiel, DE)
Henning Scholz (Berlin, DE)
Paul Tafforeau (Grenoble, FR)
Ole S. Tendal (Copenhagen, DK)
Aline Tribollet (Kaneohe, US)
Alfred Uchman (Kraków, PO)
Geerat J. Vermeij (Davis, US)
Klaus P. Vogel (Frankfurt, DE)
Eva Walpersdorf (Aberdeen, UK)
Mark A. Wilson (Wooster, US)

Last but certainly not least, we are indebted to Sonja-B. Löffler (Erlangen, Germany) for her commitment in verifying and formatting the vast amount of references as well as for proof reading manuscripts, and André Freiwald (Erlangen, Germany), chief editor of this book series, for his valuable conceptual suggestions and proof comments. Sibylle Noè (Erlangen, Germany) kindly provided the cover inset image, a ~100 µm wide SEM image of an etched hardround (Porcupine Seabight, off Ireland, 740 m water depth) featuring a sponge chip which is bioeroded by a microboring fungus – 'bioerosion cubed'. We are very pleased to have had the opportunity to realise this project with Springer-Publishing as a helpful and readily competent partner and we specifically like to express our sincere thanks to Christian Witschel (head of geoscience branch) and his colleague Christine Adolph.

Max Wisshak & Leif Tapanila

Erlangen, October 2007

Contents

I

Evolution and classification

The endolithic guild: an ecological framework for residential cavities in hard substrates

Leif Tapanila[1]

[1] Department of Geosciences, Idaho State University, Pocatello 83209, Idaho, USA, (tapaleif@isu.edu)

Abstract. Endolithic organisms that live inside rocks, shells and wood use two processes, bioerosion and embedment, to produce boring and bioclaustration residential cavities. In the trace fossil record, borings and bioclaustrations preserve the distinctive and often stereotypic activity of particular endolithic organisms, thereby linking important biological and ecological information to this special group of trace fossils. In order to use these fossils to study aspects of evolutionary paleoecology, an ecologically-significant framework is needed to logically group and subdivide hard substrate trace fossils.

A modified version of the guild concept can be applied to borings and bioclaustrations, which groups these cavities and their producers on the basis of biological affinity (or behavior), trophic level and habitat. In general, the endolithic guild is unified as a group of mostly soft-bodied organisms that produce shallow infaunal, stationary domichnia in hard substrates. A major subdivision of the endolithic guild roughly corresponds to the commonly used size-based divide between microborers and macroborers, but it is founded on the ecologically-significant criterion of trophic group. The autotrophic endolithic guild includes the photosynthetic microboring cyanobacteria, chlorophytes and rhodophytes, and the heterotrophic endolithic guild is dominated by macroscopic invertebrate borers and embedders, and includes bacteria, fungi and other microscopic fauna.

Primary observations on the endolithic guild suggest that (1) morphological diversity of cavities generally increased through geologic time for all endoliths, with a major Ordovician rise noted for invertebrate endoliths; (2) the class-level composition of invertebrates in the suspension feeding endolithic guild remain nearly unchanged since their rise in the Early–Middle Paleozoic until today, despite a major shift in dominant groups at the familial level that occurred in the Early Mesozoic; (3) diversity patterns within the heterotrophic endolithic guild show similar timing of onset between macroborers and embedders in the Ordovician, but differences in substrate availability may provide an ecological explanation for the decline of host-specific bioclaustrations in the Devonian relative to macroboring diversity.

Keywords. Bioerosion, classification, guild concept, ecology, paleoecology, evolution, bioclaustrations

M. Wisshak, L. Tapanila (eds.), *Current Developments in Bioerosion*. Erlangen Earth Conference Series, doi: 10.1007/978-3-540-77598-0_1,

Introduction

The need for organisms to obtain a stable residence is recorded by trace fossils in the rock record. Although burrows produced in soft sediment are important indicators of animal dwelling behaviors, the legacy of organisms living in hard substrates preserves some of the oldest evidence of life and behavior on the planet – from the prokaryotic bioalteration of volcanic glass since the Archean (Furnes et al. 2004, McLoughlin et al. this volume), to the microscopic borings of photosynthetic eukaryotes since the Mesoproterozoic (Zhang and Golubic 1987), to the large-scale borings of animals ushered in by the Cambrian Explosion (James et al. 1977). Acquiring living space within a hard substrate such as rock, bone, shell or wood is an adaptive strategy that has evolved repeatedly and independently in myriad lineages, including cyanobacteria, red and green algae, fungi, and several phyla of the Animalia (Bromley 1992). At a minimum, the endolithic lifestyle gives the selective advantage of living in a hard domicile secure from physical and biological stresses, and it widens the range of available habitats by permitting the endolith to occupy unrealized ecospace. Given these presumed advantages and a fossil record demonstrating its repeated success, understanding the tempo and mode of endolithic strategies may point to broader patterns in the role of competition and predation in the evolution of life.

Two different strategies – destructive bioerosion and constructive embedment – can be employed by an endolith to actively acquire living space in a hard substrate (Tapanila and Ekdale 2007). Whereas bioeroding organisms grind or etch into a lithic substrate to produce a boring cavity, embedding organisms settle upon the growing shell of a host organism and locally inhibit accretion of the host shell to produce a bioclaustration cavity (Bromley 1970; Palmer and Wilson 1988). Borings are identified by observing truncation of the hard material, which usually produces a cavity with a well defined, sharp contact with the substrate (Bromley 1970). Bioclaustration cavities also have a sharp contact, but because they are formed through the embedment process, the host skeleton shows a pronounced deflection immediately next to the cavity: these fossils therefore preserve direct evidence of symbiosis (Tapanila 2005).

Trace fossil cavities generally have a much greater preservation potential than the endolithic body fossil, thus borings and bioclaustrations provide the most reliable geologic record for endolithic activity. Note that other types of hard substrate fossils, such as bioimmurations and etchings, are not endolithic domiciles and so they will not be considered further in this paper: the former are passively overgrown body fossils and the latter are superficial attachment structures for epizoans (Voigt 1979; Taylor 1990; Gibert et al. 2004). Although some have argued that bioimmurations may be indistinguishable from bioclaustrations (Bertling et al. 2006), an increasing body of literature documents dozens of fossil and modern examples unequivocally demonstrating that an active inhibiting behavior (impedichnia) by the symbiont produces a distinct suite of features not observed in bioimmuration structures (Tapanila 2005; Tapanila and Ekdale 2007 and references therein).

Over a century of ichnotaxonomic research has revealed first-order patterns in the diversification of boring and bioclaustration trace fossils through time (Tapanila

2005; Wilson and Palmer 2006), but if these trace fossil data are to be used collectively with other fossil data to understand broader evolutionary patterns, e.g., the influence of faunal diversification, competition, predation, and mass extinction on endolith evolution, then a conceptual framework is needed to place these trace fossils in ecological significance. This paper will argue that a modified version of the guild concept is an ideal solution for categorizing these trace fossils and their tracemakers on the basis of ecological factors. After characterizing endolithic trace fossils as ecological entities, their potential utility in evolutionary paleoecology will be discussed in the context of their Phanerozoic record.

Endolithic fossils for evolutionary paleoecology

Documenting residences in hard materials

Evaluating the progress of endoliths through time requires stratigraphic range information for their trace fossils. Effort during the past two decades to analyze bioerosion and embedment structures in the Paleozoic record has helped to recognize the first occurrences of many endoliths once thought to be restricted to post-Paleozoic strata. For example, bivalve borings, which have a stereotypic shape of a flask (*Gastrochaenolites* Leymerie, 1842) or elongate groove (*Petroxestes* Wilson and Palmer, 1988) are now recognized from the Ordovician, Silurian and Pennsylvanian (Wilson and Palmer 1988, 1998; Ekdale et al. 2002; Tapanila and Copper 2002). Sponge borings, which form interconnected chambers and galleries (*Entobia* Bronn, 1837), are common in the Devonian, but absent again until the Late Permian (Weidlich 1996; Tapanila 2006a). Detailed analyses of Paleozoic substrates have demonstrated that many cyanobacterial and algal borings occur in assemblages very similar to those found in modern oceans (Podhalańska and Nõlvak 1995; Vogel et al. 1995; Glaub and Bundschuh 1997; Wisshak et al. this volume). Recent synoptic studies of bioclaustrations (Tapanila 2005) find that these cavities have recurrent morphologies often linked to clade-specific hosts, although identification of the symbiont is usually impossible with the exception of lingulid brachiopods (Tapanila and Holmer 2006). Together, these papers have provided a critical mass of information that will allow new analyses to examine patterns in the endolith record (Perry and Bertling 2000; Taylor and Wilson 2003; Bromley 2004; Glaub and Vogel 2004; Tapanila 2005; Glaub et al. 2007; Tapanila and Ekdale 2007; Wilson 2007; Wisshak et al. this volume).

The unity of endolithic residences

Trace fossils have a successful track record in identifying significant ecological patterns in the evolution of life, including the precipitous rise of infaunalization strategies during the Ordovician Radiation by soft-sediment burrowers and the reciprocal evolution of predator-prey systems recorded by *Oichnus* Bromley, 1981, drillholes in molluscan prey shells (Thayer 1983; Vermeij 1987; Droser and Bottjer 1989; Kelley and Hansen 1993). Residential borings and bioclaustrations (domichnia) share several important taphonomic and biologic features that make them particularly attractive subjects for evolutionary paleoecology.

Most obvious is that hard substrate trace fossils occur in materials that have a very high preservation potential in the geologic record. Except where bioerosion intensity is severe (e.g., clionaid penetrations), most hard materials will survive in the rock record, more so than the body of a typical endolith. As trace fossils in a substrate, these endolithic cavities inherently are in situ structures and therefore give unambiguous information on the relative timing and original location of their emplacement. A final important consideration of taphonomy is the quality of trace fossil preservation. Because a hard substrate is resistant to ductile deformation, morphological features of borings and bioclaustrations often retain a high degree of resolution (e.g., submicron scale for microborings), whereas the detailed preservation of burrows often is compromised by compaction or the coarseness of the substrate grain. This high degree of detail preserved by hard substrate trace fossils permits a greater level of ichnotaxonomic resolution and can help to crosslink distinctive behaviors observed in modern biotaxa with ancient trace fossil ichnotaxa.

A strong link between trace fossil and producer helps in assessing the biological and ecological role of endoliths. Trace fossils produced in hard materials may preserve a variety of behaviors, but they are dominated in number and diversity by residential cavities (Gibert et al. 2004). Domichnia in hard substrates typically are good indicators of their tracemaker because borings and bioclaustrations are morphologically stereotypic. Conservatism in cavity form issues from the series of distinct anatomical, physiological and behavioral innovations that evolved among endoliths to allow borers to mechanically and chemically penetrate hard materials, and embedders to inhibit a host's growth and initiate symbiosis. The metabolic effort required to bore or inhibit overgrowth promotes borings and bioclaustrations that have relatively simple shapes, often reflecting the cross-section or entire body shape of the endolith.

Classifying hard substrate trace fossils

Research on borings and bioclaustrations has been ongoing for more than 150 years (e.g., Bronn 1837; Wedl 1859; Duncan 1876; Clarke 1908, 1921), with much of the research effort devoted to basic systematic description and some application to biological and geological problems (see Tapanila 2005; Schönberg and Tapanila 2006). Influential papers by Neumann (1966), Golubic (1969), Bromley (1970), Warme (1977) and Warme et al. (1978) recognized the importance of these cavities in marine ecosystems and focused on identifying bioerosion and embedment as processes. Currently, only a few conceptual frameworks have been developed for the classification of hard substrate trace fossils, aside from the Linnaean-inspired ichnotaxonomical approach (see Bertling et al. 2006 and Bertling 2007 for recent reviews). A brief summary of frequently used classifications specific to hard substrates follows.

Terminology

Different approaches have been suggested in naming hard substrate fossils to make explicit the nature of the organism / substrate relationship. Golubic et al. (1981) introduced prefixes to the term endolith to distinguish borers (euendoliths) from organisms that occupy preexisting cavities. Bromley (1970) added the term paraendoliths for organisms that embed into skeletons to form bioclaustrations. Taylor and Wilson (2002) produced the most thorough treatment on the terminology of borers and embedders using the Folk (1959) approach of two- or three-part names, with the prefixes denoting the organism and its relationship to the substrate, identified by the base-word. For example, a symbiont that forms a bioclaustration in a coral is called an endozoozoan, and a sponge that bores into a disarticulated bivalve shell an endoskeletozoan. This scheme was not intended to connote any specific ecological information, but it does afford great descriptive precision compared to other terminologies.

A more simplistic terminology gained wide usage beginning in the 1960's and 1970's when scanning electron microscopy permitted detailed study on borings by microbial algae and fungi (e.g., Golubic et al. 1975). Borings were subdivided by size as microborings and macroborings. This informal subdivision, arbitrarily set at 100 μm by most authors (e.g., Glaub and Vogel 2004), has an obvious practical quality where microborings must be visualized under microscope and macroborings can be seen with the unaided eye. No equivalent size-based terminology has been developed for bioclaustrations. The informal size classification of hard substrate trace fossils is not designed to imply any particular significance beyond a descriptive term, but as discussed later, it does carry some unintended ecological significance.

Behavior

Seilacherian ethologic categories offer a useful approach for categorizing trace fossils based on the purpose of their formation (Seilacher 1953, 1964; Bromley 1996). In a review of bioerosion trace fossils, Gibert et al. (2004) noticed a lower diversity of behaviors recorded in hard substrates compared to soft substrates, including domichnia, pascichnia, equilibrichnia, praedichnia, and fixichnia. Since their study, two additional ethologic categories are recognized in hard substrates: impedichnia, a special type of domichnia produced by bioclaustration-making animals (Tapanila 2005), and calichnia, for the recently discovered larval borings in dinosaur bones (Roberts et al. 2007). Examination of the evolutionary development of particular ethologies could be interesting; however, using only ethologic categories might be too broad and the results might be difficult to interpret.

Depositional environment

The ichnofacies concept, originally conceived by Seilacher (1953, 1964), identifies recurrent trace fossil assemblages and ethologic categories that typify particular depositional environments (see MacEachern et al. 2007). Ichnofacies defined by the presence of hard substrates were later added to describe environments with hardgrounds and rocky coastlines (*Trypanites*, *Gnathichnus* and *Entobia* ichnofacies: Frey and Seilacher 1980; Bromley and Asgaard 1993a) and woodgrounds (*Teredolites* ichnofacies: Bromley et al. 1984). An examination of ichnofacies through time can give insight on how marine depositional environments and their endolithic communities have changed during the Phanerozoic (e.g., Palmer 1982; Wilson and Palmer 1992; Gibert et al. 1998; Johnson and Baarli 1999; Taylor and Wilson 2003). At their best, ichnofacies analysis can benefit from community-level comparisons through time that are restricted to specific depositional environments (e.g., the *Trypanites* ichnofacies in hardgrounds), but other substrate-based ichnofacies (e.g., *Teredolites* ichnofacies) have highly variable environments of deposition. A further drawback of restricting one's scope to ichnofacies is that it may overlook endolithic innovations occurring away from the ichnofacies being studied, e.g., a study on *Trypanites* ichnofacies in hardgrounds overlooks the endolithic communities that colonize reefs, biostromes and bioclasts on the muddy subtidal seafloor (Tapanila et al. 2004a).

The guild concept

The guild is an ecological concept to categorize organisms that exploit their environment and resources in the same manner. Originally defined by Root (1967) for modern ecosystems, the guild concept was modified slightly by Bambach (1983) for applications in the fossil record. Bambachian guilds are defined by three unifying ecological criteria: the bauplan (as defined by class-level taxonomy), food source (trophic group), and the specific living habitat (space utilization). Examples of Bambachian guilds include, 'spired grazing gastropods' or 'erect, epibyssate suspension feeding bivalves' (Bambach 1983). Using this concept for fossil biotaxa, Bambach related the Phanerozoic increase in species richness to a simultaneous increase in guilds through time. This increased utilization of ecospace was permitted by the evolution of more ecologically plastic taxa, and it can be recognized in the progression of Sepkoski's (1981) three major evolutionary faunas, i.e., the Cambrian, Paleozoic and Modern (see Fig. 1).

Trace fossil data are absent from Bambach's (1983) analyses, as he examined only the body fossil record dominated by skeletonized marine animals. Bridging this gap, Bromley (1990, 1996) extended the guild concept to soft sediment trace fossils, coining the term 'ichnoguilds' to distinguish groupings of trace fossils from Bambach's organism guilds. Ichnoguilds were developed primarily as a tool to analyze the complexity of an ichnocoenosis (Ekdale and Bromley 1991; Bromley and Asgaard 1993b) by grouping trace fossils on the basis of ethology (= bauplan), food source, and level of tiering (= space utilization). Examples of Bromley's ichnoguilds include the 'vagile, shallow-tier deposit feeding *Planolites* ichnoguild' and the 'stationary, deep-tier suspension-feeder *Skolithos-Ophiomorpha* ichnoguild.' Using this concept, Buatois et al. (1998) were first to use ichnoguilds in evolutionary paleoecology by documenting the historical exploitation of continental ecospace. With some minor modifications, the guild concept is fully applicable to endolithic communities, not just their trace fossils, and this approach may serve as an ideal ecological framework to study their evolutionary history.

Fig. 1 Integration of body and trace fossil data to assess adaptive strategies employed by marine faunas through the Phanerozoic. **A** Summary of Bambach's (1983: fig. 3) assessment of adaptive strategies based only on body fossil data of diverse organisms classes. Three digit code denotes number of classes employing each adaptive strategy during the Cambrian (1st number), Middle / Upper Paleozoic (2nd) and Mesozoic / Cenozoic (3rd). **B** Assessment of the 'suspension feeding, shallow, passive infaunal' adaptive strategy based on endolithic trace fossils and the inferred biological affinities (Vermeij 1987; Bromley 1992; Glaub and Vogel 2004; Tapanila 2005). When including classes of macroboring (1 6 7) and embedding taxa (0 3 3) with Bambach's body fossil data (0 2 3), the combined number of different classes

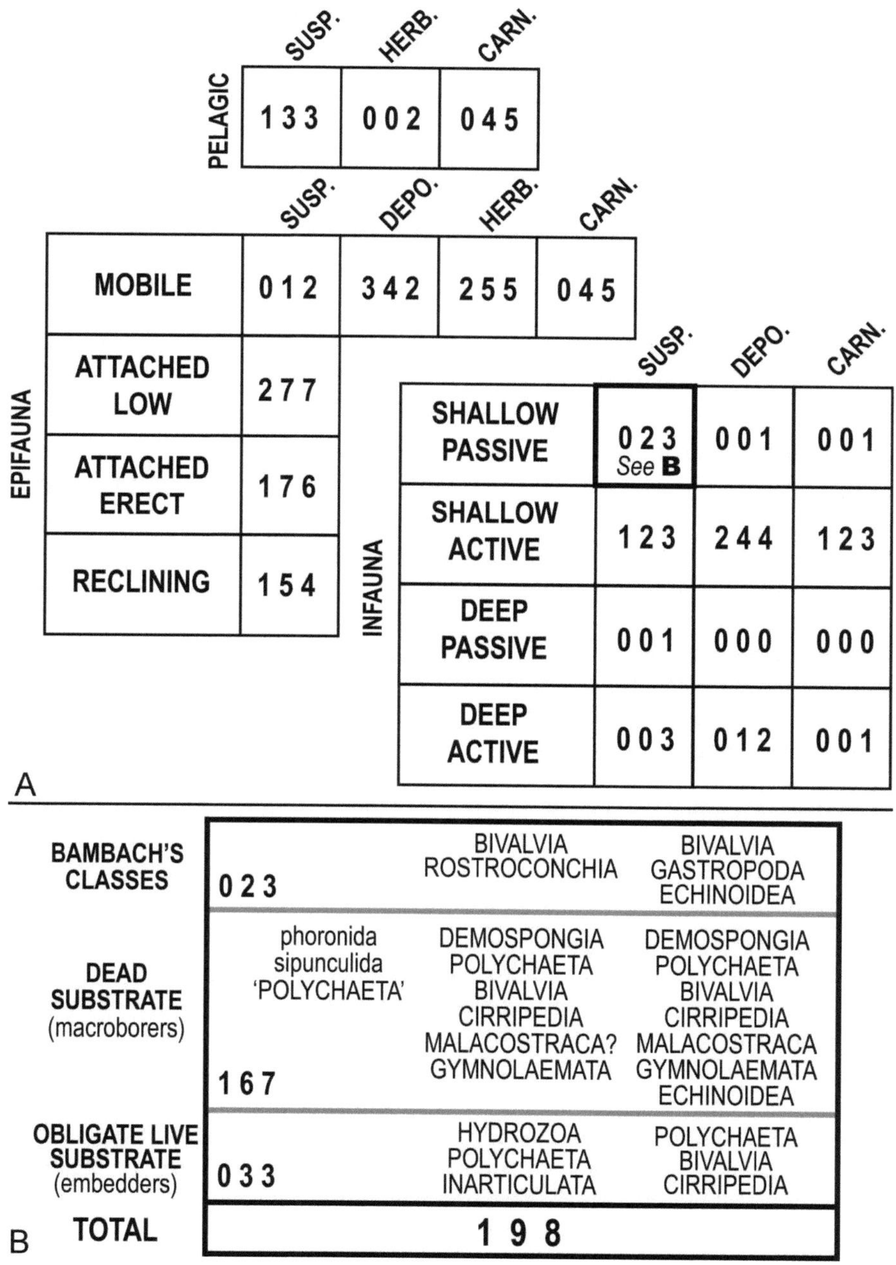

employing the suspension feeding, shallow, passive infaunal adaptive strategy is (1 9 8). Macroboring taxa listed in three columns corresponds, from left to right, to the Cambrian, Middle / Upper Paleozoic, and Mesozoic / Cenozoic intervals. Taxa listed in lower-case type were not included in the tabulation as their body fossil record was not diverse enough to be used in Bambach's class counts. Malacostraca? refers to possible producer of *Sanctum* Erickson and Bouchard, 2003. Susp = Suspension; Depo = Deposit; Herb = Herbivore; Carn = Carnivore

Table 1 Comparison of criteria used in subdividing endoliths into guilds based on Bambach (1983), Bromley (1996) and this study. See text for elaboration

Guild style	Bambach 1983	Bromley 1996	This paper
Name	Bambachian Guild	Bromleyan Ichnoguild	Endolithic Guild
Bauplan	• Linnaean Class: Demospongia	• Ethology: domichnia • Vagility: stationary	• Affinity: sponge • Ethology: domichnia • Vagility: stationary
Food source	Suspension feeding	Suspension feeding	Heterotrophic, suspension feeding
Habitat	Shallow, passive infauna	Shallow tier	Shallow tier endolithic boring
Example: *Entobia* (ichno)guild	Shallow infaunal suspension feeding demosponge	Stationary, shallow tier suspension-feeder domichnia	Shallow tier suspension feeding boring sponge

The endolithic guild

Guild analysis can be used as a means to unify biological and ichnological data within an ecological framework. For this reason I have opted to remove the 'ichno-' prefix in my discussion of the endolithic guild despite Bromley's (1996) caution that his trace fossil ichnoguilds differ significantly from Bambachian organism guilds. The overarching purpose for the guild concept, whether using body or trace fossils, is to identify a biotaxon or ichnotaxon as an ecological entity so that it maybe compared and contrasted to others. The criteria used to define the identity of the ecological entity are tantamount regardless of the type of fossil, and so Bambachian guilds and Bromleyan ichnoguilds should be comparable: there is no need for duplicity if the purpose is to unify fossils as ecological (not phylogenetic) entities. The three-part definition of a guild can equally describe endolithic trace fossils as it can define endolithic organisms.

Bauplan

The bauplan (structural bodyplan) describes the intrinsic qualities of the organism that allows it to function in its environment, from its physiology and growth to its activity and reproductive strategy. Bambachian guilds make an explicit phylogenetic definition for the bauplan using class-level systematics, contrary to the intentions of Root (1967). Bambach (1983: 728) justified his phylogenetic groupings on the basis that baupläne among classes are significantly different and that classes are easily recognized in the fossil record. While this justification has merit, it ignores the possibility of linking polyphyletic groups within the same bauplan, and therefore Bambach's resulting analyses are not purely ecological, but rather carry an unwanted phylogenetic component at the class-level.

The design of a guild classification is to mitigate phylogenetics – and augment ecological factors – in the paleoenvironmental or evolutionary analysis. This is not a problem for trace fossils, where the confidence in assessing the affinity of the tracemaker often is limited to higher biotaxonomic levels (e.g., above the ordinal level). In retooling the bauplan category for trace fossil data, Bromley (1990, 1996) chose ethology (e.g., domichnia, repichnia, etc.) and the vagility of the burrower (e.g., transitory or stationary). It is this different means of categorizing bauplan that distinguishes Bromley's ichnoguilds from Bambach's guilds (Table 1). For example, following Bromley's definition of bauplan, borings and bioclaustrations could both be described as stationary domichnia. Whereas if the endoliths themselves are emphasized, i.e., the bauplan sensu Bambach, then borings and bioclaustrations could be categorized on the basis of inferred biological affinity. Inferences of this kind are routinely made for macro- and microborings based on a wealth of neoichnological information and the conservative, stereotypic forms of borings that usually indicate biological affinity (Bromley 1970; Taylor and Wilson 2003; Glaub and Vogel 2004). Bauplan categories such as 'sponge borer' or 'chlorophyte microborer' group endolithic organisms at crude levels of taxonomy and so reduce the influence of phylogeny on guild analysis. In modern oceans, many simple macroborings and most bioclaustrations are produced by polychaetes and sipunculans, but in the fossil record they can only be attributed to unknown 'worms' – a desirable term that conveys no phylogenetic information and epitomizes the objective of the bauplan criterion in grouping organisms with shared physical constraints.

Resource requirements

Among modern endoliths, and presumably among ancient ones, trophic strategies are rather limited within autotrophy and heterotrophy. Most groups of microborers are photosynthetic, including the cyanobacteria, red algae and green algae (Golubic 1969). Little is known about the biology of the earliest microborers in volcanic glass, but they appear to have oxidized iron for nutrients, i.e., chemotrophs (Furnes et al. 2004, McLoughlin et al. this volume). Saprophytic fungi and some bacteria are the smallest borers among endolithic heterotrophs. The majority of euendolithic invertebrate animals are suspension feeders, from the microscopic bryozoans and sponges, to the larger borers and embedders that produce residential cavities (Bromley 1970). Important exceptions include parasitic foraminifera (e.g., Neumann and Wisshak 2006), deposit feeding borers (gastropods and some polychaetes: Hutchings this volume), and motile borers such as urchins that produce residential cavities (*Circolites* Mikuláš, 1992), but use suspension feeding and grazing strategies both within and away from their home cavity (e.g., Asgaard and Bromley this volume). In general, two major divisions among the endoliths are recognized to mostly coincide with the practical 100 μm size classification: microscopic autotrophs and macroscopic heterotrophs, with the inclusion of microscopic fungi, bacteria, sponges and bryozoans.

Space utilization

By definition, all endolithic organisms share the same general habitat, living in a hard substrate. The specific type of hard substrates occupied by particular endoliths is restricted by the method for cavity formation. Mechanical borers tend to occur in more types of substrates than do borers that require a chemical agent to excavate, for example phosphatic materials may be bored by pholadid bivalves, but not by clionaid sponges (Tapanila et al. 2004b). Bioclaustrations are exclusively produced in living skeletal substrates. Therefore occupation of an endolithic habitat is a useful unifying criterion for endoliths, but more precisely these organisms succeed in a variety of different hard materials within the broader marine environment.

In general, the endolithic guild is unified as a group of mostly soft-bodied organisms that produce shallow infaunal, stationary domichnia in hard substrates. It is argued that endoliths comprise two major superguilds (coarsely defined guilds: Bambach 1983), the autotrophic endoliths and the heterotrophic endoliths. Although superguilds are not desirable in a paleocommunity analysis because they under-represent resource utilization, they may be useful in examining the overall trends in the evolution of the endolithic lifestyle. The next section describes some first-order observations that can be made about the invertebrates of the heterotrophic endolithic guild.

Primary observations on the endolithic guild

Adaptive strategies and evolutionary faunas

To develop his rationale for using guild analysis, Bambach (1983: fig. 3) depicted the adaptive strategies employed by the three great evolutionary faunas (EFs) as recognized by Sepkoski (1981). The evolutionary faunas are statistically grouped classes of animals that show similar patterns of richness, and succeed one another through the Phanerozoic as the dominant marine constituents. For example, the Trilobita-dominated Cambrian EF (which also includes the Inarticulata, Monoplacophora, Hyolitha and a subset of the Polychaeta) is succeeded by the Articulata-dominated Paleozoic EF in the Ordovician, which in turn is succeeded by the Bivalvia- and Gastropoda-dominated Modern EF in the Triassic (Sepkoski 1981). Bambach's examination of adaptive strategies focused on three intervals of time, roughly corresponding to the turnover of the three EFs during the Cambrian, the Middle - Upper Paleozoic, and the Mesozoic - Cenozoic. For each time interval, Bambach listed classes of biotaxa on the basis of their trophic group and habitat. These groupings essentially represent superguilds that show an overall increase in the kinds of adaptive strategies used by marine animals and an increase in the diversity of classes that use these strategies through each major time interval. Underlining the elegance and significance of his findings, Bambach's (1983) adaptive strategy diagram has received wide popularity, including reproduction in several recent texts on paleontology (e.g., Prothero 2004: 124; Foote and Miller 2007: 266)

Figure 1A summarizes Bambach's (1983: fig. 3) adaptive strategy diagram and adds new information based on endolithic trace fossil data (Fig. 1B). In each box, the number of classes using a particular adaptive strategy is depicted for each of the three time intervals (refer to Bambach's original diagram for the identity of each class). For example, Bambach (1983) determined that the number of classes belonging to the pelagic suspension feeders include one in the Cambrian, three in the Middle - Upper Paleozoic, and three in the Mesozoic - Cenozoic (coded 1 3 3). Unfortunately, Bambach's depiction of infaunal exploitation is limited to the body fossil record, especially skeletonized organisms. His shallow passive infauna (suspension feeders, deposit feeders and carnivores) includes a low diversity of classes. For example, the suspension feeding group lists zero classes in the Cambrian, two classes (Bivalvia, Rostroconchia) in the Middle - Upper Paleozoic, and three classes (Bivalvia, Echinoidea and Gastropoda) in the Mesozoic - Cenozoic (coded 0 2 3 in Fig. 1A). When the classes of suspension feeding macroborers and embedders are added (i.e., the suspension feeding endolithic guild), the total diversity of invertebrates using the 'suspension feeding passive, shallow infaunal' strategy increases by at least 13 classes (total 1 9 8: Fig. 1B).

The greatest advance in this adaptive strategy clearly occurs after the Cambrian time interval, and includes the Ordovician Bioerosion Revolution (Wilson and Palmer 2006), which contributes at least six classes of macroboring infauna, excluding the bioeroding Sipunculida and Phoronida. Bambach did not include these latter two classes in his analysis due to their low diversity in the body fossil record, and so they are here excluded from the adaptive strategy count to remain comparable with Bambach's values. The addition of embedding infauna during the Paleozoic gives the impression that infaunal suspension feeding during this time (9 classes) exceeded the diversity of classes using this strategy in the Mesozoic - Cenozoic (8 classes), however it should be noted that a compilation of post-Paleozoic bioclaustrations is, as yet, incomplete.

Remarkably, the composition of invertebrate classes that make up the suspension feeding endolithic guild (i.e., nearly all invertebrates of the heterotrophic endolithic superguild) appears to have changed very little since their rise in the Early–Middle Paleozoic when the Paleozoic EF replaced the Cambrian EF as the dominant marine fauna. Researchers of fossil macrobioerosion have long recognized that a major shift in the dominant borers and their abundance occurs in the Triassic–Jurassic interval (Taylor and Wilson 2003). These changes most notably include an increase in bivalve borings in rock and wood substrates, an increase in clionaid and other sponge borings, and the rise of boring echinoids. But despite changes to the endolithic community structure at lower taxonomic levels – often regarded as a response to the increased predation pressures of the Mesozoic Marine Revolution (Vermeij 1987) – new classes of organisms do not appear to attempt or succeed in adopting the endolithic strategy (except for echinoids). Perhaps those classes of invertebrates that adopted endolithic strategies in the Paleozoic represent all (or nearly all) animals with a bauplan (i.e., physiology, anatomy, behavior) capable of efficiently exploiting endolithic habitats, or perhaps other anti-predatory strategies such as skeletal armoring and burrowing were more effective options for non-endolithic classes.

From this exercise it is evident that ecological analyses based solely on the body fossil record (e.g., Bambach 1983; Bambach et al. 2007; Bush et al. 2007) significantly risk under-representing the history of shallow infaunal exploitation, and most likely this bias affects analyses of all infaunal guilds.

Endolithic guild diversity of the Paleozoic

Synoptic studies of the total (gamma) diversity of trace fossils through time generally demonstrate an increase through the Paleozoic (e.g., Mángano and Droser 2004). Counting the total diversity of borings and bioclaustrations is hindered by the variability in how trace fossils are named (see discussion of ichnotaxobases by Bertling 2007) and is biased by the different availability and study dedicated to each interval of the geologic record, but some primary patterns appear to be robust in spite of these obstacles. Recent compilations on the stratigraphic ranges of macroborings and bioclaustrations show a general increase in these fossils through time (Taylor and Wilson 2003; Bromley 2004; Tapanila 2005). The record of microborings through time is still a work in progress (see Glaub and Vogel 2004; Wisshak et al. this volume for most recent surveys), and so these will not be described here.

A comparison of the two subgroups within the heterotrophic guild – macroborings and bioclaustrations – demonstrates the utility of guild analysis. More than 30 ichnogenera of borings and bioclaustrations from Paleozoic rocks have been identified. Current understanding of the diversity of the macroborers and embedders (Fig. 2) shows that the heterotrophic endolithic guild underwent its most significant radiation during the Middle to Late Ordovician (Tapanila 2005; Wilson and Palmer 2006). Changes in the diversity of these heterotrophic trace fossils can be related to taxonomy-independent controls. This leads to the hypothesis that a convergence in the evolution of lifestyles (e.g., the Ordovician invasion of hard substrates) is primarily a response to extrinsic controls rather than phylogenetic diversification. These ecological changes may include the availability of resources and habitat space, and they may involve an increase in interspecific competition and predation. The rise of heterotrophic endolithic guilds in the Paleozoic provides new evidence of the evolution of adaptive strategies, which at first glance appear to coincide with a general increase in animal diversity (Sepkoski 1981) and the increased development of infaunalization and tiering observed in the Ordovician (Ausich and Bottjer 1982; Thayer 1983).

Between the two heterotrophic endolithic groups, the history of bioclaustrations appears to be more complex than that for bioerosion domiciles. Although only

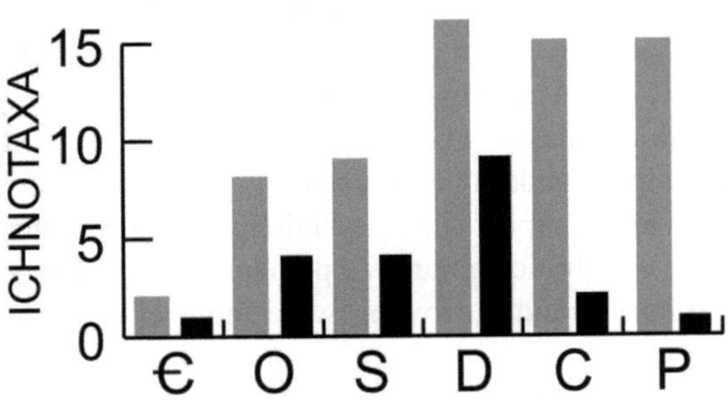

Fig. 2 Diversity of the heterotrophic endolithic guild in the Paleozoic. Macroborings (ichnogenera: grey) and bioclaustrations (ichnogenera: black) produced mostly by suspension feeding animals. Data from most recent compilations reported by Tapanila and Ekdale 2007, Wilson 2007

the Paleozoic has received systematic treatment (Tapanila 2005), it is evident that bioclaustration ichnotaxa are strongly influenced by the availability of preferred live host taxa (e.g., Tapanila 2002, 2004, 2006b). The onset of symbioses in the Late Ordovician mirrors the invasion of hard substrates by macroborers, but the Late Devonian decline in bioclaustration diversity is not matched by a decline in macroboring diversity. Tapanila (2005) argued that the collapse in bioclaustrations coincides with a Givetian decline in the preferred host coralline taxa millions of years prior to the Frasnian–Famennian mass extinction. Therefore, the requirement of a preferred living skeletal substrate might have limited the geologic range of bioclaustration ichnotaxa, whereas borers having less stringent habitat requirements might favor evolutionary conservatism. More research effort documenting the Mesozoic–Cenozoic record of bioclaustrations is needed to more thoroughly investigate this hypothesis.

Conclusions

By grouping fossils on the basis of standardized ecological criteria, the guild concept can effectively dissolve the boundary between body fossil and trace fossil data, such that both can be integrated to get a more complete sense of ecological change through time. Residential cavities formed in hard substrates are produced by a variety of organisms that utilize their environment in a similar way. As a result, these organisms share an ecological role that can be defined using the guild concept. Endoliths are divided into two superguilds, including (1) autotrophs consisting exclusively of microborers and (2) heterotrophs that include micro- and macroboring invertebrates, embedders as well as fungal and bacterial microborers.

Preliminary analyses of heterotrophic endoliths demonstrate the utility of trace fossils as ecological entities and combining them with body fossil data. A re-examination of Bambach's (1983) adaptive strategy diagram demonstrates that without considering endolithic trace fossils, the shallow infaunal category was significantly under-represented by at least 13 classes. Similarly, the context of ecology is important when studying the gamma diversity patterns of macroborings and bioclaustrations. In this case, observation of a comparable onset in the Ordovician, but contrasting diversity through the Devonian points to the efficacy of different endolithic strategies over short vs. long time periods. Future efforts to refine guild groupings and to use them in comparative and community-level studies have the potential to reveal new hypotheses on the evolution of infaunal strategies.

Acknowledgements

I thank the participants of the Fifth International Bioerosion Workshop in Erlangen for giving much needed critical feedback and suggestions to improve the ideas in this manuscript. I am indebted to reviewers K. Vogel, J.M. de Gibert and A.A. Ekdale for their challenging and constructive comments.

References

Asgaard U, Bromley RG (this volume) Echinometrid sea urchins, their trophic styles and corresponding bioerosion. In: Wisshak M, Tapanila L (eds) Current developments in bioerosion. Springer, Berlin, pp 279-304

Ausich WI, Bottjer DJ (1982) Tiering in suspension feeding communities on soft substrata throughout the Phanerozoic. Science 216:173-174

Bambach RK (1983) Ecospace utilization and guilds in marine communities through the Phanerozoic. In: Tevesz MJS, McCall PL (eds) Biotic interactions in recent and fossil benthic communities. Plenum Press, New York, pp 719-746

Bambach RK, Bush AM, Erwin DH (2007) Autecology and the filling of ecospace: key metazoan radiations. Palaeontology 50:1-22

Bertling M (2007) What's in a name? Nomenclature, systematics, ichnotaxonomy. In: Miller III W (ed) Trace fossils: concepts, problems, prospects. Elsevier, Amsterdam, pp 81-91

Bertling M, Braddy SJ, Bromley RG, Demathieu GR, Genise J, Mikuláš R, Nielsen JK, Nielsen SS, Rindsberg AK, Schlirf M, Uchman A (2006) Names for trace fossils: a uniform approach. Lethaia 39:265-286

Bromley RG (1970) Borings as trace fossils and *Entobia cretacea* Portlock, as an example. In: Crimes TP, Harper JG (eds) Trace fossils. Seel House Press, Liverpool, pp 49-90

Bromley RG (1981) Concepts in ichnotaxonomy illustrated by small round holes in shells. Acta Geol Hisp 16:55-64

Bromley RG (1990) Trace fossils, biology and taphonomy. Special topics in palaeontology 3. Unwin Hyman, London, 280 pp

Bromley RG (1992) Bioerosion: eating rocks for fun and profit. In: Maples CG, West RR (eds) Trace fossils. Short Courses Paleont 5:121-129

Bromley RG (1996) Trace fossils, biology, taphonomy and applications. Chapman and Hall, New York, 361 pp

Bromley RG (2004) A stratigraphy of marine bioerosion. In: McIlroy D (ed) The application of ichnology to palaeoenvironmental and stratigraphic analysis. Geol Soc London, Spec Publ 228:455-479

Bromley RG, Asgaard U (1993a) Two bioerosion ichnofacies produced by early and late burial associated with sea-level change. Geol Rundsch 82:276-280

Bromley RG, Asgaard U (1993b) Endolithic community replacement on a Piocene rocky coast. Ichnos 2:93-116

Bromley RG, Pemberton SG, Rahmani RA (1984) A Cretaceous woodground: the *Teredolites* ichnofacies. J Paleont 58:488-498

Bronn HG (1837) Lethaea Geognostica oder Abbildungen und Beschreibungen der für die Gebirgs-Formationen bezeichnendsten Versteinerungen. Schweizerbart, Stuttgart, 1:1-544 and plates

Buatois LA, Mángano MG, Genise JF, Taylor TN (1998) The ichnologic record of the continental invertebrate invasion: evolutionary trends in environmental expansion, ecospace utilization, and behavioral complexity. Palaios 13:217-240

Bush AM, Bambach RK, Daley GM (2007) Changes in theoretical ecospace utilization in marine fossil assemblages between the mid-Paleozoic and late Cenozoic. Paleobiology 33:76-97

Clarke JM (1908) The beginnings of dependent life. New York State Mus Bull 121:146-196

Clarke JM (1921) Organic dependence and disease: their origin and significance. New York State Mus Bull 221-222:1-113

Droser ML, Bottjer DJ (1989) Ordovician increase in extent and depth of bioturbation: implications for understanding early Paleozoic ecospace utilization. Geology 17:850-852

Duncan PM (1876) On some thallophytes parasitic within recent Madreporaria. Roy Soc London, Proc 24:238-257

Ekdale AA, Bromley RG (1991) Analysis of composite ichnofabrics: an example in uppermost Cretaceous chalk of Denmark. Palaios 6:232-249

Ekdale AA, Benner JS, Bromley RG, Gibert de JM (2002) Bioerosion of Lower Ordovician hardgrounds in southern Scandinavia and western North America. Acta Geol Hisp 27:9-13

Erickson JM, Bouchard TD (2003) Description and interpretation of *Sanctum laurentiensis*, new ichnogenus and ichnospecies, a domichnium mined into Late Ordovician (Cincinnatian) ramose bryozoan colonies. J Paleont 77:1002-1010

Folk RL (1959) Practical petrographic classification of limestones. AAPG Bull 43:1-38

Foote M, Miller AI (2007) Principles of paleontology. Freeman and Company, New York, 354 pp

Frey RW, Seilacher A (1980) Uniformity in marine invertebrate ichnology. Lethaia 13:183-207

Furnes H, Banerjee NR, Muehlenbachs K, Staudigel H, de Wit M (2004) Early life recorded in Archean pillow lavas. Science 304:578-581

Gibert JM de, Martinell J, Domènech R (1998) *Entobia* ichnofacies in fossil rocky shores, Lower Pliocene, Northwestern Mediterranean. Palaios 13:476-487

Gibert JM de, Domènech R, Martinell J (2004) An ethological framework for animal bioerosion trace fossils upon mineral substrates with proposal of a new class, fixichnia. Lethaia 37:429-437

Glaub I, Bundschuh M (1997) Comparative study on Silurian and Jurassic/Lower Cretaceous microborings. Courier Forschinst Senckenberg 201:123-135

Glaub I, Vogel K (2004) The stratigraphic record of microborings. Fossils and Strata 51:126-135

Glaub I, Golubic S, Gektidis M, Radtke G, Vogel K (2007) Microborings and microbial endoliths: geological implications. In: Miller III W (ed) Trace fossils: concepts, problems, prospects. Elsevier, Amsterdam, pp 368-381

Golubic S (1969) Distribution, taxonomy, and boring patterns of marine endolithic algae. Amer Zoologist 9:747-751

Golubic S, Perkins RD, Lukas KJ (1975) Boring microorganisms and microborings in carbonate substrates. In: Frey RW (ed) The study of trace fossils: a synthesis of principles, problems, and procedures in ichnology. Springer, New York, pp 229-259

Golubic S, Friedmann I, Schneider J (1981) The lithobiontic ecological niche, with special reference to microorganisms. J Sediment Petrol 51:475-478

Hutchings P (this volume) Role of polychaetes in bioerosion of coral substrates. In: Wisshak M, Tapanila L (eds) Current developments in bioerosion. Springer, Berlin, pp 249-264

James NP, Kobluk DR, Pemberton SG (1977) The oldest macroborers: Lower Cambrian of Labrador. Science 197:980-983

Johnson ME, Baarli BG (1999) Diversification of rocky-shore biotas through geologic time. Geobios 32:257-273

Kelley PH, Hansen TA (1993) Evolution of the naticid gastropod predator-prey system: an evaluation of the hypothesis of escalation. Palaios 8:358-375

Leymerie A (1842) Suite de mémoire sur le terrain Crétacé du department de l'Albe. Mem Geol Soc France 4:1-34

MacEachern JA, Pemberton SG, Gingras MK, Bann KL (2007) The ichnofacies paradigm: a fifty-year retrospective. In: Miller III W (ed) Trace fossils: concepts, problems, prospects. Elsevier, Amsterdam, pp 52-77

Mángano MG, Droser ML (2004) The ichnologic record of the Ordovician radiation. In: Webby BD, Droser ML (eds) The great Ordovician biodiversification event. Columbia Univ Press, New York, pp 369-379

McLoughlin N, Furnes H, Banerjee NR, Staudigel H, Muehlenbachs K, de Wit M, Van Kranendonk MJ (this volume) Micro-bioerosion in volcanic glass: extending the ichnofossil record to Archaean basaltic crust. In: Wisshak M, Tapanila L (eds) Current developments in bioerosion. Springer, Berlin, pp 371-396

Mikuláš R (1992) Early Cretaceous borings from Stramberk (Czechoslovakia). Čas Mineral Geol 37:297-312

Neumann AC (1966) Observations on coastal erosion in Bermuda and measurements of the boring rate of the sponge, *Cliona lampa*. Limnol Oceanogr 11:92-108

Neumann C, Wisshak M (2006) A foraminiferal parasite on the sea urchin *Echinocorys*: Ichnological evidence from the Late Cretaceous (Lower Maastrichtian, Northern Germany). Ichnos 13:185-190

Palmer T (1982) Cambrian to Cretaceous changes in hardground communities. Lethaia 15:309-323

Palmer TJ, Wilson MA (1988) Parasitism of Ordovician bryozoans and the origin of pseudoborings. Palaeontology 31:939-949

Perry CT, Bertling M (2000) Spatial and temporal patterns of macroboring within Mesozoic and Cenozoic coral reef systems. In: Insalaco E, Skelton PW, Palmer TJ (eds) Carbonate platform systems: components and interactions. Geol Soc London, Spec Publ 178:33-50

Podhalańska T, Nõlvak J (1995) Endolithic trace-fossil assemblage in Lower Ordovician limestones from northern Estonia. GFF 117:225-231

Prothero DR (2004) Bringing fossils to life: an introductory to paleobiology. McGraw-Hill, New York, 503 pp

Roberts EM, Rogers RR, Foreman BZ (2007) Continental insect borings in dinosaur bone: examples from the Late Cretaceous of Madagascar and Utah. J Paleont 81:201-208

Root RB (1967) The niche exploitation pattern of the blue-gray gnatcatcher. Ecol Monogr 37:317-350

Schönberg CHL, Tapanila L (2006) Bioerosion research before and after 1996 – a discussion on what has changed since the First International Bioerosion Workshop. Ichnos 13:99-102

Seilacher A (1953) Über die Methoden der Palichnologie. 1. Studien von Palichnologie. N Jb Geol Paläont Abh 96:421-452

Seilacher A (1964) Biogenic sedimentary structures. In: Imbrie J, Newell ND (eds) Approaches to paleoecology. Wiley and Sons, New York, pp 296-316

Sepkoski JJ Jr (1981) A factor analytic description of the Phanerozoic marine fossil record. Paleobiology 7:36-53

Tapanila L (2002) A new endosymbiont in Late Ordovician tabulate corals from Anticosti Island, eastern Canada. Ichnos 9:109-116

Tapanila L (2004) The earliest *Helicosalpinx* from Canada and the global expansion of commensalism in Late Ordovician sarcinulid corals (Tabulata). Palaeogeogr Palaeoclimatol Palaeoecol 215:99-110

Tapanila L (2005) Palaeoecology and diversity of endosymbionts in Palaeozoic marine invertebrates: Trace fossil evidence. Lethaia 38:89-99

Tapanila L (2006a) Devonian *Entobia* borings from Nevada, with a revision of *Topsentopsis*. J Paleont 80:760-767

Tapanila L (2006b) Macroborings and bioclaustrations in a Late Devonian reef above the Alamo Impact Breccia, Nevada, USA. Ichnos 13:129-134

Tapanila L, Copper P (2002) Endolithic trace fossils in Ordovician-Silurian corals and stromatoporoids, Anticosti Island, eastern Canada. Acta Geol Hisp 37:15-20

Tapanila L, Ekdale AA (2007) Early history of symbiosis in living substrates: trace-fossil evidence from the marine record. In: Miller III W (ed) Trace fossils: concepts, problems, prospects. Elsevier, Amsterdam, pp 339-349

Tapanila L, Holmer LE (2006) Endosymbiosis in Ordovician-Silurian corals and stromatoporoids: a new lingulid and its trace from eastern Canada. J Paleont 80:750-759

Tapanila L, Copper P, Edinger E (2004a) Environmental and substrate control on Paleozoic bioerosion in corals and stromatoporoids, Anticosti Island, eastern Canada. Palaios 19:292-306

Tapanila L, Roberts EM, Bouaré ML, Sissoko F, O'Leary MA (2004b) Bivalve borings in phosphatic coprolites and bone, Cretaceous-Paleogene, northeastern Mali. Palaios 19:565-573

Taylor PD (1990) Preservation of soft-bodied and other organisms by bioimmuration – a review. Palaeontology 33:1-17

Taylor PD, Wilson MA (2002) A new terminology for marine organisms inhabiting hard substrates. Palaios 17:522-525

Taylor PD, Wilson MA (2003) Palaeoecology and evolution of marine hard substrate communities. Earth-Sci Rev 62:1-103

Thayer CW (1983) Sediment-mediated biological disturbance and the evolution of marine benthos. In: Tevesz M, McCall P (eds) Biotic interactions in recent and fossil communities. Plenum Press, New York, pp 479-625

Vermeij GJ (1987) Evolution and escalation: a natural history of life. Princeton Univ Press, Princeton, 527 pp

Vogel K, Bundschuh M, Glaub I, Hofmann K, Radtke G, Schmidt H (1995) Hard substrate ichnocoenoses and their relations to light intensity and marine bathymetry. N Jb Geol Paläont Abh 195:49-61

Voigt E (1979) The preservation of slightly or non-calcified fossil Bryozoa (Ctenostoma and Cheilostoma) by bioimmuration. In: Larwood GP, Abbott MB (eds) Advances in bryozoology. Acad Press, London, pp 541-564

Warme JE (1977) Carbonate borers-their role in reef ecology and preservation. In: Frost SH, Weiss MP, Saunders JB (eds) Reefs and related carbonates-ecology and sedimentology. AAPG Stud Geol 4:261-279

Warme JE, Slater RA, Cooper RA (1978) Bioerosion in submarine canyons. In: Stanley DJ, Kelling G (eds) Sedimentation in submarine canyons, fans, and trenches. Dowden, Hutchinson and Ross, Pennsylvania, pp 65-70

Wedl G (1859) On the significance of the canals found in many mollusk and gastropod shells. Sitzber Kl Akad Wiss 33:451-472

Weidlich O (1996) Bioerosion in Late Permian Rugosa from reefal blocks (Hawasina Complex, Oman Mountains): implications for reef degradation. Facies 35:133-142

Wilson MA (2007) Macroborings and the evolution of marine bioerosion. In: Miller III W (ed) Trace fossils: concepts, problems, prospects. Elsevier, Amsterdam, pp 356-367

Wilson MA, Palmer TJ (1988) Nomenclature of a bivalve boring from the Upper Ordovician of the Midwestern United States. J Paleont 62:306-308

Wilson MA, Palmer TJ (1992) Hardgrounds and hardground faunas. Univ Aberystwyth, Inst Earth Stud Publ 9:1-131

Wilson MA, Palmer TJ (1998) The earliest *Gastrochaenolites* (Early Pennsylvanian, Arkansas, USA): an Upper Paleozoic bivalve boring? J Paleont 72:769-772

Wilson MA, Palmer TJ (2006) Patterns and processes in the Ordovician bioerosion revolution. Ichnos 13:109-112

Wisshak M, Seuß B, Nützel A (this volume) Evolutionary implications of an exceptionally preserved Carboniferous microboring assemblage in the Buckhorn Asphalt lagerstätte (Oklahoma, USA). In: Wisshak M, Tapanila L (eds) Current developments in bioerosion. Springer, Berlin, pp 21-54

Zhang Y, Golubic S (1987) Endolithic microfossils (Cyanophyta) from early Proterozoic stromatolites, Hebei, China. Acta Micropalaeont Sin 4:1-12

Evolutionary implications of an exceptionally preserved Carboniferous microboring assemblage in the Buckhorn Asphalt lagerstätte (Oklahoma, USA)

Max Wisshak[1], Barbara Seuß[1], Alexander Nützel[2]

[1] Institute of Palaeontology, Erlangen University, 91054 Erlangen, Germany, (wisshak@pal.uni-erlangen.de)
[2] Bayerische Staatssammlung für Paläontologie und Geologie, 80333 München, Germany

Abstract. In the Buckhorn Asphalt deposit, exceptional preservation of mollusc shells with original aragonitic mineralogy, owing to an early impregnation with migrating hydrocarbons, provides a 'preservational window' for studying a Late Palaeozoic microboring assemblage. The evaluation of thin-sections, bioclast surface features, and SEM analysis of epoxy resin casts reveals a total of 18 known ichnospecies and the new ichnotaxon *Aurimorpha varia* – reflecting the most diverse Palaeozoic microboring assemblage known to date.

The ichno-inventory is dominated by the cyanobacterial traces *Eurygonum nodosum*, *Scolecia filosa*, and *Fascichnus dactylus*, complemented by the common *Fascichnus frutex*, *Planobola macrogota* and *Cavernula coccidia*, and the chlorophyte traces *Cavernula pediculata* and *Rhopalia catenata*. Presumable borings of heterotrophs are rare albeit diverse constituents of the microboring assemblage. The ichnocoenosis composition indicates a palaeoenvironment in the shallow-euphotic zone and is in many respects 'modern'. The fact that 17 out of the 19 recorded ichnospecies are also known to neoichnology in closely similar ichnocoenoses of today's shallow-euphotic seas underlines the pronounced longevity of microendolithic taxa and promotes their value as deep-time palaeoenvironmental indicator.

Two thirds of the recorded ichnotaxa were reported for the first time from Carboniferous strata, thereby extending the known stratigraphic range of *Cavernula pediculata*, *Cavernula coccidia*, *Rhopalia catenata*, *Rhopalia spinosa*, *Scolecia serrata*, *Polyactina fastigata*, *Saccomorpha terminalis*, and *Aurimorpha varia* igen. n., isp. n. back into the Palaeozoic. This shows that at least 21 ichnospecies, including a high number of modern ichnotaxa, were established as early as the late Palaeozoic and reflects an increasing abundance and potential competition by chlorophyte algae and heterotroph microborers breaking the former strong dominance of endolithic cyanobacteria in microborer communities. The curve of Phanerozoic ichnospecies diversity suggests that the evolution of microbioerosion agents is to a certain

M. Wisshak, L. Tapanila (eds.), *Current Developments in Bioerosion*. Erlangen Earth Conference Series, doi: 10.1007/978-3-540-77598-0_2,

degree parallel to that of the macroborers with main radiations in the Ordovician and Devonian. However, according to the current state of knowledge, the end-Permian mass extinction event seemingly had little or no impact on the diversity and evolution of microborers. Thus, diversity patterns of micorborings differ from those of marine invertebrates, most of which have a minimum in standing diversity at the Permian / Triassic boundary and a general turn-over from typical Palaeozoic to modern evolutionary faunas.

Keywords. Bioerosion, microborings, *Aurimorpha varia*, evolution, Buckhorn Asphalt, lagerstätte, Pennsylvanian

Introduction

Microboring record in the Palaeozoic

Owing to diagenetic alteration, the analysis of microborings in fossil rocks is often rendered difficult or even impossible. This is particularly true for Palaeozoic carbonates, which allow insight into microboring ichnocoenoses composition only via 'preservational windows' with absence of silicification, dolomitisation or intra-granular precipitation of calcite spar cements. During the past two to three decades, several suitable Palaeozoic carbonate sequences have been tackled in this respect, setting the stage for the evaluation of evolutionary patterns in the microboring niche.

Diverse microboring assemblages were for instance reported from the Ordovician of the Cincinnati Arch in the USA (Vogel and Brett 2006), the Silurian of Gotland and the Baltic States (Glaub and Bundschuh 1997; Bundschuh 2000), from the Middle Devonian of New York State, USA (Vogel et al. 1987), and from the Permian Capitan Reef Complex in New Mexico and Texas, USA (Balog 1996, 1997). These observations are complemented by the investigations of individual microbioerosion features as for instance Early Cambrian chlorophytes (Runnegar 1985), microborings in ostracods from the Middle Ordovician of Poland (Olempska 1986), or red algal body fossils and their traces from the Silurian of Poland (Campbell et al. 1979; Campbell 1980).

Based on these data and unpublished observations, Glaub and Vogel (2004) compiled the stratigraphic record of some 50 microboring ichnotaxa and body fossils. According to their study, the first microendoliths made their appearance already in the Proterozoic (*Eohyella campbelli* – 1,500 to 1,700 Ma: Zhang and Golubic 1987) and many taxa emerged during the Ordovician and Silurian, followed by a diversity increase at the base of the Mesozoic. They showed that some 35% of all modern taxa were present as early as the Palaeozoic in closely similar morphotypes. This pronounced longevity and morphostasis of endolithic life forms is promoting their value as palaeoenvironmental indicators for instance with respect to palaeobathymetry and palaeotemperature (Glaub and Vogel 2004; Vogel and Glaub 2004).

Carboniferous microborings to date have been treated only marginally in a brief note by Vogel (1991) and the inclusion of unpublished data in tables (Glaub et al.

2001; Glaub and Vogel 2004). A particularly promising target for studying a well preserved Carboniferous microboring trace fossil assemblage is represented by the middle Pennsylvanian Buckhorn Asphalt lagerstätte in the Arbuckle Mountains in southern Oklahoma, USA. Bandel et al. (2002) reported the presence of potential cyanobacterial microborings in gastropods and bivalves from this locality and, as presented in this paper, this fossil lagerstätte preserves the most diverse Palaeozoic microboring assemblage known to date.

The Buckhorn Asphalt lagerstätte

The Buckhorn Asphalt deposit is part of the Boggy Formation which belongs to the Deese Group (lower-middle Desmoinesian, middle Pennsylvanian). It is unconformably capped by deposits of the Virgillian Ada Formation and the Vanoss Conglomerate (Brand 1987; Brown and Corrigan 1997). The strata of the Buckhorn Asphalt lagerstätte crop out at the northern flank of the Arbuckle Mountains (Fig. 1; Squires 1973; Cree 1984; Brand 1987). At present these deposits are available in an abandoned quarry, an elongated pit of a few hundred metres length. Here, Cree (1984) logged a sedimentary sequence of ~6.5 m in thickness, distinguishing 14 lithological subunits. This limited outcrop was re-opened and excavated a few years ago. According to Squires (1973), the total extent of the Buckhorn Asphalt area is estimated to cover around 30 square miles.

The Pennsylvanian shallow marine sediments exposed in the Buckhorn Asphalt deposit were infiltrated by Ordovician aged oil which later transformed to asphalt. This infiltration has happened synsedimentarily or shortly after sedimentation of the Pennsylvanian deposit when the included shell material was diagenetically still unaltered or only weakly altered. Therefore, the Buckhorn Asphalt deposit is one of the few Palaeozoic deposits which yields aragonitic shell material and this renders an unusually high information content, especially for mollusc palaeobiology. The preservation of the shelly organisms is often excellent in that they are commonly preserved with their original mineralogy and microstructure (e.g., nacre and crossed lamellar microstructures). Delicate morphological details and minute early ontogenetic shells have been reported for gastropods, bivalves and cephalopods. Even colour preservation has been reported. The fact that shells are commonly unaltered by diagenesis and are present in their original mineralogical composition allows the study of original isotope signals.

General facies and fauna of the Buckhorn Asphalt

The Buckhorn sediments consist of mixed bioclastic and siliciclastic sediments. Grain sizes vary from coarse to silt / fine sand. Wackestones, rudstones, grainstones, mudstones, and conglomerates are present (e.g., Squires 1973; own observation). The deposits are highly fossiliferous and mollusc coquinas are especially abundant. The most common fossil groups are orthoconic nautiloids and other cephalopods, gastropods, bivalves, ostracods, and foraminiferans. Plant remains and even coal deposits (Brand 1989) are also present. Due to its exceptional preservation, numerous publications on specific fossil groups (especially molluscs) and palaeobiological

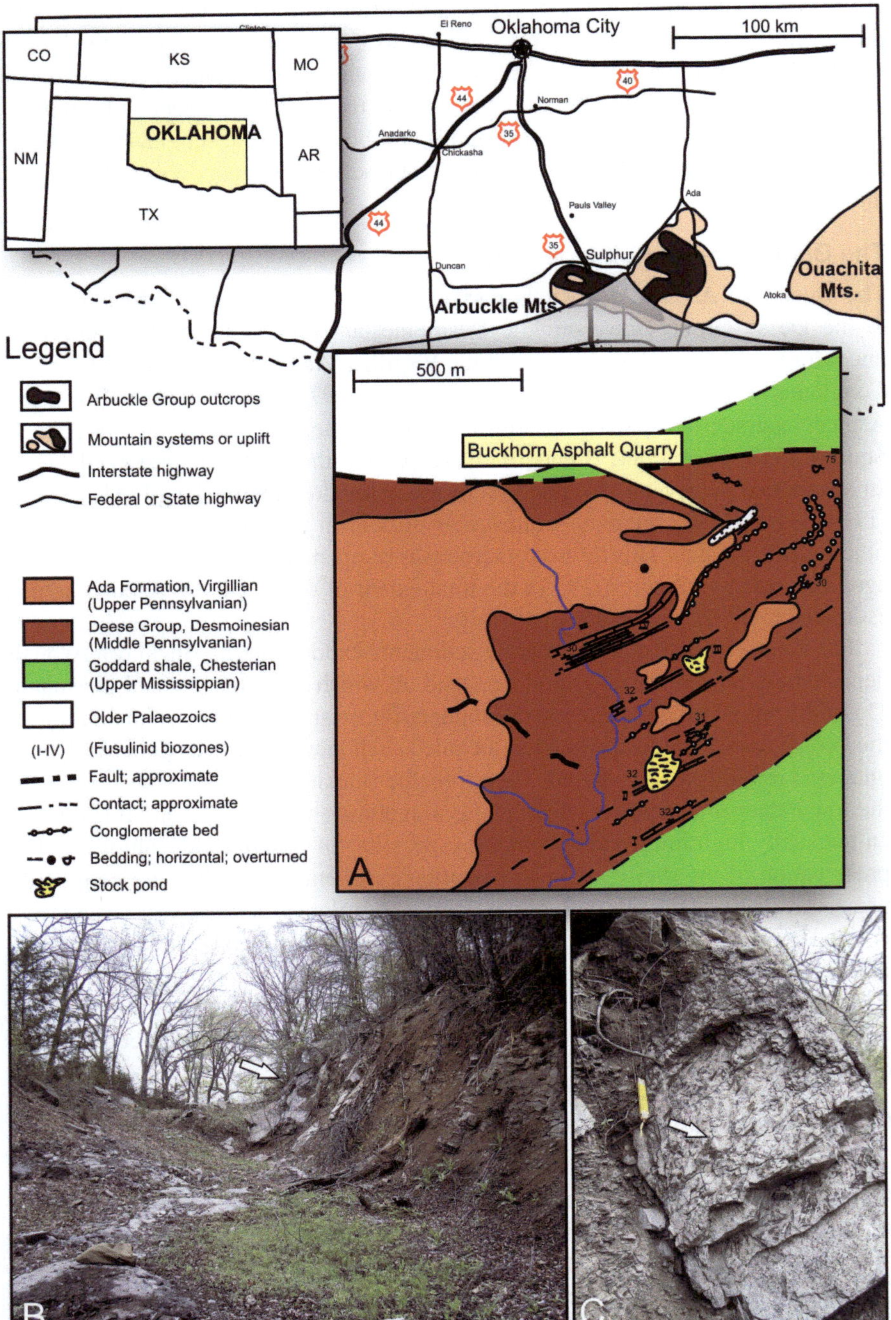

Fig. 1 A The location of the Buckhorn Asphalt lagerstätte (Boggy Formation, Desmoinesian) in the northern Arbuckle Mountains in southern Oklahoma (USA) and a geologic map of the vicinity of the quarry (modified after Squires 1973). **B** The present state of the outcrop with the main sample site (arrow). **C** The sample site (#1) with particularly bioclast-rich layers (arrow)

problems are based on material from the Buckhorn Asphalt deposit. However, a comprehensive study of the palaeoenvironment and biota is still lacking.

Mississippian deposits from Oklahoma are fully marine. During the Pennsylvanian, a progressive shallowing upward trend can be recognised and several areas were emerging above sea level. At the same time, during tectonically active phases, the formation of mountain chains (Ouachita Mts., Wichita Mts., and Arbuckle Mts.) strongly influenced the sedimentation in the study area. Cree (1984) interpreted the bioclastic deposits in the study area as turbidites and Squires (1973) as channel fills. We found no clear evidence for turbiditic deposition and would more agree with the interpretation as channel fills. It is likely that the skeletal material was transported (land plant debris, fractured bioclasts, etc.) and bioerosion may have taken place before and after this relocation. However, the commonly large size of poorly sorted shell fragments and clasts suggest that transport was not very far.

Hydrocarbon migration into the sediments of the Buckhorn Asphalt strata occurred in two pulses and started during or soon after deposition (e.g., Cree 1984). According to Brand (1989), this migration took place prior to the deposition of the Ada Conglomerate. Fossils were impregnated and cavities filled by the hydrocarbons. This process inhibited cementation and protected skeletal material against recrystallisation. Thus, the sediments have commonly only undergone minimal or incomplete diagenesis (Brand 1989; Sadd 1991). Shortly after the deposition of the Ada Conglomerate, compaction of the rocks (Squires 1973) and transformation of oil into asphalt likely took place during the Arbuckle Orogeny in the middle to late Pennsylvanian (Brand 1987; Sadd et al. 1986). The amount of asphalt in the sediments varies from 12 to 20% (Brown and Corrigan 1997) and the content of asphaltenes is as high as 78% (Sadd et al. 1986).

The salinity of the Deese Group sea-water was evaluated by analysing Sr / Na-ratios in nautiloids leading to an estimated mean salinity of 32 ppt (Brand 1987). This indicates some influx of fresh-water and is, together with plant debris, pointing towards a sedimentation close (~10 km) to the palaeoshoreline (Cree 1984). Most of the transported marine skeletal material probably formed in shallow water prior to transport. This is also corroborated by the microboring assemblages described in this paper. The water depth at the place of deposition (channels?) is uncertain. Given the large grain size especially of the conglomeratic beds, a relatively steep palaeoslope has to be assumed if the mass flow or channel fill hypothesis is correct.

Palaeotemperature estimations assessed through stable isotope analyses ($\delta^{18}O$) have to be interpreted with care and range from 14°C up to 38°C with a mean in shallow waters of 36°C and in the deeper regions 21°C (Brand 1982, 1987). These rather extreme temperature-values should be tested by future isotope analyses. According to world palaeogeographic reconstructions, the Buckhorn site was located about 5° south of the Pennsylvanian palaeoequator (Brand 1987).

The palaeoenvironment supported a diverse faunal assemblage comprising various species of gastropods, bivalves, nautiloids, fishes as well as a rich microfauna including among others foraminiferans and ostracods. Molluscs are the dominating macrofossil element (e.g., Mapes et al. 2000) and represent the target of our study. Bivalves with preserved prodissoconchs were studied by Heaney (1997) and Heaney and Yancey (1998). Occasionally, even ligaments and the organic matrix of the shell is conserved (Yancey and Heaney 2000). Some of the best preserved Palaeozoic gastropods have been reported from the Buckhorn lagerstätte (Bandel et al. 2002), showing original shell-mineralogy with preserved microstructures and micro-ornamentation. Squires (1976) and Bandel et al. (2002) reported colour patterns in naticopsid gastropods. Colour preservation is a relatively rare phenomenon in the Palaeozoic. Cephalopods from the Buckhorn Asphalt deposits have been studied by various authors. Blind (1991) studied endocameral deposits in orthoconic nautiloids. These nautiloids can often be found oriented in the sediments (Squires 1973). Kulicki et al. (2002) reported the oldest ammonoideans with preservation of original aragonite and with well-preserved ammonitellas from the Buckhorn Asphalt deposit. There are also reports on other fossil groups but the Buckhorn lagerstätte is especially important for mollusc palaeobiology. Here, we present the first comprehensive study of bioerosion and microboring assemblages of the Buckhorn Asphalt deposit.

Material and methods

Most of the sample material was gathered during a field campaign in 2002 from a selected particularly bioclast-rich layer of the exposed section in the asphalt quarry (sample set #1), which is located approximately 4.8 km south of the city of Sulphur (Fig. 1). Additional samples were collected from loose rocks in the quarry and are thus lacking a fine scale stratigraphic context (sample set #2). This material was complemented by a few mollusc samples taken more than twenty years ago in the same quarry (sample set #3).

The microborings were investigated in mollusc shells by three means: in thin-sections, via scanning electron microscopy (SEM) of surface pitting, and via SEM of epoxy resin casts.

Thin-sections were prepared after impregnation with transparent polymer resin (Araldite BY158 + Aradur 21) in a vacuum chamber and were photographed with a transmission light microscope (Zeiss Axiophot) with non-polarised light.

All further analyses required a de-asphaltation procedure utilising a suitable solvent. The sample blocks had to be broken into chunks of a maximum size of 4 cm^3 in order to fit in the cotton husks of a soxhlet extractor. Methylene chloride (CH_2Cl_2) was used as an appropriate strong solvent, which is convenient to work with as it vaporises at only 40°C. The solvent is advantageous for preserving fine detail of shell structure, pigmentation, and the isotopic signal. Depending on the asphalt content, the dissolving process took one to four days, while exchanging the methylene chloride and the husks every 24 hours. After consecutive air drying, the remaining organics in the rock matrix were removed by soaking the samples in a mixture of water and tensides for one to three days and rinsing with warm water until no more foam emerged.

After sieving the samples, the material was dried at 70°C and macro- as well as microfossils were picked for further analyses. Selected mollusc and ostracod shells were sputter-coated with gold and analysed with the aid of a (CamScan) scanning electron microscope (SEM) in order to study the fine detail of the aragonitic shells, including surficial bioerosion features.

The centrepiece of the present analysis of microboring traces was carried out utilising the approved modified cast-embedding technique based on Golubic et al. (1970, 1983). The selected samples (neritimorph gastropods, bellerophontid gastropods, unidentified bivalves and cephalopods) were placed in a vacuum chamber specifically designed for this method (Struers Epovac). The chamber allows the infiltration of several samples with a low-viscosity epoxy resin (Araldite BY158 + Aradur 21) while the samples are held under vacuum. This way, a complete infiltration of the borings is assured and trapping of air is avoided. After curing, the samples were formatted and decalcified by a treatment with diluted HCl (~5%) and consecutive rinsing in distilled water and drying at 70°C. The samples – now exhibiting the positive casts of the borings – were then sputter-coated with gold and finally analysed and photographed using SEM. In order to prevent delicate casts from collapsing and to help in recognising the actual penetration depth of the traces, some sample cross sections were only partially etched, allowing a visualisation of the traces in context of the shell matrix. The encountered traces were semiquantitatively grouped in four abundance classes (very common, common, rare, and very rare).

All trace-bearing epoxy resin casts, including the holotype material for the new ichnotaxon, are hosted in the Bayerische Staatssammlung für Paläontologie und Geologie, München (Munich, Germany) under the inventory numbers BSPG 2007 XII 1 to BSPG 2007 XII 45.

Abbreviations used in the text and figure captions: #1 = sample set #1; #2 = sample set #2; #3 = sample set #3; G = gastropod; B = bivalve; C = cephalopod; M = unidentified mollusc.

Results

Superficial bioerosion features

Many of the calcareous macro- as well as microfossils investigated by SEM yield distinct signs of bioerosion. Among the molluscs, particularly the gastropods are often carved by grooves and unroofed tunnels of microendolithic traces. Thereby, the protoconch and larval shells show the strongest density of borings in most cases (Fig. 2A-B), but also the inner surface of shells is infested, as for instance the columella of large neritimorph (Fig. 2C-D). The grooves and tunnels are usually 3 to 6 µm in width and extend in a straight (Fig. 2C-D) or curved (Fig. 2B) path for often more than 100 µm in length or form extensive networks. In contrast to the gastropods, bivalves often show corresponding microendolithic traces all over the shell and borings are not restricted to the prodissoconch (Fig. 2E-F). Cephalopod fragments rarely display bioerosion features and scaphopods were not observed having borings. Aside from the molluscs, bryozoans and ostracods (Fig. 2G-H) as well as foraminiferans are occasionally found exhibiting similar microendolithic borings, whereas echinoderms and non-calcareous vertebrate remains were lacking corresponding features. Macroborings were not observed.

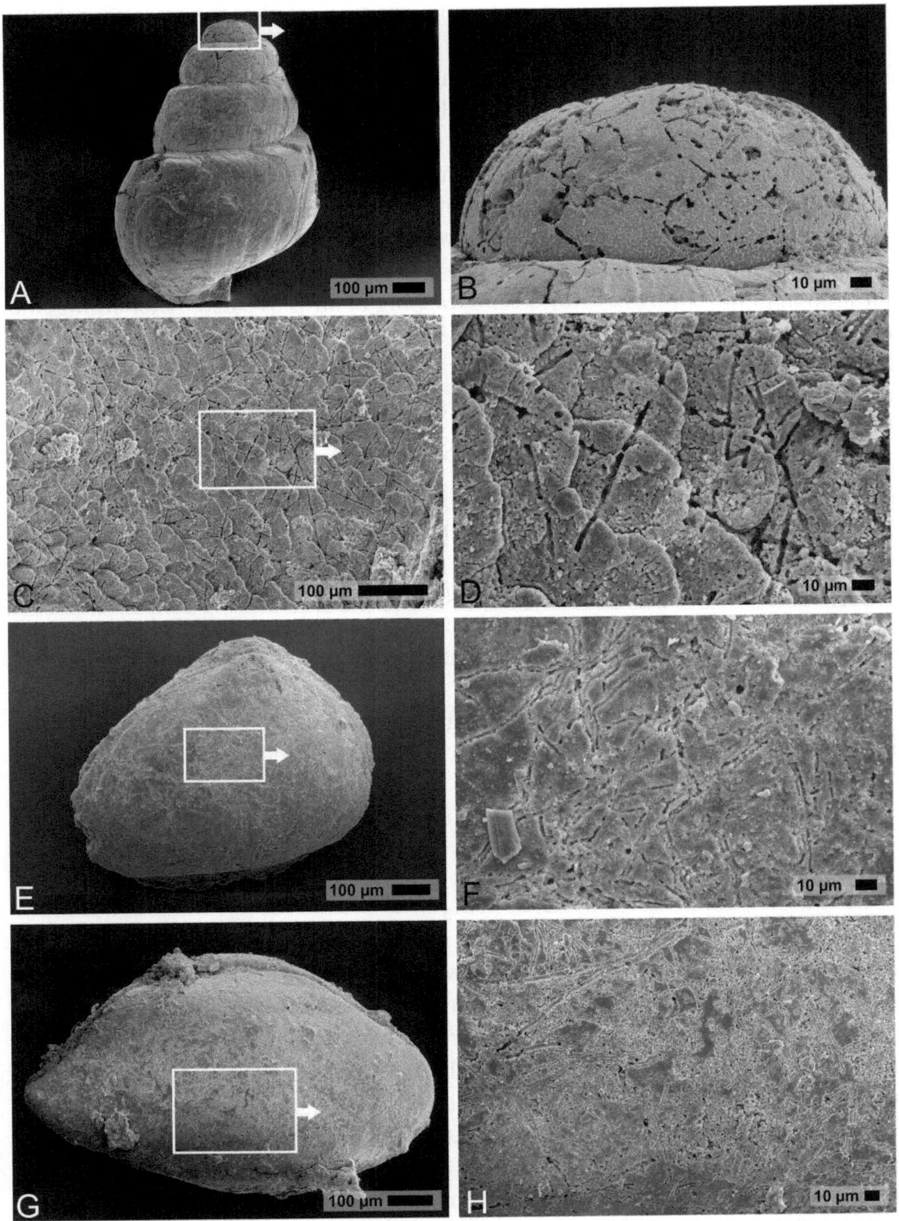

Fig. 2 Superficial bioerosion features observed via SEM on various calcareous shelly fossils. **A-B** A gastropod exhibiting grooves and unroofed tunnels of microborings particularly abundant on the larval shell (#2). **C-D** Close ups of the columella of a neritimorph gastropod featuring straight microborings (#1). **E-F** A small bivalve showing a network of corresponding microborings, which cover the whole shell (#1). **G-H** An ostracod also showing superficial microborings with tunnels of varying diameters (#1)

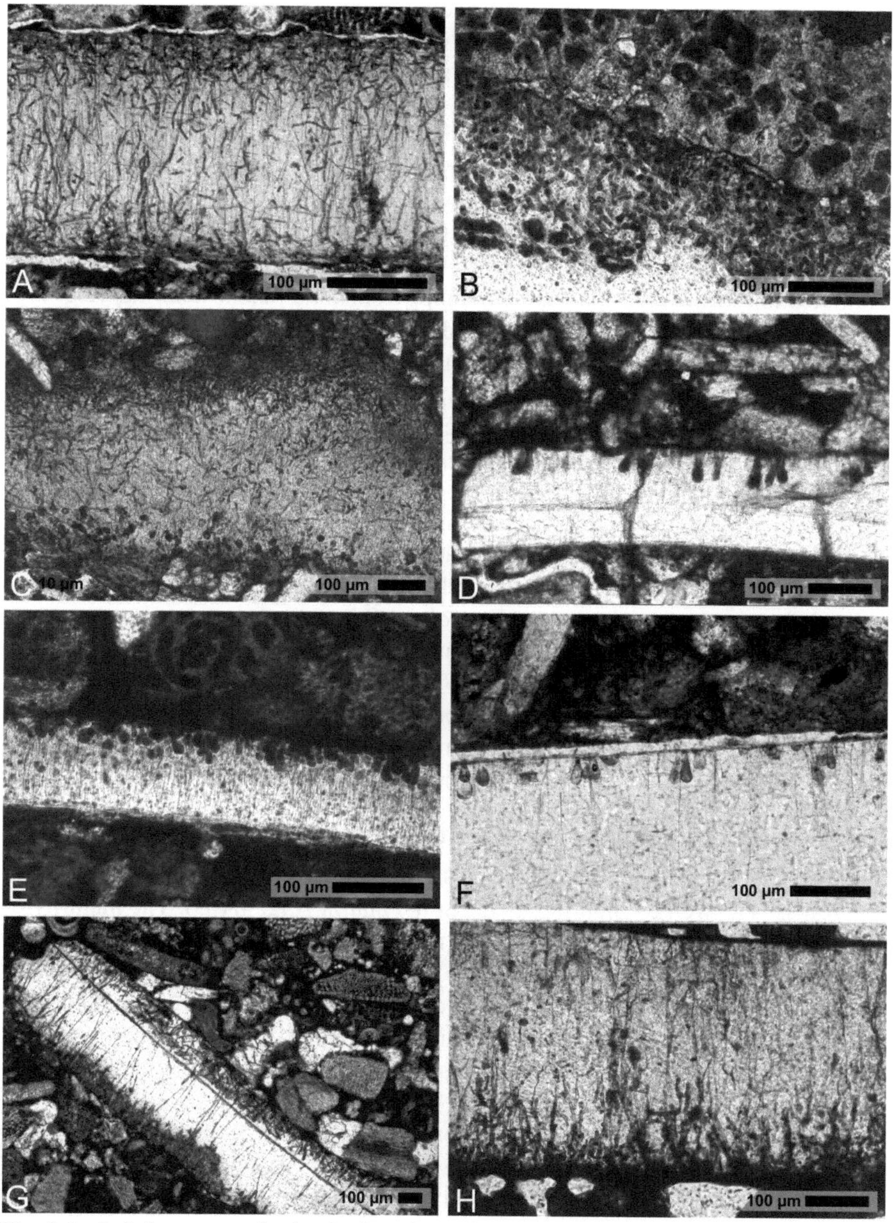

Fig. 3 Asphalt-impregnated microborings in mollusc shells observed in thin-sections under non-polarised light (#2; M). **A** Deeply penetrating *Eurygonum nodosum*. **B** Dense meshwork of *Eurygonum nodosum* in a shell encrusted by a calcareous algal colony. **C** Densely bored shell fragment. **D-F** Sack-shaped cavities of *Cavernula pediculata*. **G-H** Vertically penetrating boring systems of uncertain affinity

We refrain from an ichnotaxonomic identification of the superficial traces and leave this task to the much more reliable method of the cast-embedding technique, which reveals the true three-dimensional architecture of the traces (see below).

Bioerosion features in thin-sections

Owing to the asphalt impregnation, microendolithic borings in mollusc shells are clearly visible in thin-sections under non-polarised light (Fig. 3). In all cases, the asphalt-bearing matrix has readily filled the microborings, whereas closure of the voids by spar cements was not observed. A large percentage of bioclastic calcareous grains exhibit such bioerosion features ranging from initial, shallow borings (e.g., Fig. 3F) to shell fragments which are crowded by deeply penetrating borings (e.g., Fig. 3A, C, H). The infestation is either restricted to one side of the shell fragments (e.g., Fig. 3D) or both sides are infested, in the latter case often by different microborers (e.g., Fig. 3C, G). Some of these traces can be identified to the ichnospecies level such as the very abundant *Eurygonum nodosum* forming a branched meshwork of 6 to 10 µm thick galleries with diagnostic short swollen apophyses (Fig. 3A-C) and *Cavernula pediculata* which is characterised by sack-shaped cavities, up to 70 µm deep, connected to the substrate surface by few rhizoidal appendages (Fig. 3D-F). Some of the deeper penetrating and rarely branching galleries probably represent traces belonging to the ichnogenus *Fascichnus* (Fig. 3A). Other deeply penetrating boring systems are of uncertain affinity and can only be identified by applying the cast-embedding technique (Fig. 3G-H). With the exception of a few vertical oriented traces (Fig. 3G-H), a preferred orientation following the shell ultrastructure is not apparent. In some cases, the shell fragments were encrusted by calcareous epizoans such as bryozoans or calcareous algae (Fig. 3B).

Ichno-inventory in epoxy resin casts

The fraction of calcareous bioclasts with signs of bioerosion is high, both among the macrofossils as well as shell fragments in the matrix (Fig. 4A-B). The maturity of the ichnocoenoses recorded in the epoxy resin casts ranges from pristine substrates to densely bored mature ichnocoenoses and the range in quality of preservation is as wide. In contrast to bivalves, which show a homogenous infestation with microborers, the gastropods typically exhibit differential degrees of bioerosion with the protoconchs and larval shell being distinctly more affected by microbioerosion compared to the subsequent whorls (Fig. 4C-D).

The diverse trace fossil assemblage comprises a total of 18 known ichnospecies and one new ichnospecies (Table 1). The material of half a dozen further traces of unknown affinity is at present insufficient for ichnotaxonomic treatment and is thus omitted from this compilation. Macroborings were not encountered in the studied material.

In the following account, a brief description of the encountered ichnotaxa is given in order of relative abundance together with their known stratigraphic range and knowledge on the potential trace maker(s).

Table 1 Ichnotaxon inventory in order of abundance as recorded in epoxy resin casts taken from various molluscs, with their known or inferred Recent trace maker(s), substrate specifications (B = bivalve; G = gastropod; C = cephalopod) and relative abundance (+ + = very common; + = common; - = rare; - - = very rare)

Ichnotaxon	Potential trace maker(s)	Substrate B	Substrate G	Substrate C	Relative abund.
Eurygonum nodosum Schmidt, 1992	*Mastigocoleus testarum* Lagerheim, 1886 (cyanobacterium)	X	X	X	+ +
Scolecia filosa Radtke, 1991	*Plectonema terebrans* Bornet and Flahault, 1889 (cyanobacterium)	X	X		+ +
Fascichnus dactylus (Radtke, 1991)	e.g., *Hyella caespitosa* Bornet and Flahault, 1889 (cyanobacterium)	X	X	X	+ +
Fascichnus frutex (Radtke, 1991)	*Hyella gigas* Lukas and Golubic, 1983 (cyanobacterium)	X	X		+
Planobola macrogota Schmidt, 1992	? (cyanobacterium, alga?)	X	X	X	+
Cavernula coccidia Glaub, 1994	? (cyanobacterium, bacterium?, alga?)	X	X	X	+
Cavernula pediculata Radtke, 1991	*Gomontia polyrhiza* (Lagerheim) Bornet and Flahault, 1888 (chlorophyte)	X	X		+
Rhopalia catenata Radtke, 1991	*Phaeophila dendroides* (Crouan) Batters, 1902; *Eugomontia sacculata* Kornmann, 1960 (chlorophytes)		X	X	+
Rhopalia spinosa Radtke and Golubic, 2005	*Phaeophila dendroides* (Crouan) Batters, 1902 (chlorophyte)		X		-
Fascichnus rogus (Bundschuh and Balog, 2000)	*Hyella racemus* Al-Thukair, Golubic and Rosen, 1994; *H. conferta* Al-Thukair and Golubic, 1991 (cyanobacteria)	X	X		-
Scolecia serrata Radtke, 1991	? (bacterium?)		X		-
Saccomorpha terminalis Radtke, 1991	'*Arborella*' Zebrowski, 1937 or '*Phytophthora*' Bary, 1876 (fungi)	X	X		-
Aurimorpha varia igen. n., isp. n.	? ?		X		-
Ichnoreticulina elegans (Radtke, 1991)	*Ostreobium quekettii* Bornet and Flahault, 1889 (chlorophyte)	X	X		- -
?*Saccomorpha clava* Radtke, 1991	*Dodgella priscus* Zebrowski, 1937 (fungus)	X	X		- -
Orthogonum fusiferum Radtke, 1991	*Ostracoblabe implexa* Bornet and Flahault, 1889 (fungus)	X	X		- -
Polyactina fastigata Radtke, 1991	? ?		X		- -
?*Semidendrina pulchra* Bromley, Wisshak, Glaub and Botquelen, 2007	? (foraminiferan?)	X			- -
'*Palaeoconchocelis starmachii*' Campbell, Kasmierczak and Golubic, 1979	*Porphyra nereocystis* Anderson in Blankinship and Keeler, 1892 (rhodophyte)		X	X	- -

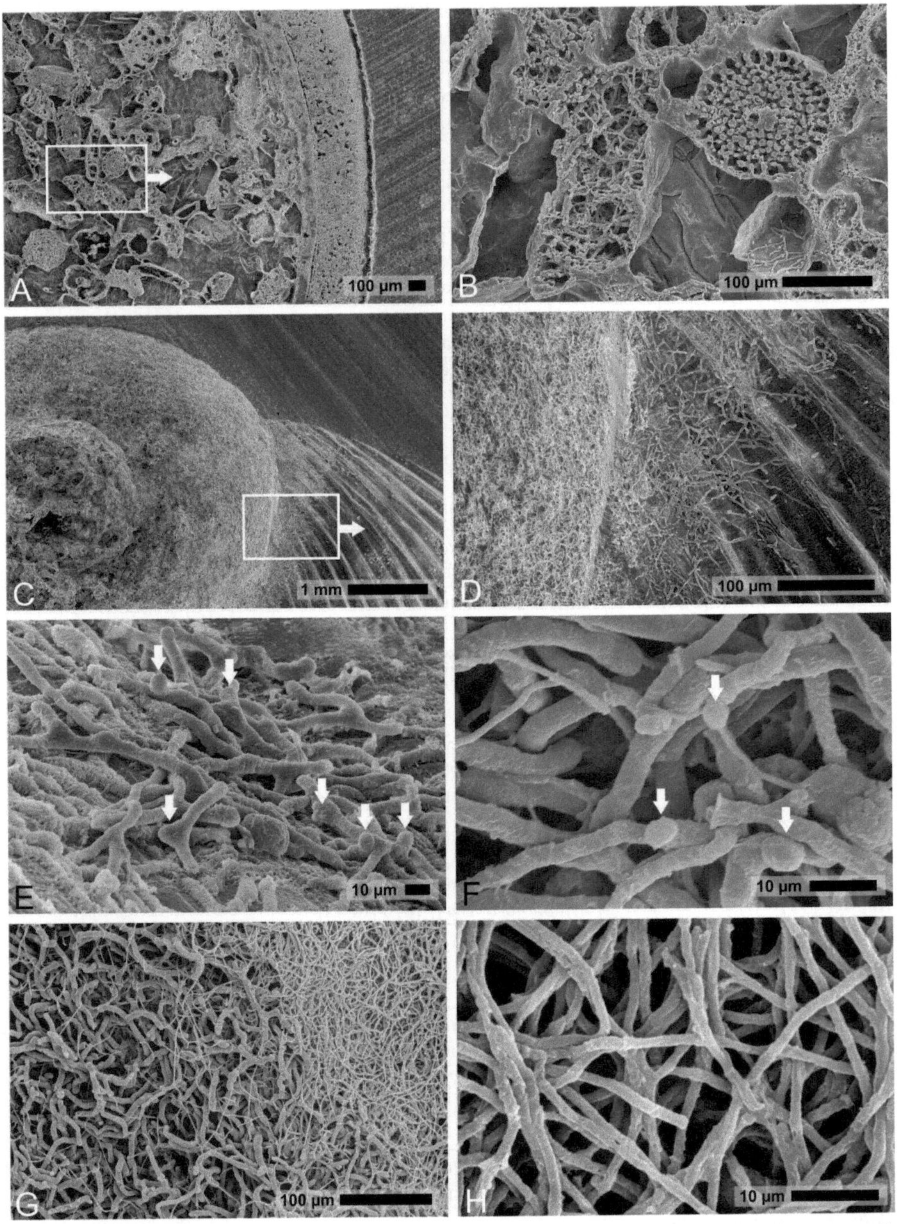

Fig. 4 A Partially etched cross-section of a bioclast-rich matrix within a neritimorph gastropod, with many components showing bioerosion features (#1; G; BSPG 2007 XII 3). **B** Close-up of an echinoid spine and densely bored shell fragments, cemented by coarse calcite spar. **C** Cast of umbilical region of a neritimorph gastropod showing a dense microborer meshwork on the larval shell as opposed to the subsequent whorl (#1; G; BSPG 2007 XII 23a). **D** Close-up of the latter transition. **E** *Eurygonum nodosum* colony with diagnostic lateral swellings indicated by arrows (#1; G; BSPG 2007 XII 23a). **F** Close-up of dense *E. nodosum* network (#1; B; BSPG 2007 XII 12a). **G** Dense pavement of *E. nodosum* to the right overlain by

Eurygonum nodosum Schmidt, 1992

Eurygonum nodosum is by far the most abundant microboring in the Buckhorn material. The borings consist of a three-dimensional network of galleries, about 4-6 μm in diameter, with dichotomous branching and diagnostic lateral nodular appendices or swellings of 4-8 μm in size (Fig. 4E-G). Bifurcations in various angles are common and the terminations of the galleries are round to blunt. The networks may extend several 100 μm into the substrate with a decreasing density and less branching points (Fig. 3A), whereas initial colonies run more closely parallel to the substrate surface (Fig. 4E).

The stratigraphic range of *Eurygonum nodosum* is restricted to the Permian to the Jurassic and the Quaternary to the Recent (Glaub and Vogel 2004) and is now supplemented by the Carboniferous.

In Recent environments, this trace is produced by the very common cyanobacterium *Mastigocoleus testarum* with thalli that bear the diagnostic lateral heterocysts (e.g., Radtke 1991; Gektidis 1997).

Scolecia filosa Radtke, 1991

The 1-2 μm uniformly thin galleries of this trace are usually found collapsed to the cast surface and overlying other traces, and thereby only apparently form curled carpets or dense networks (Fig. 4G-H). The galleries rarely bifurcate and the individual galleries can often be followed for long stretches of more than 100 μm.

Scolecia filosa is known from all periods since the Ordovician except for the Devonian (Glaub and Vogel 2004).

The modern representative of this trace is produced by the filamentous cyanobacterium *Plectonema terebrans* (e.g., Bornet and Flahault 1889; Perkins and Tsentas 1976; Radtke 1991) und *Plectonema endolithicum* (see Glaub 1994) in shallow euphotic down to dysphotic depths (maximum 370 m; Lukas 1978).

Fascichnus dactylus (Radtke, 1991)

In the present material, *Fascichnus dactylus* is by far the most common ichnospecies of this ichnogenus and is quite variable in its morphological appearance. Its thin (4-8 μm in diameter) galleries of uniform diameter penetrate the substrate in oblique to perpendicular manner and bifurcate only rarely if at all. Initial (Fig. 5B) and small (Fig. 5A) colonies show a radiating pattern from a central area of entry whereas other forms build up dense carpets of thin and deeply penetrating galleries with no individual colonies recognisable (Fig. 5C-D).

This ichnotaxon first appears in the Ordovician, reappears in the Permian and extends to the Recent (Glaub and Vogel 2004). The apparent gab in the Palaeozoic is partly closed by the present Carboniferous finding.

In modern seas, this ichnotaxon is produced by a variety of similar endolithic cyanobacteria of the genera *Hyella* and *Solentia*, most common of which is the ubiquitous *Hyella caespitosa* (e.g., Radtke 1991; Gektidis 1997).

---

collapsed *Scolecia filosa* (#1; B; BSPG 2007 XII 12a). **H** Close-up of collapsed filamentous *S. filosa* galleries (#1; B; BSPG 2007 XII 12b). (For a list of abbreviations see 'materials and methods' section)

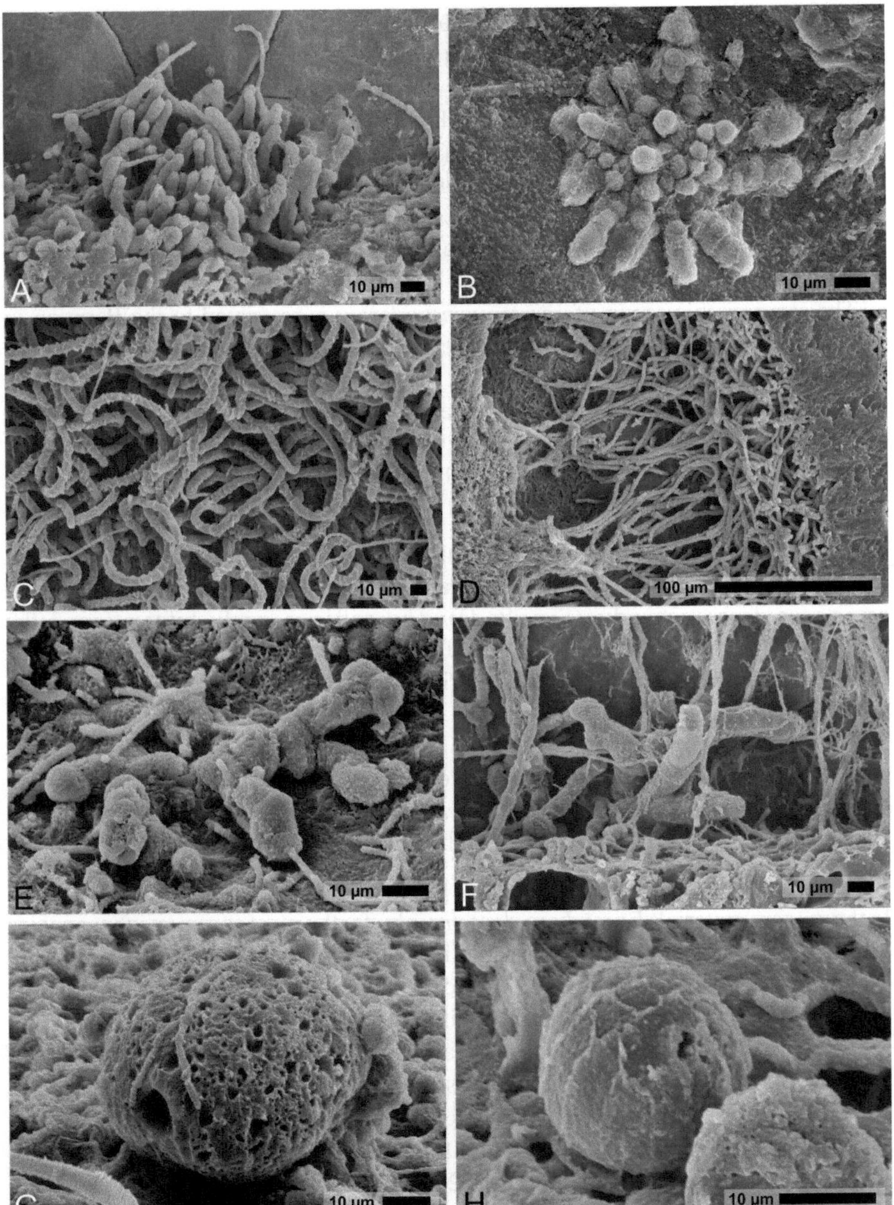

Fig. 5 A Typical cluster of *Fascichnus dactylus* in a partially etched section (#1; G; BSPG 2007 XII 34). **B** Initial radiating colony of *F. dactylus* (#1; G; BSPG 2007 XII 23b). **C** Dense pavement of elongate and unbranched *F. dactylus* (#2; B; BSPG 2007 XII 32a). **D** Partially etched cross-section with deeply penetrating slender *F. dactylus* (#2; G; BSPG 2007 XII 34). **E** Initial colony of *Fascichnus frutex* with distinct cell-like constrictions and bifurcations (#1; G; BSPG 2007 XII 6a). **F** Partially etched sample with branched *F. frutex* (#1; G; BSPG 2007 XII 14a). **G** The globular ichnospecies *Planobola macrogota* (#1; B; BSPG 2007 XII 27a). **H** Semi-spherical *P. macrogota* with diagnostic lateral substrate contact (#1; G; BSPG 2007 XII 14b)

Fascichnus frutex (Radtke, 1991)

In contrast to *Fascichnus dactylus*, this ichnospecies has shorter galleries with larger diameters of 8 to 15 µm and often displays bifurcations (Fig. 5E-F). In addition, cell-like constrictions are a common feature (Fig. 5E). Individual galleries are more or less constant in diameter or may slightly widen distally, and the terminations are round and blunt.

As for the stratigraphic range, *Fascichnus frutex* is so far the oldest representative of this ichnogenus with a record starting already in the Proterozoic represented by the corresponding body fossil of the cyanobacterium *Eohyella rectoclada* (see Green at al. 1988). Further records come from the Ordovician to Silurian and from the Permian to the Recent with a brief record gap in the Neogene (Glaub and Vogel 2004) and is now supplemented by the Carboniferous finding.

According to Radtke (1991), this ichnospecies is best compared to the Recent *Hyella gigas*, a common cyanobacterium in shallow euphotic waters worldwide (e.g., Budd and Perkins 1980; Wisshak et al. 2005).

Planobola macrogota Schmidt, 1992

Featureless spherical to semi-spherical cavities, 15-60 µm in diameter, with latitudinal substrate contact characterise *Planobola macrogota* (Fig. 5G-H). The cavities may either be isolated or clustered in moderate numbers.

Planobola macrogota is known from the Silurian, Carboniferous (this study) and Permian as well as throughout the Mesozoic to the Recent (Glaub and Vogel 2004).

The producer of *Planobola macrogota* is not known but several organisms may form such simple coccoidal cavities, as for instance unicellular cyanobacteria such as *Chroococcus* or *Cyanosaccus* (Lukas and Golubic 1981; Glaub 1994). Further possible producers are chlorophytes (sporophyte phase) or fungal zoospores (May and Perkins 1979).

Cavernula coccidia Glaub, 1994

This small (5-8 µm; much smaller than *Planobola macrogota*) and common trace fossil is characterised by globular coccoidal cavities directly fused with the substrate (Fig. 6B) or connected to the latter by a thick stalk (Fig. 6A). The cavities are found either solitary or associated in often large numbers, without interconnecting galleries (Fig. 6B).

The ichnotaxon *Cavernula coccidia* has as yet only been reported from the Jurassic type locality (Glaub 1994) and its range is herein considerabley extended to the Carboniferous.

The specific identity of the trace maker is unknown and difficult to determine but comprises principally the same options as for *Planobola macrogota* (see above) and in addition – owing to the small size of the traces – also bacteria.

Cavernula pediculata Radtke, 1991

Solitary large (20-35 µm wide and 30-70 µm deep) cavities mostly oriented perpendicular to the substrate surface and connected to the latter by few thick rhizoidal appendages (Fig. 6C-F). The cavity is variable in its morphological appearance ranging from broad dome-shaped forms to narrower dome (Fig. 6C, E)

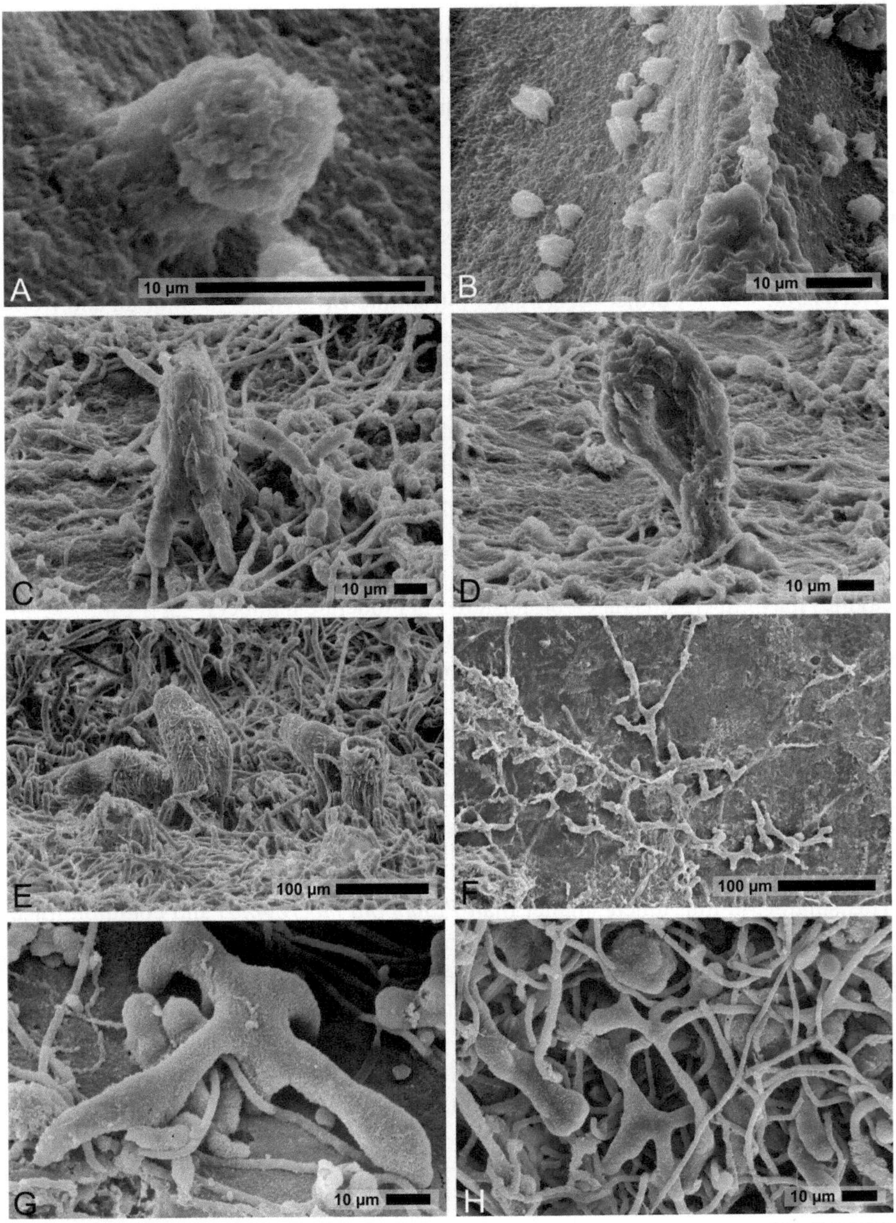

Fig. 6 A The small *Cavernula coccidia* with diagnostic vertical substrate contact (#1; G; BSPG 2007 XII 9a). **B** Clustering *C. coccidia* (#1; G; BSPG 2007 XII 20). **C** Dome-shaped cavity of *Cavernula pediculata* with rhizoidal appendages towards the substrate surface (#1; B; BSPG 2007 XII 28). **D** Slender, sack-shaped variation of *C. pediculata* (#1; B; BSPG 2007 XII 27b). **E** Several large *C. pediculata* towering a microborer meshwork (#1; B; BSPG 2007 XII 32b). **F** Substrate-parallel *Rhopalia catenata* (#1; G; BSPG 2007 XII 22). **G** Close-up of *R. catenata*, connected to the surface by rhizoidal appendages (#1; G; BSPG 2007 XII 44). **H** Microborer layer with variant of *R. catenata* lacking rhizoidal appendages (#2; G; BSPG 2007 XII 33a)

to club-shaped forms (Fig. 6D). The short rhizoids or at least their point of entry is often obscured by a carpet of other traces which are towered by *Cavernula pediculata* (Fig. 6E).

This microboring was previously not reported from the Palaeozoic but is known throughout the Mesozoic to the Recent (Glaub and Vogel 2004).

Cavernula pediculata is a trace produced by the single-celled Codiolum-stage of the chlorophyte *Gomontia polyrhiza* in shallow euphotic Recent settings (e.g., Bornet and Flahault 1889; Kornmann 1959; Radtke 1991).

Rhopalia catenata Radtke, 1991

There is a considerable morphological variability and size range of this ichnotaxon, but all representatives show a closely substrate-parallel system of branched tunnels with prominent spherical to irregular-shaped swellings at the branching points and at the terminations of short lateral side branches (Fig. 6F-H). The galleries and swellings may be connected to the substrate by thin apophyses, which, if present, are often hidden underneath the trace.

The stratigraphic range of this ichnotaxon spans all periods from the early Mesozoic to the Recent (Glaub and Vogel 2004) and is extended to the Carboniferous with the present finding.

Two Recent chlorophyte algae are known to produce *Rhopalia catenata*: *Phaeophila dendroides* and *Eugomontia sacculata.* Both species are subject to considerable morphological variability depending on environmental factors as well as on their multi-staged life cycle (Kornmann 1960; Nielsen 1972; Wilkinson 1974).

Rhopalia spinosa Radtke and Golubic, 2005

This small *Rhopalia* ichnospecies is characterised by a substrate-parallel multi-branched boring system with individual flattened chambers being interconnected by short galleries (Fig. 7A-B). Both the chambers and the interconnecting tunnels are connected to the substrate surface by short, tapering rhizoidal appendages. Individual, often elongate cavities are 6-12 μm wide and the interconnecting passages measure 2-4 μm in circular diameter.

The recently established ichnotaxon was previously reported only from the Palaeogene and the Recent (Radtke and Golubic 2005).

Rhopalia spinosa is best compared to the modern chlorophyte *Phaeophila dendroides* which is a typical representative of shallow-water microfloras from the tropics to cold-temperate settings (Radtke 1993).

Fascichnus rogus (Bundschuh and Balog, 2000)

The cauliflower-shaped colonies of this ichnospecies are about 50-90 μm in diameter and consist of densely spaced to fused radiating galleries with only distally separated rounded terminations, 4-6 μm in diameter (Fig. 7C).

According to Glaub and Vogel (2004), this ichnotaxon spans from the Silurian to the Recent with exception of record gaps in the Permian and Neogene.

Following Bundschuh and Balog (2000), the most likely trace maker of Recent representatives of this ichnospecies are the cyanobacteria *Hyella racemus* and possibly also *Hyella conferta*, both to be typically encountered in shallow tropical waters.

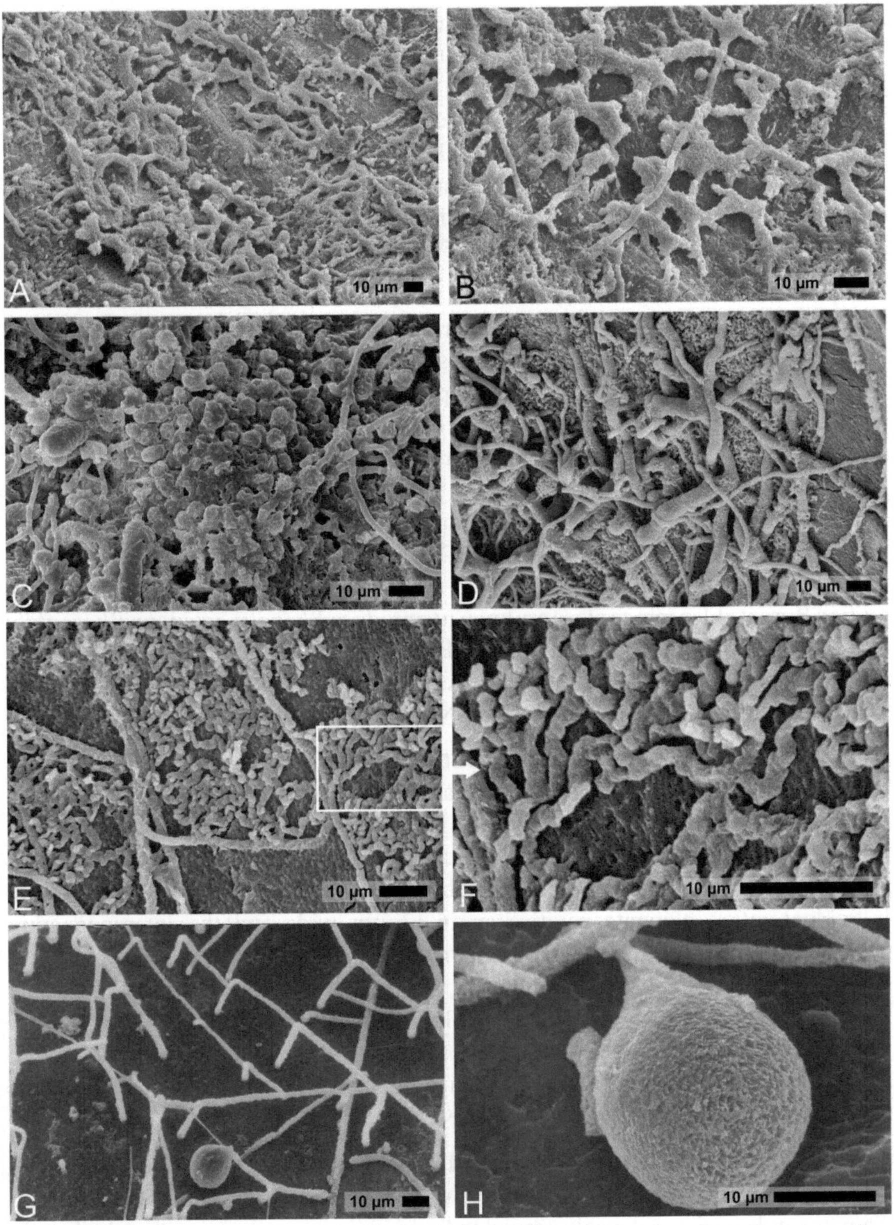

Fig. 7 **A** *Rhopalia spinosa* colony with multiple cavities connected to the substrate surface by many rhizoidal appendages (#1; G; BSPG 2007 XII 15). **B** Close-up of *R. spinosa* colony architecture (#1; G; BSPG 2007 XII 15). **C** Raspberry-shaped, semi-spherical aggregate of *Fascichnus rogus* (#1; B; BSPG 2007 XII 27c). **D** Dense pavement of *Scolecia serrata*, largely obscured by various other microborings (#2; G; BSPG 2007 XII 33b). **E** Typical closely substrate parallel colonies of *S. serrata*, apparently 'trapped' between other microborings (#2; G; BSPG 2007 XII 33b). **F** Close-up of densely-spaced, ramified and meandering *S. serrata* galleries (#2; G; BSPG 2007 XII 33b). **G** Part of a *Saccomorpha terminalis* colony

Scolecia serrata Radtke, 1991

This distinct trace was only rarely encountered and forms characteristic dense pavements close to the substrate surface consisting of irregularly shaped galleries of only 1-1.5 µm in diameter with serrate micro-sculpture (Fig. 7D-F). The galleries are flattened and follow an irregular branching or fusion pattern and often exhibit a meandering course. Thereby the networks fill gaps within the colony rather than surpassing each other. As a result, the colonies are often apparently 'trapped' in areas between other (larger) traces (Fig. 7E). Due to their minute size and very shallow penetration, these traces are often obscured by deeper tiers of the ichnocoenosis (Fig. 7D).

The known stratigraphic range of this ichnotaxon was previously restricted to the Cenozoic Era (Radtke 1991) and now reaches back to the Carboniferous.

The producer of the well known modern equivalent of this trace is still unknown but most authors suggest a bacterial origin of the trace (e.g., Zeff and Perkins 1979; Budd and Perkins 1980; Young and Nelson 1988) and in any case a heterotrophic organism given its appearance down to aphotic depths (e.g., Radtke 1991; Wisshak 2006).

Saccomorpha terminalis Radtke, 1991

The colonies of this ichnotaxon are oriented parallel in some tens of microns distance below the substrate and are connected to the latter in regular intervals by short side branches (Fig. 7G-H). However, the three-dimensional colony architecture (Fig. 7G) is often lost during the preparation due to collapse. The galleries are of uniform diameter (2-3 µm) and at some terminations bear the diagnostic spherical to pear-shaped cavities of 20-25 µm in diameter (Fig. 7H).

This ichnotaxon was previously reported from Triassic, Jurassic and Palaeogene strata (Glaub and Vogel 2004) and is also abundant in Recent settings (e.g., Wisshak et al. 2005). The Carboniferous finding extends the range back to the Palaeozoic.

The identity of the certainly heterotrophic trace maker (appearance down to aphotic depths; Wisshak et al. 2005) is not settled as yet and requires re-evaluation. Corresponding microborings were described and interpreted as fungal borings by Zebrowski (1937) under the genus *Arborella* and by Höhnk (1969) applying the name *Phytophthora*, both of which are not valid taxa of marine fungi in modern mycology.

Aurimorpha varia igen. n., isp. n.

This conspicuous microboring is not known in the literature and is herein established as a new ichnotaxon (see appendix for the formal description). It is characterised by a variably ear- to tongue-shaped morphology oriented oblique to perpendicular into the substrate (Fig. 8A-D). The individual traces are 30-150 µm wide at a thickness of less than 10 µm and extend up to 150 µm into the substrate.

Aside from the present material, related traces which may be assigned to the same ichnogenus in the future were also encountered in Palaeogene strata by Radtke

---

with galleries running with some distance parallel to the substrate surface and showing a diagnostic terminal globular swelling (#3; M). **H** Close-up of globular swelling of *S. terminalis* connected to the thin main gallery (#3; M)

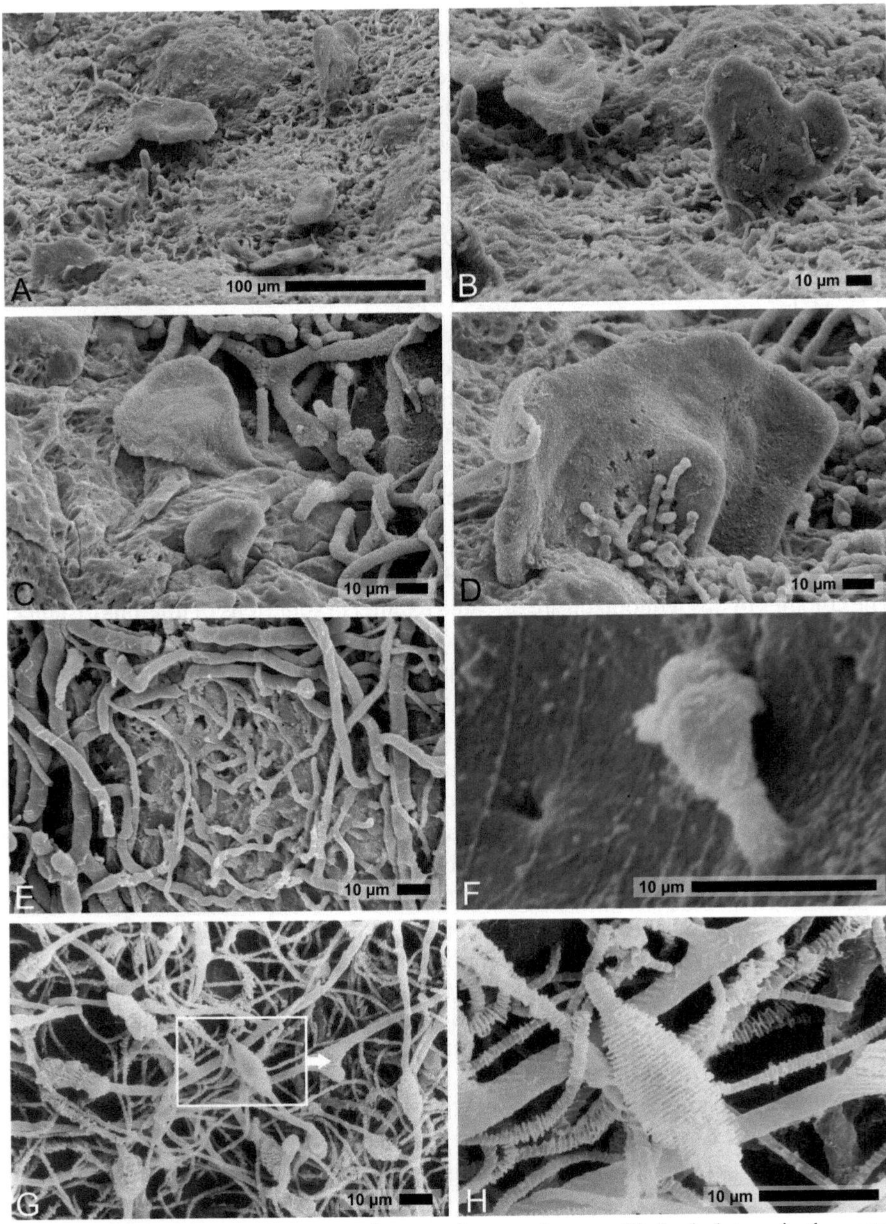

Fig. 8 **A** Five specimens of *Aurimorpha varia* igen. n., isp. n. with the holotype in the upper right (#1; G; BSPG 2007 XII 6b). **B** Part of the same cluster from a different angle with the holotype on the right. **C** Two *A. varia* in a partially etched section with their base obscured by undissolved spar matrix (#2; G; BSPG 2007 XII 36). **D** Large *A. varia* with base hidden in matrix (#2; G; BSPG 2007 XII 36). **E** The rare *Ichnoreticulina elegans* towered by *Eurygonum nodosum* (#1; B; BSPG 2007 XII 12c). **F** Sack-shaped ?*Saccomorpha clava* with a collar (#1; G; BSPG 2007 XII 4). **G** Thin galleries of *Orthogonum fusiferum* (#3; M). **H** Close-up of diagnostic spindle-shaped swelling exhibiting imprint of host-shell ultra-structure (#3; M)

(pers. comm.) and were reported from the Silurian by Bundschuh (2000) applying open nomenclature ('Piatella-Form'), implying an extended stratigraphic range of this ichnogenus from the Silurian into the early Cenozoic. From the Recent, however, corresponding traces are unknown.

The potential trace maker of this ichnotaxon remains enigmatic, leaving many options (e.g., with respect to the mode of metabolism) due to its occurrence in a shallow euphotic ichnocoenosis.

Ichnoreticulina elegans (Radtke, 1991)

In contrast to many fossil and Recent settings, this ichnospecies was encountered only very sporadically in the present material. The ichnospecies is characterised by substrate-parallel branched networks of thin (~3 μm) galleries with flattened-oval cross section (Fig. 8E) and occasional zig-zag course of second order galleries. The colonies are oriented closely parallel to the substrate surface and are thus often obscured by other traces (Fig. 8E).

Ichnoreticulina is known from all geologic periods since the Ordovician and is very common in the Mesozoic and Cenozoic (Glaub and Vogel 2004).

The siphonal alga *Ostreobium quekettii* is the well-known modern producer of this trace (e.g., Kornmann and Sahling 1980) and is most commonly encountered in the deep euphotic zone but is also found in dysphotic depths as the only representative of the chlorophytes, thus serving as key bathymetric ichnospecies (Glaub 1994).

?*Saccomorpha clava* Radtke, 1991

Only few somewhat doubtful specimens of this otherwise very common ichnospecies were encountered which showed the diagnostic club- to sack-shaped morphology (10-20 μm in length) and in one case a moderate collar around its neck (Fig. 8F), but did not display thin interconnecting substrate-parallel galleries.

Saccomorpha clava is known as early as from Ordovician carbonates, has been reported from the Carboniferous and is very common in the Mesozoic and the Cenozoic (Glaub and Vogel 2004).

In modern settings, this ichnospecies is produced by the ubiquitous marine fungus *Dodgella priscus* and is encountered at all depths but most commonly thrives under aphotic conditions (e.g., Zebrowski 1937; Zeff and Perkins 1979; Budd and Perkins 1980; Radtke 1991). *Saccomorpha clava* functions as a key ichnospecies for the aphotic zone together with *Orthogonum lineare* (Glaub 1994).

Orthogonum fusiferum Radtke, 1991

The present *Orthogonum fusiferum* are primarily perpendicular branching, substrate-parallel networks of thin (2-3 μm) galleries with occasional swellings at the branching points and diagnostic intercalar swellings of 4-7 μm in width and about 10-12 μm in length (Fig. 8G-H).

Orthogonum fusiferum is the oldest representative of this ichnogenus and was reported from Ordovician strata to the Recent with several gaps in the record in the Devonian, Permian, Cretaceous and Neogene (Glaub and Vogel 2004).

The marine fungus *Ostracoblabe implexa* with thin hyphae and distinct intercalar swellings is held reponsible for this trace (Radtke 1991).

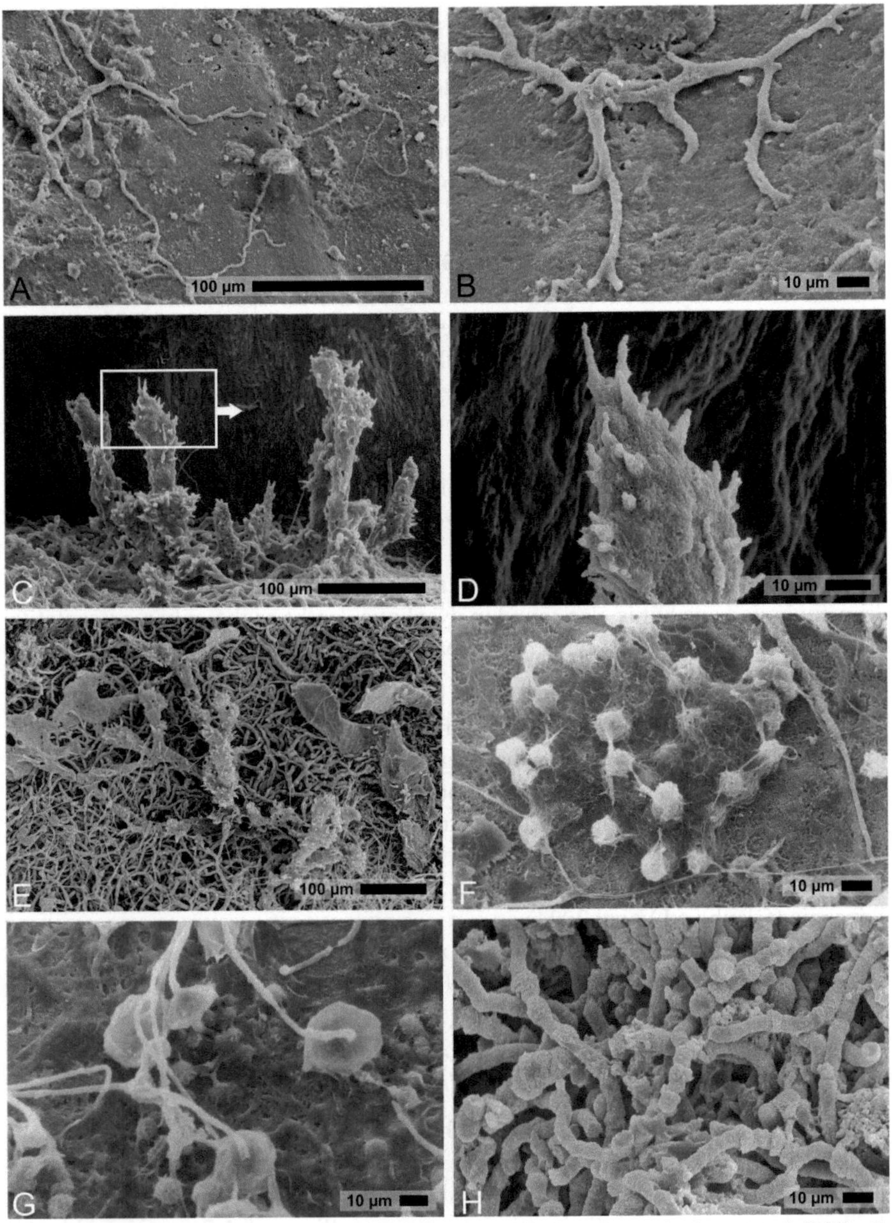

Fig. 9 **A** Tapering galleries of *Polyactina fastigata* with central area in the upper left (#1; G; BSPG 2007 XII 9b). **B** Multiply bifurcating and tapering galleries of *P. fastigata* exhibiting rhizoidal connections to the substrate surface (#1; G; BSPG 2007 XII 9b). **C** Lateral view of the enigmatic ?*Semidendrina pulchra* with several galleries oriented straight into the substrate from a horizontal dendritic main trunk (#1; B; BSPG 2007 XII 12d). **D** Close-up of swollen gallery termination with spiny protrusions (#1; B; BSPG 2007 XII 12d). **E** Dendritic architecture of ?*S. pulchra* (#1; B; BSPG 2007 XII 12d). **F** Central part of a '*Palaeoconchocelis starmachii*' with short galleries tapering towards thin galleries (#3;

Polyactina fastigata Radtke, 1991

This very rare, substrate-parallel, dendritic boring system is characterised by dichotomously ramifying galleries, emerging from a central small cavity or elongated gallery of 5-8 µm in diameter, thereby slowly tapering towards their terminations (Fig. 9A-B). In addition, the galleries are occasionally connected to the substrate surface by tapering rhizoidal appendages (Fig. 9B). Moderate swellings are often present at the bifurcations and may additionally appear along the galleries.

Besides the Recent occurrence reported by Budd and Perkins (1980), this trace has only been reported from the Palaeogene type locality (Radtke 1991), so that the present finding considerably extends the stratigraphic range back into the Palaeozoic.

The producer of the Recent equivalent ('problematic algal form B' after Budd and Perkins 1980) is yet unknown, but Radtke (1991) suggests a heterotrophic organism based on unpublished material from the Mediterranean.

?*Semidendrina pulchra* Bromley, Wisshak, Glaub and Botquelen, 2007

Only one specimen of this enigmatic dendriniform microboring was encountered (Fig. 9C-E). The trace measures 600 µm in length and 300 µm in maximum width at a penetration depth of 240 µm. Its dendritic branches display uni- or bilateral ramifications with angles of 45-55° parallel to the substrate and some of the individual branches fuse to form an anastomosing pattern (Fig. 9E). In addition to the substrate parallel galleries, several branches, some of which with perpendicular sub-branches extend straight into the substrate (Fig. 9C). All branches exhibit diameters of 15-25 µm, have slightly swollen terminations and bear many short whip-shaped protrusions (Fig. 9D). The trace is not entirely typical for this ichnospecies but still within the diagnostic limits, and the main chamber may be small or hidden beneath the surrounding microborer carpet.

Semidendrina pulchra is known from the Carboniferous, Jurassic, ?Cretaceous, Neogene and the Recent (Bromley at al. 2007).

The interpretation concerning the identity of the trace maker has most recently been reevaluated in the course of the formal establishment of this ichnospecies by Bromley et al. (2007). They neither found the previously reported foraminiferan tests in the trace's main chamber (Cherchi and Schroeder 1991) nor an agglutinated chimney described as foraminiferan body fossil on corresponding traces by Plewes et al. (1993). Nevertheless, several lines of reasoning point towards a foraminiferal origin and a naked endolithic type is found a likely candidate (Bromley et al. 2007).

'*Palaeoconchocelis starmachii*' Campbell, Kasmierczak, Golubic, 1979

Identifying this 'ichnospecies' is problematic owing to its status as a poorly defined compound ichnospecies originally described as a combination of body and trace fossil (thus set in quotation marks here) and the fact that the compound form is only rarely observed in direct association. Two morphologies with affinity to this taxon were encountered. Clusters of short round galleries of vertical to sub-vertical

M). **G** Close-up of abruptly tapering galleries and subsequent pronounced swellings of '*P. starmachii*' (#3; M). **H** String-of-pearls appearance of '*P. starmachii*' (#2; C; BSPG 2007 XII 40)

orientation, 10 to 15 µm in diameter, abruptly tapering towards uniformly thick galleries of only 2-3 µm in diameter (Fig. 9F-G) along which pronounced swellings, 3-10 µm in diameter occur (Fig. 9G-H).

The oldest occurrence and type material of '*Palaeoconchocelis starmachii*' was reported from the Silurian (Campbell et al. 1979). It is also known from the Permian and younger periods with the exception of the Cretaceous (Glaub and Vogel 2004).

This problematic trace was interpreted as the Conchocelis-stage of rhodophytes and shows close resemblance to various Recent *Porphyra nereocystis* (Campbell et al. 1979). These Bangiacaea develop conchosporangial branches (first morphotype of the present material) as well as vegetative filaments with monosporangial swellings (second morphotype) during their endolithic Conchocelis-stage (Campbell 1980).

Discussion

Ichnobathymetry

Although some material with unknown fine scale stratigraphic information was included in this analysis, the vast majority of microborings including representatives of all recorded ichnotaxa were encountered in samples which come from one confined bioclastic layer (sample set #1), allowing for a palaeobathymetric interpretation of the latter. It is a lenticular sedimentary body particularly rich in bioclasts which can tentatively be interpreted as channel deposits filled with shallow-water material based on the sediment geometry, the faunal composition (e.g., the dominance of neritimorph gastropods) and the presence of fossil plant remains (see also Squires 1973 and Bandel et al. 2002). It needs, however, to be stressed that the sedimentary succession recorded in the Buckhorn Asphalt quarry comprises very different lithologies, most likely not all reflecting the very same palaeoenvironment. For instance, the deposits which are particularly rich in orthoconic nautiloids yield only a few gastropods. But this has not been studied in detail.

The ichnocoenosis is clearly dominated by *Eurygonum nodosum*, *Scolecia filosa*, and *Fascichnus dactylus*, all of which are produced by cyanobacteria in Recent seas. The second category in relative abundance comprises further probable cyanobacterial borings as there are *Fascichnus frutex*, *Planobola macrogota*, and *Cavernula coccidia*, as well as traces today produced by chlorophyte algae (*Cavernula pediculata* and *Rhopalia catenata*). Such an ichnocoenosis composition and the fact that the cyanobacterial traces follow primarily a more vertical mode of penetration strongly imply a bathymetric position of the palaeoenvironment in the shallow euphotic zone II to III following the general ichnobathymetric scheme after Glaub (1994) and Vogel et al. (1995). All traces produced by heterotrophs (fungi, ?bacteria, ?foraminiferans) are among the rare or very rare constituents of the ichnocoenosis, further reinforcing this interpretation. Specifically *Ichnoreticulina elegans* and '*Palaeoconchocelis starmachii*' – both very common key ichnotaxa of the deep euphotic zone – are encountered only very sporadically in the present material and *Saccomorpha clava* and *Orthogonum lineare* – the corresponding key ichnotaxa for aphotic settings – were only tentatively present or completely absent, respectively.

Evolution of microendoliths

The ichnocoenosis composition in the Buckhorn Asphalt material is by all means 'modern' and is reminiscent of many Recent shallow euphotic examples such as those reported by Günther (1990) from tropical Yucatan (Mexico), by Budd and Perkins (1980) from the Puerto Rican shelf, or by Wisshak et al. (2005) from the cold-temperate Kosterfjord (Sweden). The close similarity is apparent not only in terms of the relative abundance of ichnotaxa but also in the close morphological resemblance of the individual microboring forms, with 17 out of the 19 recorded ichnospecies (90%) also known from the Recent. It is noteworthy that, as in most modern environments, the Buckhorn Asphalt has a comparatively high proportion of shells bearing signs of bioerosion – a fact that contrasts findings in other Palaeozoic strata from the Ordovician (Brett pers. comm.), Silurian (Bundschuh 2000), Devonian (Vogel et al. 1987) and Permian (Balog 1997) with mostly only low to moderate infestation rates or microboring densities. Whether this effect is facies dependent, substrate dependant (these analyses were largely based on brachiopods as host substrates) or a preservational artefact, remains to be answered.

The remarkably diverse microboring assemblage of the Buckhorn Asphalt – the most diverse Palaeozoic assemblage known to date – extends the current knowledge on the stratigraphic range of microendolithic trace fossils. Following the recent review on the stratigraphic record of microborings by Glaub and Vogel (2004), 13 of the present ichnotaxa were recorded for the first time from Carboniferous strata, nine of which (*Eurygonum nodosum, Cavernula pediculata, Cavernula*

Table 2 The Palaeozoic microboring ichnotaxa record in order of first stratigraphic appearance according to Glaub and Vogel (2004) emended by the ichno-inventory in the Buckhorn Asphalt. Body fossils related to *F. frutex* appear already in the Precambrian (arrow)

Ichnotaxon	Cambrian	Ordovician	Silurian	Devonian	Carbonif.	Permian
Fascichnus frutex	←	X	X		X (first)	X
Fascichnus dactylus		X			X (first)	X
Orthogonum fusiferum		X	X		X	
Ichnoreticulina elegans		X	X	X	X	X
Saccomorpha clava		X		X	?X	
Scolecia filosa		X	X		X	X
Fascichnus rogus			X	X	X	
Planobola macrogota			X		X (first)	X
'Palaeoconchocelis starmachii'			X		X (first)	X
Semidendrina pulchra					?X	
Eurygonum nodosum					X (oldest)	X
Cavernula pediculata					X (oldest)	
Cavernula coccidia					X (oldest)	
Rhopalia catenata					X (oldest)	
Rhopalia spinosa					X (oldest)	
Scolecia serrata					X (oldest)	
Polyactina fastigata					X (oldest)	
Saccomorpha terminalis					X (oldest)	
Aurimorpha varia					X (oldest)	

coccidia, *Rhopalia catenata*, *Rhopalia spinosa*, *Scolecia serrata*, *Polyactina fastigata*, *Saccomorpha terminalis*, and *Aurimorpha varia* igen. n., isp. n.) as the oldest record of its ichnotaxon and the latter eight for the first time in Palaeozoic strata (Table 2). Only four ichnotaxa reported by Glaub and Vogel (2004) from Carboniferous or earlier periods were not encountered in the Buckhorn Asphalt as there are *Orthogonum tripartitum*, *Orthogonum lineare*, *Polyactina araneola* and *Hyellomorpha microdendritica* (Table 3). In addition, ichnospecies of the dendrinid ichnogenera *Nododendrina*, *Platydendrina* and *Ramodendrina* (first described by Vogel, Golubic and Brett 1987), which were not considered by Glaub and Vogel (2004) based on their potential status as junior synonyms of *Clionolithes* Clarke, 1908, were included in Table 3, as well as *Pyrodendrina cupra*, established by Tapanila in this volume.

Table 3 Microboring ichnotaxa known from the Carboniferous or older strata (according to Glaub and Vogel 2004; supplemented by dendrinid ichnospecies) that are not recorded in the Buckhorn Asphalt

Ichnotaxon	Cambrian	Ordovician	Silurian	Devonian	Carbonif.	Permian
Orthogonum tripartitum		X			X	
Pyrodendrina cupra		X	X			
Polyactina araneola			X			X
Hyellomorpha microdendritica				X		
Nododendrina nodosa				X		
Platydendrina platycentrum				X		
Platydendrina convexa				X		
Ramodendrina alcicornis				X		
Ramodendrina cervicornis				X		
Orthogonum lineare					X	

What can we deduce from these updated Palaeozoic stratigraphic ranges with respect to microendolith evolution? We can conclude that the previously assumed early Mesozoic radiation of microborers with a first appearance of nine ichnotaxa (Glaub and Vogel 2004) is a sampling artefact because a highly diverse microborer spectrum including a large number of modern ichnotaxa was already established by the late Palaeozoic (Fig. 10). With an overall 21 ichnospecies in the Carboniferous, this assemblage is almost as diverse as the average Mesozoic assemblage and presently bears the highest number of first occurrences per period (11 first occurrences). This also means that during this time, at the latest, the Early Palaeozoic dominance of cyanobacteria among microendoliths had ended and an increasing abundance (competition?) of chlorophyte algae and heterotroph microborers can be recognised. The present extension of the stratigraphical range of probable chlorophyte borings (*Rhopalia catenata*, *Rhopalia spinosa*, *Cavernula pediculata*) implies an early diversification of the Phaeophilaceae and Chlorococcales, which produce Recent equivalents of these traces.

It is, however, important to keep in mind that this pattern, seen in the present state of knowledge on the stratigraphic range of microboring ichnotaxa, is prone to sampling and preservational bias – the Permian for instance has only been tackled

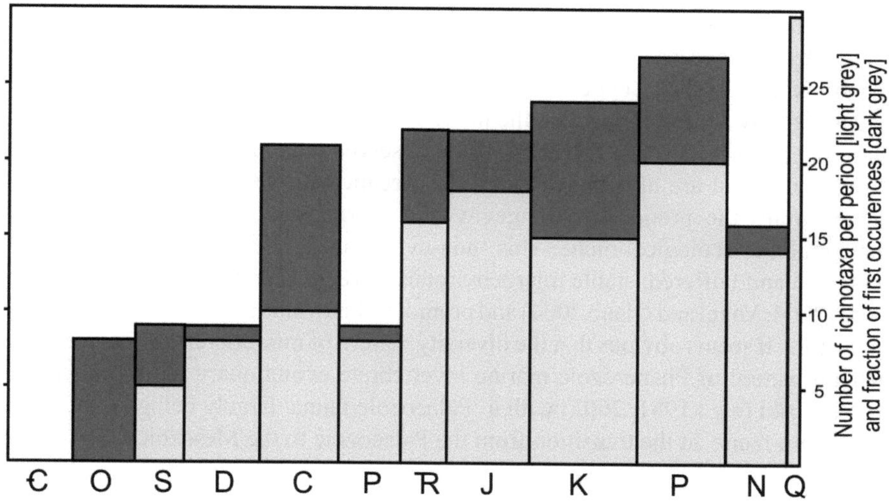

Fig. 10 Graph illustrating the number of known microboring ichnospecies per period on a relative time scale with the fraction of first occurrences (dark grey). This analysis is based upon the compilation by Glaub and Vogel (2004; ichnospecies only; *Orthogonum spinosum* considered synonym to *Orthogonum lineare*), supplemented by new ichnotaxa erected since then (*Rhopalia spinosa* Radtke and Golubic, 2005; *Eurygonum pennaforme* Wisshak, Gektidis, Freiwald and Lundälv, 2005; *Flagrichnus profundus* and *F. baiulus* Wisshak and Porter, 2006; *Semidendrina pulchra* Bromley, Wisshak, Glaub and Botquelen, 2007, *Pyrodendrina cupra* Tapanila, this volume, *Entobia mikra* and *E. nana* Wisshak, this volume, *Rhopalia clavigera* Golubic and Radtke, this volume, and *Kardopomorphos polydioryx* Beuck, López Correa and Freiwald, this volume), additionally supplemented by the dendrinid ichnospecies of the ichnogenera *Platydendrina*, *Nododendrina* and *Ramodendrina* and the new data from the Buckhorn Asphalt of the present study.

marginally (Balog 1996, 1997) and the Cambrian not at all (only in terms of body fossils). For the Ordovician, an extensive study of strata in Ohio and Kentucky (USA) is in progress and preliminary results promise a further extension of stratigraphic ranges (see Vogel and Brett 2006 for an abstract). Thus, more studies on microborers from Palaeozoic 'preservational windows' are needed for a better understanding of the bioeroding assemblages and their evolution.

The diversification pattern nevertheless provokes the speculation whether the microboring evolution parallels that of macroborings and other macroevolutionary phenomena. Currently, a strong increase of macroboring ichnogenera is recognised in the Ordovician and the Devonian (Bromley 2004; Wilson 2007). This is interpreted as the 'Ordovician Bioerosion Revolution' (sensu Wilson and Palmer 2006), being part of the 'Great Ordovician Biodiversification Event' (sensu Webby et al. 2004) and the 'Mid-Palaeozoic precursor to the Mesozoic Marine Revolution' (sensu Signor and Brett 1984). According to Wilson (2007), the 'Mesozoic Marine Revolution' (sensu Vermeij 1977) is also mirrored by a macroborer diversification as a result of enhanced predation pressure but this radiation is not as significant as the Ordovician and Devonian radiations. In contrast, microborers seem not to be affected by the 'Mesozoic Marine Revolution', which would be little plausible anyway because there seems no evident link between a predation driven escalation, mainly by crustaceans (Vermeij 1977, 1983), and microborings. Our new data from the Buckhorn Asphalt suggest that the Permian / Triassic mass extinction event had no or only a minor impact on the microboring ecological niche (see also Glaub and Vogel 2004). The fact that 90% of the observed ichnotaxa from the Buckhorn Asphalt material are also known from the Recent with very similar morphology demonstrates the pronounced longevity and morphostasis of representatives of the endolithic ecological niche. This 'non-evolution' is seen as a result of a well-protected and buffered, stable microenvironment (e.g., Campbell 1980; Glaub and Vogel 2004; Vogel and Glaub 2004) and promotes their value as palaeoenvironmental indicators. It seems obvious that the diversity pattern of microborers differs from the general pattern of Phanerozoic marine invertebrate evolutionary faunas described by Sepkoski (e.g., 1981, 2002) with a 'Palaeozoic fauna' largely being replaced by a 'Modern fauna' at the transition from the Palaeozoic to the Mesozoic.

Conclusions

A remarkably high diversity of microborings – the most diverse Palaeozoic assemblage known to date – is indicated by thin-sections and surface features of the asphalt-impregnated shelly fauna, and is unravelled in detail by SEM analysis of epoxy resin casts. A wide range of ichnocoenosis maturity and preservational quality is developed, yielding a total of 18 known ichnospecies including one new ichnotaxon, the ear-shaped *Aurimorpha varia* igen. n., isp. n.

The ichnocoenosis is clearly dominated by the cyanobacterial traces *Eurygonum nodosum*, *Scolecia filosa* and *Fascichnus dactylus* in concert with the common cyanobacterial borings *Fascichnus frutex*, *Planobola macrogota* and *Cavernula coccidia* and the chlorophyte traces *Cavernula pediculata* and *Rhopalia catenata*. Traces presumably produced by heterotroph microendoliths are diverse but rare constituents of the microboring assemblage. This ichnocoenosis composition indicates a palaeobathymetry in the shallow euphotic zone II to III following the established ichnobathymetric scheme.

This ichnocoenosis composition is 'modern' in all respects with 17 out of the 19 recorded ichnospecies also known from the Recent, and closely resembles assemblages of modern shallow euphotic seas.

The present microboring assemblage extends the known stratigraphic range of microendolithic trace fossils. A total of 13 ichnotaxa is recorded for the first time from the Carboniferous, nine of which (*Eurygonum nodosum*, *Cavernula pediculata*, *Cavernula coccidia*, *Rhopalia catenata*, *Rhopalia spinosa*, *Scolecia serrata*, *Polyactina fastigata*, *Saccomorpha terminalis*, and *Aurimorpha varia* igen. n., isp. n.) as oldest representatives of the respective ichnospecies and except for *Eurygonum nodosum* for the first time in Palaeozoic strata.

These results demonstrate that a highly diverse microborer spectrum of at least 21 ichnospecies, including a considerable number of modern ichnotaxa, were already established in the late Palaeozoic. By this time, the strong dominance of endolithic cyanobacteria was broken with increasing abundance and potential competition by chlorophyte algae and increasingly diverse heterotroph microborers.

An updated diversity curve of Phanerozoic microboring diversity according to the current state of knowledge suggests that the end-Permian mass extinction event had little or no impact on the diversity of microborers. However, it appears that microborer diversity parallels that of the macroborers with major radiations in the Ordovician ('Ordovician Bioerosion Revolution') and Devonian ('Mid-Palaeozoic precursor to the Mesozoic Marine Revolution'). Bioeroding microbiota seem not to follow the general pattern of Sepkoski's Phanerozoic marine evolutionary faunas with a 'Palaeozoic fauna' having been replaced by a 'Modern fauna' by the early Mesozoic.

Acknowledgements

We are most grateful to R. H. Mapes (Ohio University, USA) and T. E. Yancey (Texas A&M University, USA) for their commitment during the field campaign as well as K. P. Vogel (J. W. Goethe-University Frankfurt, Germany) for contributing his results based on additional material provided by R. E. Crick (University of Texas at Arlington, USA). I. Glaub (J. W. Goethe-University Frankfurt, Germany) and C. E. Brett (University of Cincinnati, USA) provided useful reviews. The de-asphaltation procedure was kindly supported by M. Brettreich and T. Schunk (Erlangen University, Germany) and thin-sections were prepared by B. Leipner-Mata (Erlangen University, Germany). This work is financially supported by the Deutsche Forschungsgemeinschaft (DFG NU 96/10-1).

References

Al-Thukair AA, Golubic S (1991) Five new *Hyella* species from the Arabian Gulf. In: Hickel B, Anagnostidis K, Komarek J (eds) Cyanophyta / Cyanobacteria - morphology, taxonomy, ecology. Arch Hydrobiol, Suppl 92, Algol Stud 64:167-197

Al-Thukair AA, Golubic S, Rosen G (1994) New endolithic cyanobacteria from the Bahama Bank and the Arabian Gulf: *Hyella racemus* sp. nov. J Phycol 30:764-769

Balog S-J (1996) Boring thallophytes in some Permian and Triassic reefs: bathymetry and bioerosion. Göttinger Arb Geol Pal Sb 2:305-309

Balog S-J (1997) Mikroendolithen im Capitan Riff Komplex (New Mexico, USA). Courier Forschinst Senckenberg 201:45-55

Bandel K, Nützel A, Yancey TE (2002) Larval shells and shell microstructures of exceptionally well-preserved Late Carboniferous gastropods from the Buckhorn Asphalt deposit (Oklahoma, USA). Senckenbergiana Lethaea 82:639-689

Bary A de (1876) Researches into the nature of the potato-fungus *Phytophthora infestans*. J R Agric Soc 12:239-268

Batters EAL (1902) A catalogue of the British marine algae. J Bot Suppl 40:1-107

Beuck L, López Correa M, Freiwald, A (this volume) Biogeographical distribution of *Hyrrokkin* (Rosalinidae, Foraminifera) and its host-specific morphological and textural variability. In: Wisshak M, Tapanila L (eds) Current developments in bioerosion. Springer, Berlin, pp 329-360

Blankinship JW, Keeler CA (1892) On the natural history of the Farallon Islands. Zoe 3:144-165

Blind W (1991) Über Anlage und Funktion von Kammerablagerungen in Orthoceren-Gehäusen. Palaeontographica A 218:35-47

Bornet E, Flahault C (1888) Note sur deux nouveaux genres d'algues perforantes. J Bot, Morot 2:161-165

Bornet E, Flahault C (1889) Sur quelques plantes vivant dans le test calcaire des mollusques. Bull Sci Bot France 36:147-179

Brand U (1982) The oxygen and carbon isotope composition of Carboniferous fossil components: sea-water effects. Sedimentology 29:139-147

Brand U (1987) Biogeochemistry of nautiloids and palaeoenvironmental aspects of Buckhorn seawater (Pennsylvanian), southern Oklahoma. Palaeogeogr Palaeoclimatol Palaeoecol 61:255-264

Brand U (1989) Aragonite-calcite transformation based on Pennsylvanian molluscs. Geol Soc Amer Bull 101:377-390

Bromley RG (2004) A stratigraphy of marine bioerosion. In: McIlroy D (ed) The application of ichnology to palaeoenvironmental and stratigraphic analysis. Geol Soc London Spec Publ 228:455-481

Bromley RG, Wisshak M, Glaub I, Boutquelen A (2007) Ichnotaxonomic review of dendriniform borings attributed to foraminifera: *Semidendrina* igen. nov. In: Miller III W (ed) Trace fossils: concepts, problems, prospects. Elsevier, Amsterdam, pp 518-530

Brown AA, Corrigan J (1997) Petroleum systems, Ardmore Basin and Arbuckle Mountains, Oklahoma. Guideb Field Trip 2, AAPG Annu Convent, Dallas Geol Soc, Dallas, pp 64-68

Budd DA, Perkins RD (1980) Bathymetric zonation and paleoecological significance of microborings in Puerto Rican shelf and slope sediments. J Sediment Petrol 50:881-904

Bundschuh M (2000) Silurische Mikrobohrspuren, ihre Beschreibung und Verteilung in verschiedenen Faziesräumen (Schweden, Litauen, Großbritannien und U.S.A.). PhD Thesis, Fachbereich Geowiss, Univ Frankfurt, 129 pp

Bundschuh M, Balog S-J (2000) *Fasciculus rogus* nov. isp., an endolithic trace fossil. Ichnos 7:149-152

Campbell SE (1980) *Palaeoconchocelis starmacchii,* a carbonate boring microfossil from the Upper Silurian of Poland (425 million years old): implications for the evolution of the Bangiaceae (Rhodophyta). Phycologia 19:25-36

Campbell S, Kazmierczak J, Golubic S (1979) *Palaeoconchocelis starmachii* gen. n., sp. n., an endolithic rhodophyte (Bangiaceae) from the Silurian of Poland. Acta Palaeont Pol 24:405-408

Cherchi A, Schroeder R (1991) Perforations branchues dues à des Foraminifères cryptobiotiques dans des coquilles actuelles et fossiles. C R Acad Sci Paris 312:111-115

Clarke JM (1908) The beginnings of dependent life. New York State Mus Bull 121:146-169

Cree SB (1984) A biostratigraphic study of the asphalt-bearing limestones of Pennsylvanian age in the Arbuckle Mountains, Oklahoma. Unpubl Master Thesis, Univ Texas, Arlington, 82 pp

Gektidis M (1997) Vorkommen, Ökologie und Taxonomie von Mikrobohrorganismen in ausgewählten Riffbereichen um die Insel Lee Stocking Island (Bahamas) und One Tree Island (Australien). PhD Thesis, Univ Frankfurt, 276 pp

Glaub I (1994) Mikrobohrspuren in ausgewählten Ablagerungsräumen des europäischen Jura und der Unterkreide (Klassifikation und Palökologie). Courier Forschinst Senckenberg 174:1-324

Glaub I, Bundschuh M (1997) Comparative study on Silurian and Jurassic/Lower Cretaceous microborings. Courier Forschinst Senckenberg 201:123-135

Glaub I, Vogel K (2004) The stratigraphic record of microborings. Fossils and Strata 51:126-135

Glaub I, Vogel K, Gektidis M (2001) The role of modern and fossil cyanobacterial borings in bioerosion and bathymetry. Ichnos 8:185-195

Golubic S, Radtke G (this volume) The trace *Rhopalia clavigera* isp. n. reflects the development of its maker *Eugomontia sacculata* Kornmann, 1960. In: Wisshak M, Tapanila L (eds) Current developments in bioerosion. Springer, Berlin, pp 95-108

Golubic S, Brent G, LeCampion T (1970) Scanning electron microscopy of endolithic algae and fungi using a multipurpose casting-embedding technique. Lethaia 3:203-209

Golubic S, Campbell S, Spaeth C (1983) Kunstharzausgüsse fossiler Mikroben-Bohrgänge. Präparator 29:197-200

Green JW, Knoll AH, Swett K (1988) Microfossils from oolites and pisolites of the Upper Proterozoic Eleonore Bay Group, Central east Greenland. J Paleont 62:835-852

Günther A (1990) Distribution and bathymetric zonation of shell-boring endoliths in recent reef and shelf environments: Cozumel, Yucatan (Mexico). Facies 22:233-262

Heaney MJ (1997) Pennsylvanian aged bivalves (Mollusca) from the Buckhorn Asphalt quarry of south-central Oklahoma: taxonomy and systematics. Unpubl PhD Thesis, Texas A&M Univ, College Stn, Texas, 108 pp

Heaney MJ, Yancey TE (1998) Conocardioid molluscs from the Buckhorn Asphalt quarry of south-central Oklahoma: rostroconchs or rostroconch homeomorphic bivalves. GSA Abstr Programs 30(3):7

Höhnk W (1969) Über den pilzlichen Befall kalkiger Hartteile von Meerestieren. Ber Dt Wiss Komm 20:129-140

Kornmann P (1959) Die heterogene Gattung *Gomontia*. I. Der sporangiale Anteil, *Codiolum polyrhizum*. Helgoländer Wiss Meeresunters 6:229-238

Kornmann P (1960) Die heterogene Gattung *Gomontia*. II. Der fädige Anteil, *Eugomontia sacculata* nov. gen. nov. spec. Helgoländer Wiss Meeresunters 7:59-71

Kornmann P, Sahling P-H (1980) *Ostreobium quekettii* (Codiales, Chlorophyta). Helgoländer Wiss Meeresunters 34:115-122

Kulicki C, Landmann NH, Heaney MJ, Mapes RH, Tanabe K (2002) Morphology of the early whorls of goniatites from the Carboniferous Buckhorn Asphalt (Oklahoma) with aragonitic preservation. Abh Geol Bundesanst 57:205-224

Lagerheim G (1886) Notes sur le *Mastigocoleus*, nouveau genre des algues marines de l'ordre des Phycochromacées. Notarisia 1:65-69

Lukas KJ (1978) Depth distribution and form among common microboring algae from the Florida continental shelf. Geol Soc Amer, Abstr Programs 10:448

Lukas KJ, Golubic S (1981) New endolithic cyanophytes from the North Atlantic Ocean. I. *Cyanosaccus piriformis* gen. et sp. nov. J Phycol 17:224-229

Lukas KJ, Golubic S (1983) New endolithic cyanophytes from the North Atlantic Ocean. II. *Hyella gigas* Lukas & Golubic sp. nov. from the Florida continental margin. J Phycol 19:129-136

Mapes RH, Mapes G, Yancey TE, Liu ZH (2000) Fossil plants from the Buckhorn lagerstätte (Late Carboniferous - Desmoinesian) in southern Oklahoma. GSA Annu Meet 30(7): A-15

May JA, Perkins RD (1979) Endolithic infestation of carbonate substrates below the sediment-water interface. J Sediment Petrol 49:357-378

Nielsen R (1972) A study of the shell-boring marine algae around the Danish island Læsø. Bot Tidsskr 67:245-269

Olempska E (1986) Endolithic microorganisms in Ordovician ostracod valves. Acta Palaeont Pol 31:229-236

Perkins RD, Tsentas CI (1976) Microbial infestations of carbonate substrates planted on the St. Croix shelf. Geol Soc Amer Bull 87:1615-1628

Plewes CR, Palmer TJ, Haynes JR (1993) A boring foraminiferan from the Upper Jurassic of England and Northern France. J Micropalaeont 12:83-89

Radtke G (1991) Die mikroendolithischen Spurenfossilien im Alt-Tertiär West-Europas und ihre palökologische Bedeutung. Courier Forschinst Senckenberg 138:1-185

Radtke G (1993) The distribution of microborings in molluscan shells from Recent reef environments at Lee Stocking Island, Bahamas. Facies 29:81-92

Radtke G, Golubic S (2005) Microborings in mollusc shells, Bay of Safaga, Egypt: morphometry and ichnology. Facies 51:118-134

Runnegar B (1985) Early Cambrian endolithic algae. Alcheringa 9:179-182

Sadd JL (1991) Tectonic influences on carbonate deposition and diagenesis, Buckhorn Asphalt, Deese Group (Desmoinesian), Arbuckle Mountains, Oklahoma. J Sediment Petrol 61:28-42

Sadd JL, Peterson N, Kendall C (1986) Origin of Buckhorn Asphalt, Deese Group (Pennsylvanian), central Arbuckle Mountains, Oklahoma. AAPG Bull 70:643

Schmidt H (1992) Mikrobohrspuren ausgewählter Faziesbereiche der tethyalen und germanischen Trias (Beschreibung, Vergleich und bathymetrische Interpretation). Frankfurter Geowiss Arb A 12:1-228

Sepkoski JJ Jr (1981) A factor analytic description of the Phanerzoic marine fossil record. Paleobiology 7:36-53

Sepkoski JJ Jr (2002) A compendium of fossil marine animal genera. Bull Amer Paleont 363:1-560

Signor PW, Brett CE (1984) The mid-Paleozoic precursor to the Mesozoic marine revolution. Palaeobiology 10:229-245

Squires RL (1973) Burial environment, diagenesis, mineralogy, and Mg & Sr contents of skeletal carbonates in the Buckhorn Asphalt of middle Pennsylvanian age, Arbuckle Mountains, Oklahoma. Unpubl PhD Thesis, California Inst Technol, Pasadena, 184 pp

Squires RL (1976) A colour pattern of *Naticopsis* (*Naticopsis*) *wortheniana*, Buckhorn Asphalt deposit, Oklahoma. J Paleont 50:349-350

Tapanila L (this volume) The medium is the message: imaging a complex microboring (*Pyrodendrina cupra* igen. n., isp. n.) from the early Paleozoic of Anticosti Island, Canada. In: Wisshak M, Tapanila L (eds) Current developments in bioerosion. Springer, Berlin, pp 123-146

Vermeij GJ (1977) The Mesozoic marine revolution: evidence from snails, predators and grazers. Paleobiology 3:245-258

Vermeij GJ (1983) Traces and trends of predation, with special reference to bivalved animals. Palaeontology 26:455-465

Vogel K (1991) Comment on: Delle Phosphatic Member: an anomalous phosphatic interval in the Mississipian (Osagean-Meramecian) shelf sequence of central Utah. Newsl Stratigr 24:109-110

Vogel K, Brett CE (2006) Microendoliths: early evolution of ichnotaxa and ichnocoenoses and their ecological tolerances. In: Wisshak M, Löffler S-B, Schulbert C, Freiwald A (eds) 5th Int Bioerosion Workshop, Programme Abstr, Erlangen, pp 13-15

Vogel K, Glaub I (2004) 450 Millionen Jahre Beständigkeit in der Evolution endolithischer Mikroorganismen? Sitzber Wiss Ges Univ Frankfurt 42:1-42

Vogel K, Golubic S, Brett CE (1987) Endolith associations and their relation to facies distribution in the Middle Devonian of New York State, U.S.A. Lethaia 20:263-290

Vogel K, Bundschuh M, Glaub I, Hofmann K, Radtke G, Schmidt H (1995) Hard substrate ichnocoenoses and their relations to light intensity and marine bathymetry. N Jb Geol Paläont Abh 195:49-61

Webby BD, Paris F, Droser M, Percival IG (2004) The great Ordovician biodiversification event. Columbia Univ Press, New York, 484 pp

Wilkinson M (1974) Investigations on the autecology of *Eugomontia sacculata* Kornmann, a shell-boring alga. J Exp Mar Biol Ecol 16:19-27

Wilson MA (2007) Macroborings and the evolution of marine bioerosion. In: Miller III W (ed) Trace fossils: concepts, problems, prospects. Elsevier, Amsterdam, pp 356-367

Wilson MA, Palmer TJ (2006) Patterns and process in the Ordovician bioerosion revolution. Ichnos 13:109-112

Wisshak M (this volume) Two new dwarf *Entobia* ichnospecies in a diverse aphotic ichnocoenosis (Pleistocene / Rhodes, Greece). In: Wisshak M, Tapanila L (eds) Current developments in bioerosion. Springer, Berlin, pp 213-234

Wisshak M (2006) High-latitude bioerosion: the Kosterfjord experiment. Lect Notes Earth Sci 109: 1-202

Wisshak M, Porter D (2006) The new ichnogenus *Flagrichnus* - a palaeoenvironmental indicator for cold-water settings? Ichnos 13:135-145

Wisshak M, Gektidis M, Freiwald A, Lundälv T (2005) Bioerosion along a bathymetric gradient in a cold-temperate setting (Kosterfjord, SW Sweden): an experimental study. Facies 51:93-117

Yancey TE, Heaney MJ (2000) Carboniferous praecardioid bivalves from the exceptional Buckhorn Asphalt biota of south-central Oklahoma, USA. In: Harper EM, Taylor JD, Crame JA (eds) The evolutionary biology of the Bivalvia. Geol Soc Spec Publ 177:291-301

Young HR, Nelson CS (1988) Endolithic biodegradation of cool-water skeletal carbonates on Scott shelf, northwestern Vancouver Island. Canad J Sediment Geol 60:251-267

Zebrowski G (1937) New genera of Cladochytriaceae. Ann Missouri Bot Gard 23:553-564

Zeff ML, Perkins RD (1979) Microbial alteration of Bahamian deep-sea carbonates. Sedimentology 26:175-201

Zhang Y, Golubic S (1987) Endolithic microfossils (Cyanophyta) from early Proterozoic stromatolites, Hebei, China. Acta Micropaleont Sin 4:1-12

Appendix – Systematic ichnology

Aurimorpha **igen. n.**

Diagnosis: Ear- to tongue-shaped flat microboring cavities, widening but not thickening from an sub-circular to elongate area of entry and entering the substrate obliquely to perpendicular.

Differential diagnosis: The trace is clearly distinguished from all other microboring ichnotaxa. There is only some superficial similarity with the macroboring ichnogenus *Caulostrepsis* Clarke, 1908, which, however, exhibits a more clearly defined bent marginal tunnel interconnected by a vane and is much larger in dimension.

Etymology: auris, Latin = ear; *-morpha*, Latin = -shaped

Type ichnospecies: Aurimorpha varia isp. n.

Aurimorpha varia isp. n.
Fig. 8A-D

Diagnosis: Ear- to tongue-shaped microboring cavities of variable, occasionally incised outline, widening but not thickening from an elongate area of entry. The thickness of the tongue is uniform or slightly thinning towards its centre. The traces are entering the substrate in oblique to perpendicular orientation.

Etymology: varia, Latin = varied

Type material, locality and horizon: Epoxy resin cast hosted in the Bayerische Staatssammlung für Paläontologie und Geologie, München (BSPG 2007 XII 6b) with several specimen including the holotype (Fig. 8A upper right and Fig. 8B right) taken from the outside of a neritimorph gastropod shell. Three further traces are found in a partially etched epoxy cast (BSPG 2007 XII 36) taken from an undetermined gastropod fragment. Buckhorn Asphalt Quarry, Oklahoma, USA. Buckhorn Asphalt / Boggy Formation / Deese Group / Desmoinesian / middle Pennsylvanian.

Description: The individual traces are 30-150 µm wide at a thickness of less than 10 µm and extend up to 150 µm obliquely to perpendicular into the substrate. Thereby the cavity is widening from an elongate point of entry, which is in epoxy resin casts often obscured by other microborings. The outline of the boring is variably ear- to tongue-shaped or irregular-shaped with incisions. The central area of the traces is often thinner than the margin, which is in any case less than 10 µm thick. The traces appear solitary or clustered in moderate numbers.

Stratigraphic range: Middle Pennsylvanian

Remarks: There is one trace reported in the literature with some similarity to *Aurimorpha* and potentially assignable to the latter as further ichnospecies in the future, as there is the 'Piatella-Form' reported by Bundschuh (2000: 69-70, table 10, figs. 1-3). This trace is in contrast to *A. varia* always oriented closely parallel to the substrate and often exhibits a more pronounced slender proximal part. Further and better preserved traces of this form were encountered by Radtke in the Palaeogene of the Paris Basin (pers. comm.).

Enigmatic organisms preserved in early Ordovician macroborings, western Utah, USA

Jacob S. Benner[1], Allan A. Ekdale[2], Jordi M. de Gibert[3]

[1] Department of Geology, Tufts University, Medford, MA 02155, USA, (jacob.benner@tufts.edu)

[2] Department of Geology and Geophysics, University of Utah, Salt Lake City, UT 84112, USA

[3] Departament d'Estratigrafia, Paleontología i Geociències Marines, Universitat de Barcelona, 08028 Barcelona, Spain

Abstract. Macroborings in the Lower Ordovician Fillmore Formation, western Utah, USA, occasionally contain fossil remains of enigmatic organisms. In the most complete specimens a common morphology can be observed. The calcified body wall of the animal is vase-shaped, mimicking the shape of the boring itself. An ovoid body leads up to a neck that contains either a single or double cylinder near the aperture of the boring. The incomplete preservation of the specimens is not sufficient to identify the biological affinity of the organism at this time, but a review of potential groups is warranted. While such groups as barnacles, bivalves, mitrates, and a host of worm-like forms are potential boring inhabitants, none fit what is known of the morphology of the specimens from Utah. Regardless, recognition and future identification of these animals will lead to a greater understanding of complex hardground trophic systems during the Ordovician Bioerosion Revolution.

Keywords. Bioerosion, macroboring, hardground, Ordovician Bioerosion Revolution, Fillmore Formation

Introduction

No body fossils representing organisms that are obviously capable of macroboring behavior have been found in rocks prior to the Pennsylvanian, and macroborings themselves are not abundant in rocks earlier than that age. Macroborings identified as *Gastrochaenolites* Leymerie, 1842 occur in the Lower Ordovician Fillmore Formation at Skull Rock Pass (SRP), western Utah, USA (Ekdale et al. 2002; Benner et al. 2004). Several specimens of these macroborings contain the presumed body fossils of enigmatic organisms. This is only the second known occurrence of an animal fossilized inside a pre-Pennsylvanian macroboring (Wilson and Palmer 1998).

Because of the physically and chemically erosive environments in which hardground dwellers live, it is unlikely that a borer without a thickly shelled body would

M. Wisshak, L. Tapanila (eds.), *Current Developments in Bioerosion*. Erlangen Earth Conference Series, doi: 10.1007/978-3-540-77598-0_3,

survive. While it is unknown whether the organism was responsible for excavating the boring or was a secondary inhabitant, it is clear that these specimens represent an important source of new information regarding hard substrate infaunalization during the Ordovician Bioerosion Revolution (Wilson and Palmer 2006).

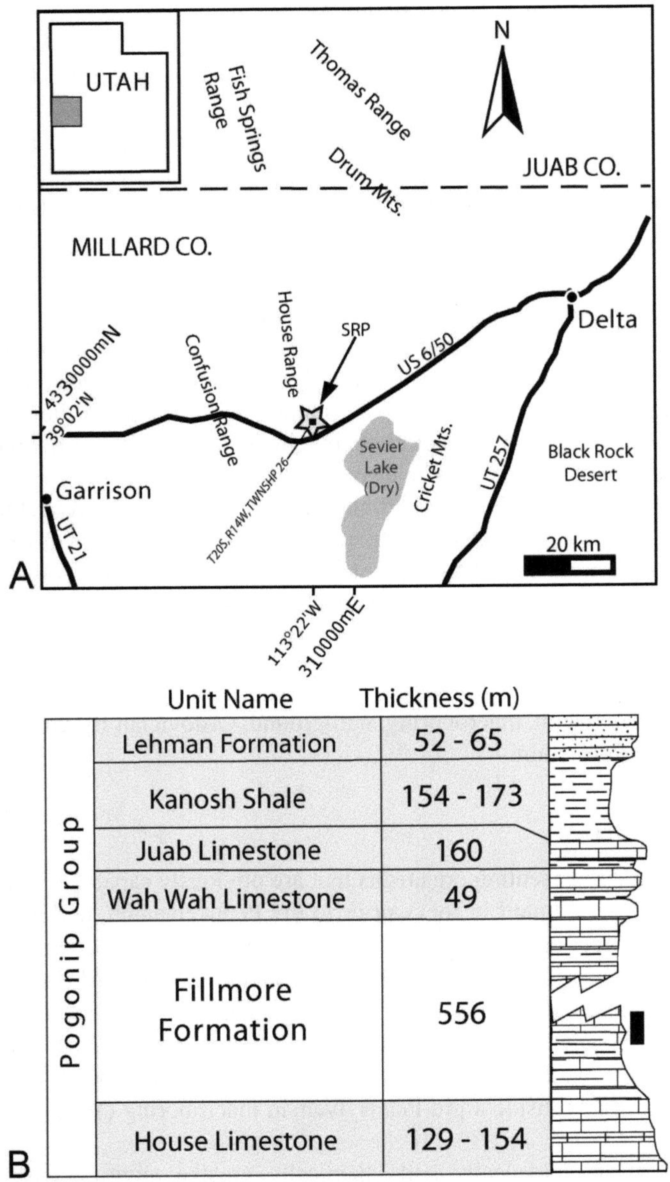

Fig. 1 Location and stratigraphy of the study area in the Fillmore Formation. **A** Pyramid Section locality at Skull Rock Pass, Utah. **B** Generalized stratigraphy of the Lower Ordovician Fillmore Formation showing the location of the Pyramid Section (black bar). Modified from Hintze (1973)

Geological setting

Macroborings containing the enigmatic organisms discussed in this paper occur in hardground surfaces in the Lower Ordovician (Ibexian) Fillmore Formation at SRP in Millard County, Utah (Fig. 1A). At the study locality, nearly 100 m of limestone and shale are exposed in outcrop, arranged in meter-scale shallowing-upward cycles (Hintze 1973; Dattilo 1993; Evans et al. 2003) (Fig. 1B). Surfaces interpreted as hardgrounds are present on many different rock types, including nodular lime mudstone, sponge-algal mud mounds, and flat pebble conglomerates (Dattilo 1993; Benner et al. 2004).

Description of body fossils within borings

In situ body fossils were found inside borings in four limestone cobbles from SRP hardground surfaces (J. S. Benner 2002, Masters Thesis Collection, Department of Geology and Geophysics, University of Utah). These surfaces were examined in the field for the presence of borings (Benner 2002; Benner et al. 2004), but it was not until portions of hardgrounds were investigated in the lab that organisms were found inside some of the borings. Because of the small sample size, the full morphology and identity of these organisms is not yet known, making it difficult to ascertain whether they were the organisms responsible for the borings or merely were secondary inhabitants utilizing the excavation for living space (i.e., nestlers). In borings that preserve the most complete body fossils, a common feature is the presence of singular (Fig. 2A-B) or paired (Figs. 2C-D, 3A-B) calcareous tubes, occasionally within a sheath of calcite (Fig. 2A-B). Diameter of the tubes is between 0.5 and 0.75 mm. When the inside of the boring is exposed due to erosion of the cobble surface, a larger portion of the body fossil can be observed at a deeper, wider point within the boring (Figs. 2A-B, 3C-D). The main body of the fossil is vase-shaped, with an outer 'body wall' preserved as a thin layer of calcite. Typically, the interior of the body wall contains bioclasts in a micrite matrix (Figs. 2A, C, 3C). Longitudinal and transverse sections through the body are available. In one longitudinal section (Fig. 3C-D), the tube or neck portion of the body leads from a funnel-shaped aperture of the boring to the main body cavity. The posterior (deepest) portion of the body is tapered, and the whole body fits tightly within the boring. In transverse section (Fig. 3E-F) it appears that the body is bilaterally symmetrical with axes that give an elliptical aspect, a shape that is common to the *Gastrochaenolites* borings described from SRP (Benner et al. 2004). An idealized reconstruction of the boring and the body fossil show that the boring chamber and body are elliptical, while at the hardground surface, the cross-sectional aspect of the aperture of the boring and tubes of the body fossil are circular (Fig. 4).

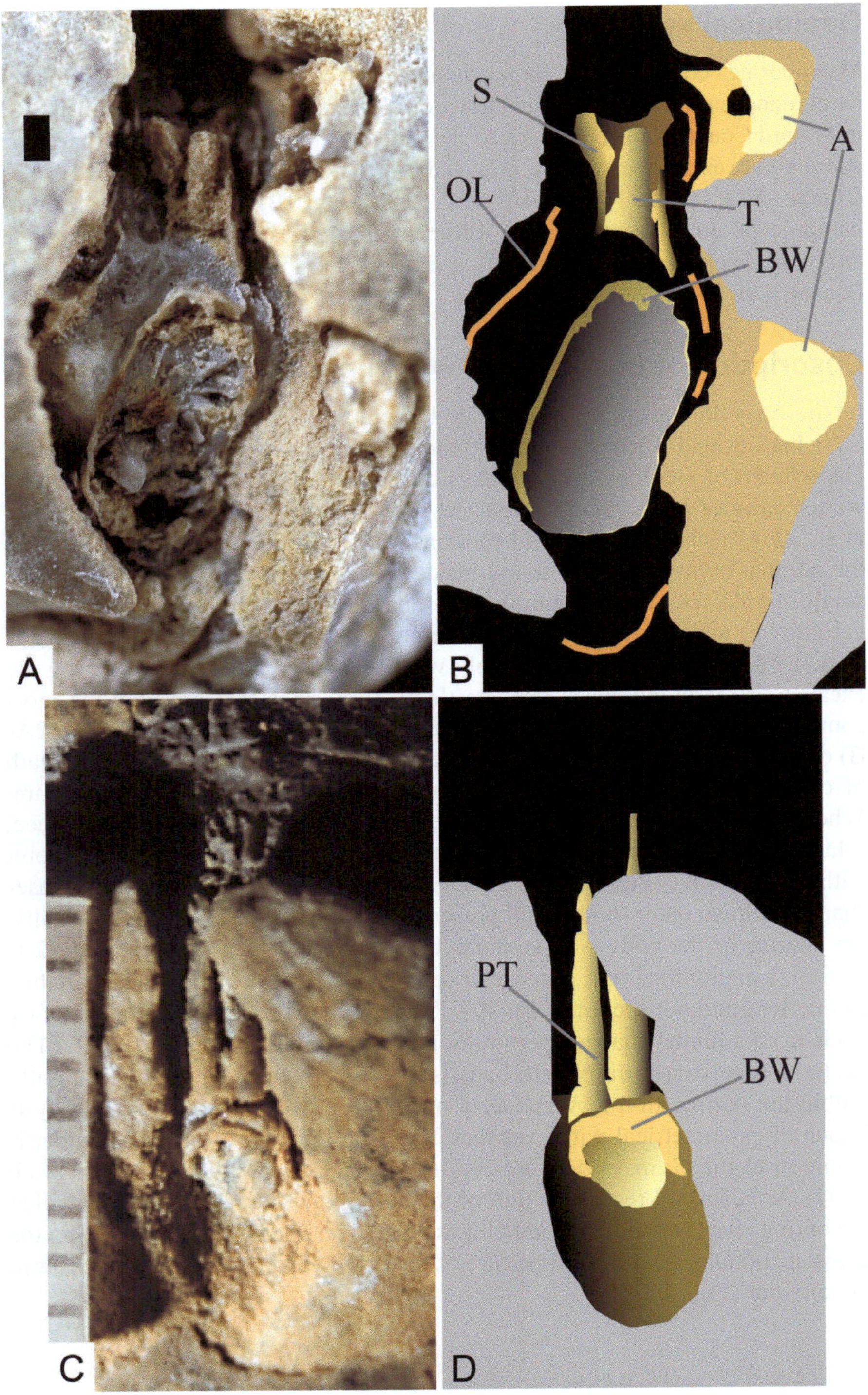
S
A
OL
T
BW
A
B
PT
BW
C
D

Discussion

Marine invertebrates bore into rock in response to a variety of ecological and physical pressures (Bromley 1970, 1978; Warme 1975, 1977), such as predation avoidance, protection from high-energy waves / currents, and specialized feeding habits. Infaunal borers occupy important niches in benthic marine ecosystems, often in the lower tier of suspension feeding communities, and the boring strategy has proved to be a very successful one (Bromley 1970, 1978; Warme 1975, 1977; Jones and Pemberton 1988; Ekdale and Bromley 2001). Protection provided by a lithic substrate has resulted in success in an extremely competitive trophic structure. The borings in Fillmore Formation hardgrounds represent the earliest known attempts at a *Gastrochaenolites*-type strategy, and they provide evidence for more complex hardground trophic structures than were previously thought to exist in the Early Ordovician.

It has been asserted that evolution of new body plans during the Ordovician biotic radiation was driven by the ability to exploit new habitats, including those at and below the sediment-water interface (Bambach 1985). The rise in the abundance and diversity of epifaunal suspension feeders in Lower Ordovician hardgrounds (e.g., Wilson et al. 1989, 1992) that were exploiting similar food resources could have led to the evolution of the *Gastrochaenolites*-type behavior. If this were the case, the advent of the *Gastrochaenolites*-type behavior may have been a precursor to the creation of a new body plan more suited to hard substrate habitat (Palmer 1982; Bambach 1985).

Evidence is scant that the Fillmore body fossils are the remains of the animals that were responsible for creating the *Gastrochaenolites* borings. The tight fit of the body within the boring may indicate either the bioerosional modification of the original boring to fit the body of a borer, or the restricted growth of the body of a nestler (e.g., Savazzi 1994). However, because the potential exists to identify the organism inside the boring, consideration of certain groups is necessary, including those that are not known from direct body fossil evidence in the Fillmore Formation.

The borings described here exhibit some similarities to those produced by cirripeds, particularly acrothoracican barnacles (Ahr and Stanton 1973). The acrothoracicans may have their earliest ancestor in the Cambrian, but this group is only known in fossil form since the Devonian (Rodriguez and Gutschick 1970, 1977). What is known about the morphology of the boring made by these

Fig. 2 Photos of enigmatic organisms in two bored cobbles from the Fillmore Formation at Skull Rock Pass with corresponding labeled diagrammatic sketches. For descriptions see text. Specimens held in research collections in the Department of Geology and Geophysics, University of Utah. **A-B** Longitudinal section through boring showing body inside with attached tube. Body outline is a thin layer of calcite and the interior is filled with bioclastic material. Note two semi-cylindrical features to the right of the individual oriented normal to the plane of the photo. These represent transverse sections through the necks of two other individuals. Scale bar = 1 mm. **C-D** Oblique view of partially eroded specimen. Body with attached paired tube is visible. Scale in mm. For this and all subsequent figures: BM = boring margin, OL = outer lining, BW = body wall, B = body, T = single tube, PT = paired tubes, S = sheath (around tubes), A = aperture, PS = plane of symmetry

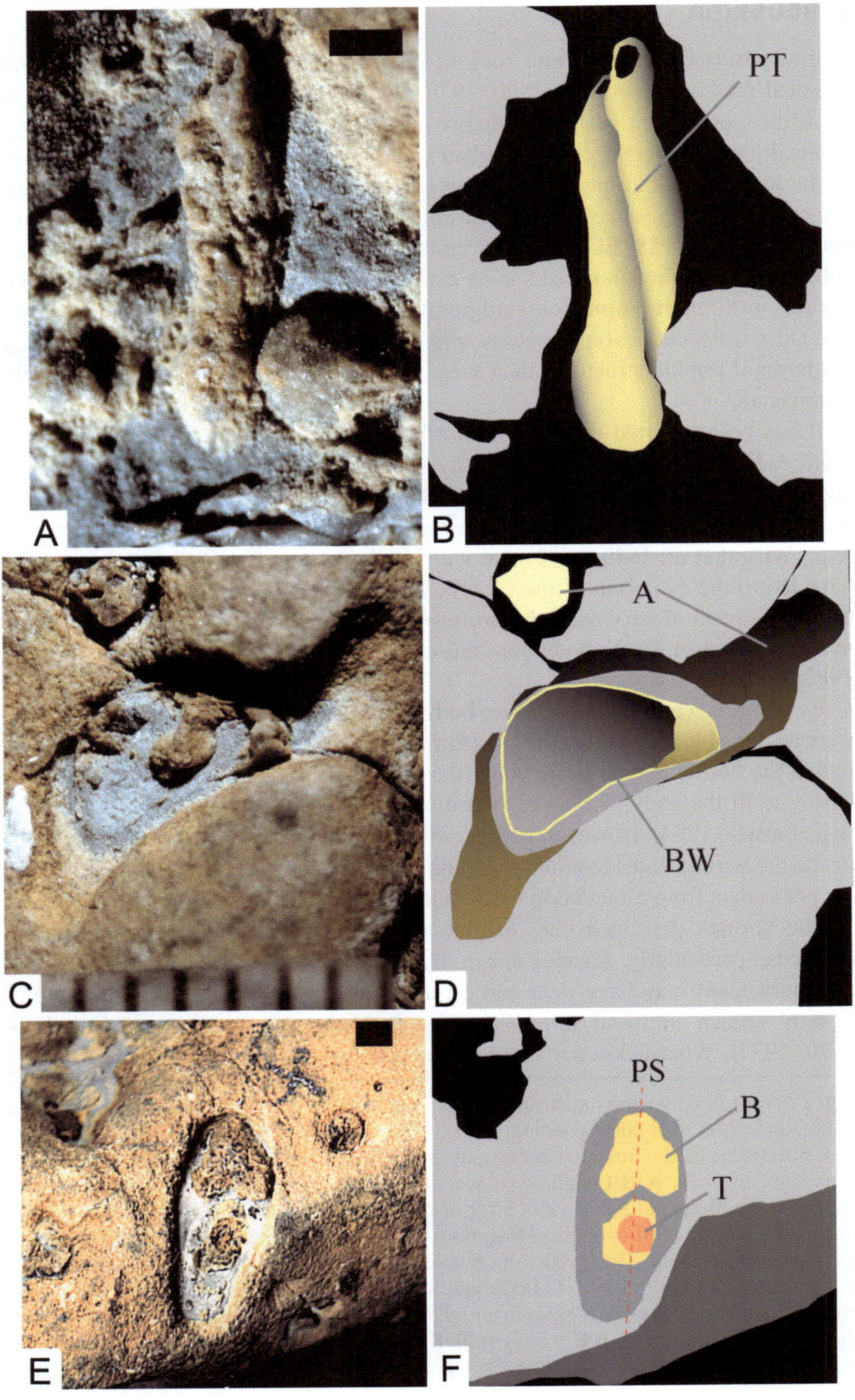
A
B
PT
C
D
A
BW
E
F
PS
B
T

animals makes it unlikely that it is a producer of *Gastrochaenolites* (e.g., Seilacher 1969; Rodriguez and Gutschick 1977). Acrothoracican borings do not have the characteristic neck of *Gastrochaenolites*, which is a key feature of the Fillmore Formation borings and their inhabitants. Borings emplaced into hard substrates by *Lithotrya* Ellis and Solander, 1786 are cylindrical to subcylindrical in shape and taper near the bottom (e.g., Dineen 1990). These borings, if preserved in the fossil record, should be ascribed to the ichnogenus *Trypanites* Mägdefrau, 1932 and are not representative of the borings from the Fillmore Formation. In addition, no evidence exists for cirripedian skeletal material in the borings or elsewhere in the entire Fillmore Formation.

Bivalves are the most common producers of *Gastrochaenolites*. Except for the Upper Ordovician occurrence of mytilaceans in Ohio (Whitfield 1893; Pojeta and

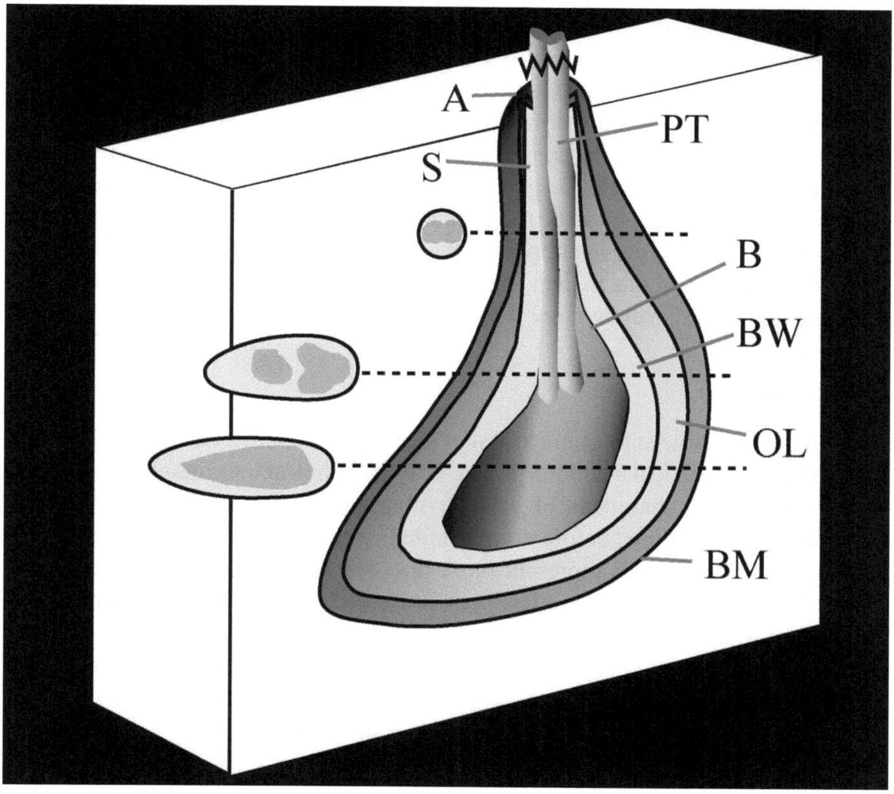

Fig. 4 Idealized rendering of the body inside the boring showing three cross-sections through the body along the length of the boring. (See Fig. 2 for abbreviations)

Fig. 3 Photos of enigmatic organisms, cont. **A-B** Set of paired tubes leading out of the surface of a bored cobble. Scale bar = 1 mm. **C-D** Longitudinal section through partially eroded boring showing body outlined by thin calcite lining leading up (to the right) to a funnel-shaped boring aperture. Scale in mm. **E-F** Transverse section through body inside boring. Scale bar = 1 mm. (See Fig. 2 for abbreviations)

Palmer 1976; Wilson and Palmer 1988), no other bivalves capable of boring into indurated substrates have been identified in Paleozoic rocks. Pojeta (1971) and Bambach (1985) suggested the existence of siphonate bivalves which eventually gave rise to epifaunal, nonsiphonate forms as early as the Upper Cambrian. It is possible that the infaunal, boring life habit of bivalves developed early in the siphonate forms, and if so, it may have allowed some types of bivalves to produce the borings in the Fillmore Formation. The thin calcite lining that outlines the body of each organism is reminiscent of the calcite lining secreted by some bivalve borers (e.g., by some mytilids and gastrochaenids), but that alone is not sufficient for body fossil identification.

Mitrate (stylophoran echinoderm) bioclasts have been found in the Fillmore Formation (Sprinkle and Geunsberg 1995). The stylophoran body plan and inferred life habits offer no indication of the ability to penetrate hard substrates, but some workers interpret them as soft substrate infauna (e.g., Sutcliffe et al. 2000). The enigmatic body fossils in the Fillmore Formation lack any indication of a plated skeletal structure, but their overall shape approximates that of an inverted mitrate.

Recent *Gastrochaenolites* are known to have been made by a variety of organisms, including some worm-like forms, so barnacles, bivalves and mitrates are not the only animals to be considered in this discussion. Sipunculans (e.g., Rice 1969) and some polychaetes (e.g., Bromley 1978: fig. 8) can create vase-shaped biogenic structures that would be referred to the ichnogenus *Gastrochaenolites*, and they may have been a prevalent part of the hardground communities in the Lower Ordovician. However, due to their low fossilization potential, it is unlikely that their identification would ever be more than speculative.

Conclusions

While it is impossible at this time to assign the enigmatic body fossils in the Fillmore Formation to a particular animal taxon, it is nonetheless important to recognize their existence as inhabitants of the macroborings. More specimens need to be discovered and examined before more detailed interpretations can be made. It may be that these body fossils represent more than one of an as-yet undescribed soft-bodied hardground dweller that has been preserved under unusual circumstances. Regardless of biologic affinity, these animals are evidence of a more complex hardground community than previously known, and their discovery lends credence to the recognition of significant evolutionary advances that occurred early during the Ordovician Bioerosion Revolution (Ekdale and Bromley 2001).

Acknowledgments

Improvements to the manuscript were made by reviewers M. Wilson and K. Kleemann, and editors L. Tapanila and M. Wisshak. B. Dattilo first introduced the authors to the Skull Rock Pass section.

References

Ahr WM, Stanton RJ (1973) The sedimentologic and paleoecologic significance of *Lithotrya*, a rock-boring barnacle. J Sediment Petrol 43:20-23

Bambach RK (1985) Classes and adaptive variety: The ecology of diversification in marine faunas through the Phanerozoic. In: Valentine JW(ed) Phanerozoic diversity patterns: profiles in macroevolution. Princeton Univ Press, Princeton, Pacific Division AAAS, San Francisco, pp 191-254

Benner JS (2002) Ichnology and cyclic stratigraphy of the Lower Ordovician Fillmore Formation at Skull Rock Pass, Millard County, western Utah. MSc Thesis Univ Utah, Salt Lake City, 115 pp

Benner JS, Ekdale AA, Gibert JM de (2004) Macroborings (*Gastrochaenolites*) in Lower Ordovician hardgrounds of Utah: Sedimentologic, paleoecologic, and evolutionary implications. Palaios 19:543-550

Bromley RG (1970) Borings as trace fossils and *Entobia cretacea* Portlock, as an example. In: Crimes TP, Harper JC (eds) Trace fossils. Geol J Spec Issue 3:49-90

Bromley RG (1978) Bioerosion of Bermuda reefs. Palaeogeogr Palaeoclimatol Palaeoecol 23:69-197

Dattilo BF (1993) The Lower Ordovician Fillmore Formation of western Utah: storm-dominated sedimentation on a passive margin. Brigham Young Univ Geol Stud 39:71-100

Dineen JF (1990) Burrowing rates of *Lithotrya dorsalis* (Cirripedia: Thoracica) in Jamaica. Bull Mar Sci 47:656-662

Ekdale AA, Bromley RG (2001) Bioerosional innovation for living in carbonate hardgrounds in the Early Ordovician of Sweden. Lethaia 34:1-12

Ekdale AA, Benner JS, Bromley RG, Gibert JM de (2002) Bioerosion of Lower Ordovician hardgrounds in southern Scandinavia and western North America. Acta Geol Hisp 37:9-13

Ellis J, Solander D (1786) The natural history of many curious and uncommon zoophytes, collected from various parts of the globe. White and Son, London, 208 pp

Evans KR, Miller JF, Dattilo BF (2003) Sequence stratigraphy of the Sauk Sequence: 40th anniversary field trip in western Utah. In: Swanson TW (ed) Western Cordillera and adjacent areas. Geol Soc Amer Field Guide 4:17-35

Hintze LF (1973) Lower and Middle Ordovician stratigraphic sections in the Ibex area, Millard County, Utah. Brigham Young Univ Geol Stud 20:3-36

Jones B, Pemberton SG (1988) *Lithophaga* borings and their influence on the diagenesis of corals in the Pleistocene Ironshore Formation of Grand Cayman Island, British West Indies. Palaios 3:3-21

Leymerie A (1842) Suite de mémoire sur le terrain Crétacé du département de l'Aube. Mém Soc Géol France 5:1-34

Mägdefrau K (1932) Über einige Bohrgänge aus dem unteren Muschalkalk von Jena. Paläont Z 14:150-160

Palmer TJ (1982) Cambrian to Cretaceous changes in hardground communities. Lethaia 15:309-323

Pojeta J (1971) Review of Ordovician pelecypods. US Geol Surv Prof Paper 695, 46 pp

Pojeta J, Palmer TJ (1976) The origin of rock boring in mytilacean pelecypods. Alcheringa 1:167-179

Rice M (1969) Possible boring structures of sipunculids. Amer Zoologist 9:803-812

Rodriguez J, Gutschick RC (1970) Late Devonian-Early Mississippian ichnofossils from western Montana and northern Utah. In: Crimes TP, Harper JC (eds) Trace fossils. Geol J Spec Issue 3:407-438

Rodriguez J, Gutschick RC (1977) Barnacle borings in live and dead hosts from the Louisiana limestone (Famennian) of Missouri. J Paleont 51:718-724
Savazzi E (1994) Functional morphology of burrowing and boring organisms. In: Donovan SK (ed) The palaeobiology of trace fossils. Wiley and Sons, London, pp 43-82
Seilacher A (1969) Paleoecology of boring barnacles. Amer Zoologist 9:705-719
Sprinkle J, Guensburg TE (1995) Origin of echinoderms in the Paleozoic Evolutionary Fauna; the role of substrates. Palaios 10:437-453
Sutcliffe OE, Südkamp WH, Jefferies RPS (2000) Ichnological evidence on the behaviour of mitrates: two trails associated with the Devonian mitrate *Rhenocystis*. Lethaia 33:1-12
Warme JE (1975) Borings as trace fossils, and the processes of marine bioerosion. In: Frey RW (ed) The study of trace fossils: a synthesis of principles, problems, and procedures in ichnology. Springer, New York, pp 181-227
Warme JE (1977) Carbonate borers; their role in reef ecology and preservation. AAPG Stud Geol 4:261-279
Whitfield RP (1893) New genera and species of burrowing fossil bivalve shell. Contr Paleont Ohio, Rep Geol Surv Ohio 7:492-493
Wilson MA, Palmer TJ (1988) Nomenclature of a bivalve boring from the Upper Ordovician of the midwestern United States. J Paleont 62:306-308
Wilson MA, Palmer TJ (1998) The earliest *Gastrochaenolites* (Early Pennsylvanian, Arkansas, USA): An upper Paleozoic bivalve boring? J Paleont 72:769-772
Wilson MA, Palmer TJ (2006) Patterns and processes in the Ordovician bioerosion revolution. Ichnos 13:109-112
Wilson MA, Palmer TJ, Guensburg TE, Finton CD (1989) Sea-floor cementation and the development of marine hard-substrate communities: new evidence from Cambro-Ordovician hardgrounds in Nevada and Utah. Geol Soc Amer Abstracts with Programs 21:253-254
Wilson MA, Palmer TJ, Guensburg TE, Finton CD, Kaufman LE (1992) The development of an Early Ordovician hardground community in response to rapid sea-floor calcite precipitation. Lethaia 25:19-34

II

Spectrum of bioerosive biota

The boring microflora in modern coral reef ecosystems: a review of its roles

Aline Tribollet[1]

[1] Hawaii Institute of Marine Biology, Kaneohe HI 96744, USA, (aline@hawaii.edu)

Abstract. Euendolithic microorganisms (boring microflora) – cyanobacteria, algae and fungi – colonize all carbonate substrates in modern coral reefs and are distributed worldwide. Recent studies showed that in dead carbonate, they are important primary producers and are fed upon by various excavating invertebrate and vertebrate grazers, contributing greatly to biodestruction processes (bioerosion) and sedimentation. Additionally, it has been shown that in some live calcifying organisms, they either inflict damages to live tissues or provide a benefit to the host, depending on the euendolithic community involved (parasitic or mutualistic relationships). Based on those recent studies, the following question is raised: Are euendoliths key organisms in the functioning and maintenance of coral reefs? Reviewed literature includes studies on (1) the mechanisms used by euendoliths to penetrate into carbonate substrates (production of acids or chelating fluids; use of the products of photosynthesis / respiration and / or calcium pumps), (2) their roles in reef bioerosion and sedimentation (major roles), (3) their metabolism (important rates of production), (4) their interactions with their live hosts (symbiosis, mutualism and / or parasitism) and (5) the effects of various environmental factors such as eutrophication, sedimentation and rising atmospheric pCO_2 on euendolith activities. The review concentrates on modern coral reef ecosystems.

Keywords. Boring microflora, bioerosion, environmental factors, euendoliths, modern coral reefs, production, symbiosis

Introduction

Boring autotrophic and heterotrophic microorganisms or euendoliths (boring microflora) actively penetrate (biochemical dissolution) into the interior of the hard substrates they inhabit (Golubic et al. 1981). In contrast, the chasmoendoliths colonize fissures and cracks in substrates while the cryptoendoliths colonize structural cavities within porous substrates, including spaces produced and vacated by euendoliths (Golubic et al. 1981). Some euendoliths can show an epilithic or chasmoendolithic stage in their life cycle (Golubic et al. 1975; Kobluk and Risk 1977; Jones and Goodbody 1982). The term 'endolith' includes the euendoliths, chasmoendoliths and cryptoendoliths.

M. Wisshak, L. Tapanila (eds.), *Current Developments in Bioerosion*. Erlangen Earth Conference Series, doi: 10.1007/978-3-540-77598-0_4,

Euendoliths live in various substrates such as carbonates (Schneider and Torunski 1983; Mao Che et al. 1996; Tribollet and Payri 2001; Tribollet and Golubic 2005), glass (Jones and Goodbody 1982), apatite (Königshof and Glaub 2004) and granite (De Los Rios et al. 2005). They are observed in terrestrial (Friedmann et al. 1988; Ascaso et al. 1998; Golubic and Schneider 2003), freshwater (Schneider et al. 1983; Anagnostidis and Pantazidou 1988) and marine ecosystems including the Adriatic Sea (Ghirardelli 2002), the Mediterranean Sea (Le Campion-Alsumard 1979), cold-temperate coasts (Wisshak et al. 2005), Antarctic (De Los Rios et al. 2005) and tropical waters (Le Campion-Alsumard et al. 1995a; Radtke et al. 1997a; Perry 1998; Tribollet 2007). On marine limestone coasts, carbonate rock surfaces can be infested by more than half a million euendolithic filaments per square cm (Schneider and Le Campion-Alsumard 1999).

Euendoliths have long geological histories, having been identified within substrates from Mesoproterozoic (Zhang and Golubic 1987), Neoproterozoic (Knoll et al. 1986) and Paleozoic strata (Klement and Toomey 1967; Campbell 1980; Golubic et al. 1980; Königshof and Glaub 2004). The first record of euendolithic cyanobacteria was found in 1.5 billion year old stromatolite rocks in China (Zhang and Golubic 1987). As a result, they are believed to have played a major role in the production and destruction of carbonate, including reef framebuilders and sediments, over long periods of geological time. The well fossilized microborings of euendoliths are used as proxies in paleoecological and paleobathymetrical studies (Chazottes et al. 1995; Radtke et al. 1997b; Perry and MacDonald 2002; Glaub and Vogel 2004; Radtke and Golubic 2005) as their diversity, distribution and abundance depend on substrates and environmental conditions (Gektidis 1999; Radtke et al. 1997a; Vogel et al. 2000; Tribollet 2007). They are also used for ichnological and ichnotaxonomic determinations (Glaub and Vogel 2004; Chacón et al. 2006). Those studies are based on comparisons between modern and fossilized microborings. Thus, it is important to better understand euendolith ecological characteristics as well as their metabolism in modern environments in order to improve our knowledge on the past environmental conditions and to better predict the future.

Coral reefs are maintained when constructive forces (reef framebuilders growth and calcification) are balanced with destructive forces (mainly bioerosion including boring microflora, macroborers and grazers). These ecosystems are increasingly threatened by anthropogenic and natural stresses such as overfishing, sedimentation, eutrophication, rising seawater surface temperature and atmospheric pCO_2 (Edinger et al. 1998, 2000; Kleypas et al. 1999; Langdon et al. 2000; Szmant 2002; Hallock 2005). These factors increase coral bleaching events, the rate of coral mortality and can favor shifts from coral-dominated reefs to algal-dominated reefs (Mumby et al. 2001; Szmant 2002; Wilkinson 2004). More and more dead substrates are, therefore, available for colonization by euendoliths and other agents of bioerosion. When bioerosion rates exceed accretion rates, the degradation of reef framework is accelerated (Hutchings 1986; Sheppard et al. 2002). Reefs are then even more subject to damage by cyclones and storms (Harmelin-Vivien 1994) putting in jeopardy the survival of coastal and insular human populations.

Recently, a few studies showed the necessity of studying simultaneously constructive and destructive processes in coral reefs in order to estimate their state of health under different environmental conditions (Edinger et al. 2000; Risk et al. 2001; Tribollet and Golubic 2005; Harborne et al. 2006; Mallela and Perry 2007). Euendoliths play a role in those different processes but have received much less attention than other organisms involved, such as corals, macroborers and grazers. Boring microflora is well known to colonize live and dead carbonate substrates although colonization is more intense in dead ones (Le Campion-Alsumard et al. 1995a; Radtke et al. 1997a; Perry 1998; Tribollet and Payri 2001; Tribollet 2007). Through their activity, they are known to inflict damages to their live hosts or to provide benefits to them, to be important primary producers in dead substrates, to prepare substrates for macroborers, to attract excavating invertebrate and vertebrate grazers in a combined bioerosion activity, to enhance fine sediment production and to alter sediment grains (Kobluk and Risk 1977; Hutchings 1986; Bruggemann et al. 1994; Chazottes et al. 1995; Bentis et al. 2000; Perry 2000; Fine and Loya 2002; Tribollet et al. 2002; Tribollet and Golubic 2005; Chacón et al. 2006; Tribollet et al. 2006a). But what are the contributions of euendoliths in those processes at the reef scale and under various environmental conditions?

To my knowledge there is no review summarizing the different roles played by boring microflora in modern coral reef ecosystems. Hutchings (1986) published a review on bioerosion processes including the erosive activity of boring microflora, while Radtke et al. (1997b) provided a bibliographic overview of micro- and macrobioerosion. More recently Schneider and Le Campion-Alsumard (1999) provided a review on processes of construction and destruction of carbonates by aquatic cyanobacteria including marine euendolithic cyanobacteria. The purpose of the present paper is to highlight the main functions played by boring microflora in coral reefs and their consequences on the functioning of those ecosystems. The variability of euendolith activities in space and over time will also be discussed and perspectives will be suggested. The present contribution does not pretend to be exhaustive. It is based on 162 references and highlights the most recent studies on the topic.

Diversity of the boring microflora

Boring microflora comprises cyanobacteria, algae (chlorophytes, rhodophytes) and fungi. Their taxonomy has been studied for decades (Bornet and Flahault 1889; Lukas 1974; Le Campion-Alsumard 1979; see also literature review by Kobluk and Kahle 1977) but new species are still being discovered, especially since the emergence of molecular tools (Al-Thukair and Golubic 1991; Chacón et al. 2006). Recently, Komárek and Anagnostidis (1999, 2005) published a determination manual on cyanobacterial diversity (Oscillatoriales) including endolithic cyanobacteria. Kendrick et al. (1982), Porter and Lingle (1992), Bentis et al. (2000) and Golubic et al. (2005) provided information on euendolithic fungi but their taxonomy remains relatively unknown.

The most common genera of autotrophic euendoliths observed in coral reef substrates around the world are the chlorophytes *Ostreobium* Bornet and Flahault, 1889 and *Phaeophila* Hauck, 1876 (Fig. 1A, 1B), the rhodophyte *Porphyra* Agardh, 1824 (Conchocelis stage) and the cyanobacteria *Hyella* Bornet and Flahault, 1888, *Mastigocoleus* Lagerheim, 1886 (Fig. 1C), *Plectonema* Thuret in Gomont, 1892 (Fig. 1D) and *Solentia* Ercegovic, 1927 (Raghukumar et al. 1991; Chazottes et al. 1995; Le Campion-Alsumard et al. 1995a; Gektidis and Golubic 1996; Golubic et al. 1996; Radtke et al. 1997a; Perry 1998; Gektidis 1999; Vogel et al. 2000; Tribollet and Payri 2001; Tribollet 2007). Lukas (1973) has documented the ubiquitous nature of the siphonale *Ostreobium quekettii* Bornet and Flahault, 1889 infestation in corals, occurring at 100% frequency in Atlantic and Pacific corals. However, there might be more than one species of *Ostreobium* involved (Lukas 1974). The phylogeny of this chlorophyte (Bryopsidale) is poorly known (Woolcott et al. 2000) and seems to show different lineages (pers. comm. H. Verbruggen).

Species composition of euendolithic communities varies between live and dead substrates. In the skeleton of live corals and crustose coralline algae, the dominant euendoliths are *Ostreobium quekettii, Plectonema terebrans* Bornet and Flahault, 1889 and fungi (Lukas 1974; Le Campion-Alsumard et al. 1995a; Tribollet and Payri 2001). However, Conchocelis were found very abundant in some skeletons of live corals from Caribbean coral reefs (Laborel and Le Campion-Alsumard 1979). The species composition of microbial euendoliths in those live substrates is a result of a selection in favor of oligophotic, positively phototrophic, fast-growing taxa which can follow the accretion rate of skeletons. Such an infestation occurs at the early stage of the calcifying organism's growth, and starts from the substrate of fixation. Following the death and denudation of corals and crustose coralline algae, a new colonization by euendoliths occurs at the surface of skeletons. Thus, the species composition of communities changes dramatically and a succession of communities is observed over time (Gektidis 1999). Short-lived opportunistic pioneer species such as *Mastigocoleus testarum* Lagerheim, 1886 and *Phaeophila dendroides* Crouan (Batters, 1902) are observed in substrates within a few days (Hutchings 1986; Le Campion-Alsumard et al. 1995a; Gektidis 1999). Then after 6 to 12 months, euendolithic communities become dominated by the low-light specialists (sciaphile species), *O. quekettii* and *P. terebrans,* as well as by heterotrophic fungi (Chazottes et al. 1995; Le Campion-Alsumard et al. 1995a; Gektidis 1999; Tribollet and Golubic 2005; Tribollet 2007). Those communities are called 'mature communities' (Le Campion-Alsumard et al. 1995a; Gektidis 1999; Tribollet 2007). Changes in species composition of euendolithic assemblages in dead substrates is primarily due to light availability which varies with depth (bathymetry) and over time due to the growth of epilithic organisms providing shade to euendoliths (Gektidis 1999; Tribollet 2007). Chazottes et al. (2002) showed that *M. testarum* is particularly abundant in dead *Porites lobata* Dana, 1846 under turf algae while the assemblage of *P. terebrans*, *M. testarum* and *O. quekettii* prevails under crustose coralline algae and phaeophytes (*Lobophora* Agardh (Womersley, 1987)).

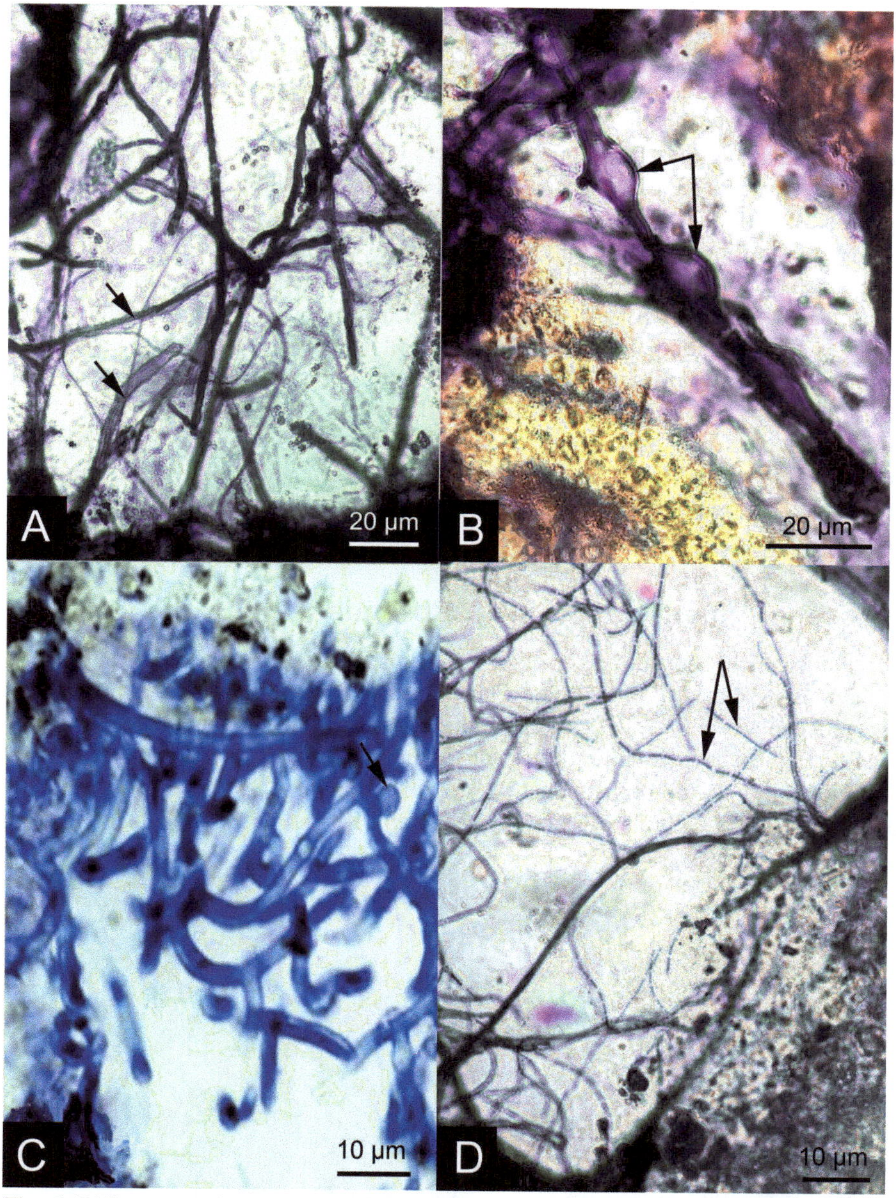

Fig. 1 Different species of euendoliths stained with Toluidine blue and determined on petrographic thin sections of the dead coral *Porites lobata*. **A** Filaments of the chlorophyte *Ostreobium quekettii* (black arrows). Thin filaments penetrating into filaments of the chlorophyte are fungal hyphae. **B** Filaments of the chlorophyte *Phaeophila dendroides* (arrows show cells). **C** Filaments of the cyanobacterium *Mastigocoleus testarum* (arrow shows a heterocyste). **D** Filaments of the cyanobacterium *Plectonema terebrans* (arrows show cells)

Differences in community species composition may result from the effects of grazing pressure. Grazers remove the surface of substrates, more or less intensively, creating new surfaces available for colonization by pioneer euendoliths and allowing light to penetrate deeper inside substrates. Crustose coralline algae and phaeophytes are less palatable than turf algae for grazers (Bruggemann et al. 1994; Chazottes et al. 2002), therefore substrates covered by those epilithic organisms are less grazed, which limits the colonization by pioneer species such as *M. testarum*. When grazing is too intense, immature assemblages of euendoliths do not develop and only mature communities are observed penetrating deeper inside coral skeletons (Schneider and Torunski 1983; Tribollet 2007). The effects of environmental factors such as eutrophication and freshwater inputs on species composition of euendolith communities are, however, relatively unknown. Ercegovic (1930) observed that *Mastigocoleus testarum* does not tolerate high variation in salinity. The effects of those factors as well as the interactions between grazers, epilithic organisms and euendolithic assemblages need to be more studied in order to better understand the spatial and temporal variability of the boring microflora activities (erosion and production).

Process of penetration into substrates

It is not clear why euendoliths evolved to live inside hard substrates. Paleontologists cannot answer clearly this question but the same hypothesis implied for decades is: unoccupied ecological niche providing protection against sunlight, grazers and desiccation (Schneider and Le Campion-Alsumard 1999).

Euendolithic autotrophs penetrate rapidly into new available substrate by dissolving chemically its crystals (Tudhope and Risk 1985; Fig. 2A). They leave

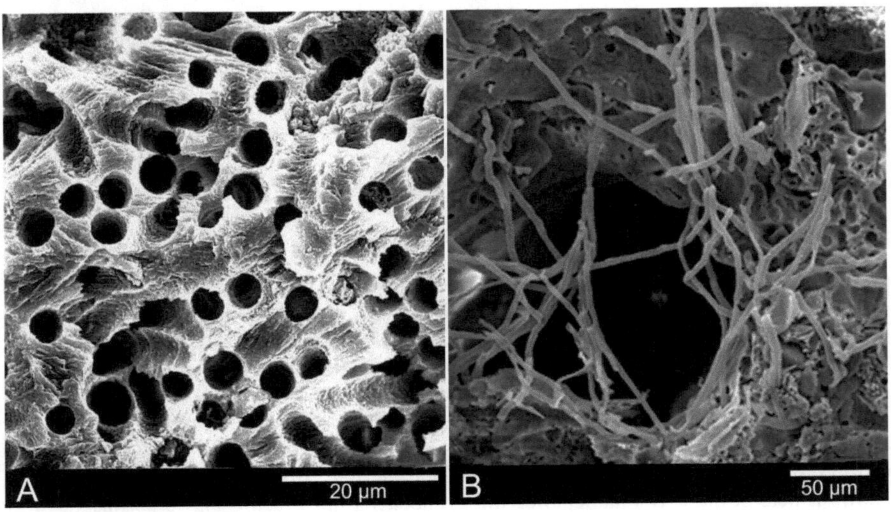

Fig. 2 Pictures of the surface of dead corals taken by SEM, showing: **A** Traces of microbioerosion due to filaments of *Ostreobium quekettii*. **B** Filaments of *Ostreobium quekettii* penetrating into coral skeleton

microborings (1-100 μm in diameter) which conform to the shape of their thalli (Fig. 2B). Crystals of substrates show a specific arrangement around microborings suggesting a precisely controlled excavating process (Alexandersson 1975; Golubic et al. 1975; Le Campion-Alsumard 1975, 1979). This process is poorly understood. It has often been suggested that substrate dissolution results from the production of acid or chelating fluids at the apical cell of euendolithic filaments (Le Campion-Alsumard 1975; Golubic et al. 1984), and even involves organelles (Alexandersson 1975). The dissolution process in coral reefs occurs in waters saturated or supersaturated with respect to aragonite and calcite, and is thermodynamically unfavorable. Excavation must thus be performed at the cost of cellular energy.

Recently, Garcia-Pichel (2006) suggested other mechanisms to explain dissolution of carbonate substrates by euendolithic cyanobacteria. Those mechanisms are (1) a temporal separation of photosynthetic and boring activities during the daily cycle (dissolution due to the CO_2 produced during respiration at night), (2) a spatial separation of photosynthetic and boring activities, and (3) the use of calcium pumps. A combination of those mechanisms is possible but the use of calcium pumps seems the most probable mechanism. Garcia-Pichel (2006) showed that an active transport of Ca^{2+} from the apical cell of euendolithic filaments to their trailing end would make dissolution thermodynamically favorable around the apical cell while interstitial pH is high due to photosynthesis. This is consistent with the known range of bored substrates including aragonite, calcite, granite and hydroxylapatite, and with precipitation of micrite and brucite observed around euendolithic filaments at the surface of dead substrates. Kobluk and Risk (1977) showed that micrite is precipitated inside and around dead filaments of *Ostreobium quekettii* after they protruded at the surface of dead carbonate substrates (filaments partially epilithic). Those authors could not explain the process involved in the micrite precipitation but suggested that *'something peculiar in the composition of Ostreobium filaments and probably of some others, facilitates precipitation'*. If euendolithic chlorophytes possess Ca^{2+} pumps like euendolithic cyanobacteria, the Ca^{2+} pumped at the apical cell and released at the other extremity of filaments (i.e., at the surface of carbonate substrates) may create the conditions necessary to induce precipitation of carbonates (micrite) at ambient pH. Le Campion-Alsumard (1978) found higher calcium contents in euendolithic cyanobacterial filaments as compared with epilithic forms. Moreover, Nothdurft et al. (2005) suggested that euendoliths allow precipitation of brucite in scleractinian corals while ambient seawater of coral reefs is undersaturated in brucite. The model proposed by those authors to explain such a precipitation follows: (1) semiconfined intracorallum spaces contain dead organic matter and live filaments of cyanobacteria, algae and fungi, (2) photosynthetic removal of CO_2 lowers the pCO_2 and increases the pH, leading to aragonite precipitation on organic matter or synthaxially on exposed skeletal aragonite, (3) aragonite precipitation lowers Ca^{2+} activity, thereby increasing the local Mg / Ca ratio, (4) increased pH due to degradation of organic matter (possibly by bacteria or fungi) produces excess of OH^-, leading to localized brucite precipitation. The increase of Mg / Ca ratio may in fact result from the decrease of Ca^{2+} in the boring microenvironment due to calcium pumps, thus creating favorable conditions for brucite precipitation under high pH. Elevated pH (>9) has been measured inside coral heads by Risk and Kramer (1981) during the day while lower pH was quantified under dark conditions by Risk and Muller (1983) due to endoliths respiration. Nevertheless, it is possible that other mechanisms are involved in substrate dissolution by euendoliths. In Antarctic desertic sandstones (arkosic / feldspar), euendoliths (in lichen: *Plectonema*) seem to penetrate by secreting oxalic acids, which require the mobilization of inorganic nutrients such as iron and potassium (Edwards et al. 1997; Russell et al. 1998). Euendolithic fungi, which are heterotrophic, penetrate both mineral substrates but also organic matter using digestive enzymes (Golubic et al. 2005). Processes involved in substrate dissolution by euendoliths as a whole may vary depending on taxa and species, the type of substrate and environmental conditions.

Distribution of the boring microflora

Type of substrates

Euendoliths penetrate all kinds of reef carbonate substrates such as live and dead corals (Le Campion-Alsumard et al. 1995a, 1995b; Tribollet 2007), crustose coralline algae (Tribollet and Payri 2001), calcifying macroalgae such as *Halimeda* Lamouroux, 1812 (Perry 2000), foraminifera (Golubic et al. 1984; Perry 1998), shells (Radtke 1993; Mao Che et al. 1996; Radtke and Golubic 2005), sediments (Tudhope and Risk 1985) and limestone (Le Campion-Alsumard 1979; Reaka-Kudla et al. 1996; Vogel et al. 2000). Infestation of dead substrates is always higher than in their live counterparts due to the new colonization of their surface by euendoliths (Le Campion-Alsumard et al. 1995a; Tribollet and Payri 2001). Moreover, Perry (1998) showed that coral sediments are more infested than those of molluscs, foraminifera, *Halimeda*, crustose coralline algae and echinoids. The structure, porosity and skeletal mineralogy of calcifying organisms influence euendolith infestation (Perry 1998; Chacón et al. 2006).

Depth of penetration in substrates – light dependence

In live and accreting reef framebuilders, such as corals and coralline algae, phototrophic as well as heterotrophic euendoliths are located beneath the layer of live tissue and keep up with the calcification rates of their hosts. For example, the chlorophyte *Ostreobium quekettii* usually forms mm-thick green colored bands beneath the surface of massive coral skeletons (Le Campion-Alsumard et al. 1995a), and some filaments have been observed in contact with polyps (Kanwisher and Wainwright 1967; Schlichter et al. 1995). Heterotrophic fungi maintain their presence close to the polyps, as they attack both algae and the polyps (Le Campion-Alsumard et al. 1995b; Bentis et al. 2000; Priess et al. 2000).

In dead corals, the depth of penetration of euendolithic filaments varies depending on the species. Species requiring high light intensities such as *Mastigocoleus testarum* do not penetrate deep into substrates (<1 mm in dead *Porites lobata*;

Chazottes et al. 1995; Tribollet 2007). In contrast, oligophotic species such as *Plectonema terebrans* and *Ostreobium quekettii* show a depth of compensation of 2 to 4 mm in dead *Porites lobata* (Chazottes et al. 1995; Tribollet 2007). Heterotrophic fungal filaments are independent of light and use the euendolithic chlorophytes and the residual organic matter in coral skeletons as a source of food (Bentis et al. 2000; Golubic et al. 2005; Tribollet 2007). The depth of penetration of euendolithic filaments in dead substrates correspond to their bathymetric distribution. Euendoliths are found from the intertidal and wave spray zones of marine environments to abyssal depths (see Golubic et al. 1984, 2005), yet individual species are restricted to special environments and water depths due to their light requirements. *M. testarum* is always found in shallow waters (Gektidis 1999; Vogel et al. 2000) while *O. quekettii* and *P. terebrans* can be observed at more than 300 m depth (Lukas 1978; Kiene et al. 1995). Diversity of euendoliths is higher in shallow waters than at depth (Perry 1998, 2000; Gektidis 1999). In general, the bathymetric distribution of euendolithic assemblages in coral reefs has been well documented and will not be detailed here (see Golubic et al. 1975; Kiene et al. 1995; Radtke et al. 1996; Radtke et al. 1997a; Gektidis 1999; Hoffmann 1999; Vogel et al. 2000).

Factors affecting the depth of penetration of euendolithic filaments in substrates

Depth of penetration of euendolithic filaments varies from species to species but also with the nature of the substrate and environmental conditions. For example, in shallow waters, depths of penetration are lower in bivalve shells than in coral skeletons (<0.1 mm; Mao Che et al. 1996; Perry 1998). This probably results from the difference in substrate architecture, porosity and respective translucency of the substrate. It was shown that at 10 m depth, the depth of penetration of euendoliths in experimental blocks of *Porites lobata* increased from inshore to offshore reefs on the northern Great Barrier Reef (GBR); inshore reefs showed turbid and eutrophic waters and a low grazing pressure. This gradient was particularly evident for *O. quekettii* (0.6 to 4.1 mm with distance from shore; Tribollet 2007). Development of euendolithic communities at inshore reefs was probably limited by the entrapment of sediments within epilithic turf algal filaments, thus restricting the light availability. Depths of penetration of euendoliths also increased over a three year period due to the increase of grazing pressure (see Schneider and Torunski 1983) and the presence of larger holes made by macroborers (they increased the amount of light penetrating deep inside substrates). This trend was less pronounced for *M. testarum*, which reaches its depth of compensation (photosynthesis = respiration) more rapidly than the oligophotic euendoliths (after one year). Tribollet (2007) showed that the depth of penetration of euendolithic filaments is the main variable affecting rates of microbioerosion in dead substrates exposed more than one year to bioerosion; the level of infestation at the surface of substrates being stable. This first study implies the necessity of studying more the effects of biotic and abiotic factors on the euendolith vertical distribution inside substrates over long periods of time to better understand the spatial and temporal variability of microbioerosion.

Boring microflora: roles in processes of bioerosion and sedimentation

Microbioerosion and sedimentation

By dissolving calcium carbonate (microbiocorrosion), the boring microflora recycles ions of calcium and carbonate which are used by calcifying organisms for their growth and calcification. Euendoliths also disintegrate the substratum into smaller particles which are then more susceptible to dissolution and abrasion (Schneider and Le Campion-Alsumard 1999). The pattern of boring is not random and determines the size of sediments produced by grazers or physical erosion (Schneider and Torunski 1983). Fine-grain sediments are usually produced with a size below 125 µm (most of the material falls into the 20-63 µm class). Part of the new sediment produced by bioerosion contributes to the sandy areas of the seafloor whereas another part is entrapped and cemented into the coral reef structure (Perry 2000). The rest is exported into the open ocean. About half of the sediment produced in shallow areas may be exported (Hubbard et al. 1981; Eakin 1996).

Sediment grains on the sea floor of shallow marine environments are colonized by euendolithic communities (Perry 1998), which contribute to their destruction by dissolving calcium carbonate (Kobluk and Risk 1977; Schneider and Le Campion-Alsumard 1999; Perry 2000). Euendoliths participate greatly to early diagenesis of sediments (Perry 1998, 2000). Tudhope and Risk (1985) estimated microbioerosion rates of 0.35 kg $CaCO_3$ dissolved per $m^{-2} y^{-1}$ in sediments at Davies Reef (Australia). Similar rates were found in dead coral skeletons of *Porites lobata* on an oligotrophic reef at Moorea in French Polynesia (0.14 kg $m^{-2} y^{-1}$; Chazottes et al. 1995). But rates found in this substrate vary greatly depending on grazing pressure and other factors. Microbioerosion rates quantified by Chazottes et al. (2002) at different reefs of La Reunion Island after one year of exposure were lower than 0.07 kg $m^{-2} y^{-1}$ because of intense grazing by echinoids on substrates (3.5 kg $m^{-2} y^{-1}$). Those estimated on the northern Great Barrier Reef were much higher (0.13 kg $m^{-2} y^{-1}$ at inshore turbid reefs to 1.3 kg $m^{-2} y^{-1}$ at offshore oligotrophic reefs), with grazing primarily by fish (0.5 kg $m^{-2} y^{-1}$; Tribollet and Golubic 2005; Tribollet 2007). Echinoids are known to be more efficient grazers than fishes (Hutchings 1986; Reaka-Kudla et al. 1996; Harborne et al. 2006).

Rates of microbioerosion measured in dead carbonate substrates using microscopy techniques and image analysis (see Chazottes et al. 1995; Tribollet and Golubic 2005) are 'residual' rates (underestimated rates), i.e., the difference measured after the removal of euendolith-affected layers by grazers. Since the carbonate loosened by microbial euendoliths and removed by grazers cannot be quantified, the observed rates are underestimated relative to the total rates of microbial bioerosion. Interactions between grazers and euendoliths have been described by several authors (see Schneider and Torunski 1983; Chazottes et al. 1995; Tribollet and Golubic 2005). The latter authors showed a positive correlation between grazing and microbioerosion rates at a large spatial scale (inshore-offshore transect along 200 km of the GBR). Phototrophic euendoliths provide a renewable food source for grazers (e.g., gastropods, echinoids and fishes). They make substrates attractive for grazers and by weakening the surface of substrates they facilitate grazing rates. On the other hand, by constantly removing the surface of substrates, grazers extend the depth to which the light can penetrate in substrates and therefore, the depth of compensation of euendoliths (Schneider and Torunski 1983). But low rates of 'residual' microbioerosion prevail when grazing pressure is too intense and catches up with microbioerosion. However, such conditions have a regulatory effect by requiring re-colonization and re-establishing of a primary trophic level for grazers. On the other hand, low rates of microbioerosion were quantified at inshore reefs on the northern Great Barrier Reef because waters at those reefs were turbid (Tribollet and Golubic 2005). Lots of particles in suspension and a thin layer of mud (2-3 mm) deposited at the surface of the studied blocks of *Porites lobata*, limited the development of euendolithic communities while it allowed the important development of macroborers which were mostly filter feeders. Grazing rates were also low at those reefs, with population of fishes (Russ 1984) and their food source (epilithic and endolithic flora) being less abundant (Tribollet and Golubic 2005). Rates of microbioerosion vary also among substrates because infestation by euendoliths depends on this factor as seen earlier (see also Vogel et al. 2000; Zubia and Peyrot-Clausade 2001; Carreiro-Silva et al. 2005; Tribollet and Golubic 2005). In mussel shells, infestation by euendoliths can be so intense that it causes sub-lethal or lethal effects on mussel populations inducing changes in growth and reproduction output (see example of *Perna perna* on the coast of South Africa: Kaehler and McQuaid 1999).

Finally, rates of microbioerosion increase over time but not linearly nor proportionally – they are cumulative over time – as shown by Chazottes et al. (1995) and Tribollet and Golubic (2005) in experiments carried out over two to three years. In higher latitudes there are probable significant seasonal differences in bioerosion rates which need to be assessed in more detail. In order to create a model of microbioerosion, and more generally a model of bioerosion (= macrobioerosion + microbioerosion + grazing), carbonate destruction by agents of bioerosion has to be studied under different environmental conditions and over long time periods.

Effects of eutrophication on microbioerosion rates

A few attempts were made to study the effects of eutrophication (high concentration of nutrients) on microbioerosion but results are controversial. Kiene (1997) and Vogel et al. (2000) reported inconclusive results on the effect of eutrophication on microbioerosion of five month-old dead substrates in microatolls on the Great Barrier Reef (ENCORE experiment). Kiene (1997) suggested that boring microflora were already saturated in nutrients. But, this may also have resulted from the design of the experiment, the residence time of seawater charged with nutrients inside microatolls being very low (microatolls were flushed at each tide), as well as to grazing pressure (data were lacking). In contrast, Chazottes et al. (2002) found an increase of microbioerosion under eutrophic conditions at La Reunion Island but again the direct effect of this factor was not evident. Under eutrophic conditions, substrates were covered by macroalgae such as *Lobophora variegata* Lamouroux in Oliveira (Womersley, 1987) and encrusting coralline algae. Those algae are less attractive to grazers. Thus, high rates of microbioerosion may be due to low grazing pressure and protection from epilithic algae. Inversely, Zubia and Peyrot-Clausade (2001) found higher rates of microbioerosion in dead skeletons of acroporids under eutrophic conditions outside damselfish territories than inside damselfish territories. In this study, the time of exposure to bioerosion was unknown and could have influenced rates of microbioerosion; branches of coral may have been exposed longer to bioerosion outside damselfish territories than those inside territories.

Other studies attempted to show that eutrophication affects microbioerosion but only the biomass of euendoliths mixed with epiliths was studied (Le Bris et al. 1998; Pari et al. 1998, 2002). Microbioerosion rates are due to the accumulation of borings of euendoliths over time. Measuring the biomass of boring microflora at one particular time gives an idea of the euendolithic community present in substrates at that specific time and not of the succession of different communities over time. Moreover, measurements of boring microflora biomass are not accurate as they include the biomass of the coral residual organic matter and of small heterotrophic organisms.

Only Carreiro-Silva et al. (2005) really showed that microbioerosion increases under high concentrations of nutrients (Glovers Reef, Belize) on a short period of time (49 days), and that this effect varies with grazing intensity. Inorganic nutrients were the strongest factors increasing microbioerosion by a factor of 10 in the absence of macrograzers (0.06 kg m^{-2} $CaCO_3$ eroded after 49 days under fertilized conditions versus 0.0007 kg m^{-2} after 49 days under unfertilized conditions). Microbioerosion rates were of course lower as they were residual when macrograzers had access to substrates (shells). In this experiment, nutrients were released slowly underneath substrates and not in the water column as in the ENCORE experiment on the Great Barrier Reef (Kiene 1997). This suggests that boring microflora may respond only to nutrient inputs in their close environments and are probably not influenced by nutrient pulses in the water column (problem of the boundary layer at the surface of substrates, dilution in the water column and water residence time). Further studies should be carried out to better understand the impact of eutrophication on microbioerosion rates. How are euendolithic communities affected by eutrophication (changes of community species composition and metabolism)? Are there any thresholds of nutrient concentrations? Is the increase of microbioerosion rates under eutrophic conditions maintained over time?

Effects of other factors on microbioerosion rates

Eutrophication is a major issue in conservation of coral reefs but it is not the only one. One important question being increasingly asked by coral reef scientists is how the rise of atmospheric partial pressure of CO_2 will affect coral reefs? By 2065-2100, the atmospheric pCO_2 should be doubled according to the forecast by the Intergovernmental Panel on Climate Change (Houghton et al. 1996). Several studies showed that calcification rates of reef framebuilders will decrease by 30-40% on average by 2065 due to the decrease in the aragonite saturation state (Borowitzka 1981; Kleypas et al. 1999; Hallock 2005; Tribollet et al. 2006b). Kleypas et al. (2001) concluded that a shift in saturation state projected for 2065 *'will shift many reefs from a net carbonate accumulation to net carbonate loss'*. Halley et al. (2005) showed that sediment dissolution increases under elevated concentration of CO_2. They suggested that this may result from the activity of boring microflora colonizing sediments but no direct proof has been provided. The question is: do euendoliths have the potential to buffer coral reef seawater under elevated pCO_2? Langdon et al. (2000) suggested that dissolution of carbonates would not compensate the acidification of coral reef seawater but considered only a maximal rate of dissolution measured by night. Anderson et al. (2003) suggested similarly that dissolution will not be enough to buffer waters of shallow environments including reefs. But those authors used a simplified model with only two major domains: surface-water and pore-water sediment domains. First of all, euendoliths probably dissolve carbonate by day and by night using different mechanisms as seen earlier (Ca^{2+} pumps by day and products of respiration by night). Secondly, boring microflora dissolve all carbonate substrates in coral reefs and not just sediments. It has been shown recently that those microorganisms are one of the main agents of bioerosion in coral reefs. Chazottes et al. (1995) showed that they are the main agents at the early stage of the bioerosion process (≤1 year) studying simultaneously all agents, i.e., microborers, macroborers and grazers at the reef of Tiahura (Moorea, French Polyensia). Tribollet et al. (2002) and Tribollet and Golubic (2005) confirmed this result at a larger scale on the northern Great Barrier Reef and showed that the important role of boring microflora is maintained over time. After a 3 years experiment, euendoliths are still contributing between 20% and 40% to total bioerosion (total bioerosion of more than 6 kg m^{-2} after three years at oligotrophic reefs). Tribollet (2007) found that the chlorophyte *Ostreobium quekettii* was the main agent of microbioerosion in blocks of *Porites lobata* exposed more than a year to bioerosion on the northern GBR, dissolving more than 1 kg of $CaCO_3$ m^{-2} y^{-1}. A similar role for this euendolith was suggested before by Le Campion-Alsumard et al. (1995a) in live colonies of *Porites lobata* as it is the most abundant euendolith. At the scale of coral reefs, *Ostreobium quekettii* is probably the most important agent of microbioerosion and therefore, plays a crucial role in reefal carbonate dissolution as a whole as it is ubiquitous in all kinds of carbonate substrates, in shallow and deep waters, and all around the world. This role may be even greater in the near future as the number of substrates available for colonization increases as a consequence of the rise of reef framebuilder mortality rate, due to diseases (coral and crustose coralline diseases), predators (e.g., Crown-of-Thorns starfish), sedimentation, rising sea surface temperature and many other factors (Goreau et al. 1998; Wilkinson 2004; Hallock 2005; Voss and Richardson 2006). The effects of elevated atmospheric pCO_2 and other factors such as rising sea surface temperature and UV on euendolith activities should receive more attention in the next few years if the overall effect of those factors on coral reef functioning is to be understood. Such research is a prime topic with biological and geological implications.

Boring microflora: important primary producers?

In 1955, Odum and Odum were the first to suggest that euendolithic oxygenic phototrophs are one of the main primary producers in coral reefs. Their assumption was based on endolithic biomass (including euendolithic biomass) and chlorophyll a concentration quantified in live corals (endolith biomass represented 81% of the total biomass of live corals on average). Their conclusion when extrapolated to the reef scale assumes a 100% cover by healthy growing corals, which is rarely the case. In contrast, Kanwisher and Wainwright (1967), Shibata and Haxo (1969) and Shashar and Stambler (1992) suggested a rather limited contribution of euendoliths in live corals to primary production amounts to less than 4% of the total activity of live corals (see also Schlichter et al. 1997). These authors measured the oxygen production of euendoliths in live coral colonies maintained in tanks or chambers, the absorption spectra of those phototrophs and the spectral distribution using various sensors. In order to assess the metabolism of euendoliths in the interior of coral skeleton, it is essential to measure the available light.

Light intensity inside substrates

To date, measurements of light intensity inside substrates were only carried out in live corals. In those substrates, the amount of light reaching euendoliths is limited by (i) the absorption of incidental light by zooxanthellae in the coral tissue, (ii) internal architecture of the coral skeleton and (iii) water depth. Enriquez et al. (2005) showed that coral endosymbionts (zooxanthellae) of *Porites branneri* Rathbun, 1887 are able to collect more than 85% of solar radiation with one order magnitude less pigment density than terrestrial leaves. This is due to multiple scattering by the skeleton which enhances the local light field (absorption increased). This would leave less than 15% of light for euendoliths. Only 1% of the Photosynthetically Active Radiation (PAR) was estimated to reach the layer of euendoliths within the skeleton of the coral *Favia pallida* Dana, 1846 in Australia by Halldal (1968) and Shibata and Haxo (1969); 0.1% of light between 340 nm and 680 nm penetrates into coral skeleton, as well as 1% and 2% at 700 nm and 720 nm, respectively. Schlichter et al. (1997) estimated that 3.9% of white, 3.8% of blue, 3.6% of green, and 6.4% of red light penetrate skeletons of deep corals in the Gulf of Aqaba. But in this case, light absorption by live coral tissues only was measured (skeleton of samples was removed), which may introduce an error in the calculation of the amount of light reaching euendoliths. Direct measurements of light intensity and wavelengths reaching the layer of euendoliths, and in general endoliths, in skeletons of live and dead corals, and other live and dead carbonate substrates, for different water depths are lacking. However, the information thus far available shows that euendoliths live in extremely shaded environments.

Euendoliths are adapted to low light intensities

To be able to live inside carbonate substrates, euendoliths present a certain number of adaptations. They have specific cellular ultrastructures such as complex and thick sheaths, well developed thylakoids and abundant cytoplasmic storage substances

(Le Campion-Alsumard 1976, 1979) and are able to absorb red and far red (peak at 730 nm) wave lengths using different forms of chlorophyll. Fork and Larkum (1989) and Koehne et al. (1999) showed that the genus *Ostreobium* including *O. quekettii,* possesses red-shifted chlorophylls. The antenna size is approximately 340 chlorophyll molecules (chlorophyll a and b) per PS II reaction center, which is significantly larger than in higher plants (~240). Additionally, euendoliths have high carbonic anhydrase and catalase activities (Shashar and Stambler 1992). The first enzyme transforms HCO_3^- into CO_2 (Weis et al. 1989) which is used by the RUBISCO for photosynthesis. During the day, euendoliths produce large amounts of O_2 (210% inside the skeleton of *Porites compressa* Dana, 1846 versus ambient seawater) which turns into H_2O_2 (Shashar and Stambler 1992). This hydrogen peroxide may inhibit CO_2 fixation (H_2O_2 is easily fixed by RUBISCO), destroy algal cellular membranes, damage their DNA and generate even more reactive oxygen species via metalcatalyzed reactions. The catalase catalyzes the reduction of H_2O_2 into H_2O and O_2. Such a catalase activity is a protecting mechanism for euendolithic filaments against reactive oxygen species; mechanisms which are also found in corals and their endosymbionts (Lesser and Shick 1989).

Those mechanisms and cellular ultrastructures change with environmental conditions. Koehne et al. (1999) showed that red-shifted chlorophylls (chlorophylls absorbing wavelengths >700 nm) are more abundant in euendolithic thalli deep inside skeletons than close to the white light at the surface of substrates. Shashar and Stambler (1992) and Schlichter et al. (1997) showed that concentration of pigments in *Ostreobium* vary depending on the illumination of the habitat (shaded versus lighted) and depth. With increasing depth, the ratio of chlorophyll b to chlorophyll a increases indicating improvement of light harvesting under low light conditions (Schlichter et al. 1997). Concentration of chlorophyll a can also vary with the type of substrate and species of coral. Endolithic chlorophyll a is much more abundant in dead corals (59 μg cm^2 which convert to 12 μg mL^{-1} of substrate; Tribollet et al. 2006a) than in healthy or bleached corals (4.6 μg cm^2 and 15 μg cm^2, respectively; Fine and Loya 2002) but less abundant than in bivalve shells (20 μg mL^{-1} of substrate; Raghukumar et al. 1991).

Euendolithic phototrophs seem to saturate at low light intensities which is another adaptation to their extreme environment. Shashar and Stambler (1992) suggested that euendoliths in live *Porites compressa* collected in 1-2 m depth are saturated at less than 160 μE m^{-2} s^{-1} external light, which corresponds to 1 μE m^{-2} s^{-1} inside skeleton (Shibata and Haxo 1969). Schlichter et al. (1997) suggested a light compensation of 10 μE m^{-2} s^{-1} for *O. quekettii* within skeletons of deep corals and a light saturation of 35-40 μE m^{-2} s^{-1}. In *Oculina patagonica* De Angelis d'Ossat, 1908 from shallow shaded waters along the coast of Israel, Fine et al. (2004) suggested that euendoliths saturate at 50 μE m^{-2} s^{-1} in healthy colonies (coral tissues + zooxanthellae) while they saturate at around 200 μE m^{-2} s^{-1} in bleached colonies (bleaching >40% of the colony), and no photoinhibition was observed at 850 μE m^{-2} s^{-1}. Those authors found that euendoliths photoacclimate to increased irradiance during a coral bleaching event. Fine et al. (2006) also showed euendolith photoacclimation in bare skeletons after corals were attack by the white syndrome (disease). However, in the case of a rapid bleaching event, euendoliths are photoinhibited by combined high light intensities and temperatures (Fine et al. 2005). This delays the recovery process of those phototrophs (Fine et al. 2005). The outcome of euendoliths in bleached coral seems to depend on two major parameters: (1) the microhabitat the coral lives in (back reef, reef crest, etc.) and (2) the response time of the coral host and its endosymbionts to the stress and recovery. Finally, Vooren (1981) found that endoliths, including *O. quekettii* and *P. terebrans*, were saturated at an external irradiance of 200 $\mu E\ m^{-2}\ s^{-1}$ in dead corals. All those studies – while different techniques and substrates were used – show the same pattern: euendoliths are light saturated at low light intensities and are able to optimize the absorption of wavelengths available inside coral skeletons.

Production rates of endoliths (including euendoliths)

Although absorption spectra and spectral distribution of endolithic phototrophs in various substrates and under different environmental conditions need more investigation, there are a few indications that endoliths may be one of the main primary producers in coral reef environments. First of all, Bruggemann et al. (1994) showed that grazing fish feed preferentially on epilithic turf and endolithic algae. Endoliths are one of the main sources of food for grazers so they must be highly productive (Schneider and Torunski 1983; Hutchings 1986; Bruggemann et al. 1994; Chazottes et al. 1995; Tribollet and Golubic 2005). Schlichter et al. (1995) and Fine and Loya (2002) also showed that euendolithic communities can cover up to 30% of coral when bleached, which suggest that they are highly productive under those conditions. In the live hydrozoan *Millepora* (1-20 m depth), Bellamy and Risk (1982) found large volumes of oxygen trapped inside skeletons (69% of oxygen, 31% of molecular nitrogen and race CO_2) and attributed this production to euendolithic phototrophs. They compared volumes of O_2 in the morning and in the afternoon and found that there was net community photosynthesis (excess of O_2 produced over 24h). In dead substrates, Wanders (1977) determined that endoliths are the most important primary producers within the sparse vegetation on dead coral limestone. Wanders (1977) suggested that it might be *'an important feature in coral reefs that boring algae are present everywhere in the limestone substrate and whenever a primary producer disappears, its task as a primary producer is replaced by the boring algae which are present in the limestone beneath it'*. Vooren (1981) found similar results in natural substrates colonized by *Ostreobium* from Curaçao at 10-25 m depth (net photosynthesis ranging between 0.4-0.8 $g\ C\ m^{-2}\ d^{-1}$). Recently, Tribollet et al. (2006a) quantified that endoliths contribute at least 40% of the overall productivity of experimental blocks of the coral *Porites lobata* in shallow waters at Kaneohe Bay, Hawaii (3 m depth), with rates of net photosynthesis averaging 2.3 $g\ C\ m^{-2}\ d^{-1}$. Casareto et al. (2004) found the same participation (40%) in natural coral rubble in shallow waters at Bora Reef (Japan). Those values are probably underestimated as epilithic organisms such as encrusting coralline algae and bivalves, which are also colonized by endoliths, have been scraped off to quantify endolithic metabolism in dead substrates. The metabolism of endoliths quantified by Casareto et al. (2004) and Tribollet et al. (2006a) is comparable to that of turf algae and live corals (Kinsey

1985; Adey 1998). It is interesting to note that reef community production rates estimated in very diverse coral reef environments always fall within the same range (2-10 g C m^{-2} d^{-1} on average; Kinsey 1985). Endolithic metabolism might be one of the factors creating certain homogeneity in coral reef production, as they are ubiquitous, and show similar rates of production as corals and turf algae.

Spatial and temporal variability of production rates

The significance of the role played by endoliths in coral reef productivity may increase as coral bleaching events and the rate of coral mortality increase. But what are the effects of environmental factors such as rising sea surface temperature, rising atmospheric pCO_2, eutrophication and terrestrial inputs on boring microfloral metabolism? So far, rare studies investigated the effects of environmental factors on the endolithic metabolism.

Tribollet et al. (2006a) showed that production of endoliths varied along an inshore-offshore transect in Kaneohe Bay; the inshore and intermediate reefs showing low wave energy, more eutrophic and turbid waters than the offshore reef (high water motion, oligotrophic and clear water). Dead substrates from the offshore reef were covered mostly by encrusting coralline algae while inshore substrates were covered by turf algae and bivalves. Hydrodynamic energy is one major factor controlling the epilithic algal cover on dead substrates (Adey 1998) and influences the thickness of the boundary layer at the surface of substrates (Williams and Carpenter 1998). Those factors could explain why endolithic production was higher at the offshore reef. Moreover, endolithic productivity in dead corals and coral rubble is probably enhanced by eutrophication because Carreiro-Silva et al. (2005) showed that locally rates of microbioerosion in those substrates increase under such conditions. Microbioerosion or dissolution of calcium carbonate increases as a result of the rise of euendolithic filament growth. Studies of the direct effects of different concentrations of nutrients on euendolithic metabolism in live and dead carbonate substrates have to be carried out over time if the variability in space and time of total bioerosion rates is to be understood.

Tribollet et al. (2006b) also studied the effects of elevated atmospheric pCO_2 (750 ppm) on endolithic metabolism which included euendolithic metabolism. No effect was highlighted under controlled conditions over a three month period. Tribollet et al. (2006b) hypothesized that endoliths as a whole, under such conditions, may have been already at saturation or limited by nutrients in tanks. Further experiments are needed to determine precisely the effects of this factor on euendolithic metabolism separately and therefore, on rates of dissolution of carbonate substrates.

Fine and Loya (2002) and Fine et al. (2004) showed that euendolithic metabolism increases with irradiance when corals experience bleaching. However, when bleaching is too rapid, euendolithic filaments are photoinhibited. Their metabolism is reduced by 80% due to an elevated temperature combined with high light intensities. Euendolithic communities are more diverse, abundant and efficient in in dead substrates, thus it would be interesting to determine the effects of those combined factors on their metabolism in dead substrates and over time.

Interactions between euendoliths and their live hosts

Mutualistic relationship

One of the most intriguing questions being asked increasingly by coral reef researchers is why some coral species, mostly massive and encrusting forms, survive bleaching events, while others do not. Coral bleaching corresponds to the disruption of the symbiotic relationship between the coral host and its algal symbionts (zooxanthellae) following in general an increase of the sea surface temperature (Hoegh-Guldberg 1999). As a consequence, corals loose their main source of energy putting at risk their main biological functions (e.g., reduced proteins and lipids, lowered growth rate, lower calcification and repair capabilities, and termination of gametogenesis; see Fine and Loya 2002). There may be more than one reason why some corals survive bleaching and others do not, but one has been given recently involving boring microflora (Fine and Loya 2002). The boring microflora and live corals may have a mutualistic relationship, helping corals survive better during bleaching events. Back in 1955, Odum and Odum suggested for the first time a symbiotic relationship between phototrophic euendoliths and live corals (ectosymbiosis). Those authors hypothesized that nutrients of coral metabolism readily diffuse through the porous skeleton to euendoliths, and polyps provide to them a protection against competitors, browsing and intense sunlight. From phototrophic euendoliths, corals would benefit in return from diffusion of organic substances, but no evidence of that was provided by Odum and Odum (1955). Later works carried out on this topic since confirmed or contradicted the hypothesis proposed by Odum and Odum and are presented below.

Shashar et al. (1997) stated that it is an advantage for phototrophic euendoliths to live inside skeletons of live corals as the latter provide mycosporine-like amino-acids (MAAs) which absorb most of UV-A and UV-B. Those authors were unable to find MAAs in euendolithic filaments suggesting that either: (a) those compounds are not synthesized by euendoliths because they are not induced by environmental conditions or (b) they do not have the ability to produce them, which forces euendoliths to live in protected niches such as coral skeletons. They also suggested that live corals provide a protection for euendoliths against predation by grazers because grazing fishes feed rarely on live corals (Rojtan and Lewis 2005). It is also plausible that euendoliths use nutrients derived from coral metabolism and degradation of residual organic matter within the skeleton. Such a short circuit of nutrient supply schemes have been shown in other closed environments such as microbial mats (Paerl and Pinckney 2005).

Schlichter et al. (1995) carried out the first study showing a transfer of photoassimilats produced by phototrophic euendoliths (*O. quekettii*) to coral tissues of the azooxanthellate coral *Tubastrea micranthus* Ehrenberg, 1834 in a shallow reef of the Red Sea, using ^{14}C incubations. Fixation of ^{14}C was light dependent. Those authors showed that euendolithic filaments are in contact with coral tissues and that within 24h-48h algal precursors are integrated into coral lipids. An 'ectosymbiotic relationship' was suggested with an intracolonial recycling of nutrients. Risk and Muller (1983) quantified high concentrations of nutrients in live colonies of *P. lobata* from the Great Barrier Reef and attributed those concentrations to euendolith metabolism. Ferrer and Szmant (1988) estimated that euendoliths may satisfy 55% to 65% of N required by live corals for balanced growth. Pore-waters extracted from skeletons of *Montastrea annularis* Ellis and Solander, 1786 showed concentrations of 9 μM, 0.3 μM, and 0.4 μM of NH_4^+, NO_3^-, and PO_4^{3-}, respectively, over ambient seawater concentrations. Shashar et al. (1994) and Casareto et al. (2004) also showed that the uptake of N_2 by endolithic communities is very high. The mutualistic relationship between boring microflora, especially *Ostreobium quekettii*, and azooxanthellate corals may support their distribution into deeper waters.

Recently, Fine and Loya (2002) highlighted an ectosymbiotic relationship between phototrophic euendoliths and the zooxanthellate coral *Oculina patagonica* from the oriental Mediterranean Sea. They found that euendoliths bloom in bleached corals (increase of biomass and pigments concentration) within 2 weeks and transfer photoassimilats to coral tissues. This is an alternative source of energy for live coral tissues which could explain the low mortality rate of *O. patagonica* following bleaching events. Such external symbiosis may be advantageous to the coral over endosymbiosis with zooxanthellae during periods of environmental changes (e.g., elevated seawater temperature). This source of energy is however insufficient for reproduction (Fine et al. 2001) but may help bleached coral to better survive and to recover from stress. In zooxanthellate corals from the Great Barrier Reef, Fine et al. (2005) observed a similar bloom of phototrophic euendoliths after bleaching events. These euendoliths photoacclimated well to their new conditions except when they were exposed too quickly to high temperature and light intensities. This study suggests that the role of phototrophic euendoliths in the resilience of corals when exposed to stresses such as bleaching (and associated rise of sea surface temperature) may vary greatly with environmental conditions and over time. This has to be investigated more.

Parasitic relationship

While phototrophic euendoliths seem to be beneficial for their live hosts, heterotrophic euendoliths often have a parasitic relationship with their hosts (Bentis et al. 2000; Golubic et al. 2005), or are associated with coral diseases (Alker et al. 2001). In live corals such as *Porites lobata*, euendolithic fungi are observed in newly deposited skeletal spines (pali), demonstrating that they are able to keep up with coral skeleton accretion (Le Campion-Alsumard et al. 1995b). Also, fungi were observed penetrating into live coral tissues of different coral species (Pacific and Atlantic oceans; Bentis et al. 2000; Golubic et al. 2005). The damage to corals caused by fungal penetration remains unknown but to defend themselves, coral tissues form pearl-like deposits of carbonate (or 'cones') around fungal hyphae to prevent further penetration (Le Campion-Alsumard et al. 1995b; Bentis et al. 2000). But as polyps respond by producing cones, fungal hyphae continue to penetrate along the axis of the defence structure (Bentis et al. 2000). Recently, Domart-Coulon et al. (2004) showed that fungal infestation extends coral cell survival of

Pocillopora damicornis (Linnaeus, 1758) up to 2 days, by producing a transient cytoprotective effect, selectively enhancing the survival of skeletogenic cell types. This may result from the activation of cell metabolism or inhibition of death. Such a result supports previous observations of cone repair production around hyphae. Additionally, euendolithic fungi may be involved in the formation of skeletal growth tissue anomalies in *Porites compressa* in Kaneohe Bay (Domart-Coulon et al. 2006). Fungi engaged in lichenization with phototrophic euendoliths may have mutually beneficial effect (Le Campion-Alsumard et al. 1995b; Golubic et al. 2005).

Factors that directly or indirectly impair skeletogenesis in corals – such as elevated sea surface temperature (Hoegh-Guldberg and Salvat 1995), coastal eutrophication (Marubini and Davies 1996) and increased atmospheric CO_2 (Kleypas et al. 1999) – could compromise the ability of corals to defend themselves against euendolithic fungal attacks. Bentis et al. (2000) suggested that coral's inability to deposit enough repair aragonite, in conjunction with factors that may increase the frequency of coral-fungal interactions, could increase the regularity with which hyphae are successful in reaching the skeletal surface of actively growing corallites. Domart-Coulon et al. (2004) also suggested that the contact of fungal hyphae with coral soft tissue may cause tissue sloughing. All those effects are likely to be deleterious to the host coral's health. This may be even more dramatic because corals may not be able to defend themselves against other pathogens (Kushmaro et al. 1997) or to invest energy in other processes such as tissue regeneration or recovery from bleaching (Meesters and Bak 1995). In regard to the increased concern over the health of coral reefs and coral diseases, with an attempt to diagnose and to identify their causative agents, more attention needs to be paid to euendolithic fungi.

Conclusions

In order to better understand the functioning of coral reef ecosystems and how they will respond to global change and anthropogenic activities in the near future, it appears critical to study rapidly the different roles played by the boring microorganisms at various spatial and temporal scales, as they had biological and geological implications for millions of years. This review summarizes evidence that they are key organisms in coral reef functioning and maintenance. The preliminary studies reported here, all tend to show that boring microflora are major primary producers and agents of bioerosion dissolving large quantities of calcium carbonate with a potential in buffering seawater, recycling nutrients and producing metabolites allowing survival of corals when bleached. High productivity of microbial euendoliths in live coral is in support of a balance between constructive and destructive processes, and appears to be beneficial to the health of the reef as a whole, whereas the high productivity of euendoliths on coral rubble may feed the grazers that destroy the reef in the process. Additionally, a few studies show that some euendoliths may be involved in coral diseases and may weaken coral health, rendering them more susceptible to death. Those studies opened new questions, especially on the effects of environmental factors such as eutrophication, rising sea surface temperature, UV and atmospheric CO_2 on euendolith metabolism.

Acknowledgements

I would like to thank Dr Max Wisshak and Dr Leif Tapanila for offering me the opportunity to send a contribution for the book volume on 'Current Developments in Bioerosion'. I also thank the reviewers, Prof Stjepko Golubic and Dr William Kiene, for their comments which helped improving the manuscript.

References

Adey WH (1998) Review. Coral reefs: algal structured and mediated ecosystems in shallow, turbulent, alkaline waters. J Phycol 34:393-406

Agardh CA (1824) Systema algarum. Berling, Lund, 312 pp

Alexandersson ET (1975) Marks of unknown carbonate-decomposing organelles in cyanophyte borings. Nature 254:237-238

Alker AP, Smith GW, Kiho K (2001) Characterization of *Aspergillus sydowii* (Thom et Church), a fungal pathogen of Caribbean sea fan corals. Hydrobiol 460:105-111

Al-Thukair AA, Golubic S (1991) New endolithic cyanobacteria from the Arabian Gulf. I. *Hyella immanis* sp. nov. J Phycol 27:766-780

Anagnostidis K, Pantazidou A (1988) *Hyella kalligrammos* sp. nov., *Hyella maxima* (Geitl.) comb. nov., and other freshwater morphotypes of the genus *Hyella* Born. et Flah. (Chroococcales, Cyanophyceae). Algol Stud 50/53:227-247

Anderson AJ, Mackenzie FT, May Ver L (2003) Solution of shallow-water carbonates: An insignificant buffer against rising atmospheric CO_2. Geol Soc Amer 31:513-516

Ascaso C, Wierzchos J, Castello R (1998) Study of the biogenic weathering of calcareous litharenite stones caused by lichen and endolithic microorganisms. Int Biodeterior Biodegrad 42:29-38

Batters EAL (1902) A catalog of the Bristish marine algae. J Bot, Brit and Foreign 40 (Suppl):1-107

Bellamy N, Risk MJ (1982) Coral gas: oxygen production in *Millepora* on the Great Barrier Reef. Science 215:1618-1619

Bentis CJ, Kaufman L, Golubic S (2000) Endolithic fungi in reef-building corals (Order: Scleractinia) are common, cosmopolitan, and potentially pathogenic. Biol Bull 198:254-260

Bornet E, Flahault C (1888) Note sur deux nouveaux genres d'algues perforantes. J Bot Morot 2:161-165

Bornet E, Flahault C (1889) Sur quelques plantes vivant dans le test calcaire des mollusques. Bull Soc Bot France 36:147-177

Borowitzka MA (1981) Photosynthesis and calcification in the articulated coralline algae *Amphiroa anceps* and *A. foliaceaa*. Mar Biol 62:17-23

Bruggemann JH, van Oppen MJH, Breeman AN (1994) Foraging by the spotlight parrotfish *Sparisoma viride*. I. Food selection in different, socially determined habitats. Mar Ecol Prog Ser 106:41-55

Campbell SE (1980) *Palaeoconchocelis starmachii*, a carbonate boring microfossil from the upper Silurian of Poland (425 millions years old): implications for the evolution of the Bangiaceae (Rhodophyta). Phycologia 19:25-36

Carreiro-Silva M, McClanahan TR, Kiene WE (2005) The role of inorganic nutrients and herbivory in controlling microbioerosion of carbonate substratum. Coral Reefs 24:214-221

Casareto B, Suzuki Y, Ishikawa Y, Kurosawa K (2004) Benthic primary production and N_2 fixation in a fringing coral reef at Miyako Island, Okinawa, Japan. Proc 10th Int Coral Reef Symp, Okinawa, Japan, Abstr, p 192

Chacón E, Berrendero E, Garcia-Pichel F (2006) Biological signatures of microboring cyanobacterial communities in marine carbonates from Cabo Rojo, Puerto Rico. Sediment Geol 185:215-228

Chazottes V, Le Campion-Alsumard T, Peyrot-Clausade M (1995) Bioerosion rates on coral reefs: interaction between macroborers, microborers and grazers (Moorea, French Polynesia). Palaeogeogr Palaeoclimatol Palaeoecol 113:189-198

Chazottes V, Le Campion-Alsumard T, Peyrot-Clausade M, Cuet P (2002) The effects of eutrophication-related alterations to coral reef communities on agents and rates of bioerosion (Reunion Island, Indian Ocean). Coral Reefs 21:375-390

Dana JD (1846) Notice of some genera of *Coclopacea.* Amer J Sci Arts 1:225-230

De Angelis d'Ossat G (1908) Altri Zoantari del Terziario della Patagonia. An Mus Nac Buenos Aires 3:93-102

De Los Rios A, Wierzchos J, Sancho LG, Green TGA, Ascaso C (2005) Ecology of endolithic lichens colonizing granite in continental Antarctica. Lichenologist 37:383-395

Domart-Coulon IJ, Sinclair CS, Hill RT, Tambutté S, Purevel S, Ostrander GK (2004) A basidiomycete isolated from the skeleton of *Pocillopora damicornis* (Scleractinia) selectively stimulates short-term survival of coral skeletogenic cells. Mar Biol 144:583- 592

Domart-Coulon IJ, Traylor-Knowles N, Peters E, Elbert D, Downs CA, Price K, Stubbs J, McLaughlin S, Cox E, Aeby G, Brown PR, Ostrander GK (2006) Comprehensive characterization of skeletal tissue growth anomalies of the finger coral *Porites compressa.* Coral Reefs 25:531-543

Eakin CM (1996) Where have all the carbonates gone? A model comparison of calcium carbonate budgets before and after the 1982-1983 El Niñio at Uva Island in the eastern Pacific. Coral Reefs 15:109-119

Edinger EN, Jompa J, Limmon GV, Widjatmoko W, Risk MJ (1998) Reef degradation and biodiversity in Indonesia: effects of land-based pollution, destructive fishing practices, and changes through time. Mar Poll Bull 36:617-630

Edinger EN, Limmon GV, Jompa J, Widjatmoko W, Heikoop J, Risk M (2000) Normal coral growth rates on dying reefs: Are coral growth rates good indicators of reef health? Mar Poll Bull 40:404-425

Edwards HGM, Russell NC, Wynn-Williams DD (1997) Fourier transform raman spectroscopic and scanning electron microscopic study of cryptoendolithic lichen from Antarctica. J Raman Spectrosc 28:685-690

Ehrenberg CG (1834) Beiträge zur physiologischen Kenntniss der Corallenthiere im allgemeinen, und besonders des Rothen Meeres, nebst einem Versuche zur physiologischen Systematik derselben. Abh K Akad Wiss Berlin, Phys Math, 1832:225-380

Ellis J, Solander D (1786) The natural history of many curious and uncommon zoophytes. White and Son, London, 208 pp

Enriquez S, Méndez ER, Iglesias-Prieto R (2005) Multiple scattering on coral skeletons enhances light absorption by symbiotic algae. Limnol Oceanogr 50:1025-1032

Ercogovic A (1927) Tri nova roda litofiskih cijanoiceja sa jadranske obale. Acta Bot 2:78-84

Ercegovic A (1930) Sur quelques types peu connus de Cyanophycées lithophytes. Arch Protistenkd 71:361-373

Ferrer LM, Szmant AM (1988) Nutrient regeneration by the endolithic communities in coral skeletons. Proc 6th Int Coral Reef Symp, Townsville, Australia, 3:1-4

Fine M, Loya Y (2002) Endolithic algae: an alternative source of photoassimilates during coral bleaching. Proc Roy Soc London 269:1205-1210

Fine M, Zibrowius H, Loya Y (2001) *Oculina patagonica,* a non lessepsian scleractinian coral invading the Mediterranean Sea. Mar Biol 138:1195-1203

Fine M, Steindler L, Loya Y (2004) Endolithic algae photoacclimate to increased irradiance during coral bleaching. Mar Fresh Res 55:115-121

Fine M, Meroz-Fine E, Hoegh-Guldberg O (2005) Tolerance of endolithic algae to elevated temperature and light in the coral *Montipora monasteriata* from the southern Great Barrier Reef. J Exper Biol 208:75-81

Fine M, Roff G, Ainsworth TD, Hoegh-Guldberg O (2006) Phototrophic microendoliths bloom during coral "white syndrome". Coral Reefs 25:577-581

Fork DC, Larkum ADW (1989) Light harvesting in the green alga *Ostreobium* sp., a coral symbiont adapted to extreme shade. Mar Biol 103:381-385

Friedmann EI, Hua M, Ocampo-Friedmann R (1988) Cryptoendolithic lichen and cyanobacterial communities of the Ross Desert, Antarctica. Polarforschung 58:215-259

Garcia-Pichel F (2006) Plausible mechanisms for the boring on carbonates by microbial phototrophs. Sediment Geol 185:205-213

Gektidis M (1999) Development of microbial euendolithic communities: the influence of light and time. Bull Geol Soc Denmark 45:147-150

Gektidis M, Golubic S (1996) A new endolithic cyanophyte / cyanobacterium: *Hyella vacans* sp. nov. from Lee Stocking Island, Bahamas. Nova Hedwigia, Beih 112:93-100

Ghirardelli LA (2002) Endolithic microorganisms in live and dead thalli of coralline red algae (Corallinales, Rhodophyta) in the northern Adriatic Sea. Acta Geol Hisp 37:53-60

Glaub I, Vogel K (2004) The stratigraphic record of microborings. Fossils Strata 51:126-135

Golubic S, Schneider J (2003) Microbial endoliths as internal biofilms. In: Krumbein WE, Dornieden T, Volkmann M (eds) Fossil and Recent biofilms. Kluwer, Dordrecht, pp 249-263

Golubic S, Perkins RD, Lukas KJ (1975) Boring microoganisms and microborings in carbonate substrates. In: Frey RW (ed) The study of trace fossils. Springer, New York, pp 229-259

Golubic S, Knoll AH, Ran W (1980) Morphometry of late Ordovician microbial borings. Tech Summa Abstr AAPG Bull 64:713

Golubic S, Friedmann I, Schneider J (1981) The lithobiontic ecological niche, with special reference to microorganisms. Sediment Geol 51:475-478

Golubic S, Campbell SE, Drobne K, Cameron B, Balsam WL, Cimerman F, Dubois L (1984) Microbial endoliths: a benthic overprint in the sedimentary record, and a paleobathymetric cross-reference with foraminifera. J Paleontol 58:351-361

Golubic S, Hernandez-Marine M, Hoffmann L (1996) Developmental aspects of branching in filamentous Cyanophyta / Cyanobacteria. Algol Stud 83:303-329

Golubic S, Radtke G, Le Campion-Alsumard T (2005) Endolithic fungi in marine ecosystems. Trends Microbiol 13:229-235

Gomont M (1892) Monographie des Oscillariees (Nostocacees homocystees) Ann Sci Nat Bot 15:263-368

Goreau TJ, Cervino J, Goreau M, Hayes R, Hayes M, Richardson L, Smith G, DeMeyer K, Nagelkerken I, Garzon-Ferrera J, Gil D, Garrison G, Williams EH, Bunkley-Williams L, Quirrolo C, Patterson K, Porter JW, Porter K (1998) Rapid spread diseases in Caribbean coral reefs. Rev Biol Trop 46 (Suppl 5):157-171

Halldal P (1968) Photosynthetic capacities and photosynthetic action spectra of endozoic algae of the massive coral *Favia*. Biol Bull 134: 411-424

Halley RB, Yates KK, Brock JC (2005) South Florida coral-reef sediment dissolution in response to elevated CO_2. Proc 10th Int Coral Reef Symp, Okinawa, Japan, Abstr, p 178

Hallock P (2005) Global change and modern coral reefs: New opportunities to understand shallow-water carbonate depositional processes. Sediment Geol 175:19-33

Harborne AR, Mumby PJ, Micheli F, Perry CT, Dahlgren CP, Holmes KE, Brumbaugh DR (2006) The functional value of Caribbean coral reef, seagrass and mangrove habitats to ecosystem processes. Adv Mar Biol 50:57-189

Harmelin-Vivien ML (1994) The effects of storms and cyclones on coral reefs: a review. J Coast Res Spec Issue 12:221-231

Hauck F (1876) Verzeichnis der im Golf von Triest gesammelten Meeralgen (Fortsetzung). Österr Bot Z 26:24-26 and 54-57

Hoegh-Guldberg O (1999) Climate change coral bleaching and the future of the world's coral reefs. Greenpeace, 27 pp, http://cmbc.ucsd.edu/content/1/docs/hoegh-gulberg.pdf

Hoegh-Guldberg O, Salvat B (1995) Periodic mass-bleaching and elevated sea-level temperatures: bleaching of outer-reef slope communities in Moorea, French Polynesia. Mar Ecol Prog Ser 121:181-190

Hoffmann L (1999) Marine cyanobacteria in tropical regions: diversity and ecology. Europ J Phycol 34:371-379

Houghton JT, Meira Filho LG, Callander BA, Harris N, Kattenberg A, Maskell K (1996) Climate change 1995. The science of climate change. Cambridge Univ Press, Cambridge, 306 pp

Hubbard DK, Sadd JL, Miller AI, Gill IP, Dill RF (1981) The production, transportation and deposition of carbonate sediments on the insular shelf of St Croix, U.S. Virgin Island. West Indies Lab Rep Contrib N°76

Hutchings PA (1986) Biological destruction of coral reefs. Coral Reefs 4:239-252

Jones B, Goodbody QH (1982) The geological significance of endolithic algae in glass. Canad J Earth Sci 19:671-678

Kaehler S, McQuaid CD (1999) Lethal and sublethal effects of phototrophic endoliths attacking the shell of the intertidal mussel *Perna perna*. Mar Biol 13:497-503

Kanwisher JW, Wainwright SA (1967) Oxygen balance in some reef corals. Biol Bull 133:378-390

Kendrick B, Risk MJ, Michaelides J, Bergman K (1982) Amphibious microborers: bioeroding fungi isolated from live corals. Bull Mar Sci 32:862-867

Kiene WE (1997) Enriched nutrients and their impact on bioerosion: results from ENCORE. Proc 8th Int Coral Reef Symp, Panama, 11:897-902

Kiene WE, Radtke G, Gektidis M, Golubic S, Vogel K (1995) Factors controlling the distribution of microborers in Bahamian reef environments. Facies 32:145-188

Kinsey DW (1985) Metabolism, calcification and carbon production. I. Systems level studies. Proc 5th Int Coral Reef Congr, Tahiti, 4:505-526

Klement KW, Toomey DF (1967) Role of bluegreen alga *Girvanella* in skeletal grain destruction and lime-mud formation in the lower Ordovician of west Texas. J Sediment Petrol 37:1045-1051

Kleypas JA, Buddemeier RW, Archer D, Gattuso J-P, Langdon C, Opdyke BN (1999) Geochemical consequences of increased atmospheric carbon dioxide on coral reefs. Science 284:118-120

Kleypas JA, Buddemeier RW, Gattuso JP (2001) The future of coral reefs in an age of global change. Int J Earth Sci 90:426-437

Knoll AH, Golubic S, Green J, Swett K (1986) Organically preserved microbial endoliths from the late Proterozoic of East Greenland. Nature 321:865-857

Kobluk DR, Kahle CF (1977) Bibliography of the endolithic (boring) algae and fungi and related geologic processes. Bull Canad Petrol Geol 25:208-223

Kobluk DR, Risk M (1977) Calcification of exposed filaments of endolithic algae, micrite envelope formation and sediment production. J Sediment Petrol 47:517-528

Koehne B, Elli G, Jennings RC, Wilhelm C, Trissl HW (1999) Spectroscopic and molecular characterization of a long wavelength absorbing antenna of *Ostreobium* sp. Biochim Biophys Acta 1412:94-107

Königshof P, Glaub I (2004) Traces of microboring organisms in Palaeozoic conodont elements. Geobios 37:416-424

Komárek J, Anagnostidis K (1999) Cyanoprokaryota 1. Chroococcales. In: Ettl H, Gärtner G, Heynig H, Mollenhauer D (eds) Süsswasserflora von Mitteleuropa. Spektrum Akad Verl, Fischer, Jena, 19:1-548

Komárek J, Anagnostidis K (2005) Cyanoprokaryota 2. Oscillatoriales. In: Büdel B, Gärtner G, Krienitz L, Schagerl M (eds) Süsswasserflora von Mitteleuropa. Elsevier, Spektrum Akad Verl, München, 19:1-759

Kushmaro A, Rosenberg E, Fine M, Loya Y (1997) Bleaching of coral *Oculina patagonica* by *Vibrio* AK-1. Mar Ecol Prog Ser 147:159-165

Laborel J, Le Campion-Alsumard T (1979) Infestation massive du squelette de coraux vivants par des Rhodophycées de type Conchocelis. C R Acad Sci Paris 288:1575-1577

Lagerheim G (1886) Notes sur le *Mastigocoleus*, nouveau genre des algues marines de l'ordre des Phycochromacées. Notarisia 1:65-69

Lamouroux JVF (1812) Extrait d'un mémoire sur la classification des polypiers coralligènes non entièrement pierreux. Nouv Bull Sci Soc Philomat Paris 3:181-188

Langdon C, Takahashi T, Marubini F, Atkinson MJ, Sweeney C, Aceves H, Barnet H, Chipman D, Goddard J (2000) Effect of calcium carbonate saturation state on the calcification rate of an experimental coral reef. Global Biogeochem Cycles 14:639-654

Le Bris S, Le Campion-Alsumard T, Romano J-C (1998) Caractéristiques du feutrage algal des récifs coralliens de Polynésie française soumis à différentes intensités de bioérosion. Oceanolog Acta 21:695-708

Le Campion-Alsumard T (1975) Etude expérimentale de la colonisation d'éclats de calcite par les cyanophycées endolithes marines. Cah Biol Mar 16:177-185

Le Campion-Alsumard T (1976) Etude préliminaire sur l'écologie et l'ultrastructure d'une cyanophycée Chrooccocale endolithe. J Microscop Biol Cell 26:53-60

Le Campion-Alsumard T (1978) Les cyanophycées endolithes marines: systématique, ultrastructure, écologie et biodestruction. Thèse Univ Aix-Marseille II, 198 pp

Le Campion-Alsumard T (1979) Les cyanophycées endolithes marines. Systématique, ultrastructure, écologie et biodestruction. Oceanolog Acta 2:143-156

Le Campion-Alsumard T, Golubic S, Hutchings PA (1995a) Microbial endoliths in skeletons of live and dead corals: *Porites lobata* (Moorea, French Polynesia). Mar Ecol Prog Ser 117:149-157

Le Campion-Alsumard T, Golubic S, Priess K (1995b) Fungi in corals: symbiosis or disease? Interaction between polyps and fungi causes pearl-like skeleton biomineralization. Mar Ecol Prog Ser 117:137-147

Lesser MP, Shick JM (1989) Photoadaptation and defenses against oxygen toxicity in zooxanthellae from natural populations of symbiotic cnidarians. J Exp Mar Biol Ecol 134:129-141

Linnaeus C (1758) Systema naturae per regna tria naturae, secundum classes, ordines, genera, species, cum characteribus, differentiis, synonymis, locis. Tomus I. Editio decima, reformata. Laurentii Salvii, Holmiae, 824 pp

Lukas KJ (1973) Taxonomy and ecology of the endolithic microflora of reef corals, with a review of the literature on endolithic microphytes. PhD Diss, Univ Rhode Island, 159 pp

Lukas KJ (1974) Two species of the chlorophyte genus *Ostreobium* from skeletons of Atlantic and Caribbean reef corals. J Phycol 10:331-335

Lukas, KJ (1978) Depth distribution and form among common microboring algae from the Florida continental shelf. Geol Soc Amer, Abstr Programs 10:1-448
Mallela J, Perry CT (2007) Calcium carbonate budgets for two coral reefs affected by different terrestrial runoff regimes, Rio Bueno, Jamaica. Coral Reefs 26:129-145
Mao Che L, Le Campion-Alsumard T, Boury-Esnault N, Payri C, Golubic S, Bézac C (1996) Biodegradation of shells of the black pearl oyster, *Pinctada margaritifera* var. *cumingii*, by microborers and sponges of French Polynesia. Mar Biol 126:509-519
Marubini F, Davis PS (1996) Nitrate increases zooxanthellae population density and reduces skeletogenesis in corals. Mar Biol 127:319-328
Meesters E, Bak R (1995) Age related deterioration of a physiological function in the branching coral *Acropora palmata*. Mar Ecol Prog Ser 121:203-209
Mumby PJ, Chisholm JRM, Edwards AJ, Clark CD, Roark EB, Andrefouet S, Jaubert J (2001) Unprecedented bleaching-induced mortality in *Porites* spp. at Rangiroa Atoll, French Polynesia. Mar Biol 139:183-189
Nothdurft LD, Webb GE, Buster NA, Holmes CW, Sorauf JE (2005) Brucite microbialites in living coral skeletons: indicators of extreme microenvironments in shallow-marine settings. Geol Soc Amer 33:169-172
Odum HT, Odum EP (1955) Trophic structure and productivity of a windward coral reef community on Eniwetak atoll. Ecol Monogr 25:291-320
Paerl HW, Pinckney JL (2005) A mini-review of microbial consortia: Their roles in aquatic production and biogeochemical cycling. Microbial Ecol 31:225-247
Pari N, Peyrot-Clausade M, Le Campion-Alsumard T, Huthcings PA, Chazottes V, Golubic S, Le Campion J, Fontaine MA (1998) Bioerosion of experimental substrates on high islands and atoll lagoons (French Polynesia) after two years of exposure. Mar Ecol Prog Ser 166:119-130
Pari N, Peyrot-Clausade M, Hutchings PA (2002) Bioerosion of experimental substrates on high islands and atoll lagoons (French Polynesia) during 5 years of exposure. J Exp Mar Biol Ecol 276:109-127
Perry CT (1998) Grain susceptibility to the effects of microboring: implication for the preservation of skeletal carbonates. Sedimentology 45:39-51
Perry CT (2000) Factors controlling sediment preservation on a north Jamaican fringing reef: a process-based approach to microfacies analysis. J Sediment Res 70:633-648
Perry CT, MacDonald IA (2002) Impacts of light penetration on the bathymetry of reef microboring communities: implications for the development of microendolithic trace assemblages. Palaeogeogr Palaeoclimatol Palaeoecol 186:101-113
Porter D, Lingle WL (1992) Endolithic thraustrochytrid marine fungi from planted shell fragments. Mycologia 84:289-299
Priess K, Le Campion-Alsumard T, Golubic S, Gadel F, Thomassin BA (2000) Fungi in corals: black bands and density-banding of *Porites lutea* and *P. lobata* skeleton. Mar Biol 136:19-27
Radtke G (1993) The distribution of microborings in molluscan shells from recent reef environments at Lee Stocking Island, Bahamas. Facies 29:81-92
Radtke G, Golubic S (2005) Microborings in mollusc shells, Bay of Safaga, Egypt: morphometry and ichnology. Facies 51:118-134
Radtke G, Le Campion-Alsumard T, Golubic S (1996) Microbial assemblages of the bioerosional "notch" along tropical limestone coasts. Algol Stud 83:469-482
Radtke G, Le Campion-Alsumard T, Golubic S (1997a) Microbial assemblages involved in tropical coastal bioerosion: an Atlantic-Pacific comparison. Proc 8th Int Coral Reefs Symp, Panama, 2:1825-1830

Radtke G, Hofmann K, Golubic S (1997b) A bibliographic overview of micro- and macroscopic bioerosion. In: Betzler C, Hüssner H (eds) Vogel-Festschrift. Courier Forschinst Senckenberg 201:307-340

Raghukumar C, Sharma S, Lande V (1991) Distribution and biomass estimation of shell boring algae in the intertidal at Goa, India. Phycologia 30:303-309

Rathbun R (1887) The crab, lobster, crayfish, rock lobster, shrimp, and prawn fisheries. In Goode GB (ed) The fisheries and fishery industries of the United States. Section V, History and Methods of the Fisheries II. U.S. Government Printing Office, Washington, pp 627-839

Reaka-Kudla ML, Feingold JS, Glynn W (1996) Experimental studies of rapid bioerosion of coral reefs in the Galapagos Islands. Coral Reefs 15:101-107

Risk MJ, Kramer JR (1981) Water chemistry inside coral heads: determination of pH, Ca and Mg. Proc 4th Int Coral Reef Symp, Manila, Abstr, p 54

Risk MJ, Muller HR (1983) Porewater in coral heads: evidence for nutrient regeneration. Limnol Oceanogr 28:1004-1008

Risk MJ, Heikoop JM, Edinger EN, Erdmann MV (2001) The assessment 'toolbox': community-based reef evaluation methods coupled with geochemical techniques to identify sources of stress. Bull Mar Sci 69:443-458

Rojtan RD, Lewis SM (2005) Selective predation by parrotfishes on the reef coral *Porites astreoides*. Mar Ecol Prog Ser 305:193-201

Russ G (1984) Distribution and abundance of herbivorous grazing fishes in the Central Great Barrier Reef. I. Levels of variability across the entire continental shelf. Mar Ecol Prog Ser 20:23-34

Russell NC, Edwards HGM, Wynn-Williams DD (1998) FT-Raman spectroscopic analysis of endolithic microbial communities from Beacon sandstone in Victoria Land, Antarctica. Antarct Sci 10:63-74

Shashar N, Stambler N (1992) Endolithic algae within corals – life in an extreme environment. J Exp Mar Biol Ecol 163:277-286

Sharshar N, Cohen Y, Loya Y, Sar N (1994) Nitrogen fixation (acetylene reduction) in stony corals: Evidence for coral-bacteria interactions. Mar Ecol Prog Ser 111:259-264

Sharshar N, Banaszak AT, Lesser MP, Amrami D (1997) Coral endolithic algae: Life in a protected environment. Pacif Sci 51:167-173

Shibata K, Haxo FT (1969) Light transmission and spectral distribution through epi- and endozoic algal layers in the brain coral, *Favia*. Biol Bull 136:461-468

Schlichter D, Zscharnack B, Krisch H (1995) Transfer of photoassimilates from endolithic algae to coral tissue. Naturwissenschaften 82:561-564

Schlichter D, Kampmann H, Conrady S (1997) Trophic potential and photoecology of endolithic algae living within coral skeletons. PSZN. Mar Ecol 18:299-317

Schneider J, Le Campion-Alsumard T (1999) Construction and destruction of carbonates by marine and freshwater cyanobacteria. Europ J Phycol 34:417-426

Schneider J, Torunski H (1983) Biokarst on limestone coasts, morphogenesis and sediment production. Mar Ecol 4:45-63

Schneider J, Schröder HG, Le Campion-Alsumard T (1983) Algal micro-reefs – Coated grains from freshwater environments. In: Peryt TM (ed) Coated grains. Springer, Berlin Heidelberg, pp 284-298

Sheppard CRC, Spalding M, Bradshaw, Wilson S (2002) Erosion vs recovery of coral reefs after 1998 El Niño: Chagos reefs, Indian Ocean. Ambio 31:40-48

Szmant AM (2002) Nutrient enrichment on coral reefs: Is it a major cause of coral reef decline? Estuaries 25:743-766

Tribollet A (2007) Dissolution of dead corals by euendolithic microorganisms across the northern Great Barrier Reef (Australia). Microbial Ecol DOI: 10.1007/s00248-007-9302-6

Tribollet A, Golubic S (2005) Cross-shelf differences in the pattern and pace of bioerosion of experimental carbonate substrates exposed for 3 years on the northern Great Barrier Reef, Australia. Coral Reefs 24:422-434

Tribollet A, Payri C (2001) Bioerosion of the crustose coralline alga *Hydrolithon onkodes* by microborers in the coral reefs of Moorea, French Polynesia. Oceanolog Acta 24:329-342

Tribollet A, Decherf G, Hutchings PA, Peyrot-Clausade M (2002) Large-scale spatial variability in bioerosion of experimental coral substrates on the Great Barrier Reef (Australia): importance of microborers. Coral Reefs 21:424-432

Tribollet A, Langdon C, Golubic S, Atkinson M (2006a) Endolithic microflora are major primary producers in dead carbonate substrates of Hawaiian coral reefs. J Phycol 42:292-303

Tribollet A, Atkinson M, Langdon C (2006b) Effects of elevated pCO_2 on epilithic and endolithic metabolism of reef carbonates. Global Change Biol 12:2200-2208

Tudhope AW, Risk MJ (1985) Rate of dissolution of carbonate sediments by microboring organisms, Davies Reef, Australia. J Sediment Petrol 55:440-447

Vogel K, Gektidis M, Golubic S, Kiene WE, Radtke G (2000) Experimental studies on microbial bioerosion at Lee Stocking Island, Bahamas and One Tree Island, Great Barrier Reef, Australia: implications for paleoecological reconstructions. Lethaia 33:190-204

Vooren CM (1981) Photosynthetic rates of benthic algae from the deep coral reef of Curaçao. Aquat Bot 10:143-154

Voss JD, Richardson LL (2006) Nutrient enrichment enhances black band disease progression in corals. Coral Reefs 25:569-576

Wanders JBW (1977) The role of the benthic algae in the shallow reef of Curaçao (Netherlands Antilles). III. The significance of grazing. Aquat Bot 3:357-390

Weis VM, Smith GJ, Muscatine L (1989) A CO_2 supply mechanism in zooxanthellate cnidarians: role of carbonic anhydraze. Mar Biol 100:195-202

Wilkinson C (2004) Status of coral reefs of the world: 2004. Austral Inst Mar Sci, Townsville, Australia, 557 pp

Williams SL, Carpenter RC (1998) Effects of unidirectional and oscillatory flow on nitrogen fixation (acetylene reduction) in coral reef algal turfs, Kaneohe Bay, Hawaii. J Exp Mar Biol Ecol 226:293-316

Wisshak M, Gektidis M, Freiwald A, Lundälv T (2005) Bioerosion along a bathymetric gradient in cold-temperate setting (Kosterfjord, SW Sweden): an experimental study. Facies 51:93-117

Womersley HBS (1987) The marine benthic flora of southern Australia. Part II. South Austral Gov Print Div, Adelaide, pp 1-481

Woolcott GW, Knoller K, King RJ (2000) Phylogeny of the Bryopsidaceae (Bryopsidales, Chlorophyta): cladistic analyses of morphological and molecular data. Phycologia 39:471-481

Zhang Y, Golubic S (1987) Endolithic microfossils (cyanophyta) from early Proterozoic stromatolites, Hebei, China. Acta Micropaleont Sin 4:1-12

Zubia M, Peyrot-Clausade M (2001) Internal bioerosion of *Acropora formosa* in Réunion (Indian Ocean): microborer and macroborer activities. Oceanolog Acta 24:251-262

The trace *Rhopalia clavigera* isp. n. reflects the development of its maker *Eugomontia sacculata* Kornmann, 1960

Stjepko Golubic[1], Gudrun Radtke[2]

[1] Biological Science Center, Boston University, Boston MA 02215, USA, (golubic@bu.edu)
[2] Hessisches Landesamt für Umwelt und Geologie, 65203 Wiesbaden, Germany

Abstract. Complex boring patterns often reflect the complexity of life cycle of the euendoliths that produce them. They are illustrated here by different stages in the development of the euendolithic ulotrichacean chlorophyte *Eugomontia sacculata* reconstructed on the basis of its complex trace in the shells of *Mya arenaria* in brackish waters of the Baltic Sea at Gdansk, Poland. The trace consists of different types of boring morphologies as distinctive from one another as many traces specific to separate organisms. Because they occur associated, they may be misinterpreted as separate members of an ichnocoenosis. We propose to describe them as parts of a complex trace instead, because they are based on the genetic program of a single organism, but expressed in different proportions at different stages of its development. A new ichnospecies, *Rhopalia clavigera* isp. n., is described to characterize these traces.

Keywords. Complex traces, development, *Eugomontia*, ichnofossils, microborings, *Rhopalia clavigera*, salinity

Introduction

Most microbial euendoliths are known from modern marine environments with very few described and reported from freshwater or brackish settings. In contrast, many fossil microborings, which have modern counterparts similar to the borings of marine euendoliths, were described from paleoenvironments interpreted as brackish (Pomerol and Feugueur 1986; Radtke 1991). In the study of paleoenvironments there are few sedimentary clues to estimate past salinities, although the salinity is generally considered an important parameter responsible for selective separation of marine and freshwater taxa of plants and animals. Boring traces preserve well in the fossil record and show relatively high morphological specificity which, in addition to their small size, makes them good potential environmental indicators (Vogel et al. 2000; Glaub et al. 2007). The effect of salinity on endolith distribution has been studied using experimentally exposed substrates in the waters of the Skagerrak where

M. Wisshak, L. Tapanila (eds.), *Current Developments in Bioerosion*. Erlangen Earth Conference Series, doi: 10.1007/978-3-540-77598-0_5,

salinities fluctuated with the change of currents between normal marine waters of the North Sea and brackish waters of the Baltic Sea; these substrates have all recruited assemblages of marine euendoliths (Wisshak et al. 2005; Wisshak 2006).

The objective of this study was to explore the extension of marine euendoliths into the brackish ranges of stable, consistently low salinities below 10 ppt of NaCl in the Baltic Sea at Gdansk, Poland. Our study in a low salinity and low diversity environment had an additional benefit by providing numerous euendolithic populations of a single taxon at different stages of its development.

We report here the prevalence of complex borings produced by the chlorophyte *Eugomontia sacculata* Kornmann, 1960 in naturally exposed shells of *Mya arenaria* Linnaeus, 1758. The organism exhibited conspicuous dominance producing coherent internal biofilms (sensu Golubic and Schneider 2003) comprised of networks of intertwined filaments. The organism produced a number of colonies at different stages of development. Its complex boring system is described as a new ichnofossil. *Eugomontia sacculata* was found euendolithically in carbonate substrates with a wide distribution, mostly reported from northern latitudes (e.g., Wilkinson and Burrows 1970; Wilkinson 1974; Akpan and Farrow 1984; Akpan 1986; Burrows 1991; Tittley et al. 2005) but also from the Mediterranean (Gallardo et al. 1993) and the Red Sea (Gektidis et al. 2007). Nielsen (1987) described a second species *Eugomontia stelligera* from cultures isolated from shells in New Zealand, which has not been observed in natural settings.

Materials and methods

A collection of dead shells of the bivalve *Mya arenaria*, which have been exposed to euendolith colonization in shallow brackish waters on the coast of the Baltic Sea at Gdansk, was examined for the presence and distribution of microborings and resident microbial euendoliths. Shells were studied for the presence of discoloration by phototrophic euendoliths using dissecting microscope. Forty shells and shell fragments from two different collection sites were separated for further analysis. The areas affected by endoliths in each sample were fragmented and sub-sampled. A ca. 10 mm^2 shell fragment from the same area was analyzed using different methods: (a) The surface of a shell fragment was analyzed using scanning electron microscope (SEM). (b) Another fragment was dissolved in 3% HCl exposing a coherent biofilm of euendoliths, which was observed under dissecting microscope, cut in half and mounted on microscope slides, each half with the other side up, so that both sides of the endolithic biofilm could be examined. (c) A fragment was treated with acetone and impregnated with slow polymerizing Araldite resin (Fluka, Bregenz, Austria), then excised and decalcified with dilute HCl (Golubic et al. 1983). The resin casts were observed by SEM. The boring replicas were measured using in-scale photomicrographs and Sigma-Scan Image software (SigmaScan, San Raphael, CA). The measurements of euendolith colonies at different stages of development were carried out and compared with the resin casts of their borings with 25-50 measurements for each of the following parameters: Minimum and maximum width of filaments observed by light microscopy at 400 x magnification and of cylindrical tunnels measured from SEM images; width and length (= depth of penetration) of largest sporangial swellings and casts of their boreholes, respectively, were measured in each of the colonies. Fully developed swellings extend perpendicular to the substrate surface. The results were expressed as mean ± standard deviation (number of measurements) and graphed as cross diagrams with arms of one standard deviation and the mean at the intersection. The distribution of measurements was tested for normalcy, and the distance of two standard deviations from the mean included most of the individual data.

Results

Shells of the bivalve *Mya arenaria* collected in shallow waters of the Baltic coast were consistently dominated by the euendolithic ulotrichacean chlorophyte *Eugomontia sacculata*, accompanied by the cyanobacteria *Hyella* Bornet and Flahault, 1888 and *Cyanosaccus* Lukas and Golubic, 1981 as minor components. *Eugomontia* Kornmann, 1960 was resident in most shells and formed complex boring systems at different stages of development. The accompanying cyanobacteria occurred in minor patches mostly separate from the colonies of *Eugomontia.* The identification was made by correlating the morphology of the borings with that of the resident euendoliths extracted from the same substrate fragment. Developmental sequences were reconstructed from resin casts by comparing a large number of boring systems arranged by size.

Eugomontia formed two different basic types of borings which prevail at different times of its development (Fig. 1). The endolithic portions of the thallus normally originate at point locations where the organism penetrates the surface of the shell from where it spreads by growing filaments directly beneath the surface of the substrate. Filaments radiate in all direction, forming a shallow circular colony. The resin replicas reveal a dense system of ramified tunnels with the highest density in the parts proximal to the point of entry, gradually decreasing toward the periphery (Fig. 1A). The second stage consists of the development of endolithic sacs with reproductive function (sporangia). The sacs start as simple spindle-shaped widening of the tunnels, but soon become asymmetric and then elongate perpendicular to the filaments, oriented toward the interior of the substrate. Less common are slight terminal swellings of filaments on short lateral branches. At this stage, the tunnels without widening are limited to the colony margin (Fig. 1B) where the expansion of the colony slowly continues. The development of sacs, possibly triggered by an environmental stimulus, is initiated in the central (oldest) region of the colony, proximal to the point of entry, from where it rapidly spreads toward the periphery, so that ultimately the entire colony including the margins are characterized by sac formation (Fig. 1C). The proportion of the colony covered by sacs varies from system to system, although the largest sacs and their highest density are found in the proximal regions of the colony.

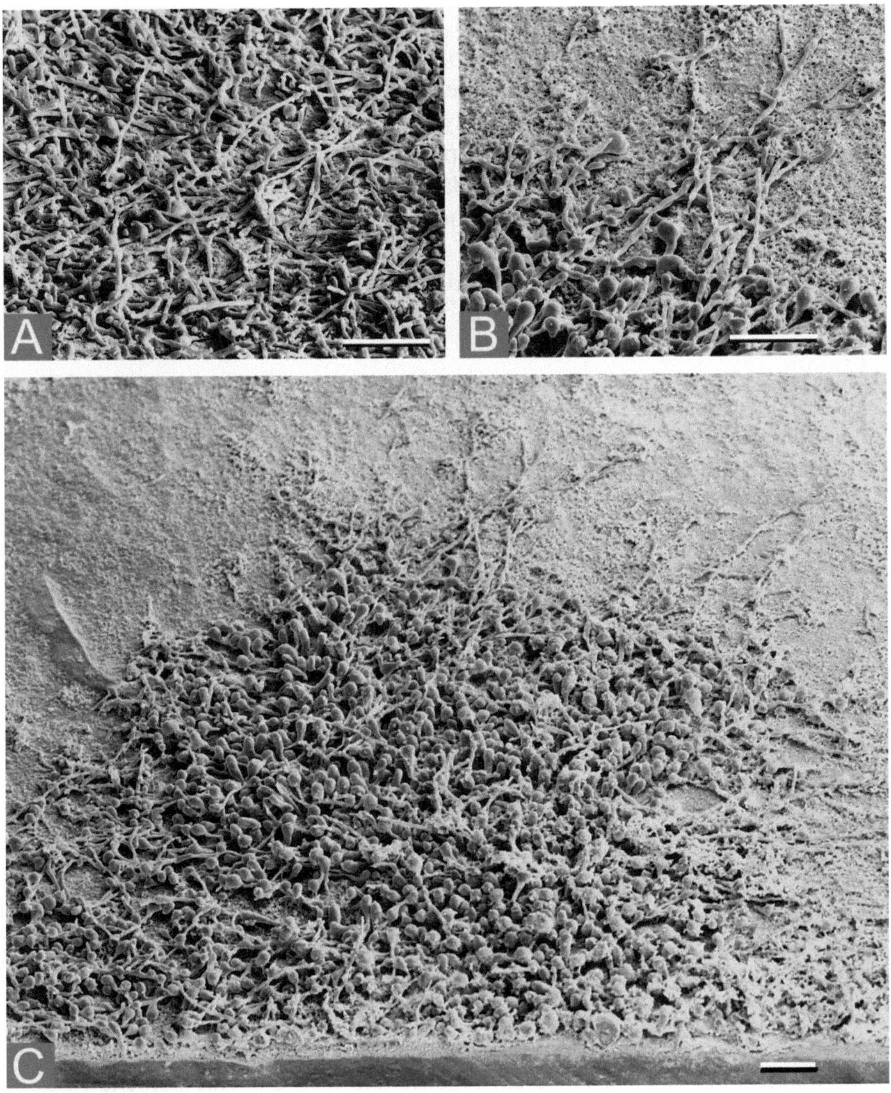

Fig. 1 Complex boring system of *Rhopalia clavigera* isp. n., produced by *Eugomontia sacculata*. **A** Early development of the boring is dominated by branched cylindrical tunnels produced by septate filaments radiating from a center outward. **B** Development of sporangial swelling advances to the margins of the system, leaving only exploratory filaments on the edges of the colony. **C** A mature colony with club-shaped enlargements extending perpendicular to the filaments and substrate surface. All scale bars: 100 μm

The tunnels are cylindrical 3 to 6 μm in diameter, with the mean ± standard deviation of 4.5 ± 0.8 μm (25 measurements), whereas the fully developed sacs are clearly elongated toward the interior of the substrate, 21.2 ± 2.9 μm wide and 42.1 ± 5.4 μm (25 measurements) long (Fig. 2). Displayed are maximum and minimum diameters for each tunnel, whereas the sacs were measured and displayed as length vs. width.

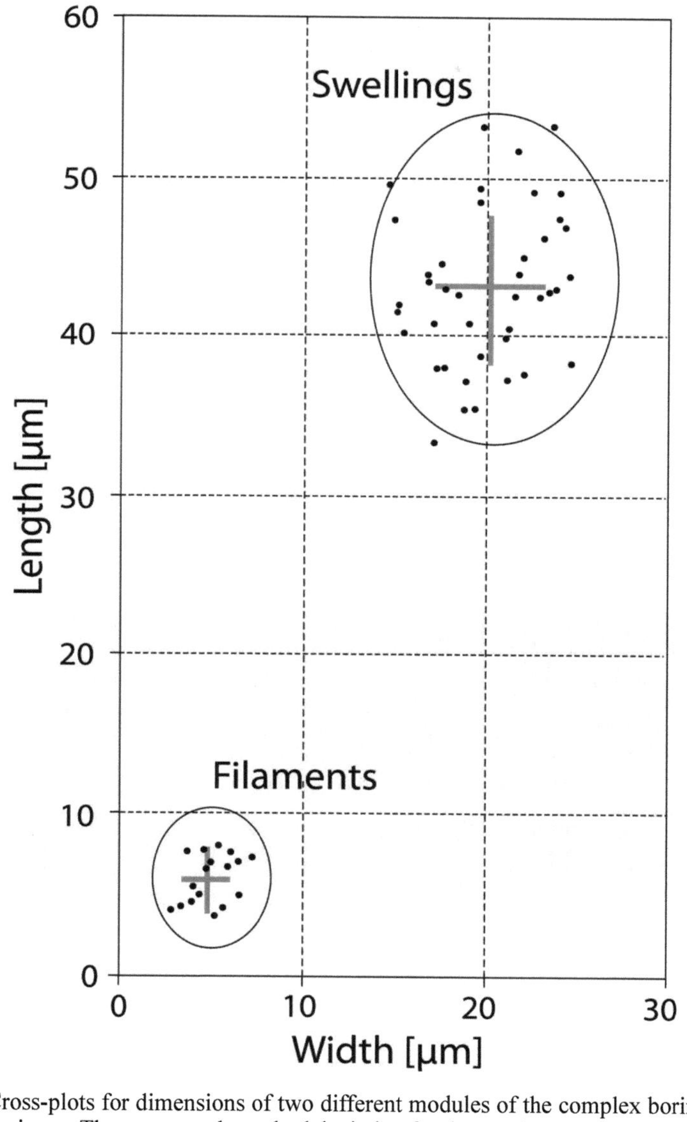

Fig. 2 Cross-plots for dimensions of two different modules of the complex boring *Rhopalia clavigera* isp. n. The means and standard deviation for the maximum vs. minimum diameters of the branched tunnels (lower left) and for the length vs. width of the swellings (upper right). Note that the field covering 2 standard deviations from the mean (oval) includes 95% of the sample

The relationship between the euendolith *Eugomontia sacculata*, its differentiation in filaments and sporangia and the morphology of its boring system is summarized in Figure 3. Filaments extracted by dissolution of the substrate (Fig. 3A) correspond in detail with the resin casts of their borings, including pointed tips (Fig 3B). Sporangial swellings are modifications of filaments, which remain attached to each other (Fig. 3C). Both are faithfully replicated in the casts (Fig. 3D). The sequence

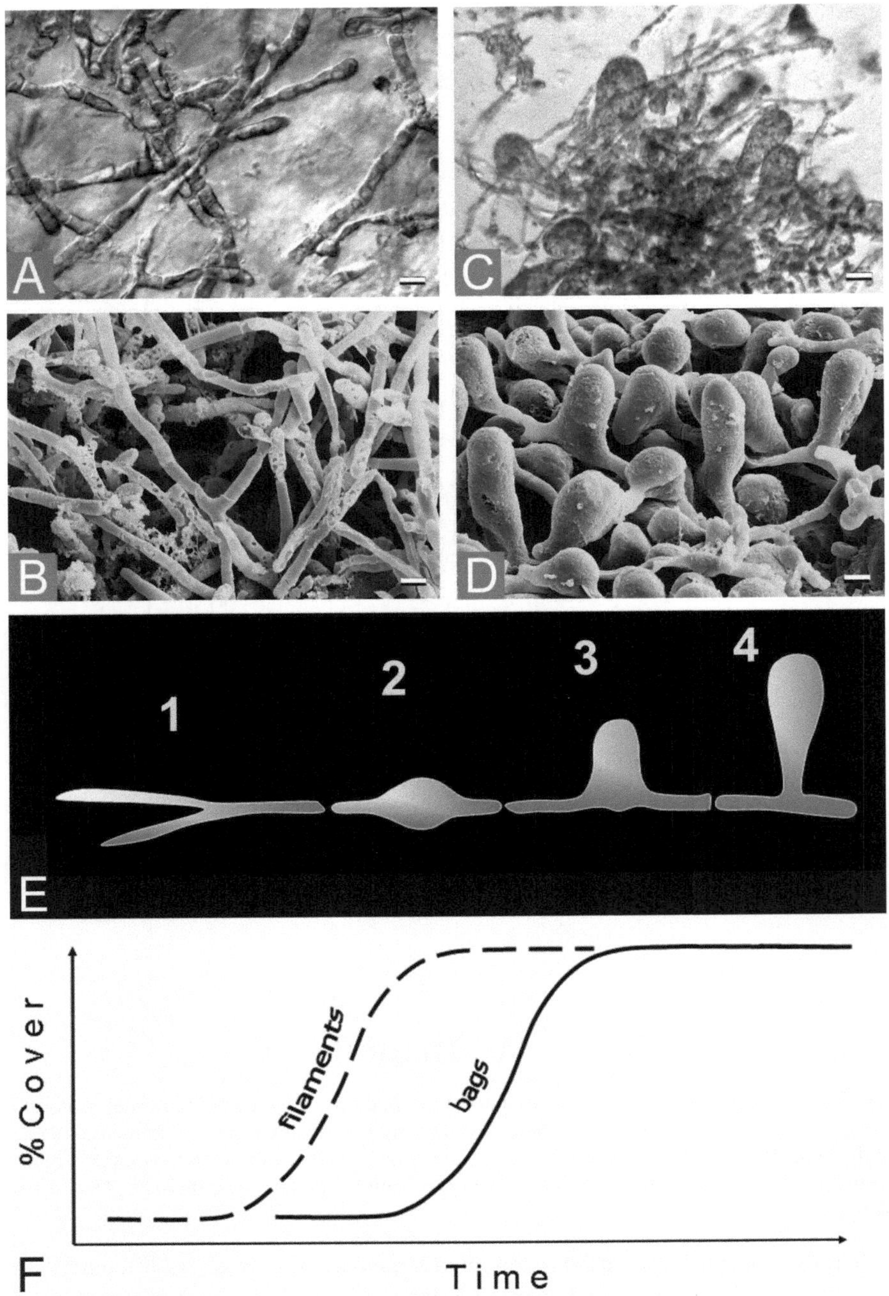

Fig. 3 The euendolith *Eugomontia sacculata* and its boring *Rhopalia clavigera* isp. n. **A** Branched septate filaments. **B** Resin cast of filamental tunnels. **C** Cluster of sporangial swellings connected to filaments. **D** Resin cast of a cluster of fully developed sacs containing sporangial swellings. **E** Reconstruction of differentiation (from 1 to 4) and colony expansion (from 4 to 1). **F** Temporal model in the development and expansion of the two modules. All scale bars: 10 μm

of developmental stages is presented schematically in Figure 3E, starting with branched tunnels, which undergo lateral swellings that develop into upright bags. In this diagram the growth of the boring system proceeds from right to left, while the differentiation in form and function proceeds from left to right. The estimated timing of the changes in boring patterns and their relative rates are presented diagrammatically (Fig. 3F).

In conclusion, the borings of *Eugomontia sacculata* reflect the changes in the course of the development of the organism. As a result, the shape of the boring includes very different elements ranging from filaments to sporangial sacs. The sacs expand toward the interior of the substrate, resulting in an upright posture of the resin replica. As the sporangial swellings mature, the sacs expand toward the substrate surface, producing a tube (see Kornmann 1960) and form a perforation in the shell where the spores are released.

Systematic ichnology

Ichnogenus *Rhopalia* Radtke, 1991

Original diagnosis: Club-shaped to spherical swellings connected with a system of tunnels (after Radtke 1991, translated from German).

Rhopalia clavigera isp. n.

Fig. 4B

1990 *Eugomontia sacculata* – Günther, 237, plate 54, fig. 6
1991 *Rhopalia catenata* – Radtke, 77, sample Bo7/41, plate 9, fig. 3; [partim]

Etymology: *claviger* (Latin) = club-carrying

Type material, locality and horizon: The holotype as shown in Figure 4B is a resin cast of borings in a shell fragment of *Callista* Poli, 1791 from the Paris Basin at Venteuil, France (Middle Lutetian, Middle Eocene). Collection of the Institute of Geology-Paleontology, University of Frankfurt a. M., Germany, inventory number Bo 7/26 (see also Radtke 1991: location marked in fig. 6).

Diagnosis: Borings comprised of two basic morphotypes and the transitions between them. The system spreads by branching tunnels, which radiate underneath the substrate surface from a single point of entry. The system deepens starting in the central portions by developing elongate lateral swellings perpendicular to the radiating tunnels and to the surface of the substrate.

Description: The boring starts from a single point of entry and expands radial forming a shallow network system which is circular in outline and expanding. The initial growth produces cylindrical tunnels, later expanding into deeper penetrating sacs, which empty to the substrate surface. The boring consists of two principal

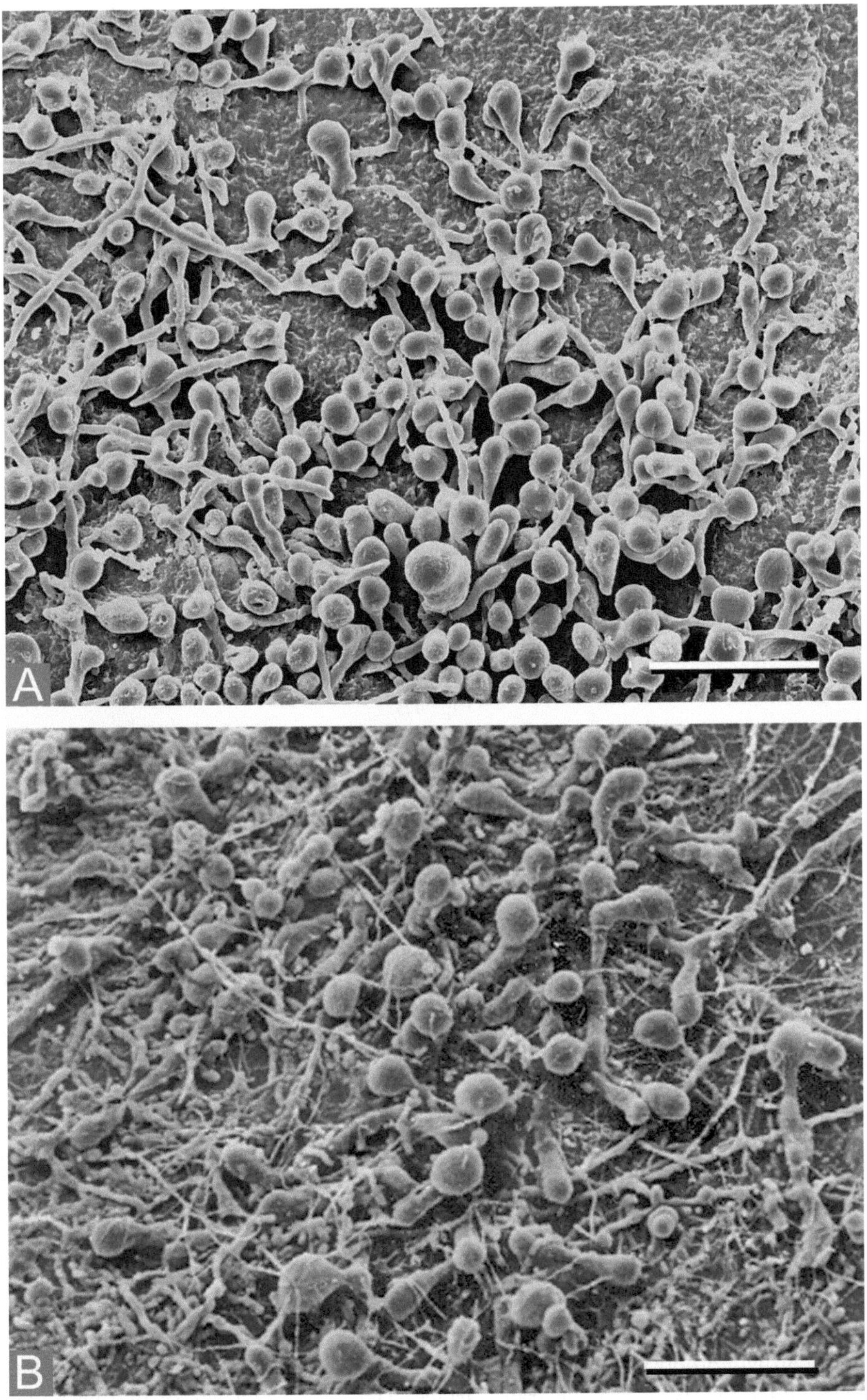
A
B

modules connected with transitional forms. **Module 1 – tunnels**: Tunnels cylindrical in cross section of relatively constant diameter of 4.5 ± 0.8 μm (25 measurements) branching at various angles, formed by an intertwined network of septate filaments. This module predominates in the early stages of growth and is responsible for lateral expansion of the system. The mature systems retain this early module at the periphery of the colony as exploratory filaments (Figs. 1A-B, 2 lower left, 3C). **Module 2 – sacs**: Sacs are series of elongated, club-shaped lateral expansions from the tunnels toward the interior of the substrate and perpendicular to the tunnels and the substrate surface. They contain sporangial swellings of the endolith. Sacs originate from lateral expansions of each cell in the filament, bounded by narrow interconnections at the positions of filament cross walls. Sac formation starts from the colony center and expands toward the periphery (Figs. 1C, 2 upper right, 3B).

Remarks and differential diagnosis: *Rhopalia clavigera* isp. n. is a complex system of microborings that tightly outlines the filaments and reproductive sacs of the ulotrichacean chlorophyte *Eugomontia sacculata* Kornmann, 1960 and reflects the changes in the development of the latter, including the timing and rates characteristic of this development. The borings are characterized by several features that serve as differential diagnosis toward other traces of the same ichnogenus: Frequent conical tips of the filaments, fairly constant diameters along the cylindrical tunnels, narrow variability range of tunnel diameters and narrow interconnections between swellings (mostly <5 μm). These properties clearly distinguish *Rhopalia clavigera* from *Rhopalia catenata* Radtke, 1991. Lack of spiny attachments to the surface and elongated, inward oriented rather than isodiametric swellings comprise the distinction from *Rhopalia spinosa* Radtke and Golubic, 2005.

Discussion

Development and differentiation in form and function

Microborings frequently conform to the shape of the euendoliths that produce them and reflect their specific boring behavior. However, the shape of the microborer as well as its behavior may change in the course of the development of the organism. As a consequence, borehole morphology may undergo considerable variation or a complete change that requires qualitative and quantitative assessments. We suggest that a biometric evaluation of spatial and temporal borehole variability would improve the definition of ichnotaxa. Comparison between resident euendoliths and their borings helps to distinguish between properties specific to different biotaxa in an assemblage, from those that characterize parts of the same biotaxon or stages in the development of the same biotaxon.

In the case of the microboring *Rhopalia clavigera* and its producer *Eugomontia sacculata*, each of the borehole morphotypes illustrated in Fig. 3E (numbered 1

Fig. 4 Comparison of modern and ancient *Rhopalia clavigera* isp. n. **A** Club-shaped swellings distributed up to the margin in a late stage of a modern *Rhopalia clavigera* in a *Mya arenaria* shell. **B** Holotype of *Rhopalia clavigera* isp. n. Middle Eocene, Paris Basin. Scale bars: 100 μm

through 4) is sufficiently distinct to be described as a separate ichnofossil. Their co-occurrence may then be misinterpreted as an ichnocoenosis, particularly when the material available for comparison is limited as it is often the case with fossils. A confusion concerning the euendolith identity is another problem and is a part of the history of the study of *Eugomontia sacculata*, which is still not completely resolved.

The organism was initially described under the name *Gomontia polyrhiza* (Lagerheim) Bornet and Flahault, 1888, which included an earlier described taxon *Codiolum polyrhizum* Lagerheim, 1885 under the assumption that both constituted parts of the same organism (Bornet and Flahault 1888, 1889). The work on cultures by Kylin (1935) and later by Kornmann (1959, 1960) showed that the two endolithic morphs are phases in the life cycles of two different organisms, which required formal renaming of the taxon to *Eugomontia sacculata* Kornmann, 1960. The name *Gomontia polyrhiza* has since been used for Lagerheim's organism, because 'Codiolum' turned out to be a developmental stage in the life cycle of several ulotrichacean algae. Recent studies using 18S rRNA gene sequencing discovered that genera *Monostroma* Thuret, 1857, *Collinsiella* Setchell and Gardner, 1903, *Gomontia* and *Eugomontia*, which all have euendolithic phases, form a phylogenetically tight cluster, so that they might be considered the same genus (O'Kelly et al. 2004). Exept for the stage of dissemination by zoospores and gametes, *Eugomontia sacculata* appears to be the only species in this group that is euendolithic throughout its entire life cycle, including sporophytic and gametophytic generations.

Timing of cell differentiation and boring behavior

Boring activity is initiated from single points of entry by rapid radial growth of septate endolithic filaments which produce cylindrical tunnels. High density of borings is maintained by filament branching. As the colonies spread horizontally, they may interfere with each other, causing a deeper penetration into the substrate in older (proximal) parts of the colony. In the second phase of the development, the peripheral expansion of the colony slows down, while the cells along the filament increase in diameter between the septa and grow sac-like outgrowths, initially symmetrical and later oriented toward the interior of the substrate. The differentiation into sacs starts in the oldest parts of the colony, close to the point of the entry into the substrate, and the maturation progresses outward so as to catch up with the peripheral growth (Fig. 1C). As a consequence of this timing, young colonies may be predominantly comprised of cylindrical tunnels whereas older colonies are entirely made of sacs up to the periphery (Fig. 4A). The life cycle of *Eugomontia sacculata* includes an isomorphic alternation of haploid and diploid generations, both producing sporangial swellings, although possibly with different timing of the onset and rates of progression (Kornmann 1960).

Similarity and convergent evolution

The ability to penetrate carbonate substrate and to continue to live inside it evolved in a number of different microorganisms, including members of Cyanobacteria, Chlorophyta and Rhodophyta among phototrophs and several endoliths of fungal affinity (reviewed by Wisshak 2006). Because of the selective pressure associated with the endolithic mode of life, which is common to all euendoliths, similar morphological features among unrelated organisms may be explained by convergent evolution. These include the adaptations facilitating substrate colonization, nutrient supply, exchange of metabolic products and reproduction. Occupation of new substrate by euendoliths requires only a small free surface area, because the endoliths continue to grow underneath the substrate surface, thus avoiding competition by the epilithic community. Accordingly, the early stages are often inconspicuous, limited to a small, spore-sized perforation on the substrate surface. Once inside, the phototrophic euendoliths stay close to the substrate surface and have to cope with the shading by epiliths. Maintenance of repeated contacts with the surface is required for nutrient supply and release of metabolic products, and separate perforations are made for the release of reproductive cells. Accordingly, many unrelated euendoliths radiate from the initial points of entry, produce shallow borings, and maintain repeated contacts with the substrate surface and the surrounding water. Because of similarities due to adaptation to common conditions, it is recommended that differential diagnoses be included in descriptions of ichnotaxa.

Studying complex traces

Euendolithic organisms form a spatial network or an internal biofilm (Golubic and Schneider 2003) which is inaccessible to direct observation because of the opacity of the carbonate substrate. This spatial network, however, collapses following dissolution of the substrate, which makes an assessment of the biological identity of the organisms difficult. Casting of borings by polymerizing resins followed by SEM observations enables the perception of borings in their three-dimensional arrangement, which does not include the resident microorganisms. A parallel study applying complementary methods to subsamples has produced optimal results in a combined study of both the traces and their makers.

Traces exhibiting different degrees of complexity are produced by both prokaryotic and eukaryotic euendoliths. Among phototrophic prokaryotes, relatively simple boring systems of *Fascichnus* (Radtke, 1991) produced by species of *Hyella* may vary in size over a wide range, possibly reflecting taxonomic diversification (Radtke and Golubic 2005). Yet, one of these species, *Hyella vacans* Gektidis and Golubic, 1996 shows a progressive change from binary fission of cells during the initial boring phase to multiple fission of the reproductive phase, at an accelerated rate in which the reproduction 'catches up' with the apical growth (Gektidis and Golubic 1996). The process is similar to that illustrated here for the eukaryotic *Eugomontia*, but does not affect borehole morphology. Simple borings are also produced by filamentous cyanobacteria *Plectonema terebrans* Bornet and Flahault, 1889 and *Schizothrix perforans* (Ercegović) Geitler, 1927, both recently reclassified within the genus *Leptolyngbya* Anagnostidis and Komárek, 1988 (see Komárek and Anagnostidis 2005). They produce narrow tunnels similar to those made by hyphae of euendolithic fungi. More complex filamentous cyanobacteria show cell differentiation into carbon-fixing and nitrogen-fixing (heterocysts) cells. Among euendoliths, these are represented by the *Mastigocoleus testarum* Lagerheim, 1886 and its trace *Eurygonum nodosum* Schmidt, 1992 (see Glaub et al. 2007: fig. 21.2-5 and Günther 1990: plate 52 fig. 6 for illustrations). In the present contribution we have dealt with a higher degree of complexity that generally characterizes eukaryotic phototrophs, exhibiting changes in the course of their life cycle, particularly in cases of alternation of generations of anisomorphic haploid and diploid phases. The life cycle of endolithic rhodophytes (Conchocelis-stages) is even more complex (Conway and Cole 1977; Campbell 1980).

The study of microborings in modern settings has the advantage of the frequent encounter of microborers in residence, which enables not only a correlation of ichnotaxa with corresponding biotaxa, but also a study of euendolith differentiation in form and function during their development, which includes their boring behavior. Such analyses reveal predictable sequences and even timing of events leading to morphological complexity of traces. It is, therefore, preferable that the documented biological coherence be acknowledged by a formal description of a complex trace (see Miller 2007). Morphologically different elements of a complex trace can then be described as 'modules', which identify distinct but spatially or temporally connected morphological elements of a complex trace. Such a procedure was applied for the siphonal chlorophyte *Ostreobium quekettii* Bornet and Flahault, 1889 and its trace *Ichnoreticulina elegans* (Radtke, 1991) by Radtke and Golubic (2005). Similar ichnotaxonomic procedures identified as phases in development have been used in descriptions of other complex traces such as *Entobia* Bronn, 1837 produced by boring sponges, supported by identification of skeletal spicules of the resident euendolith (Bromley and D'Alessandro 1984, 1989). Fossil traces are often incomplete or fragmentary. Identifying more than one module of a trace adds to the degree of confidence in interpretation, an assessment which is generally recommended in the work on traces.

Euendoliths and salinity

The present study identified marine euendoliths that inhabit permanently brackish ranges with salinities of 10 ppt and less, although with a more pronounced dominance of few taxa. This observation is consistent with the general rule that extreme environments tend to be characterized by a reduced diversity, where few successful biotaxa are represented by a large number of individuals. We hypothesize, therefore, that microbial euendoliths evolved in the marine environment, where their diversity is highest, and experience low salinities and freshwater conditions as extreme. However, their tolerance to low salinities is considerable under fluctuating conditions (see Wisshak 2006) as well as in stable brackish conditions as shown in the present study. These findings suggest that further research is required to establish the lower salinity limits of marine euendolithic taxa and their traces.

Acknowledgments

This research and participation at the Bioerosion Symposium 2006 in Erlangen has been supported by the Alexander von Humboldt Foundation, Germany. M. Wisshak and A. Munnecke are thanked for the invitation and stimulating discussions. SEM analysis has been carried out at the Institute for Geology-Paleontology of the

University of Frankfurt a. M. We thank O. Sagert for technical help. Professor K. Vogel is thanked for his support of this work. We thank K. Palinska, University of Oldenburg, Germany and A. Pawlowska, Institute of Oceanography, Gdynia for field observations and collection and for the materials used in this study.

References

Akpan EB (1986) Depth distribution of endolithic algae from the Firth of Clyde: Implications for delineation and subdivision of the photic zone. J Mar Biol Assoc UK 66:269-275

Akpan EB, Farrow GE (1984) Shell-boring algae on the Scottish continental shelf: Identification, distribution, bathymetric zonation. Trans Roy Soc Edinburgh 75:1-12

Anagnostidis K, Komárek J (1988) Modern approach to the classification system of cyanophytes. 3. Oscillatoriales. Arch Hydrobiol, Suppl 80:327-472

Bornet E, Flahault C (1888) Note sur deus nouveaux genres d'algues perforantes. J Bot 2:161-165

Bornet E, Flahault C (1889) Sur quelques plantes vivant dans le test calcaires des mollusques. Bull Soc Bot France 36:147-179

Bromley RG, D'Alessandro A (1984) The ichnogenus *Entobia* from the Miocene, Pliocene and Pleistocene of southern Italy. Riv Ital Paleont Stratigr 90:227-296

Bromley RG, D'Alessandro A (1989) Ichnological study of shallow marine endolithic sponges from the Italian coast. Riv Ital Paleont Stratigr 95:279-340

Bronn HG (1837) Lethaea Geognostica oder Abbildungen und Beschreibungen der für die Gebirgs-Formationen bezeichnendsten Versteinerungen. Schweizerbart, Stuttgart, 1:1-544 and plates

Burrows EM (1991) Seaweeds of the British Isles. Vol 2. Chlorophyta. Nat Hist Mus Publ, London, 238 pp

Campbell S (1980) *Palaeoconchocelis starmachii,* a carbonate boring microfossil from the Upper Silurian of Poland (425 million years old): Implications for the evolution of the Bangiaceae (Rhodophyta). Phycologia 19:25-36

Conway E, Cole K (1977) Studied in the Bangiaceae: structure and reproduction of the conchocelis of *Porphyra* and *Bangia* in culture (Bangiales, Rhodophyceae). Phycologia 16:205-216

Gallardo T, Gómez Garreta A, Ribera MA, Cormaci M, Furnari G, Giaccone G, Boudouresque C-F (1993) Check-list of Mediterranean Seaweeds, II. Chlorophyceae Wille s.l. Bot Mar 36:399-421

Geitler L (1927) Über Vegetationsfärbungen in Bächen. Biologia generalis 3:791-814

Gektidis M, Golubic S (1996) A new endolithic cyanophyte/cyanobacterium: *Hyella vacans* sp. nov. from Lee Stocking Island, Bahamas. Nova Hedwigia, Beih 112:91-98

Gektidis M, Dubinsky S, Goffredo S (2007) Microendoliths of the shallow euphotic zone in open and shaded habitats at 30° N – Eilat, Israel - paleoecological implications. Facies 53:43-55

Glaub I, Golubic S, Gektidis M, Radtke G, Vogel K (2007) Microborings and microbial endoliths: Geological implications. In: Miller W (ed) Trace fossils: concepts, problems, prospects. Elsevier, Amsterdam, pp 368-381

Golubic S, Schneider J (2003) Microbial endoliths as internal biofilms. In: Krumbein WE, Dornieden T, Volkmann M (eds) Fossil and Recent biofilms. Kluwer, Dordrecht, pp 249-263

Golubic S, Campbell SE, Spaeth C (1983) Kunstharzausgüsse fossiler Mikroben-Bohrgänge. Präparator 29:197-200

Günther A (1990) Distribution and bathymetric zonation of shell-boring endoliths in recent reef and shelf environments: Cozumel, Yucatan (Mexico). Facies 22:233-262

Komárek J, Anagnostidis K (2005) Cyanoprokaryota 2. Teil Oscillatoriales. In: Büdel B, Gärtner G, Krienitz L, Schlager M (eds) Süßwasserflora von Mitteleuropa. Fischer, Jena, 759 pp

Kornmann P (1959) Die heterogene Gattung *Gomontia*. I. Der sporangiale Anteil, *Codiolum polyrhizum*. Helgoländer Wiss Meeresunters 6:229-238

Kornmann P (1960) Die heterogene Gattung *Gomontia*. II. Der fädige Anteil, *Eugomontia sacculata* nov. gen. nov. spec. Helgoländer Wiss Meeresunters 7:59-71

Kylin H (1935) Über einige kalkbohrende Chlorophyceen. K Fysiogr Sälsk Förh 5:1-19

Lagerheim G (1885) *Codiolum polyrhizum* n. sp. Övers K Vetensk Akad Förh 42:21-31

Lagerheim G (1886) Notes sur le *Mastigocoleus*, nouveau genre des algues marines de l'ordre des Phycochromacées. Notarisia 1:65-69

Linnaeus C (1758) Systema naturae per regna tria naturae, secundum classes, ordines, genera, species, cum characteribus, differentiis, synonymis, locis. Tomus I. Editio decima, reformata. Laurentii Salvii, Holmiae, 824 pp

Lukas KJ, Golubic S (1981) New endolithic cyanophytes from the North Atlantic Ocean: I. *Cyanosaccus piriformis* gen. et sp. nov. J Phycol 17:224-229

Miller W III (2007) Complex trace fossils. In: Miller W (ed) Trace fossils: concepts, problems, prospects. Elsevier, Amsterdam, pp 458-465

Nielsen R (1987) Marine algae within calcareous shells from New Zealand. New Zealand J Bot 25:425-438

O'Kelly CJ, Wysor B, Bellows WK (2004) *Collinsiella* (Ulvophyceae, Chlorophyta) and other ulotrichalean taxa with shell-boring sporophytes form a monophyletic clade. Phycologia 43:41-49

Poli JX (1791) Testacea Utriusque Siciliae eorumque historia et anatome 1. Regio Typographeio, Parma, pp. 1-74 and plates

Pomerol C, Feugueur L (1986) Bassin de Paris. Guides géologiques régionaux. Masson, Paris, 222 pp

Radtke G (1991) Die mikroendolithischen Spurenfossilien im Alt-Tertiär West-Europas und ihre palökologische Bedeutung. Courier Forschinst Senckenberg 138:1-185

Radtke G, Golubic S (2005) Microborings in mollusk shells, Bay of Safaga, Egypt: Morphometry and ichnology. Facies 51:125-141

Schmidt H (1992) Mikrobohrspuren ausgewählter Faziesbereiche der tethialen und germanischen Trias (Beschreibung, Vergleich und bathymetrische Interpretation). Frankfurter Geowiss Arb A 12:1-228

Setchell WA Gardner NL (1903) Algae of northwestern America. Univ California Publ Bot 1:165-418

Tittley I, Nielsen R, Gunnarsson K (2005) Relationships of algal floras in North Atlantic Islands (Iceland, the Faroes, the Shetlands, the Orkneys). BIOFAR Proc 41:33-52

Thuret MG (1857) Deuxième note sur la fécondation des Fucacées. Ann Soc Nat 7:34

Vogel K, Gektidis M, Golubic S, Kiene WE, Radtke G (2000) Experimental studies on microbial bioerosion at Lee Stocking Island, Bahamas and One Tree Island, Great Barrier Reef, Australia: implications for paleoecological reconstructions. Lethaia 33:190-204

Wilkinson M (1974) Investigations on the autecology of *Eugomontia sacculata* Kornmann, a shell-boring alga. J Exp Mar Biol Ecol 16:19-27

Wilkinson M, Burrows EM (1970) *Eugomontia sacculata* Kornmann in Great Britain and North America. Brit Phycol J 5:235-238

Wisshak M (2006) High-latitude bioerosion: the Kosterfjord experiment. Lect Notes Earth Sci 109:1-202

Wisshak M, Gektidis M, Freiwald A, Lundälv T (2005) Bioerosion along a bathymetric gradient in a cold-temperate setting (Kosterfjord, SW Sweden): an experimental study. Facies 51:93-117

Colonisation and bioerosion of marine bivalve shells from the Baltic Sea by euendolithic cyanobacteria: an experimental study

Anna M. Pawłowska[1], Katarzyna A. Palińska[2], Halina Piekarek-Jankowska[1]

[1] Department of Marine Geology, Institute of Oceanography, University of Gdansk, 81-378 Gdynia, Poland, (ocepaw@ocean.univ.gda.pl)
[2] Department of Geomicrobiology, ICBM, Carl von Ossietzky University, 26111 Oldenburg, Germany

Abstract. The destructive activity of different euendolithic cyanobacteria on marine bivalve shells typical for the Baltic Sea area, such as *Mytilus edulis*, *Mya arenaria*, *Cerastoderma glaucum*, and *Macoma balthica*, was studied in experimental settings for four months. Specifically the cyanobacterial strains of *Phormidium* sp., *Microcoleus chthonoplastes*, *Anabaena flos-aquae* and *Synechococcus* sp. – all of which not classically considered as euendolithic – were cultured under controlled conditions in the laboratory to determine their optimal environmental conditions. Each bivalve species was maintained in four experimental settings with different calcium carbonate contents in both sediment and sea water. After four months of exposure, pieces of shells were collected, thin-sectioned, and examined using scanning electron microscopy (SEM) and light microscopy. The cyanobacterial strains of *Phormidium* sp. and *Microcoleus chthonoplastes* exhibited the fastest (after two weeks) colonisation of *Mya arenaria* and *Cerastoderma glaucum* shells and formed a dense biofilm in systems with decalcified sand and carbonate-free water. Shells of *Mytilus edulis* were colonised only in sites where the periostracum was damaged suggesting periostracal protective properties against the destruction of calcium carbonate substrate by dissolution and bioerosion. The examination of SEM images and cross-sections revealed *Phormidium* sp. to be an active euendolith that produces prominent traces under laboratory conditions whereas signs of boring activity for *Microcoleus chthonoplastes*, *Anabaena flos-aquae* were inconclusive and *Synechococcus* sp. did not contribute to bioerosion.

Keywords. Cyanobacteria, bioerosion, biofilm, calcium carbonate, bivalve, experiment

Introduction

Once marine bivalves die, their shells, like the skeletons of dead marine calcifying organisms such as crustose coralline algae (Tribollet and Payri 2001; Ghirardelli 2002) and corals (Le Campion-Alsumard et al. 1995), are exposed to various

M. Wisshak, L. Tapanila (eds.), *Current Developments in Bioerosion.* Erlangen Earth Conference Series, doi: 10.1007/978-3-540-77598-0_6,

physical, chemical, and biological processes of destruction (Smith and Nelson 2003). Whereas physical processes contribute to the breakdown of carbonate substrates into pieces and sediment grains, chemical and biological processes induce dissolution or precipitation of calcium carbonate depending on the seawater conditions. The chemical dissolution or precipitation of $CaCO_3$ depends on the seawater pH which, with seawater temperature, salinity and water column pressure, influences the saturation state of seawater with respect to aragonite and calcite (concentrations of Ca^{2+} and CO_3^{2-}; Golubic and Schneider 1979; Morse 2003). Biological processes affecting dead carbonate substrates include bioerosion, micritisation, and bioturbation that has important impact on geochemical processes occurring at the sediment / water interface (Kobluk and Risk 1977; Smith and Nelson 2003).

Most studies that investigated the effects of carbonate constructive and destructive processes were conducted in marine subtropical and tropical environments (Hutchings 1986; Schneider and Le Campion-Alsumard 1999). In comparison, knowledge of processes affecting carbonate substrates in the temperate realm is limited. Alexandersson (1974) suggested that chemical dissolution due to carbonate undersaturation in seawater is the driving mechanism for carbonate degradation in cold waters.

Recently, attempts have been made to investigate carbonate bioerosion processes in cool-water marine ecosystems (Glaub et al. 2002; Wisshak et al. 2005; Wisshak 2006). Glaub et al. (2002) studied the distribution of microendoliths in molluscan substrates in relation to light availability in various North Atlantic settings. The bathymetric scheme of index ichnocoenoses, which was established earlier in fossil and tropical settings, was successfully applied for the high-latitude environments in this process. Extensive work on both micro- and macrobioerosion along the bathymetric gradient from the Swedish Kosterfjord was carried out by Wisshak et al. (2005) and Wisshak (2006). They presented a detailed inventory of micro- and macroboring communities recorded in different carbonate substrates such as Iceland spar, bivalve shells, and fragments of cold-water corals. Overall bioerosion rates estimated in this experiment (ranging from 2 to 218 $g/m^2/y$ with mean value of 37 $g/m^2/y$) were 10 to 100 times lower (microbioerosion 2 to 10 times lower) than those quantified in tropical environments in previous experiments (reviewed in Wisshak 2006). The northern Great Barrier Reef in Australia for instance is bioeroded at rates of 180 to 2,190 $g/m^2/y$ (Tribollet and Golubic 2005).

Bioerosion is a complex process combining biological corrosion and abrasion (Pia 1937). Biological corrosion results from the action of the euendolithic flora (Golubic et al. 1981) and some macroborers (Hutchings 1986) which dissolve carbonate substrates while biological abrasion results from the action of grazers that mechanically remove pieces of carbonate substrates while feeding on epilithic and endolithic floras (Schneider and Torunski 1983; Bruggemann et al. 1994).

Several authors, including Golubic and Schneider (1979), Schneider and Le Campion-Alsumard (1999), Tribollet and Payri (2001), Glaub (2004), Radtke and Golubic (2005), and Wisshak (2006), have described the erosive activity of euendolithic cyanobacteria in carbonate substrates from various environments (terrestrial, freshwater, and marine environments). The euendolithic microorganisms (represented mainly by chlorophytes, cyanobacteria and fungi) are considered to be among the most important agents in bioerosion (Tribollet and Golubic 2005). They (especially the chlorophyte *Ostreobium quekettii* Bornet and Flahault, 1889) can dissolve more than 1 kg of $CaCO_3/m^2/y$ in dead corals from oligotrophic waters on the northern Great Barrier Reef in Australia (Tribollet and Golubic 2005). Euendolithic cyanobacteria colonise all kinds of carbonate substrates including marine bivalve shells (Mao Che et al. 1996; Kaehler 1999; Pantazidou et al. 2006; Wisshak 2006). Most of these studies focused on low-latitude environments with the exception of the Kosterfjord experiment carried out by Wisshak et al. (2005) in Sweden. These investigations conducted in situ indicate the dominance of the following cyanobacteria microborers *Hyella caespitosa* Bornet and Flahault, 1889, *Mastigocoleus testarum* Lagerheim, 1886, *Plectonema terebrans* Bornet and Flahault, 1889, and *Solentia achromatica* Ercegovic, 1932 (see Young and Nelson 1988; Glaub et al. 2002; Wisshak 2006).

The marine cyanobacteria *Phormidium* (Kützing ex Gomont) Anagnostidis and Komárek, 1988, *Microcoleus chthonoplastes* (Thuret) Gomont, 1892, *Anabaena flos-aquae* (Brébisson) Bornet and Flahault, 1888, and *Synechococcus* Nägeli, 1849 are common members of benthic microbial communities in cold-water environment such as the Baltic Sea and North Sea (Gerdes et al. 1986; Pliński and Komárek 2007) but have not been reported as being euendolithic.

Shells of bivalves such as *Mytilus edulis* Linnaeus, 1758, *Mya arenaria* Linnaeus, 1758, *Cerastoderma glaucum* (Poiret, 1789), and *Macoma balthica* (Linnaeus, 1758) are the major source of calcium carbonate in siliciclastic sediments of the southern Baltic Sea. The aim of this study was to prove whether typical representatives of cyanobacterial microbial communities of the Baltic Sea are able to colonise and bioerode the calcareous bivalve shells and to determine their optimal environmental demands under controlled conditions in the laboratory. For this purpose the organisms were cultivated in different experimental systems in relation to the calcium carbonate content in both the sediment and seawater.

Materials and methods

Experimental setup

To study the biodegradation of bivalve shells from temperate waters by euendolithic cyanobacteria under different conditions (carbonate vs. carbonate-free sand, and natural vs. artificial seawater), tank experiments were performed using biogenic calcium carbonate samples obtained from bivalve shells common in the Baltic Sea: *Mytilus edulis*, *Mya arenaria*, *Cerastoderma glaucum*, and *Macoma balthica.* The shells investigated consisted of the periostracum – the outer most organic uncalcified layer, and the ostracum and the innermost hypostracum – the inorganic calcified layers (Bathurst 1976; Kobayashi and Samata 2006).

Four strains of marine cyanobacteria from the Geomicrobiology Culture Collection of the University in Oldenburg were chosen for the experiment: *Microcoleus chthonoplastes*, *Anabaena flos-aquae* OL.10, *Phormidium* sp. ('Sphere'), and

Synechococcus PCC 73109. Each strain was cultivated in four experimental systems using different media: system one contained natural coarse sand and natural sea water; system two – carbonate-free coarse sand and natural sea water; system three – carbonate-free coarse sand and ASN_{III} (artificial seawater according to Rippka et al. (1979) – this medium shows 35 psu salinity and contained 400 mM NaCl, 28 mM $MgSO_4$, 18 mM $MgCl_2$, 6.7 mM KCl, 15 mM $CaCl_2$, 9.4 mM NH_4Cl, 0.05 mM $FeSO_4$, and 0.16 mM K_2HPO_4; the carbonate content in the medium before inoculation was 0.02 g/l and the pH 7.5); and system four – carbonate-free coarse sand and modified ASN_{III} (the medium without calcium and carbonate ions). The natural sea water (pH = 8.1) was filtered and kept in cold temperature of 4°C for one month before it was used in this investigation. Salinity of seawater was 35 psu in all beakers except in those containing *Anabaena flos-aquae* OL.10 (salinity = 11 psu), because this strain was isolated from brackish water and it does not tolerate high salinity. In order to obtain carbonate-free sand the natural coarse sand was decalcified by a treatment with 10% HCl. Four previously boiled shells of each bivalve species were placed at the surface of the sand in each experimental system. Additionally, eight similar systems (four with water salinity of 35 psu and four with 11 psu) without cyanobacteria were set up as a control. All systems were set up in 400 ml beakers covered with aluminium foil, maintained at 18°C, and illuminated with Osram tungsten light tubes providing a photosynthetic photon flux density (PPFD) of 120 μmol photons/m^2/s (measured with a LICOR LI-185B quantum / radiometer / photometer equipped with a LI-190SB quantum sensor) and with a light / dark cycle of 12/12 h. The glass beakers, sand, and cultivation media were autoclaved prior to the addition of cyanobacteria and bivalve shells.

In order to control the seawater pH and to determine how the cyanobacterial growth influences its value in ambient seawater, the pH was measured in beaker seawater with a LICOR LI-185B pH meter every two weeks. Simultaneously, cyanobacterial biofilm formation on the surface of the shells and sand was monitored visually in each beaker. In order to compare the observations, three groups of biofilm formation on shell and sand surfaces were distinguished (less than 10%, 10% to 50%, and 50% to 100% of substrate surface covered by biofilm).

Sample analysis

After four months of experiment, a set of shell pieces (n = 46) covered by cyanobacterial biofilms was fixed with a 4% glutaraldehyde solution in a 0.1 M sodium phosphate buffer (pH = 7.2) and gradually dehydrated by increasing the concentration of ethanol. Then they were critical-point dried in a Blazers Union (CPF) apparatus and sputtered with gold (Blazers Union SCD 030). In parallel, another set of shell pieces (n = 46) was treated with an ultrasonic bath in order to remove the biofilm. Both sets (shell pieces with and without biofilms) were then examined with a scanning electron microscope (SEM; Zeiss DSM 940). In order to measure the depth of penetration of euendolithic filaments, 32 petrographic thin-sections of the infested shells were prepared via fixation and dehydration, embedment in Spurr's resin (Spurr 1969), cutting with a rock saw and gluing to petrographic glass slides. The thin-sections were examined under a light microscope (Zeiss).

Results and discussion

Biofilm formation

The current study shows that the investigated strains of cyanobacteria were able to colonise the surface of bivalve shells within two weeks. In beakers that contained carbonate-free sediment and modified artificial seawater (without calcium and carbonate ions), cyanobacterial growth was limited in sediments, whereas shell surfaces were covered by a thick biofilm. Furthermore, there were differences in biofilm occurrence among species of bivalves (Table 1). Initially, the growth of *Phormidium* sp. and *Microcoleus chthonoplastes* was identified only on *Mya arenaria* shells in all systems as indicated by the formation of a slime layer with living cells at the surface (Fig. 1B). After fourth months of exposure to colonisation, all the *Mya arenaria* shells were covered by dense biofilms of each strain of cyanobacteria (Table 1). A similar pattern of biofilm formation was noted on *Cerastoderma glaucum* shells (Fig. 1A, C). It was observed, however, that *Mytilus edulis* shells were weakly colonised in comparison with those of other bivalves, because its periostracum is thicker than that of the other shells. In this latter case, the biofilm was noticed only on twelve shells (Table 1). The cyanobacteria spread on shell surfaces only where the periostracum was damaged. The periostracum protects the calcified layers from being destroyed by either dissolution in an acidic environment or predation (Bathurst 1976; Knauth-Koehler et al. 1996; Mao Che et al. 1996; Kardon 1998). Harper and Skelton (1993) demonstrated the effectiveness of a thick periostracum in shells of Mytiloidea as protection against drilling

Table 1 Percentage of shell surface covered by the biofilms of different strains of cyanobacteria (total number of shells of each species subjected to colonisation by the following cyanobacteria strains was 16)

cyanobacterial strain	bivalve shell	no. of shells covered by biofilm			total
		<10%	10-50%	50-100%	
Phormidium sp.	*Mytilus edulis*	5	2	-	7
	Mya arenaria	-	-	16	16
	Macoma balthica	2	5	9	16
	Cerastoderma glaucum	-	2	14	16
Microcoleus chthonoplastes	*Mytilus edulis*	4	-	-	4
	Mya arenaria	-	1	15	16
	Macoma balthica	2	6	5	13
	Cerastoderma glaucum	-	4	12	16
Synechococcus sp.	*Mytilus edulis*	1	-	-	1
	Mya arenaria	5	4	-	9
	Macoma balthica	4	2	-	6
	Cerastoderma glaucum	5	3	-	8
Anabaena flos-aquae	*Mytilus edulis*	3	1	-	4
	Mya arenaria	1	6	9	16
	Macoma balthica	3	5	4	12
	Cerastoderma glaucum	2	4	7	13

predators. More evidence of the periostracal protective properties was presented by Kaehler (1999), who demonstrated that the successful colonisation of *Perna perna* (Linnaeus, 1758) shells by endoliths and later infestation may be dependent on the prior removal of this outer shell layer.

Moreover, the cyanobacteria under investigation indicated variation in biofilm formation as well. The initial colonisation of shells was achieved by *Phormidium* sp. in a beaker containing natural sand and natural seawater as well as in a beaker with decalcified sand and natural seawater, followed by the intensive growth of *Microcoleus chthonoplastes* in the same systems. After two weeks of incubation they created a dense biofilm consisting of cells and extracellular polymeric substances (EPS; Fig. 1B). The slightly delayed growth of *Anabaena flos-aquae* (Fig. 1D) in comparison to the previously mentioned strains was observed especially in beakers with decalcified sediment and could be due to the salinity of the medium. This strain was isolated from brackish water (7 psu), therefore the higher salinity of

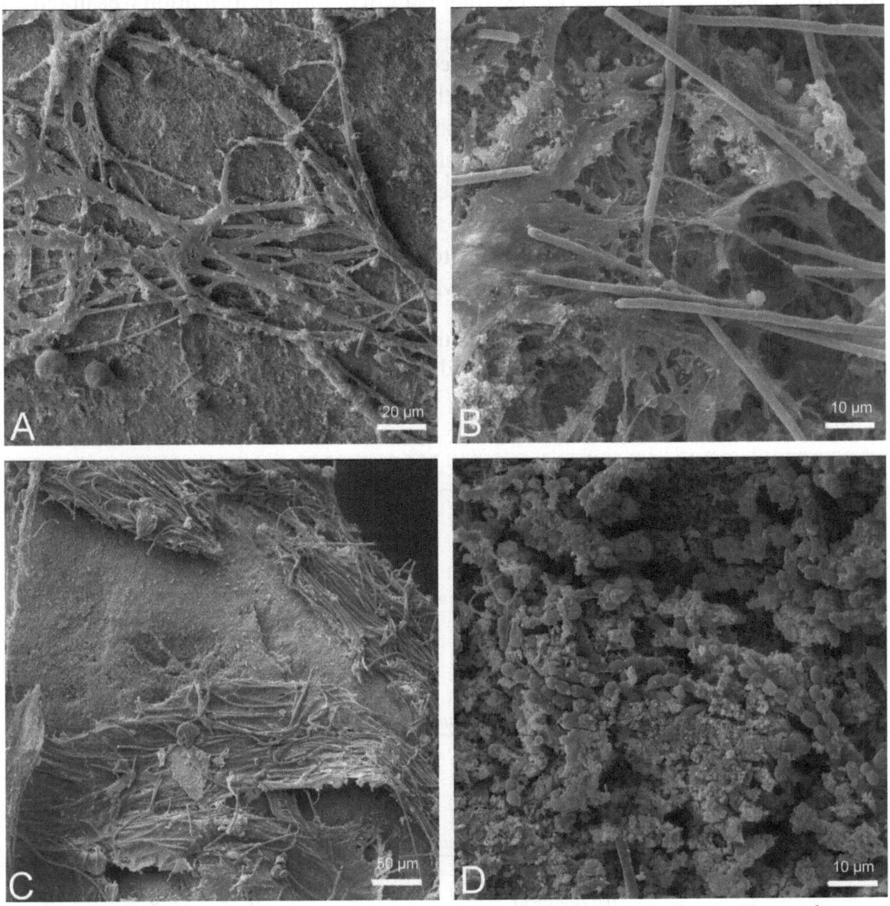

Fig. 1 SEM iamges showing cyanobacterial biofilms on bivalve shells. **A** *Phormidium* sp. on *Cerastoderma glaucum*. **B** *Microcoleus chthonoplastes* on *Mya arenaria*. **C** *Microcoleus chthonoplastes* on *Cerastoderma glaucum*. **D** *Anabaena flos-aquae* on *Mya arenaria*

the medium could have influenced the cyanobacterial development. The growth of *Synechococcus* PCC 73109 was the slowest. This species colonised shells only weakly (Table 1). This revealed that *Synechococcus* has a weaker affinity to adhere to any solid because of its planktonic way of life.

It is known from previous research that the production of EPS has a crucial impact on geochemical processes in different environments (Stone 1997). The chemical composition of EPS varies widely among different microorganisms (Stal 1994). Besides polysaccharides, EPS often contains uronic acids, which may be involved in regulation of calcification, e.g., precipitation of calcium carbonate (Kawaguchi and Decho 2002) in microbial mats and stromatolites (Feldman and McKenzie 1998; Stal 2000). Production of EPS may also contribute to carbonate corrosion (Aristovskaya 1980). Microorganisms have been shown to accelerate the dissolution of a variety of silicates and carbonates by the production of excess protons, organic ligands as well as hydroxyl (Aristovskaya et al. 1969) and extracellular polysaccharides (EPS). It was found that the formation of a cyanobacterial biofilm in marine and terrestrial environments increases the breakdown of silica and glass (Gorbushina and Palinska 1999; Brehm et al. 2005).

Boring patterns

The colonisation of bivalve shells by the different cyanobacterial strains under laboratory conditions resulted in the infestation of the carbonate substrate. SEM images of shell surfaces revealed that *Phormidium* sp. heavily infested shells. The surficial traces produced by this cyanobacterium are characterised by boreholes (Fig. 2A) and shallow branching furrows on the shell surface ranging from 20 to 60 μm in length (Fig. 2B-D). The diameter of these boreholes and furrows ranged from 2 to 5 μm. The shape of these furrows reflected the filamentous forms of the microorganisms. Filaments penetrated perpendicular to the shell surface to a depth of 140 μm (Fig. 3A-B), and were partially collapsed (Fig. 2A, C).

The way *Phormidium* sp. colonises carbonate shells is similar to that described for euendolithic cyanobacteria by Golubic et al. (1981). We thus suggest, that *Phormidium* sp. is an euendolithic organism that causes the corrosion of carbonate substrate. The boring traces left by this cyanobacterium on shell surfaces are the evidence for its destructive activity.

The other investigated strains did not exhibit any clear boring activity, since traces observed on the shell surfaces did not correlate with their shapes. However, some destructive patterns were noted on shell surfaces that had previously been covered by *Microcoleus chthonoplastes* biofilm. Long, shallow furrows of 1-2 μm in diameter were detected only on the periostracum. Shapes of these patterns are also of a filamentous form (Fig. 2E). Traces observed on shells from the systems in which *Anabaena flos-aquae* did not correlate with the shape of the organism. Several boreholes, ranging from 4-15 μm in diameter were found on the shells. These traces were probably caused by the collapse of the substrate around the boreholes (Fig. 2F). Thus, the results obtained regarding boring traces produced by *Microcoleus* and *Anabaena* are insufficient to consider these strains as being euendolithic. No boring patterns were noted on shells colonised by *Synechococcus*.

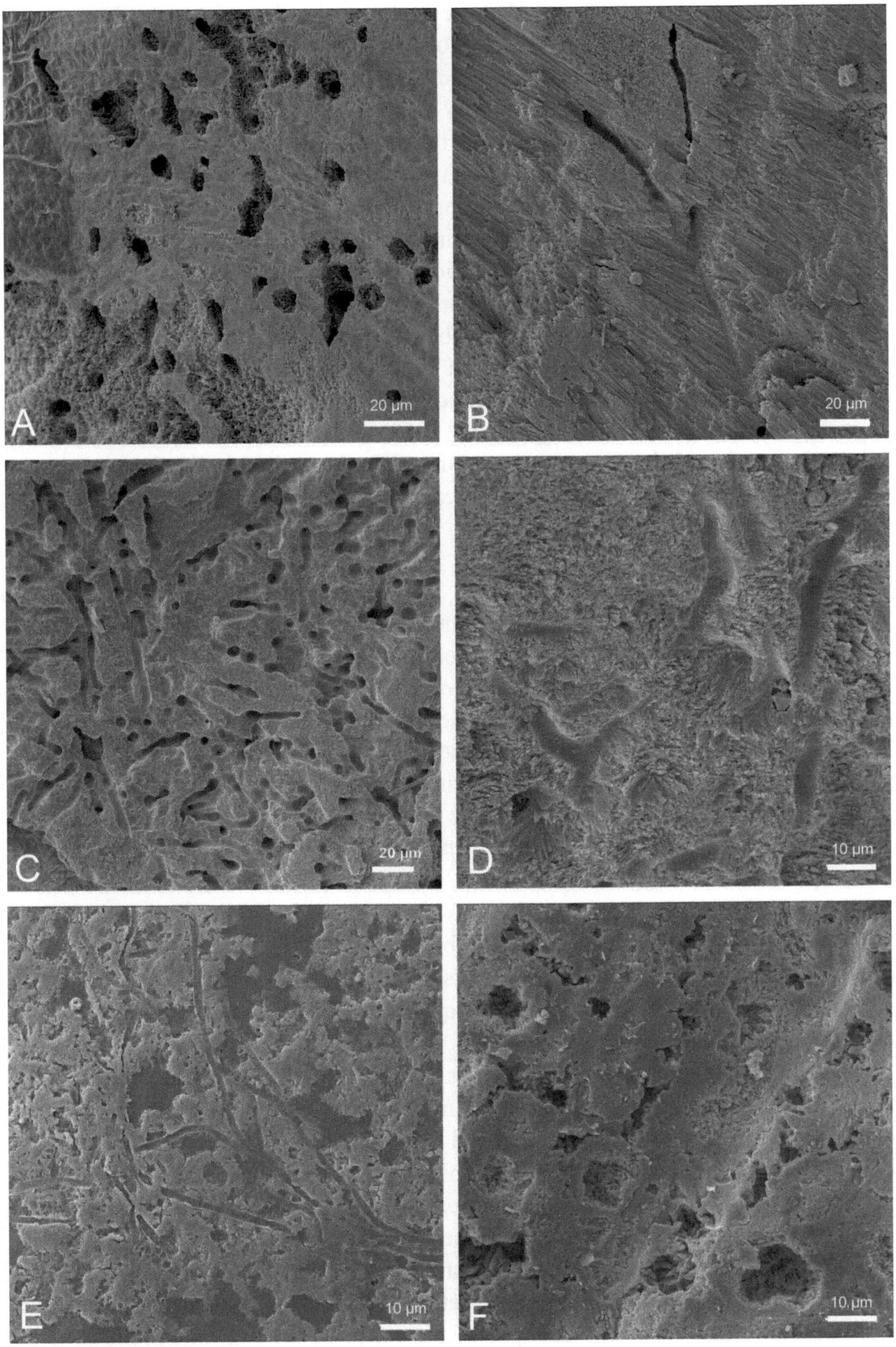

Fig. 2 SEM photographs showing borings produced by the studied euendolithic cyanobacteria in bivalve shells. **A** Partially collapsed boreholes produced by *Phormidium* sp. on the surface of *Macoma balthica*. **B** Traces of *Phormidium* sp. on the prismatic layer

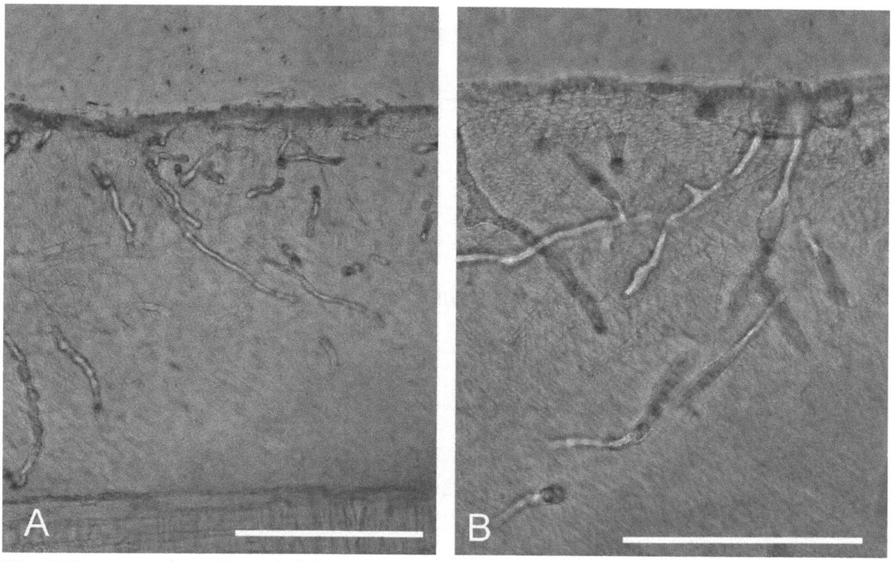

Fig. 3 Cross-sections through *Mytilus edulis* shell showing the depth of boreholes made by *Phormidium* sp. (scale bars: 100 μm)

pH index

Initially, the pH values in all beakers ranged from 6.3 in systems with decalcified sand and modified ASN_{III} to 7.9 in systems with natural sea water (Fig. 4). These lower pH values in systems four and three were probably caused by adding decalcified sand. Even though seawater pH in all systems were affected by changes of atmospheric pCO_2, the way how the cyanobacterial growth influences ambient seawater pH was determined. After the first two weeks of the experiment, the seawater pH stabilised and ranged from 7.5 to 8.0 in systems with microorganisms and 7.0-7.5 in control systems. Further changes in pH values of the sea water in beakers containing cyanobacterial strains ranged from 7.6 to 8.9 and reflected the growth of the cultures, while in control systems the range was 7.0-7.6. Increased pH correlated with intense microorganism growth. The highest pH values (8.55-8.9) were measured in beakers without carbonates both in sediment and water, containing bivalve shells densely colonised by cyanobacteria. A decrease of pH corresponded to the death of cyanobacteria. Growing organisms are expected to photosynthesise and thus increase pH up to 9, which favours $CaCO_3$ precipitation. However, in the current experiment the pH value remained relatively constant within the neutral to alkaline range in beakers containing cyanobacterial strains. Although the pH values obtained in this study correlated to the bivalve shells colonisation by cyanobacteria, they cannot be considered as any evidence of shell infestation by those microorganisms.

of *Mytilus edulis*. **C** Intense boring at the surface of *Cerastoderma glaucum* shell caused by *Phormidium* sp. **D** Network of *Phormidium* sp. micorborings at the surface of *Cerastoderma glaucum*. **E** Shallow borings of *Microcoleus chthonoplastes* on the periostracum of *Mytilus edulis*. **F** Collapsed traces produced by *Anabaena flos-aquae* at the surface of a *Mya arenaria* shell

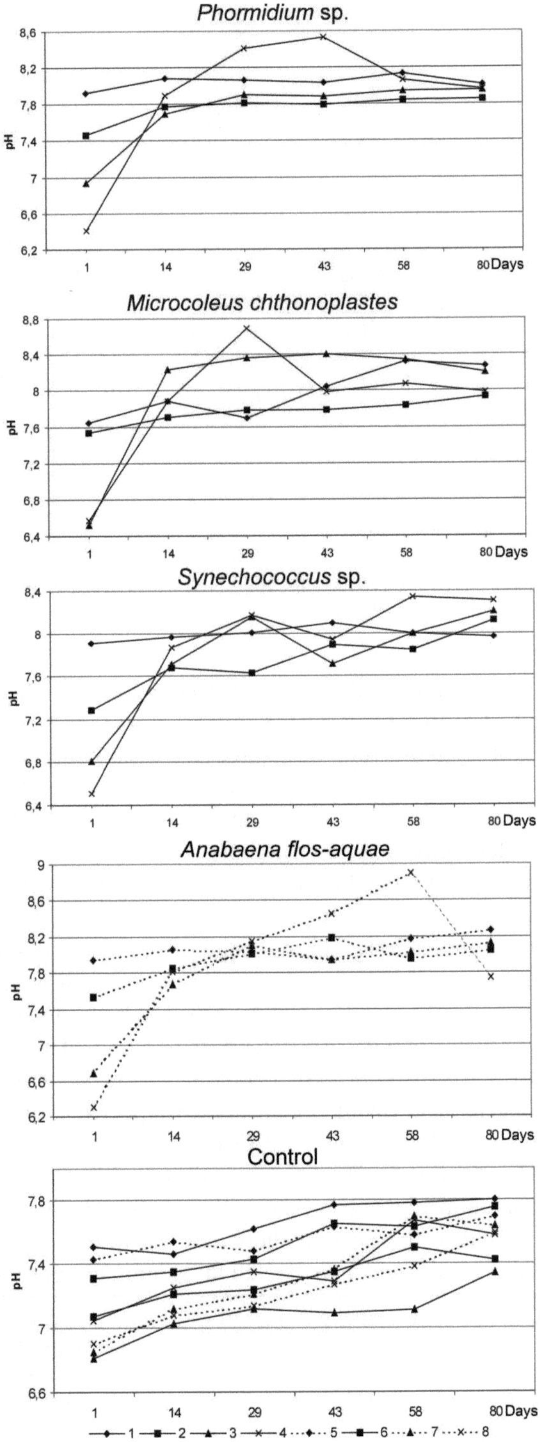
Phormidium sp.
pH
1 14 29 43 58 80 Days
Microcoleus chthonoplastes
pH
1 14 29 43 58 80 Days
Synechococcus sp.
pH
1 14 29 43 58 80 Days
Anabaena flos-aquae
pH
1 14 29 43 58 80 Days
Control
pH
1 14 29 43 58 80 Days
1 2 3 4 5 6 7 8

For example, seawater pH in beakers containing *Phormidium* sp., *Microcoleus chthonoplastes* and *Anabaena flos-aquae* were comparable to each other, but the SEM images indicated great variation in boring activity of these cyanobacteria.

There is no simple answer as to why some cyanobacteria colonise and live within carbonate substrates. Some of them, for example, may produce borings to obtain shelter against grazers or to protect themselves from drying out in less humid environments (Schneider and Le Campion-Alsumard 1999). In marine ecosystems the degree of light penetrating the water column is an important factor affecting distribution of euendolithic cyanobacteria and, consequently, microbioerosion rates (Perry and MacDonald 2002; Wisshak 2006). A description of the process how microorganisms produce borings in carbonates would provide a more complete answer to the question of colonisation. However, the cyanobacterial corrosion mechanism is not clearly understood. Several penetration agents have been described (Golubic and Schneider 1979; Schneider and Le Campion-Alsumard 1999). The acidic substances microorganisms secrete as metabolic products may change the ambient water chemistry. This change may lead to lower pH and consequently supports carbonate dissolution. The other agents thought to be responsible for shell corrosion are the acid-producing heterotrophic bacteria found in biofilm (Knauth-Koehler et al. 1996; Ferris and Lowson 1997; Schneider and Le Campion-Alsumard 1999). Furthermore, there is the 'calcium pump' mechanism described by Garcia-Pichel (2006), which is based on the active transport of calcium ions through the cyanobacterial filament. Thus, calcium uptake may lead to the undersaturation of Ca^{2+} concentrations in the interstitial space at the end of bored tunnels, which subsequently results in dissolution.

In conclusion, microorganisms used in our study preferentially colonised typical bivalve shells from the Baltic Sea in the carbonate-free experimental setting, while leaving similar shells in carbonate-containing media uncolonised and unweathered. Nevertheless, not all investigated organisms are capable of producing borings. Solely *Phormidium* sp. appears to be an euendolithic organism because it produces distinct traces, while the other strains did not reveal clear destructive activity.

Acknowledgements

We would like to thank Elizabeth Chacón and an anonymous reviewer for their constructive comments. Special thanks are extended to Sophie Martyna and Imke Notholt from the Carl von Ossietzky University in Oldenburg for their technical assistance in scanning electron microscopy. Part of the work was financed by the Polish State Committee for Scientific Research, grant No. 2P04E05729.

Fig. 4 Seawater pH changes resulted from cyanobacterial growth in different experimental systems: 1 – natural sand and water (35 psu); 2 – carbonate-free sand and natural water (35 psu); 3 – carbonate-free sand and ASN_{III} (35 psu); 4 – carbonate-free sand and modified ASN_{III} (35 psu); 5 – natural sand and water (11 psu); 6 – carbonate-free sand and natural water (11 psu); 7 – carbonate-free sand and ASN_{III} (11 psu); 8 – carbonate-free sand and modified ASN_{III} (11 psu)

References

Alexandersson ET (1974) Carbonate cementation in coralline algal nodules in the Skagerrak, North Sea: biochemical precipitation in undersaturated waters. J Sediment Petrol 44:7-26

Anagnostidis K, Komarek J (1988) Modern approach to the classification system of cyanophytes. 3-Oscillatoriales. Arch Hydrobiol, Suppl 80, Algol Stud 50-53:327-472

Aristovskaya TV (1980) Microbiology of soil formation. In: Gorbushina AA, Palinska KA (eds) Biodeteriorative processes on glass: experimental proof of the role of fungi and cyanobacteria. Aerobiologia 15:183-191

Aristovskaya TV, Daragan AY, Zykina LV, Kutuzova RS (1969) Microbiological factors in the movement of some mineral elements in the soil. Sov Soil Sci 5:538-546

Bathurst RGC (1976) Carbonate sediments and their diagenesis. Elsevier, New York, 658 pp

Bornet E, Flahault C (1888) Note sur deux nouveaux genres d'algues perforantes. J Bot, Morot 2:161-165

Bornet E, Flahault C (1889) Sur quelques plantes vivant dans le test calcaires des mollusques. Bull Soc Bot France 36:147-179

Brehm U, Gorbushina A, Mottershead D (2005) The role of microorganisms and biofilms in the breakdown and dissolution of quartz and glass. Palaeogeogr Palaeoclimatol Palaeoecol 219:117-129

Bruggemann JH, van Oppen MJH, Breeman AM (1994) Foraging by the stoplight parrotfish *Sparisoma viride*. I. Food selection in different, socially determined habitats. Mar Ecol Prog Ser 106:41-55

Ercegovic A (1932) Ekoloske i socioloske studije o litofitskim cijanoficejama sa Jugoslavenske obale Jadrana. Bull Int Acad Yougoslave Sci Math Nat 26:129-220

Feldmann M, McKenzie JA (1998) Stromatolite - thrombolite associations in a modern environment, Lee Stocking Island, Bahamas. Palaios 13:201-212

Ferris FG, Lowson EA (1997) Ultrastructure and geochemistry of endolithic microorganisms in limestone of the Niagara Escarpment. Canad J Microbiol 43:211-219

Garcia-Pichel F (2006) Plausible mechanisms for the boring on carbonates by microbial phototrophs. Sediment Geol 185:205-213

Gerdes G, Krumbein WE, Reineck H-E (1986) Mellum. Portrait einer Insel. Kramer, Frankfurt, Senckenberg B 63, 344 pp

Ghirardelli LA (2002) Endolithic microorganisms in live and dead thalli of coralline red algae (Corallinales, Rhodophyta) in the northern Adriatic Sea. Acta Geol Hisp 37: 53-60

Glaub I (2004) Recent and sub-recent microborings from the upwelling area off Mauritania (West Africa) and their implications for palaeoecology. In: McIlroy D (ed) The application of ichnology to palaeoenvironmental and stratigraphic analysis. Geol Soc London, Spec Publ 228:63-77

Glaub I, Gektidis M, Vogel K (2002) Microborings from different North Atlantic shelf areas - variability of the euphotic zone extension and implications for paleodepth reconstructions. Courier Forschinst Senckenberg 237:25-37

Golubic S, Schneider J (1979) Carbonate dissolution. In: Trudinger PA, Swaine DJ (eds) Biogeochemical cycling of mineral-forming elements. Stud Environ Sci 3:29-45

Golubic S, Friedmann I, Schneider J (1981) The lithobiontic ecological niche, with special reference to microorganisms. J Sediment Petrol 51:475-478

Gomont M (1892) Monographie des Oscillariées. Ann Sci Nat Bot 15:263-368 and 16:91-264

Gorbushina AA, Palinska KA (1999) Biodeteriorative processes on glass: experimental proof of the role of fungi and cyanobacteria. Aerobiologia 15:183-191

Harper EM, Skelton PW (1993) A defensive value of the thickened periostracum in Mytiloidea. Veliger 36:36-42

Hutchings PA (1986) Biological destruction of coral reefs. Coral Reef 4:239-252

Kaehler S (1999) Incidence and distribution of phototrophic shell-degrading endoliths of the brown mussel *Perna perna*. Mar Biol 135:505-514

Kardon G (1998) Evidence from the fossil record of an antipredatory exaptation:conchiolin layers in corbulid bivalves. Evolution 52:68-79

Kawaguchi T, Decho AW (2002) A laboratory investigation of cyanobacterial extracellular polymeric secretions (EPS) in influencing $CaCO_3$ polymorphism. J Crystal Growth 240:230-235

Knauth-Koehler K, Albers BP, Krumbein WE (1996) Microbial mineralization of organic carbon and dissolution of inorganic carbon from mussel shells (*Mytilus edulis*). Senckenbergiana Marit 26:157-165

Kobayashi I, Samata T (2006) Bivalve shell structure and organic matrix. Mater Sci Eng C 26:692-698

Kobluk DR, Risk JM (1977) Micritization and carbonate-grain binding by endolithic algae. Bull Amer Assoc Petroleum Geol 61:1069-1082

Lagerheim G (1886) Notes sur le *Mastigocoleus*, nouveau genre des algues marines de l'ordre des Phycochromacées. Notarisia 1:65-69

Le Campion-Alsumard T, Golubic S, Hutchings PA (1995) Microbial endoliths in skeletons of live and dead corals: *Porites lobata* (Moorea, French Polynesia). Mar Ecol Prog Ser 117:149-157

Mao Che L, Le Campion-Alsumard T, Boury-Esnault N, Payri C, Golubic S, Bézac C (1996) Biodegradation of shells of the black pearl oyster, *Pinctada margaritifera* var. *cumingii* by microborers and sponges of French Polynesia. Mar Biol 126:509-519

Morse JW (2003) Formation and diagenesis of carbonate sediments. In: Holland HD, Turekian KK (eds) Treatise on geochemistry. Elsevier, Amsterdam, 7:67-86

Nägeli C (1849) Gattungen einzelliger Algen, physiologisch und systematisch bearbeitet. N Denkschr Schweiz Ges Natw 8:44-60

Pantazidou A, Louvrou I, Economou-Amilli A (2006) Euendolithic shell-boring cyanobacteria and chlorophytes from the saline lagoon Ahivadolimni on Milos Island, Greece. Europ J Phycol 41:189-200

Perry CT, Macdonald IA (2002) Impact of light penetration on the bathymetry of reef microboring communities: implications for the development of microendolithic trace assemblages. Palaeogeogr Palaeoclimatol Palaeoecol 186:101-113

Pia J (1937) Die kalklösenden Thallophyten. Arch Hydrobiol 31:264-328 and 341-398

Pliński M, Komárek J (2007) Flora Zatoki Gdańskiej i wód przyległych (Bałtyk Południowy). 1. Sinice - Cyanobakterie (Cyanoprokaryota). Gdańsk Univ Publ House, Gdańsk, 186 pp

Poiret JLM (1789) Voyage en Barbarie, ou lettres écrites de l'ancienne Numidie pendant les années 1785 et 1786. La Rochelle, Paris, 315 pp

Radtke G, Golubic S (2005) Microborings in mollusk shells, Bay of Safaga, Egypt: Morphometry and ichnology. Facies 51:125-141

Rippka R, Deruelles J, Waterbury JB, Herdman M, Stanier RY (1979) Generic assignments, strain histories and properties of pure cultures of cyanobacteria. J Gen Microbiol 111:1-61

Schneider J, Torunski H (1983) Biokarst on limestone coasts, morphogenesis and sediment production. Mar Ecol 4:45-63

Schneider J, Le Campion-Alsumard T (1999) Construction and destruction of carbonates by marine and freshwater cyanobacteria. Europ J Phycol 34:417-426

Smith AM, Nelson CS (2003) Effects of early sea-floor processes on the taphonomy of temperate shelf skeletal carbonate deposits. Earth-Sci Rev 63:1-31

Spurr AR (1969) A low-viscosity epoxy resin embedding medium for electron microscopy. J Ultrastruct Res 26:31-43

Stal LJ (1994) Microbial mats: Ecophysiological interactions related to biogenic sediment stabilisation. In: Krumbein WE, Paterson DM, Stal LJ (eds) Biostabilization of sediments. Oldenburg Univ, Oldenburg, pp 41-53

Stal LJ (2000) Microbial mats and stromatolites. In: Whitton BA, Potts M (eds) The ecology of cyanobacteria: Their diversity in time and space. Kluwer Academic Publishers, Dordrecht, pp 61-120

Stone AT (1997) Reactions of extracellular organic ligands with dissolved metal ions and mineral surfaces. In: Banfield JF, Nealson KH (eds) Geomicrobiology: Interactions between microbes and minerals. Rev Mineral 35:309-344

Tribollet A, Golubic S (2005) Cross-shelf differences in the pattern and pace of bioerosion of experimental carbonate substrates exposed for 3 years on the northern Great Barrier Reef, Australia. Coral Reefs 24:422-434

Tribollet A, Payri C (2001) Bioerosion of the crustose coralline alga *Hydrolithon onkodes* by microborers in the coral reefs of Moorea, French Polynesia. Oceanolog Acta 24:329-342

Wisshak M (2006) High-latitude bioerosion: the Kosterfjord experiment. Lect Notes Earth Sci 109:1-202

Wisshak M, Gektidis M, Freiwald A, Lundälv T (2005) Bioerosion along a bathymetric gradient in a cold-temperate setting (Kosterfjord, SW Sweden): an experimental study. Facies 51:93-117

Young HR, Nelson CS (1988) Endolithic biodegradation of temperate-water skeletal carbonates by boring sponges on the Scott shelf, British Columbia, Canada. Mar Geo 65:33-45

The medium is the message: imaging a complex microboring (*Pyrodendrina cupra* igen. n., isp. n.) from the early Paleozoic of Anticosti Island, Canada

Leif Tapanila[1]

[1] Department of Geosciences, Idaho State University, Pocatello 83209, Idaho, USA, (tapaleif@isu.edu)

Abstract. Three complementary imaging techniques were used to describe a complex rosette-shaped microboring that penetrates the shells of brachiopods from the Ordovician - Silurian shallow marine limestones of Anticosti Island, Canada. *Pyrodendrina cupra* igen. and isp. n. is among the oldest dendrinid microborings and consists of shallow and deep penetrating canals that radiate from a central polyhedral chamber. The affinity of the tracemaker is unknown, but a foraminiferal origin, as proposed for some dendrinid borings, is rejected. Combining micro-CT with traditional stereomicroscopy and SEM helped distinguish and quantify fine morphological features while maintaining contextual information of the microboring within the shell substrate. Different imaging techniques inherently bias the description of microborings. These biases must be accounted for as new methods in ichnotaxonomy are integrated with past research based on different methods.

Keywords. Bioerosion, microboring, Dendrinidae, *Pyrodendrina cupra*, micro-computed tomography, ichnotaxonomy

Introduction

Studies on microbioerosion – the submillimeter excavations of algae, fungi and cyanobacteria – have shown the importance and ubiquity of this process in the destruction of marine rocks (Golubic et al. 1970; Vogel 1993). In addition to recording some of the oldest organisms on the planet (Furnes et al. 2004), fossil microboring assemblages can record paleobathymetry in terms of light penetration through the water column (i.e., photofacies), and allow for the recognition of ancient euphotic, dysphotic and aphotic zones (Vogel et al. 1987, 1995; Glaub 1994). Correlation of microboring trace fossils to their producer's biotaxon is commonly achieved by comparison to modern boring taxa (e.g., Gatrall and Golubic 1970; Campbell et al. 1979; Glaub et al. 1999; Bundschuh and Balog 2000); however, in some instances, fossil borings are difficult to relate to modern analogues. For this reason, separate nomenclature is used for trace fossil ichnotaxa and tracemaking biotaxa.

M. Wisshak, L. Tapanila (eds.), *Current Developments in Bioerosion*. Erlangen Earth Conference Series, doi: 10.1007/978-3-540-77598-0_7,

This study describes an intricate rosette-shaped (dendrinid) microboring for which no known modern analogue exists. Previous work on dendrinid borings suggest that some are made by foraminiferans, although this likely is not true for all dendrinids (Bromley et al. 2007). Owing to the complexity of the fossil, three visualization approaches were used: stereomicroscopy, scanning electron microscopy (SEM) and the newly developed method of micro-computed tomography (micro-CT).

In relating advances in technology to popular culture, the Canadian philosopher Marshall McLuhan (1964) declared *'The medium is the message'*. This adage equally applies to the taxonomist – the method used to describe a fossil imparts a bias on that description. The process of describing this dendrinid microboring with multiple techniques provided a complementary set of observations to better interpret the fossil, but also gave an opportunity to reflect on how the visualization medium might bias the ichnotaxonomic message.

Technology and bioerosion research

Technological advances have played a large role in the development of bioerosion research, especially in the identification of microborings. The advent of light microscopy in the 17th Century revealed the microscopic world to early naturalists, and eventually microborers and their cavities were described and named for the first time (Duncan 1876; see summary of early microboring research by Golubic et al. 1975). Among geological researchers, the use of 30 μm thin sections was a practical way to identify the presence of microborings and recognize their influence in sedimentary diagenesis (e.g., the process of micritization: Bathurst 1966). Visualization of microscopic structures advanced little until the 1960's with the invention of SEM, but to be effective this technique would require methods for casting microborings. Solving this problem, Golubic and others (1970) developed the embedding-casting technique, where hollow borings are cast in epoxy and the enclosing carbonate shell dissolved away with weak HCl. While this technique and its more streamlined variations (e.g., Golubic et al. 1983; Nielsen and Maiboe 2000; Wisshak 2006) have proven to be effective ways to image microborings, they require the full destruction of the host shell.

Although attempts to use x-radiography were useful in some macroboring studies (Bromley 1970; Warme 1977; Benner et al. 2004) resolution limits this technique to cavities >0.5 mm. Most recently advances in the technology and availability of computed tomography has provided an exciting new approach in paleontology (Tafforeau et al. 2006). Micro-CT is capable of rendering full three-dimensional (3D) reconstructions of complex microborings with little or no destruction to the fossil itself. The main part of this study, conducted in 2003 at the University of Utah using a cone beam x-ray microtomography system, is among the first to attempt microboring description with micro-CT, and the technique has now been successfully applied by a number of researchers (see Bromley et al. this volume).

Geologic setting

Anticosti Island preserves one of the most continuous sequences of Upper Ordovician - Lower Silurian strata in the world (Lespérance 1981), and it includes a record of the collapse and recovery of benthic communities associated with global Late Ordovician mass extinction events (Copper 1999). The limestones and minor calcareous shales of the Anticosti sequence were deposited in a mid- to distal shelf environment formed along the eastern continental margin of Laurentia at a paleolatitude of approximately 15°S (Brunton and Copper 1994). The exposed strata are subdivided into seven formations (Fig. 1; for references and summary, see Long and Copper 1994). The units contain abundant, well preserved fossils indicating an open marine, epibenthic invertebrate fauna dominated by corals, stromatoporoids, brachiopods and echinoderms. Brachiopod faunas are especially well documented, and their community and paleoenvironmental contexts are well established (Copper 1973, 1995, 1999; Jin et al. 1990, 1996; Jin and Copper 1997, 1998, 1999, 2000; Dewing 1999; Schoenhals 2000).

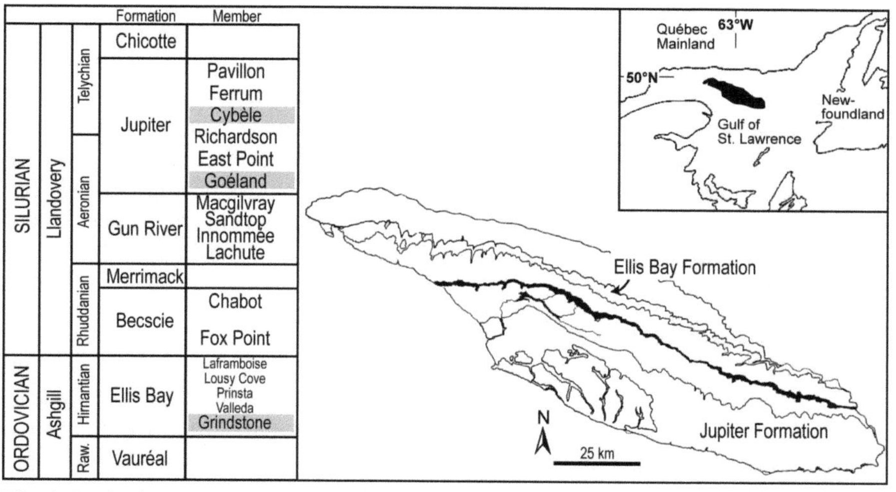

Fig. 1 Geologic map and stratigraphic section of Anticosti Island, noting formations where *Pyrodendrina* igen. n. occurs in brachiopod shells

Three brachiopod taxa in the Lower Silurian (Llandovery) Jupiter Formation contain abundant dendrinid microborings. The lower 20 m of the Goéland Member (Aeronian age) and the upper Cybèle Member (Telychian age) are both characterized by calcareous mudstone and micrite deposits and are interpreted to have been deposited in a quiet water setting below normal storm wave base (Copper and Long 1990; Copper 1995; Jin and Copper 2000). Similarly, the Upper Ordovician (Ashgill) Grindstone Member of the Ellis Bay Formation on the western end of Anticosti contains calcareous mudstones and shales deposited in a quiet water setting, and dendrinid microborings are found in a brachiopod from this member (Long and Copper 1987).

Compared to modern studies, relatively few have examined early Paleozoic microborings (e.g., Hessland 1949; Grahn 1981; Podhalańska 1984; Liljedahl 1986; Olempska 1986; Podhalańska and Nõlvak 1995; Bundshuh 2000; Bromley et al. 2007). By contrast the endolithic communities preserved in the Anticosti sequence are becoming well known (Tapanila 2002, 2004; Tapanila and Copper

2002; Tapanila et al. 2004; Ebbestad and Tapanila 2005; Tapanila and Holmer 2006) and it permits a broader view of the endoliths (i.e., embedders, macroborers, and microborers) at a pivotal time in their history (Taylor and Wilson 2003; Bromley 2004; Glaub and Vogel 2004; Tapanila 2005a). Ongoing investigations of Anticosti endolithic communities have uncovered a wide variety of microborings (Tapanila 2005b). Commonly visible by infilling pyrite, microborings were discovered in several bioclastic materials, including brachiopod shells.

Materials and methods

Brachiopod collections

Brachiopod specimens used in this study were selected from a systematic collection amassed by Dr. Paul Copper over four decades of research (originally at Laurentian University, Sudbury, now reposited at the Geological Survey of Canada [GSC], Ottawa). More than 1,000 articulated brachiopods from the Ordovician - Silurian Anticosti sequence were examined for the presence of visible micrborings using reflected light microscopy. Of these, nearly 100 specimens were found to contain dendrinid microborings filled with pyrite, iron oxides or micrite. The four microbored brachiopod taxa include *Pentamerus palaformis* Jin and Copper, 2000, *Joviatrypa brabyla* Copper, 1995, *Ehlersella davidsonii* Jin and Copper, 2000, and *Hindella* sp. Davidson, 1882.

Pentamerus palaformis (85 specimens)

Smooth shelled pentamerid with inflated biconvex valves that sometimes grew up to 10 cm in length (Jin and Copper 2000). The shell is formed of a thin (100 µm) outer lamellar layer and an internal prismatic layer that is posteriorly thickened (~1,500 µm). Articulated shells are most common from the lower 20 m of the Goéland Member (units 1-9, Copper and Long 1990) and were sometimes collected from in situ nests (beak down). They co-occur with *Joviatrypa brabyla* and in units 7-9 with tabulate corals, stromatoporoids and the calcareous dasyclad chlorophyte, *Cyclocrinites* Eichwald, 1840.

Joviatrypa brabyla (8 specimens)

Cherry-sized atrypid (~2 cm length) with a ribbed shell ornamentation (Copper 1995). The shell microstructure is dominated by a ~500 µm thick lamellar layer. Articulated shells are bed forming and co-occur with *Pentamerus palaformis* in units 7-9 of the Goéland Member.

Ehlersella davidsonii (9 specimens)

Smooth shelled pentamerid with weakly biconvex valves 4-5 cm in length (Jin and Copper 2000). Lamellar and prismatic shell microstructure is similar to *Pentamerus palaformis*, but not as thick in the posterior region (up to 500 µm thick). Articulated shells are bed forming and sometimes in situ (beak down) in the upper part of the Cybèle Member.

Hindella sp. (5 specimens)

Smooth shelled athyrid with strongly biconvex valves (1-1.5 cm length). The shell is dominated by a fibrous microstructure roughly 500 μm in thickness. Specimens are articulated and were collected in clusters along with *Eospirigerina* Boucot and Johnson, 1967, in green calcareous shales of the upper Grindstone Member of the Ellis Bay Formation, western Anticosti Island.

Visualization techniques

Stereomicroscopy

Brachiopod specimens were submerged in water and the shell was examined under stereomicroscope (Olympus 40 x and Leica MZ75 75 x) with high intensity reflected lighting with polarizing lens. Microborings filled with contrasting materials (e.g., iron sulfides, iron oxides or micrite) were readily visible using this method. Distributions of dendrinid microborings were mapped on photographs of a subset of the target brachiopods.

Scanning electron microscopy (SEM)

Two *Pentamerus* (Localities A1187, specimen no. 5 [GSC 10854] and A1187, specimen no. 8), one *Ehlersella* (Locality A1443 [GSC 10856]), one *Joviatrypa* (Locality A890) and one *Hindella* (Locality A4) were cast using the epoxy embedding technique. Epofix (Struers) two-part epoxy impregnated the microboring cavities under a low pressure vacuum (15 mm Hg), and the shell was dissolved in dilute HCl following the technique of Nielsen and Maiboe (2000). The specimens were coated with carbon and imaged by SEM with an electron beam intensity of 30 kV.

Microscopic-computed tomography (micro-CT)

A *Pentamerus* shell (Locality A1287c, specimen no. 8 [GSC 10855]) was observed to contain pyrite-filled microborings using stereomicroscopy. A small chip of brachiopod shell (5 mm x 3 mm) visibly containing pyrite-filled microborings was removed by Dremel saw from the remaining shell. Using the Konoscope 40-130 cone beam x-ray microtomography system at the University of Utah (Lin and Miller 2001), the chip was analyzed to obtain a 3D spatial reconstruction of the multiphase materials, pyrite (natural boring fill) and calcite (brachiopod substrate). X-ray emission was conducted at 130 kV with a current of 0.05 mA. A 10 mil aluminum filter was used with a focus spot of 8 μm, giving a voxel resolution of 5 μm. Exposure time per frame was set at 24 sec and 500 frames were taken for a total scanning time of 4 h 18 min. Reconstruction of the data took 3 h 35 min using a SGI Octane computing workstation (2 MIPS R12000 processors, 2 Gb of RAM). The output is a 568 x 512 pixel reconstruction of the specimen. Visualization software for the 3D reconstructions used VolSuite v.3.2.0, and 2D reconstructions were imaged using ImageJ v.1.29x. Measurements were taken directly using ImageJ for the micro-CT-imaged microborings. The original data file (available in RAW and SDT formats) and a movie file (AVI) of the microboring is archived on Pangaea® (www.pangaea.de) under doi:10.1594/PANGAEA.633815.

Systematic Ichnology

Ichnofamily Dendrinidae Bromley, Wisshak, Glaub and Botquelen, 2007
Pyrodendrina igen. n.

Type ichnospecies: Pyrodendrina cupra isp. n.

Diagnosis: Dendrinid boring system with shallow and deeply penetrating canals that radiate away from a central polyhedral node. Shallow canals are dominantly horizontal and may bifurcate. Deeply penetrating canals extend vertically in straight to slightly curved path and may bifurcate.

Differential Diagnosis: Pyrodendrina igen. n. is the only known rosette boring to have deeply penetrating distal canals, and this feature is a primary ichnotaxobase for the ichnogenus. Although these distal canals are distinctive, they may not always be preserved by the casting medium, and therefore additional criteria can be used to identify *Pyrodendrina* igen. n. Morphological differences in the central node and the shape of proximal canals can be used to distinguish *Pyrodendrina* igen. n. from *Platydendrina* Vogel, Golubic and Brett, 1987, *Ramodendrina* Vogel, Golubic and Brett, 1987, *Hyellomorpha* Vogel, Golubic and Brett 1987, *Semidendrina* Bromley, Wisshak, Glaub and Botquelen, 2007) and other dendrinid microborings. Neither *Platydendrina* nor *Ramodendrina* have the well developed central node and distal canals observed in *Pyrodendrina* igen. n. The proximal canals of *Hyellomorpha* have a swollen, rounded termination not observed in *Pyrodendrina* igen. n. Proximal canals of *Semidendrina* extend from only one side of the main chamber. An unnamed dendrinid boring from the Upper Ordovician of Sardinia (Bromley et al. 2007: fig. 32.5) has proximal canals with stubby, rounded terminations not observed in *Pyrodendrina* igen. n. and no preserved distal canals. An unnamed rosette-shaped boring (Botquelen and Mayoral 2005: fig. 2I) has superficial similarity to the proximal canals of *Pyrodendrina* igen. n., however the poor preservation and the fourfold larger size of the unnamed specimen makes comparison difficult. The size of *Pyrodendrina* igen. n., although not a formal ichnotaxobase, is distinct from other described dendrinids. The maximum proximal diameter of other rosette microborings is between 2 to 8.5 times greater than *Pyrodendrina* igen. n., with the exception of *Hyellomorpha microdendritica* Vogel, Golubic and Brett, 1987, which is barely half the diameter of *Pyrodendrina* igen. n. A similar trend holds for the diameter of the central node.

Etymology: Pyro, Latin = fire, refers both to the flame-shaped appearance of the boring cast and the fact that nearly all examples of the boring contain pyrite fill; *dendron*, Greek = tree.

Fig. 2 Proportional schematic diagram of *Pyrodendrina* igen. n. **A** Oblique vertical view of microboring placed in the context of the shell substrate, note the three parts of the microboring system. **B** Plan view of microboring showing the central node and proximal canals; inner

Pyrodendrina cupra isp. n.

Figs. 2-7

2005 'Rosette A'.- Tapanila 2005b, 172-186, fig. 7.3-7.4

Diagnosis: Rosette-shaped boring with shallow canals that bifurcate radially from a central node located below shell surface. Deep vertical canals extend radially below the central node. Canals taller than wide, with tapered terminals.

Etymology: Cupra, Latin = copper, in recognition of Dr. Paul Copper's legacy of research on Anticosti Island, including the brachiopods that contain this boring.

Type material, locality and horizon: Holotype (Fig. 3E lower right, 3F), paratype (Fig. 3E upper left, 3G) and various ontogenetic stages (Figs. 3A-D, H, 5) are cast in epoxy from *Pentamerus palaformis* shell (GSC 10854), Locality A1187, specimen no. 5, Goéland Member, Jupiter Formation, Anticosti Island. Additional reposited samples of *Pyrodendrina* igen. n. are in the shell of *Pentamerus palaformis* from Localities A1287c (GSC 10855: includes the micro-CT specimens), A1493x (GSC 10857) and A1151 (specimen no. 5: GSC 10858), and in the brachiopod shell of *Ehlersella davidsonii* from Locality A1443 (GSC 10856). *Pyrodendrina* igen. n. are known from 15 localities (summarized in Table 1) from the Grindstone Member of the Ellis Bay Formation and the Cybèle and Goéland members of the Jupiter Formation, Anticosti Island, Canada. Geological range is from Hirnantian (Ashgill, Late Ordovician) to Telychian (Llandovery, Early Silurian).

Description: The dendrinid microborings are subdivided into three parts, which occupy successive depths within the shell. Nearest the outer shell surface, the proximal part of the microboring consists of dominantly horizontal radial canals, the middle part is a central node, and the distal part (most deeply penetrating portion) consists of dominantly vertical radial canals (Figs. 2, 3F). The overall profile of the boring has an hourglass shape (Figs. 2A, 3F, 5) and in plan view the boring is oval in extent (Fig. 2B). The maximum diameter of the rosette is on average 445 μm, with a central node diameter of 105 μm. The maximum depth of penetration (distal-most extent) observed is 312 μm (see Table 2 for morphometry).

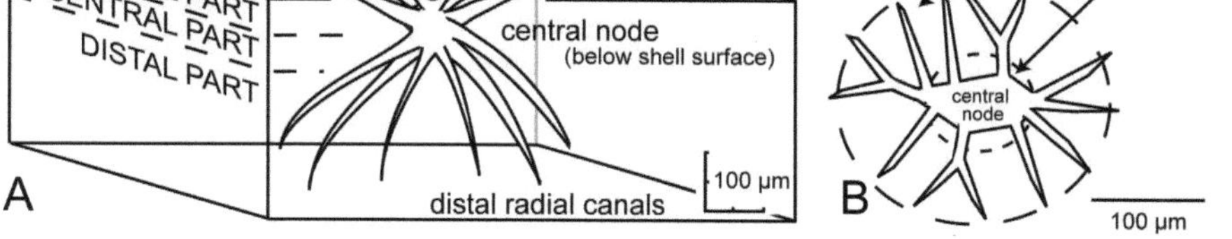

dashed line marks the medial zone where shell is not eroded above the central node and outer dashed line marks the maximum diameter of the microboring periphery. Scale bars are approximate, based on average *Pyrodendrina* igen. n. dimensions

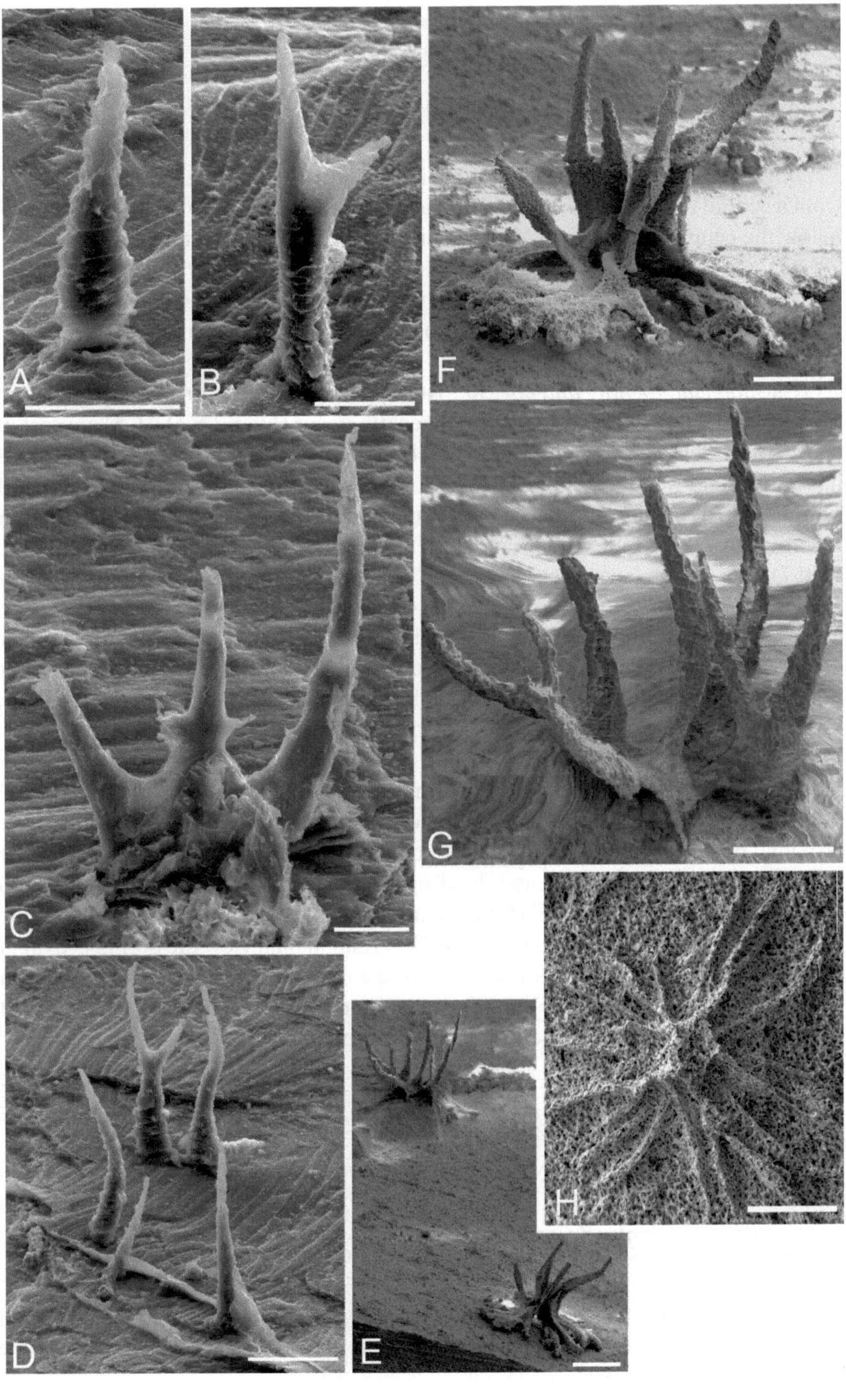
A
B
F
C
G
H
D
E

The proximal part of the microboring has the greatest lateral extent. It consists of roughly a dozen primary radial canals that may bifurcate at an angle of 33-98° to form secondary canals (Fig. 3F, H). All proximal canals are 33-50% taller than they are wide (mean width is 16 μm). The canals radiate from multiple points on the central node, roughly 50 μm below the shell surface (Figs. 3F, 5B). The length of individual canals averages 140 μm, and these bifurcate at an average distance of 44 μm away from the central node (Fig. 6). Surface apertures of the proximal canals are most abundant away from the center of the microboring, i.e., toward the periphery of the rosette (Fig. 5B). The vertical dimension of the canals enlarges toward their connection with the central node, giving them the appearance of buttresses (Fig. 3F).

The central node is polyhedral, typically having a greater horizontal (radial) dimension (80-100 μm) than vertical (50-60 μm). The node typically initiates roughly 50 μm below the shell surface and extends to a depth of 100-150 μm before branching into the distal part (Fig. 5B). Although difficult to image in SEM (see discussion), both stereomicroscopy and micro-CT confirm that the surface of shell immediately above the central node is intact. Using incident light on a dry and wetted specimen (Fig. 7), it is possible to see the reflected surface of calcite shell (Fig. 7A; no polarizer) above the pyrite-filled boring. With the polarizing filter rotated 90° to reduce reflected light, both dry (Fig. 7B) and wetted (Fig. 7C) specimens clearly show the dark surface of the pyrite-filled microboring at depth within the shell. Micro-CT scans similarly show the central node to lie below the shell surface (see white chevron symbols in Fig. 5B), giving *Pyrodendrina* igen. n. an overall hourglass-like shape. In *Pentamerus*, micro-CT demonstrates that the transition from the central node to the distal part coincides with the brachiopod shell microstructure transition from the outermost fibrous layer to the innermost thick prismatic layer (Fig. 5B).

The distal part of the microboring consists of canals that radiate distally from the apices of the polyhedral central node (Figs. 3F-G, 4G-H). Up to 9 distal canals are observed in one rosette. The canals extend in a dominantly vertical path, but may form broad curves toward their distal part. As with proximal canals, distal canals are thickest nearest the central node and they have greatest dimension in the

Fig. 3 SEM images of *Pyrodendrina cupra* igen. and isp. n. casts in the ventral valve of *Pentamerus palaformis* brachiopod shell, GSC 10854, Locality A1187-5, Goéland Member, Jupiter Formation, Anticosti Island. **A-C** Ontogenetic sequence of microboring, from solitary distal canal (A), to first branching (B), to second branching, trident phase (C). **D** Multiple early stage microborings. **E-G** Mature phase of microboring. **E** Well preserved distal part of two mature specimens (holotype and paratype), note top left (paratype) microboring occurs along a crack in the shell, obscuring the proximal part. **F** Holotype, detail of bottom right microboring in E showing well preserved proximal, central, and distal parts of the cavity. **G** Paratype, detail of top left microboring in E showing nine distal canals. **H** Image of proximal part and central node of a mature microboring. Note irregular surface of central node indicating incomplete casting of the distal part of the microboring. All images except F and H are perspective views taken at a 30° angle from the shell surface. Image H is viewed perpendicular to the shell surface, and image F is taken at a 10° angle from shell surface. Images are shown with the distal ends of the microborings pointing upward in the photos. Scale bars: A, B, C = 50 μm; D, F, G, H = 100 μm; E = 200 μm

Table 1 Locality data for brachiopods bearing *Pyrodendrina* igen. n. from Anticosti Island. (* = Number of brachiopods containing *Pyrodendrina* igen. n., with total examined brachiopods at given locality in brackets; † = contains holotype and paratype specimens. Abbreviations: S-Tely = Telychian; S-Aero = Aeronian; O-Hirn = Hirnantian)

Locality	Formation	Member	Unit	Age	Brachiopod taxon	Bored (total)*	Notes on specimens	UTM coordinates
A1092	Jupiter	Cybèle	high	S-Tely	*Ehlersella davidsonii*	4(5)		12E/7 500300:5472680
A1443	Jupiter	Cybèle	high	S-Tely	*Ehlersella davidsonii*	4(4)	#1 = GSC 10856 (cast + SEM)	12E/6 575750:5476400
A889	Jupiter	Goéland	9	S-Aero	*Joviatrypa brabyla*	3(6)		12E/1 544700:5454820
A890	Jupiter	Goéland	7, 9	S-Aero	*Joviatrypa brabyla*	2(2)	#1 Cast	12E/1 544400:5454300
A1493	Jupiter	Goéland	7, 8	S-Aero	*Pentamerus palaformis*	4(4)		12E/1 558607:5454408
A1493x	Jupiter	Goéland	7, 8	S-Aero	*Pentamerus palaformis*	6(6)	#7 = GSC 10857	12E/1 558386:5454434
A1151	Jupiter	Goéland	7, 8	S-Aero	*Pentamerus palaformis*	10(11)	#5 = GSC 10858	12E/1 558780:5454400
A1188	Jupiter	Goéland	6, 7	S-Aero	*Pentamerus palaformis*	3(3)		12E/1 542950:5451950
A1179	Jupiter	Goéland	5, 6	S-Aero	*Pentamerus palaformis*	1(3)		12E/1 543220:5452300
A1187	Jupiter	Goéland	5, 6	S-Aero	*Pentamerus palaformis*	10(11)	#5 = GSC10854† (cast + SEM) #8 cast	12E/1 544400:5454300
A1212	Jupiter	Goéland	5, 6	S-Aero	*Pentamerus palaformis*	5(17)		12E/1 542410:5451200
A1287c	Jupiter	Goéland	5, 6	S-Aero	*Pentamerus palaformis*	6(9)	#8 = GSC10855 (micro-CT)	12E/1 545260:5455040
A1211	Jupiter	Goéland	5	S-Aero	*Pentamerus palaformis*	2(4)		12E/1 542400:5451370
A1264	Jupiter	Goéland	1, 2	S-Aero	*Pentamerus palaformis*	1(5)		12E/8 545800:5456650
A4	Ellis Bay	Grindstone		O-Hirn	*Hindella* sp.	1(5)	#1 Cast	12E/13 553450:5413420

Table 2 Morphometric data for *Pyrodendrina* igen. n.

	Rosette Diam / Node Diam	Rosette Max Diam [μm]	Rosette Min Diam [μm]	Centr. Node Diam [μm]	Prox. Canal Width [μm]	Prox. Canal Length [μm]	Prox. Node to Branch [μm]	Prox. Branch Angle [degrees]	Dist. Canal Width [μm]	Dist. Canal Length [μm]
Mean	4.92	445.4	396.0	104.8	16.4	139.9	43.6	56.6	40.8	187.1
SD	0.96	60.6	57.0	46.1	5.5	43.3	15.7	19.4	13.6	31.3
Min	3.22	317.5	286.8	55.6	7.0	58.4	21.8	33.0	13.2	140.1
Max	6.24	542.1	502.7	232.9	28.8	278.1	67.8	98.0	66.3	220.3
N	14	14	11	15	70	69	24	20	13	8

vertical plane. Branching in the distal canals is less frequent than in the proximal canals. The width of the distal canals averages 41 μm at their contact with the central node and they extend distally for an average length of 187 μm to give the microboring a penetrative depth of up to 312 μm. The maximum horizontal span of the distal canals (~300 μm) generally does not exceed the diameter of the proximal rosette.

The densest concentrations of *Pyrodendrina* igen. n. (15 microborings per 5 mm^2) occur in *Pentamerus* and *Ehlersella* shells from the Jupiter Formation. The borings do not appear to exhibit a preference for posterior versus anterior or distal versus proximal positions within the valves. Furthermore, borings do not appear to be restricted to any particular part of the shell, based on the final burial orientation as preserved by geopetal cements within the brachiopod shells.

Different specimens exhibit different ontogenetic stages in the development of the microborings. The rosette microboring appears to initiate as a single vertical canal (Fig. 3A). Unbranched solitary canals are observed at a maximum depth of ~100 μm. The solitary canal continues to penetrate to a depth of ~200 μm and branches once to form a hooked appearance (Fig. 3B), although no proximal branches are evident at this point of the microboring's development. A third branch in the development of the boring produces a trident shape with a maximum penetration of ~300 μm (Fig. 3C). At the trident phase of branching, all distal canals are dominantly vertical, roughly equidistant, and connected at their branching point by a widening of adjacent canals, akin to the webbing between fingers. Proximal canals appear to originate during the trident phase, and the widened branch point of the distal canals appears to serve as the initiation of the central node. An increased number of distal and proximal canals along with a widening of the confluent central node characterize the mature phase of *Pyrodendrina* igen. n. (Fig. 3F-G). Variability in the amount of webbing between canals is noted among individual rosettes, although the variability of this feature is not considered here to be an ichnospecific taxobasis.

Examples of *Pyrodendrina* igen. n. from the Grindstone Member are recognized only by immature forms of the microboring, including single, double, and trident stages of development. No proximal portions of *Pyrodendrina* igen. n. have yet been discovered in the Ordovician of Anticosti.

Remarks: Pyrodendrina igen. n. occurs commonly in brachiopods that grew in muddy bottom, quiet water settings. These depositional environments are interpreted as being below normal storm weather wave base, but they are within the photic zone, as suggested by the co-occurrence of the photosynthetic algal fossils, *Cyclocrinites* in the Goéland Member and *Girvanella* Nicholson and Etheridge, 1878, in the Goéland and Cybèle members (e.g., Lauritzen and Worsley 1974; Riding 1975; Hammann and Serpagli 2003). The absolute water depth of these deposits has been estimated at 30-70 m on the basis of physical taphonomy (Schoenhals 2000), sedimentary structures (Copper and Long 1990), brachiopod communities (benthic assemblage 3, Jin and Copper 2000; Schoenhals 2000), and conodont communities (Zhang and Barnes 2002).

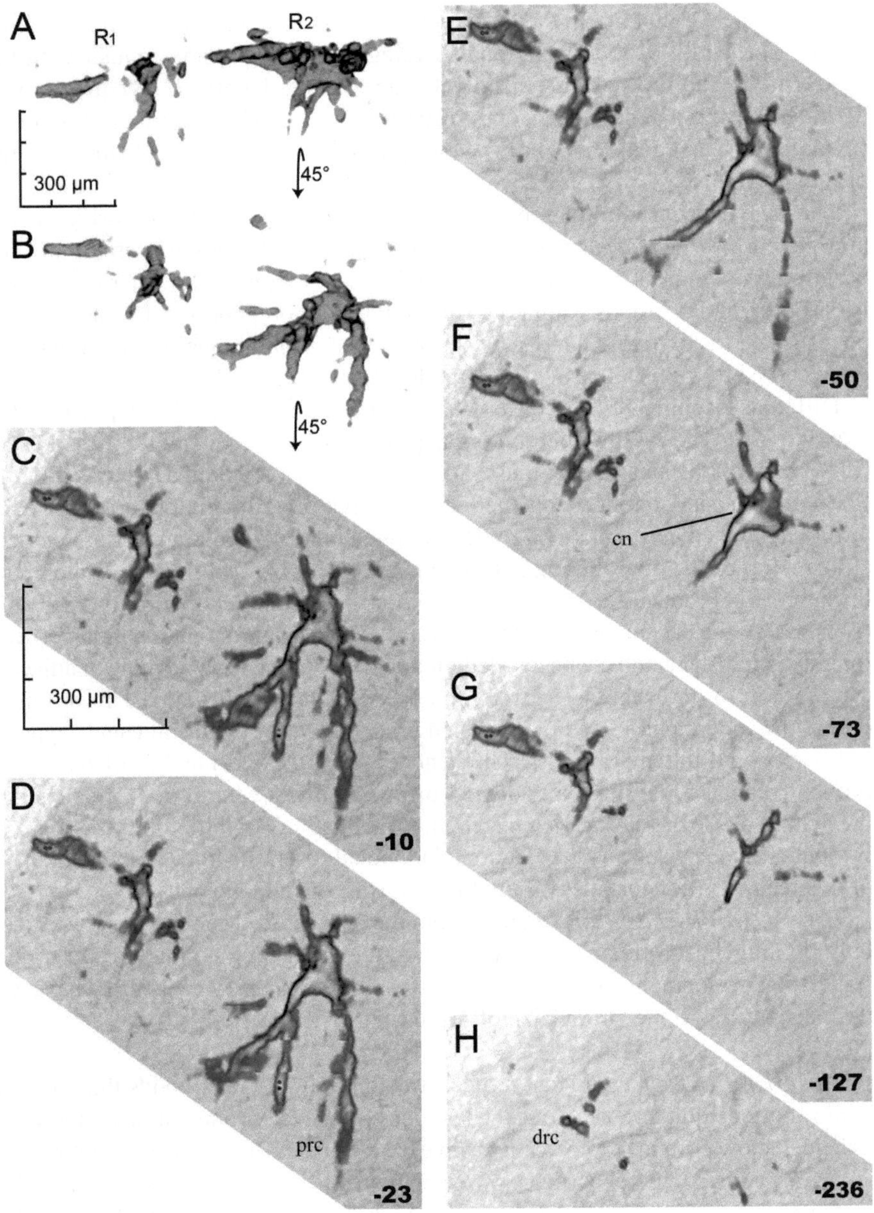

Fig. 4 Horizontal micro-CT series for two pyritized *Pyrodendrina* igen. n. microborings (left R_1, right R_2) in calcitic *Pentamerus palaformis* brachiopod shell, GSC 10855, Locality A1287c-8, Goéland Member, Jupiter Formation, Anticosti Island. **A** 3D view of microborings as seen in vertical view (as in Fig. 2). **B** Rotation toward the reader 45°. **C-H** Horizontal serial section in the plane that is parallel to the shell surface. Scale bar for C applies for images C-H. Bold numbers at bottom right indicate the depth in microns below the shell surface. Darker gray values correspond to greater density, i.e., medium gray = calcite shell; dark gray = pyrite cavity fill; prc = proximal radial canals; drc = distal radial canals; cn = central node

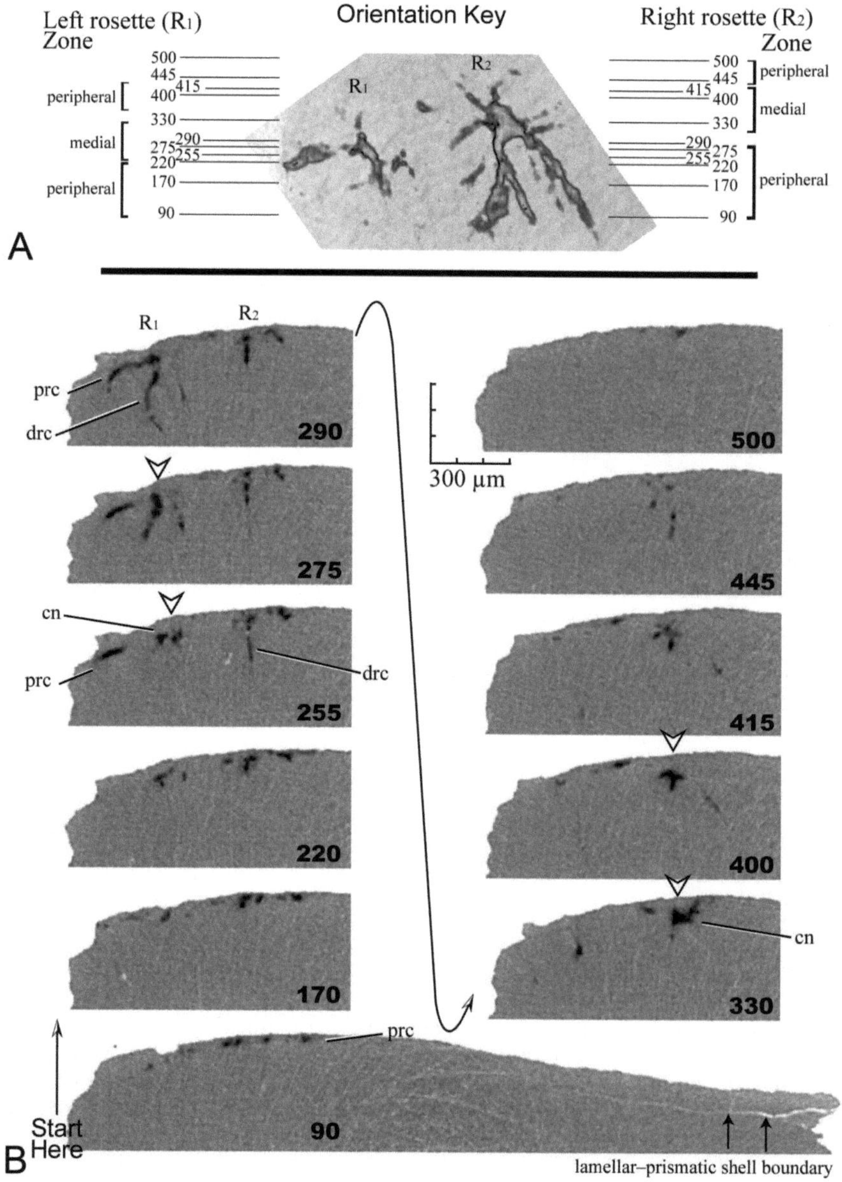

Fig. 5 Vertical micro-CT series for two pyritized *Pyrodendrina* igen. n. microborings (left R_1, right R_2) in calcitic *Pentamerus palaformis* brachiopod shell, GSC 10855, Locality A1287c-8, Goéland Member, Jupiter Formation, Anticosti Island. **A** Orientation key relating horizontal section (Fig. 4C) to the vertical planes (in microns) depicted in B; medial and peripheral zones denoted for the left and right rosette borings. **B** Vertical serial section of microborings perpendicular to the plane of the shell surface. Outer side of the shell is shown at the top of the images. Note the light-colored fracture on the right side of the shell in the 90 μm section (thin arrows), which corresponds to the lamellar-prismatic transition in the *Pentamerus* shell. Unbored shell above the central node (white chevron symbol) demonstrates an hourglass vertical profile to *Pyrodendrina* igen. n. (see Fig. 2). Abbreviations same as in Fig. 4

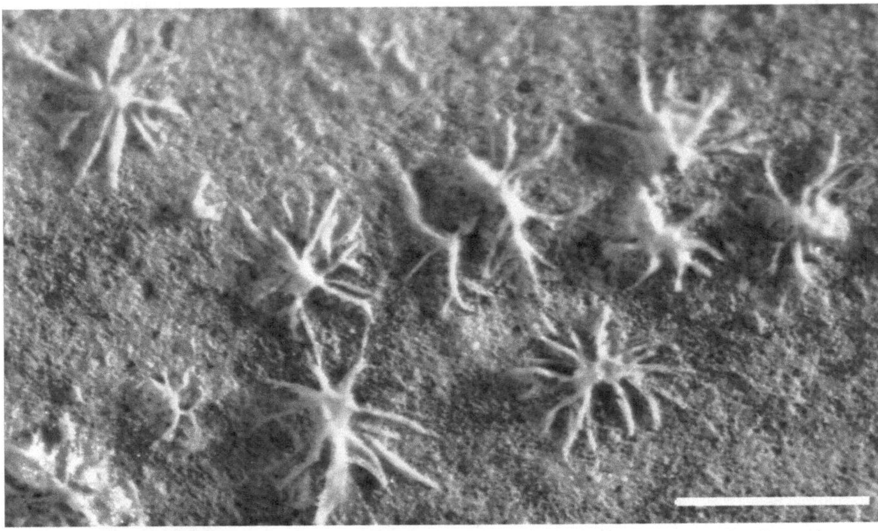

Fig. 6 Stereomicroscope image of a cluster of incompletely cast *Pyrodendrina* igen. n. microborings in the ventral valve of a *Pentamerus palaformis* brachiopod shell, GSC 10854, Locality A1187-5, Goéland Member, Jupiter Formation, Anticosti Island: Original borings filled with yellow micrite. Epoxy casting-embedding technique cast only the shallow (proximal and incomplete central) parts of the borings. Scale bar = 500 μm

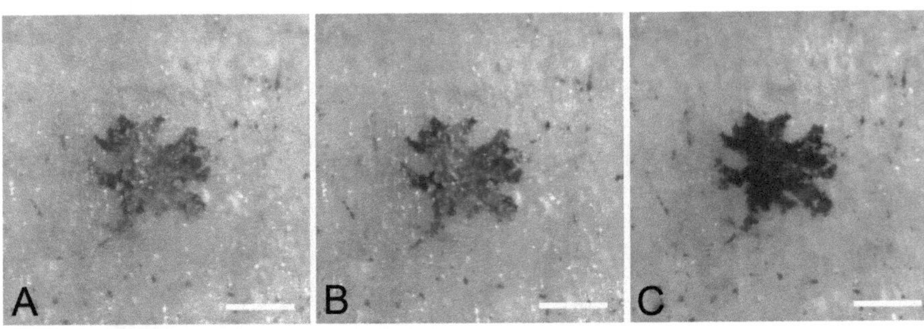

Fig. 7 Three stereomicroscope images of the same pyrite-filled *Pyrodendrina* igen. n. microboring in the ventral valve of a *Pentamerus palaformis* brachiopod shell, GSC 10858, Locality A1151, specimen no. 5, Goéland Member, Jupiter Formation, Anticosti Island. **A** Dry specimen viewed with reflected light, no polarizing filter. **B** Dry specimen viewed with reflected light, with polarizing filter rotated 90°. **C** Wet specimen viewed with reflected light, with polarizing filter rotated at 90°. See text for explanation. All scale bars = 500 μm

The microboring investigated in this study belongs to the ichnofamily Dendrinidae, a distinctive, enigmatic family of borings that has a dominantly radial disposition. The eponymous *Dendrina* Quenstedt, 1848, was the first of the dendrinid borings to receive widespread attention, owing largely to their common and conspicuous penetrations in the transluscent calcite of Mesozoic belemnite rostra (e.g., Mägdefrau 1937; Radwański 1972; Hofmann and Vogel 1992). More recently the stratigraphic record of dendrinids is recognized from the Late Ordovician to the modern (Vogel et al. 1987; Plewes et al. 1993; Hofmann 1996; Bundschuh 2000) according to the most recent systematic review by Bromley et al. (2007). Sponges, algae and foraminifera have all been posited as tracemakers for dendrinids; however, a consensus among researchers has not been reached (e.g., Clarke 1908; Solle 1938; Kornmann 1959; Bernard-Dumanois and Delance 1983; Vogel et al. 1995; Bromley et al. 2007). Similarly, it is difficult to assign a tracemaker for *Pyrodendrina* igen. n., as there exists no known modern analogue for this fossil microboring. The possibility that *Pyrodendrina* igen. n. was produced by a foraminifer is considered here unlikely. All dendrinid borings allied to foraminifera (Cherchi and Schroeder 1991; Plewes et al. 1993) include a conspicuous club-shaped pit adjacent to the branching canals. No such feature is found in *Pyrodendrina* igen. n. It is evident that *Pyrodendrina* igen. n. occurs within the photic zone, but without further evidence, it cannot be determined whether it was formed by an autotrophic or heterotrophic borer.

The morphology of *Pyrodendrina* igen. n. may be affected in part by the microstructure of the brachiopod shell. Micro-CT demonstrates that the transition from proximal to distal parts of *Pyrodendrina* igen. n. roughly correspond to the junction of the outer lamellar layer with the deeper prismatic layer of the *Pentamerus* shell, which occurs at roughly 100 μm below the surface.

The discovery of *Pyrodendrina* igen. n. adds to the growing understanding of Ordovician - Silurian endolithic communities. Data acquired during an intensive survey of macroborings and bioclaustrations in coral and stromatoporoid substrates

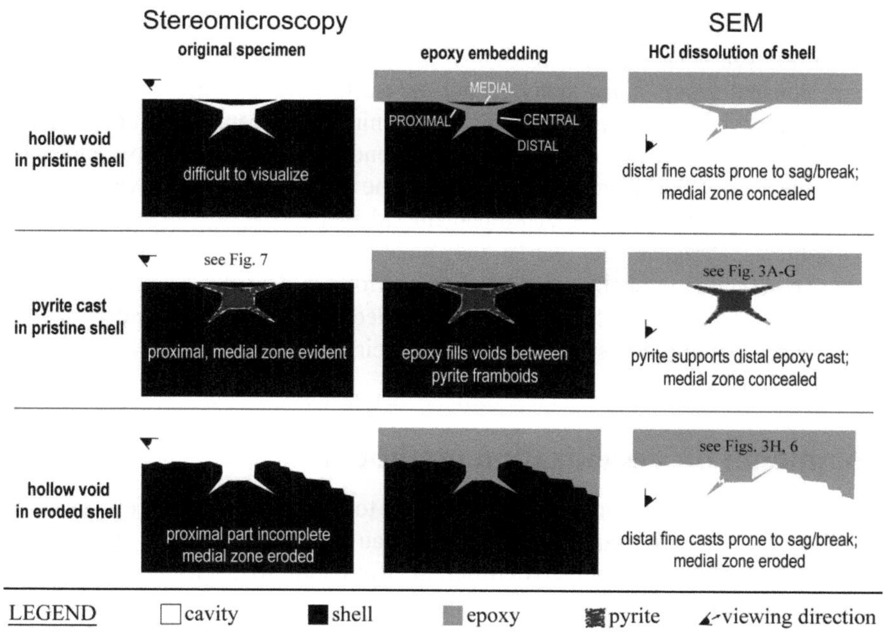

Fig. 8 Schematic diagram of the biases encountered when visualizing complex microborings using stereomicroscopy and SEM. Variables in taphonomy (natural fill and preservation quality) and laboratory methods (epoxy embedding and method of visualization) influence how and what is observed in a fossil specimen. Example figures are noted for some of the illustrated scenarios

(Tapanila 2001) allows for a comparison of trends in macroborings with those among microborings, especially for the Goéland Member. Ten of the localities from the Goéland Member that contain *Pyrodendrina*-bored brachiopods also contain stromatoporoids and corals that have been bored by macroscopic organisms (24 of 27 specimens are bored: Tapanila 2001). These corals and stromatoporoids are frequently penetrated by *Trypanites* Mägdefrau, 1932, borings, rarely by *Petroxestes* Wilson and Palmer, 1988, and some contain lingulid bioclaustrations, *Klemmatoica* Tapanila and Holmer, 2006, (see Tapanila and Copper 2002; Tapanila and Holmer 2006). Intensity of macrobioerosion at these localities is among the highest recorded in the Anticosti sequence, ranging from moderate to high (up to 20 *Trypanites* borings per 4 cm^2: Tapanila et al. 2004). At the member scale, the Goéland is among the most intensely macrobored deposits in the Anticosti sequence, having 65% of its corals and stromatoporoids macrobored, compared to the 40% average for the entire Anticosti sequence (Tapanila et al. 2004; Ebbestad and Tapanila 2005). The frequency of macrobioerosion increases exponentially with increased substrate height, a function of substrate exposure time to the water column, i.e., bioerosion is primarily limited by sediment burial (Tapanila et al. 2004; Ebbestad and Tapanila 2005). A low accumulation rate of sediments during the deposition of the lower nine units of the Goéland Member likely contributed to the diverse record of endoliths, including *Pyrodendrina* igen. n., in these strata.

Glaub and Vogel (2004) have summarized the stratigraphic record of microborings and demonstrated the morphological conservatism of these fossils over the past 500 million years. Dendrinid shaped microborings, such as *Pyrodendrina* igen. n., are an unusual group, since many do not appear to have modern representatives, and this has led to the current ambiguity concerning the tracemakers of dendrinids. The Anticosti specimens demonstrate that dendrinids were common endoliths during the Early Silurian and possibly during the Late Ordovician. Given the high boring density observed in Goéland pentamerid shells, it is possible that in some environments, *Pyrodendrina* igen. n. may have had a substantial deleterious effect on the preservation of shelly material. Future efforts in describing and quantifying ancient bioerosion, especially from the Early Paleozoic, will help our understanding of how evolving biological systems influenced their environment and the preservation of the fossil record.

Comments on visualization methodologies

The addition of micro-CT to the paleontologist's toolbox will no doubt revolutionize how fossils are described and analyzed (Tafforeau et al. 2006), but it is important to consider how this new method gives different information so that it can be integrated with data acquired by older, tried-and-true methods.

Basic stereomicroscopy provides a cheap and immediate tool to survey large collections of shells at a wide range of magnifications. While it is nondestructive, stereomicroscopy only provides an external view of the proximal (shallowest) penetrations of the substrate and it requires sufficient contrasting fill (e.g., pyrite) to be effective. These problems make it difficult to use stereomicroscopy alone in diagnosing and describing microborings. The use of thin-sections is not much better. Most microscopic borings have complex 3D geometries that are difficult to assess within a 30 µm-thick sheet of substrate, resulting in ambiguous ichnotaxonomic descriptions (e.g., putative Paleozoic 'sponge' borings: Lindström 1979; Weidlich 1996).

Coupled with SEM, epoxy casting is widely regarded as the standard tool among microboring researchers and represents the type material for most microboring ichnotaxa. Epoxy casting of microborings requires a modest investment of time per sample, although once a lab is properly outfitted the time factor diminishes. The quality of epoxy casts in fossil material are highly dependent on adequate porosity and permeability of the cavity fill, a problem that is significant in substrates whose primary mineralogy is recrystallized or replaced (Fig. 8). Biases in the epoxy casting method can hamper observation of both proximal and distal parts of a microboring structure. In a typical preparation, an epoxy-casted microboring is viewed from the distal or lateral side, and almost never from the proximal view. This means that shallow structures located toward the center of a boring are obscured, e.g., the funnel-shaped zone that is not bored in the shallow medial part of *Pyrodendrina* igen. n. Distal parts of microborings can be obscured by the epoxy casting method because of incomplete infiltration of the epoxy into the deepest, thinnest parts of the cavity (e.g., no distal parts cast in Figs. 3H, 6). Even when distal components of the microboring are casted, the acid dissolution phase of the method causes thin distal parts of the flexible epoxy cast to collapse on its side (e.g., Wisshak and Porter 2006: figs. 2, 3). Partial dissolution of the substrate can help with this problem, but will also hide parts of the cast. In this study, distal parts of *Pyrodendrina* igen. n. were completely cast because the framboidal pyrite natural casts provided sufficient permeability for epoxy infiltration and provided rigidity to the fine distal canals of the boring such that they did not collapse (Figs. 3F-G, 8). In specimens not containing pyrite, the distal parts of *Pyrodendrina* igen. n. are not typically cast and instead the central node is truncated with an irregular distal surface (Figs. 3H, 6). It is possible that other dendrinid microborings studied with epoxy casts suffer from incomplete casting of the deepest parts of the boring. Natural and epoxy casts figured by Vogel et al. (1987: fig. 4A) and Bromley et al. (2007: fig. 32.5b) show central nodes that abruptly terminate with an irregular distal surface reminiscent of the incomplete cast of *Pyrodendrina* igen. n. in Figure 3H. The potential absence of distal parts due to biases in the casting method should make the ichnologist cautious in relying on these parts as the sole diagnostic ichnotaxobase.

Micro-CT provides full 3D renderings of the microboring while maintaining context within the substrate (Figs. 4-5). This is important for recognizing the relationship between proximal and distal parts of *Pyrodendrina* igen. n. relative to the microstructural transition of the brachiopod shell (90 µm slice in Fig. 5B). Micro-CT also has an advantage over SEM and stereomicroscopy in allowing the object to be viewed in 3D through a full 360° range (Figs. 4-5), including the ability to make serial sections through the object through any plane. In the analysis of *Pyrodendrina* igen. n., it is evident that the apertures are only common toward the periphery of the boring, a feature that is supported by stereomicroscopy, but not accessible by means of SEM. Distal parts of the boring are clearly visible with micro-CT. When viewed as a 2D serial section, the morphologies imaged by micro-CT can be readily quantified using free software such as ImageJ. Another advantage of micro-CT is its ability to record the most deeply penetrating parts of the microboring.

Micro-CT poses some limitations on the types of specimens that can be effectively imaged. The size of the boring and the resolution to which it is imaged are important considerations when using micro-CT. Most microborings require a minimum of 10 μm resolution in order to identify morphological features. Increased scanning resolution (e.g., 5-10 μm) requires an increase in scanned frames per specimen, which increases the total scanning time and computer reconstruction time, and ultimately increases the cost. Fortunately, many compact microborings (<500 μm in diameter) have an ideal size that allows them to be imaged at the maximum micro-CT resolution of 5 μm. Given a proper size, the microboring also must be filled with a material that has contrasting density relative to the surrounding substrate. Iron sulfides and oxides are ideal, although detrital grains (e.g., siliciclastics) or pore / air space provide enough contrast to detect the morphology of the boring, though it is important to recognize that these natural casting media, especially pyrite and silica, often fill the microboring cavity incompletely (Kobluk and Risk 1977). An additional limitation particular to the cone beam x-ray microtomography system at the University of Utah was the maximum sample size that could be introduced into the machine. Minor destruction of the original shell was necessary to reduce its size in order to image the microboring by micro-CT. This may limit the use of certain micro-CT systems on museum specimens, although it is not as destructive as the SEM method. Lastly, the operational costs associated with micro-CT are substantially greater than for SEM or stereomicroscopy, but as with any technology, costs tend to decrease with time.

The micro-CT, SEM and stereomicroscopy techniques provide a variety of approaches to help analyze microborings, each with its own advantages and limitations. Used together, the techniques allow for a more complete account of the microboring, from their centimeter-scale distribution on the host substrate to their micron-scale morphologic features.

Conclusions

The Paleozoic microboring, *Pyrodendrina cupra* igen. and isp. n., is among the oldest dendrinid microborings currently known, and it occurs commonly in strata that were deposited in quiet water carbonate settings within the photic zone. It is most abundant in shells of large pentamerid brachiopods from the Goéland Member where it co-occurs with abundant *Trypanites* and rare *Petroxestes* macroborings. Micro-CT scanning is a useful method for imaging microborings and can reveal morphologic and contextual information about the substrate that often is difficult to recognize using either SEM or stereomicroscopy techniques alone. Awareness of the biases and benefits to each visualization method will help integrate observations that use new technologies with those of the past.

Acknowledgments

Thank you to A.A. Ekdale and F.R. Brown (Univ Utah) for encouraging and facilitating this research. Brachiopod material and locality information was generously provided by P. Copper (Laurentian Univ). I thank C.L. Lin for his technical assistance with the micro-CT analysis and D. Harding for his assistance on SEM (Univ Utah). The Department of Geology and Geophysics and the College of Mines and Earth Sciences at the University of Utah are acknowledged for supporting this research through financial aid and use of equipment. Earlier versions of this manuscript benefited from comments by A.A. Ekdale, J.M. deGibert, and M. Wisshak. Constructive reviews by A. Botquelen and G. Radtke greatly helped to clarify the systematic description of the fossil.

References

Bathurst RGC (1966) Boring algae, micrite envelopes and lithification of molluscan biosparites. Geol J 5:15-32

Benner JS, Ekdale AA, Gibert JM de (2004) Macroborings (*Gastrochaenolites*) in Lower Ordovician hardgrounds of Utah: sedimentologic, paleoecologic, and evolutionary implications. Palaios 19:543-550

Bernard-Dumanois A, Delance J-H (1983) Microperforations par algues et champignons sur les coquilles des "marnes à *Ostrea acuminata*" (Bajocien supérieur) de Bourgogne (France), relations avec le milieu et utilization paléobathymétrique. Geobios 16:419-429

Botquelen A, Mayoral E (2005) Early Devonian bioerosion in the Rade de Brest, Armorican Massif, France. Palaeontology 48:1057-1064

Boucot AJ, Johnson JG (1967) Silurian and Upper Ordovician atrypids of the genera *Plectatrypa* and *Spirigerina*. Norsk Geol Tidsskr 47:79-101

Bromley RG (1970) Borings as trace fossils and *Entobia cretacea* Portlock, as an example. In: Crimes TP, Harper JG (eds) Trace fossils. Seel House Press, Liverpool, pp 49-90

Bromley RG (2004) A stratigraphy of marine bioerosion. In: McIlroy D (ed) The application of ichnology to palaeoenvironmental and stratigraphic analysis. Geol Soc London, Spec Publ 228:455-479

Bromley RG, Wisshak M, Glaub I, Botquelen A (2007) Ichnotaxonomic review of dendriniform borings attributed to foraminiferans: *Semidendrina* igen. nov. In: Miller III W (ed) Trace fossils: concepts, problems, prospects. Elsevier, Amsterdam, pp 518-530

Bromley RG, Beuck L, Taddei Ruggiero E (this volume) Endolithic sponge versus terebratulid brachiopod, Pleistocene, Italy: accidental symbiosis, bioclaustration and deformity. In: Wisshak M, Tapanila L (eds) Current developments in bioerosion. Springer, Berlin, pp 361-368

Brunton FR, Copper P (1994) Paleoecologic, temporal, and spatial analysis of Early Silurian reefs of the Chicotte Formation, Anticosti Island, Quebec, Canada. Facies 31:57-80

Bundschuh M (2000) Silurische Mikrobohrspuren, ihre Beschreibung und Verteilung in verschiedenen Faziesräumen (Schweden, Litauen, Großbritannien und U.S.A.). PhD Thesis, Fachbereich Geowiss, Univ Frankfurt, 129 pp

Bundschuh M, Balog S-J (2000) *Fasciculus rogus* nov. isp., an endolithic trace fossil. Ichnos 7:149-152

Campbell S, Kazmierczak J, Golubic S (1979) *Palaeoconchocelis starmachii* gen. n., sp. n., an endolithic rhodophyte (Bangiaceae) from the Silurian of Poland. Acta Palaeont Pol 24:405-408

Cherchi A, Schroeder R (1991) Perforations branchues dues à des Foraminifères cryptobiotiques dans des coquilles actuelles et fossiles. C R Acad Sci Paris 312:111-115

Clarke JM (1908) The beginnings of dependent life. New York State Mus Bull 121:146-196

Copper P (1973) The type species of *Lissatrypa* (Silurian Brachiopoda). J Paleont 47:70-76

Copper P (1995) Five new genera of Late Ordovician-Early Silurian brachiopods from Anticosti Island, Eastern Canada. J Paleont 69:846-862

Copper P (1999) Brachiopods during and after the Late Ordovician mass extinctions on Anticosti Island, E Canada. Acta Univ Carolinae - Geol 43:207-209

Copper P, Long DGF (1990) Stratigraphic revision of the Jupiter Formation, Anticosti Island, Canada: a major reference section above the Ordovician-Silurian boundary. Newsl Stratigr 23:11-36

Davidson T (1882) British fossil Brachiopoda. Devonian and Silurian supplements. Monogr Palaeontologr Soc 5:1-1134

Dewing K (1999) Late Ordovician and Early Silurian strophomenid brachiopods of Anticosti Island, Quebec, Canada. Palaeontogr Canad 17:1-143

Duncan PM (1876) On some thallophytes parasitic within recent Madreporaria. Roy Soc London, Proc 24:238-257

Ebbestad JOR, Tapanila L (2005) Non-predatory borings in *Phanerotrema* (Gastropoda, Early Silurian), Anticosti Island, Québec, Canada. Palaeogeogr Palaeoclimatol Palaeoecol 221:325-341

Eichwald KE von (1840) Über das silurische Schichtensystem Estlands. Bull Acad Sci St. Pétersbourg 7:1-210

Furnes H, Banerjee NR, Muehlenbachs K, Staudigel H, de Wit M (2004) Early life recorded in Archean pillow lavas. Science 304:578-581

Gatrall M, Golubic S (1970) Comparative study on some Jurassic and Recent endolithic fungi using scanning electron microscope. In: Crimes TP, Harper JG (eds) Trace fossils. Seel House Press, Liverpool, pp 167-178

Glaub I (1994) Mikrobohrspuren in ausgewählten Ablagerungsräumen des europäischen Jura und der Unterkreide (Klassifikation und Palökologie). Courier Forschinst Senckenberg 174:1-324

Glaub I, Vogel K (2004) The stratigraphic record of microborings. Fossils and Strata 51:126-135

Glaub I, Balog S-J, Bundschuh M, Gektidis M, Hofmann K, Radtke G, Schmidt H, Vogel K (1999) Euendolithic cyanobacteria/cyanophyta and their traces in earth history. In: Charpy L, Larkum AWD (eds) Marine Cyanobacteria. Bull Inst Océanogr Monaco, Spec Issue 19:135-142

Golubic S, Brent G, LeCampion T (1970) Scanning electron microscopy of endolithic algae and fungi using a multipurpose casting-embedding technique. Lethaia 3:203-209

Golubic S, Perkins RD, Lukas KJ (1975) Boring microorganisms and microborings in carbonate substrates. In: Frey RW (ed) The study of trace fossils: a synthesis of principles, problems, and procedures in ichnology. Springer, New York, pp 229-259

Golubic S, Campbell S, Spaeth C (1983) Kunstharzausgüsse fossiler Mikroben-Bohrgänge. Präparator 29:197-200

Grahn Y (1981) Parasitism on Ordovician Chitinozoa. Lethaia 14:135-142

Hammann W, Serpagli E (2003) The algal genera *Ischadites* Murchison, 1839 and *Cyclocrinites* Eichwald, 1840 from the Upper Ordovician Portixeddu Formation of SW Sardinia. Boll Soc Paleont Ital 42:1-29

Hessland I (1949) Investigations of the Lower Ordovician of the Siljan District, Sweden, II: Lower Ordovician penetrative and enveloping algae from the Siljan District. Bull Geol Inst Univ Uppsala 33:409-424

Hofmann K (1996) Die mikro-endolithischen Spurenfossilien der borealen Oberkreide Nordwest-Europas und ihre Faziesbeziehungen. Geol Jb, A, 136:3-153

Hofmann K, Vogel K (1992) Endolithische Spurenfossilien in der Schreibkreide (Maastricht) von Rügen (Norddeutschland). Z Geol Wiss 20:51-65

Jin J, Copper P (1997) *Parastrophinella* (Brachiopoda): its paleogeographic significance at the Ordovician/Silurian boundary. J Paleont 71:369-380

Jin J, Copper P (1998) *Kulumbella* and *Microcardinalia* (*Chiastodoca*) new subgenus, Early Silurian divaricate stricklandiid brachiopods from Anticosti Island, eastern Canada. J Paleont 72:441-453

Jin J, Copper P (1999) The deep-water brachiopod *Dicoelosia* King, 1850, from the Early Silurian tropical carbonate shelf of Anticosti Island, Eastern Canada. J Paleont 73:1042-1055

Jin J, Copper P (2000) Late Ordovician and Early Silurian pentamerid brachiopods from Anticosti Island, Quebec, Canada. Palaeontogr Canad 18:1-140

Jin J, Caldwell WGE, Copper P (1990) Evolution of the Early Silurian rhynchonellid brachiopod *Fenestrirostra* in the Anticosti Basin of Quebec. J Paleont 64:214-222

Jin J, Long DGF, Copper P (1996) Early Silurian *Virgiana* pentamerid brachiopod communities of Anticosti Island, Québec. Palaios 11:597-609

Kobluk DR, Risk MJ (1977) Algal borings and framboidal pyrite in Upper Ordovician brachiopods. Lethaia 10:135-143

Kornmann P (1959) Die heterogene Gattung *Gomontia* I. Der sporangiale Anteil, *Codiolum polyrhizum*. Helgoländer Wiss Meeresunters 6:229-238

Lauritzen Ø, Worsley D (1974) Algae as depth indicators in the Silurian of the Oslo region. Lethaia 7:157-161

Lespérance PJ (1981) International Union of Geological Sciences, Subcommission on Silurian Stratigraphy, Ordovician-Silurian Boundary Working Group, Field Meeting, Anticosti-Gaspé, Québec, 1981. Vol I: Guidebook. Dépt Géol, Univ Montréal, 56 pp

Liljedahl L (1986) Endolithic micro-organisms and silicification of a bivalve fauna from the Silurian of Gotland. Lethaia 19:267-278

Lin CL, Miller JD (2001) A new cone beam X-ray microtomography facility for 3D analysis of multiphase materials. Second World Congress on Industrial Process Tomography, Hannover, Germany, 29th-31st August 2001, pp 98-109

Lindström M (1979) Probable sponge borings in Lower Ordovician limestone of Sweden. Geology 7:152-155

Long DGF, Copper P (1987) Stratigraphy of the Upper Ordovician upper Vaureal and Ellis Bay formations, eastern Anticosti Island, Quebec. Canad J Earth Sci 24:1807-1820

Long DGF, Copper P (1994) The Late Ordovician-Early Silurian carbonate tract of Anticosti Island, Gulf of St. Lawrence, Eastern Canada. GAC/MAC, Joint Annu Meet, Waterloo, Ontario. 20-25 May, 1994. Field Trip Guideb B4, 70 pp

Mägdefrau K (1932) Über einige Bohrgänge aus dem unteren Muschelkalk von Jena. Paläont Z 14:150-160

Mägdefrau K (1937) Lebensspuren fossiler 'Bohr'-Organismen. Beitr Natkd Forsch Südwestdeutschland 2:54-67

McLuhan M (1964) Understanding media: the extensions of man. McGraw-Hill, New York, 365 pp

Nicholson HA, Etheridge R (1878) A monograph of the Silurian fossils of the Girvan District in Ayrshire. Edinburgh – London, 341 pp

Nielsen JK, Maiboe J (2000) Epofix and vacuum: an easy method to make casts of hard substrates. Palaeont Electron, 3(1), Art 2, 10 pp. http://palaeo-electronica.org/2000_1/epofix/issue1_00.htm

Olempska E (1986) Endolithic microorganisms in Ordovician ostracod valves. Acta Palaeont Pol 31:229-236

PANGAEA (accessed August 16th, 2007) Publishing network for geoscientific and environmental data. http://www.pangaea.de/

Plewes CR, Palmer TJ, Haynes JR (1993) A boring foraminiferan from the Upper Jurassic of England and Northern France. J Micropalaeont 12:83-89

Podhalańska T (1984) Microboring assemblage in Lower/Middle Ordovician limestones from northern Poland. N Jb Geol Paläont Mh 1984:497-511

Podhalańska T, Nõlvak J (1995) Endolithic trace-fossil assemblage in Lower Ordovician limestones from northern Estonia. GFF 117:225-231

Quenstedt FA (1845-1849) Petrefaktenkunde Deutschlands. 1, Abh 1, Cephalopoden. Fues, Tübingen, 580 pp

Radwański A (1972) Remarks on the nature of belemnicolid borings *Dendrina*. Acta Geol Pol 22:257-263

Riding R (1975) *Girvanella* and other algae as depth indicators. Lethaia 8:173-179

Schoenhals TJ (2000) Paleoecological constraints on Early Silurian *Pentamerus* brachiopod communities of Anticosti Island, Quebec. Unpubl PhD Thesis, Univ West Ontario, London, Canada, XX pp

Solle G (1938) Die ersten Bohr-Spongien im europäischen Devon und einige andere Spuren. Senckenb Lethaea 20:154-178

Tafforeau P, Boistel R, Boller E, Bravin A, Brunet M, Chaimanee Y, Cloetens P, Feist M, Hoszowska J, Jaeger J-J, Kay RF, Lazzari V, Marivaux L, Nel A, Nemoz C, Thibault X, Vignaud P, Zabler S (2006) Applications of x-ray synchrotron microtomography for non-destructive 3D studies of paleontological specimens. Appl Phys A 83:195-202

Tapanila L (2001) Bioerosion in Late Ordovician and Early Silurian tropical carbonate settings of Anticosti Island, Québec, Canada. Unpubl MSc Thesis, Laurentian Univ, Sudbury, Canada, 152 pp

Tapanila L (2002) A new endosymbiont in Late Ordovician tabulate corals from Anticosti Island, eastern Canada. Ichnos 9:109-116

Tapanila L (2004) The earliest *Helicosalpinx* from Canada and the global expansion of commensalism in Late Ordovician sarcinulid corals (Tabulata). Palaeogeogr Palaeoclimatol Palaeoecol 215:99-110

Tapanila L (2005a) Palaeoecology and diversity of endosymbionts in Palaeozoic marine invertebrates: Trace fossil evidence. Lethaia 38:89-99

Tapanila L (2005b) Paleoecology and evolutionary significance of hard substrate trace fossils. Unpubl PhD Thesis, Univ Utah, 204 pp

Tapanila L, Copper P (2002) Endolithic trace fossils in Ordovician-Silurian corals and stromatoporoids, Anticosti Island, eastern Canada. Acta Geol Hisp 37:15-20

Tapanila L, Holmer LE (2006) Endosymbiosis in Ordovician-Silurian corals and stromatoporoids: a new lingulid and its trace from eastern Canada. J Paleont 80:750-759

Tapanila L, Copper P, Edinger E (2004) Environmental and substrate control on Paleozoic bioerosion in corals and stromatoporoids, Anticosti Island, eastern Canada. Palaios 19:292-306

Taylor PD, Wilson MA (2003) Palaeoecology and evolution of marine hard substrate communities. Earth-Sci Rev 62:1-103

Vogel K (1993) Bioeroders in fossil reefs. Facies 28:109-114

Vogel K, Golubic S, Brett CE (1987) Endolith associations and their relation to facies distribution in the Middle Devonian of New York State, U.S.A. Lethaia 20:263-290

Vogel K, Bundschuh M, Glaub I, Hofmann K, Radtke G, Schmidt H (1995) Hard substrate ichnocoenoses and their relations to light intensity and marine bathymetry. N Jb Geol Paläont Abh 195:49-61

Warme JE (1977) Carbonate borers-their role in reef ecology and preservation. In: Frost SH, Weiss MP, Saunders JB (eds) Reefs and related carbonates-ecology and sedimentology. AAPG Stud Geol 4:261-279

Weidlich O (1996) Bioerosion in Late Permian Rugosa from reefal blocks (Hawasina Complex, Oman Mountains): implications for reef degradation. Facies 35:133-142

Wilson MA, Palmer TJ (1988) Nomenclature of a bivalve boring from the Upper Ordovician of the Midwestern United States. J Paleont 62:306-308

Wisshak M (2006) High-latitude bioerosion: the Kosterfjord experiment. Lect Notes Earth Sci 109:1-202

Wisshak M, Porter D (2006) The new ichnogenus *Flagrichnus* - a paleoenvironmental indicator for cold-water settings? Ichnos 13:135-145

Zhang S, Barnes CR (2002) Late Ordovician-Early Silurian (Ashgillian-Llandovery) sea level curve derived from conodont community analysis, Anticosti Island, Québec. Palaeogeogr Palaeoclimatol Palaeoecol 180:5-32

Micro-computed tomography for studies on *Entobia*: transparent substrate versus modern technology

Christine H. L. Schönberg[1], Greg Shields[2]

[1] Centre of Marine Studies, University of Queensland, St. Lucia, QLD 4072, Australia; current address: Institut für Bio- und Umweltwissenschaften, Carl von Ossietzky Universität Oldenburg, 26111 Oldenburg, Germany, (christine.schoenberg@uni-oldenburg.de)

[2] Australian Key Centre for Microscopy and Microanalysis, University of Sydney, 2006 Sydney, Australia

Abstract. Endolithic bioerosion is difficult to analyse and to describe, and it usually requires damaging of the sample material. Sponge erosion (*Entobia*) may be one of the most difficult to evaluate, as it is simultaneously macroscopically inhomogeneous and microstructurally intricate. We studied the bioerosion traces of the two Australian sponges *Cliona celata* and *Cliona orientalis* with modern technology: high resolution X-ray micro-computed tomography. Micro-CT allows non-destructive visualisation of live and dead structures in three dimensions and was compared to traditional microscopic methods. Micro-CT and microscopy showed that *C. celata* bioerosion was more intense in the centre and branched out in the periphery (21 vs. 9% substrate removed). In contrast, *C. orientalis* produced a dense, even meshwork and caused an overall more intense erosion pattern than *C. celata* (48 central vs. 42% marginal substrate removed). Extended pioneering filaments were not usually found at the margins of the studied sponge erosion, but branches ended abruptly or tapered to points. Results obtained with micro-CT were similar in quality to observations from transparent optical spar under the dissecting microscope. Microstructures could not be resolved as well with micro-CT as anticipated. Even though sponge scars and sponge chips were easily recognisable on maximum magnification micro-CT images, they lacked the detail that is available from SEM. Other drawbacks of micro-CT involve high costs and presently limited access. Even though micro-CT cannot presently replace traditional techniques such as epoxy resin casts viewed by SEM, we obtained valuable information. Especially for the possibility to measure endolithic pore volumes, we regard micro-CT as a very promising tool that will continue to be optimised. A combination of different methods will produce the best results in the study of *Entobia*.

Keywords. *Entobia*, *Cliona celata*, *Cliona orientalis*, bioerosion, traces, ichnology, micro-computed tomography

---

M. Wisshak, L. Tapanila (eds.), *Current Developments in Bioerosion*. Erlangen Earth Conference Series, doi: 10.1007/978-3-540-77598-0_8,

Introduction

Quantitative and qualitative studies on endolithic bioerosion have always presented problems. Most methods for the quantification of bioerosion are destructive and usually allow only a two-dimensional assessment. They include sectioning or digestion of entire samples to facilitate point-counts, two-dimensional transcription, evaluation of light permeability, production of corrosion casts and extracting the tissue of endolithic organisms (see also summary in table 1 in Schönberg 2001). This is undesirable in rare or valuable samples such as fossils or type specimens. Moreover, more accurate methods than those listed above are sometimes necessary, especially for qualitative observations. Scanning electron microscopy (SEM) allows a high level of resolution to evaluate the quality of traces in three dimensions, but also is at least partly destructive and is mostly used for details rather than for overviews (Hoeksema 1983; Calcinai et al. 2003, 2004; Borchiellini et al. 2004). Traditional, medical X-radiography can provide non-destructive means for a good general overview of the erosion trace in two dimensions, which is suitable for relatively flat substrates such as bivalve shells (e.g., Bromley and Tendal 1973; Wesche et al. 1997). More complex substrates such as gastropod shells or three-dimensional erosion patterns cannot adequately be visualised with traditional radiography. Conventional, medical techniques of computerised axial tomography (CAT scanning) allow the viewing of the distribution through the object by producing a series of computer-generated, two-dimensional 'slices', but this has not routinely been used to produce continuous images of internal structures or to quantify bioerosion (e.g., Becker and Reaka-Kudla 1997; Schönberg 2001).

In context of the above considerations, studies on endolithic sponges and their traces (*Entobia* Bronn, 1837) probably are the most difficult. Only a few sponges excavate large cavities entirely filled by sponge tissue (e.g., most species of *Aka* (Johnson, 1899), *Cliothosa* Topsent, 1905, and a few species of *Cliona* Grant, 1826 such as *Cliona ampliclavata* Rützler, 1974). In almost all other cases the sponges produce smaller, three-dimensional, multiple and discontinuous cavities, which are connected by smaller ducts and canals, often forming networks (e.g., Bromley and D'Alessandro 1989). Such structures are very difficult to visualise and especially cause problems when attempting to provide a quantitative estimate of the affected volume. To date, most studies have considered such porous sponge excavations as coherent bodies, including substrate bridges and columns remaining in the sponge tissue (e.g., Schönberg 2001 and references therein; Schönberg 2003, 2006). Rützler (1975) and Bergman (1983) showed that even in dense sponge bioerosion only 40-50% of the substrate is removed. Therefore, the above procedure may result in overestimates. Radiographic methods can occasionally produce more accurate results, but the structure of porous substrate, e.g., coral, presents a serious problem: natural pores of the corals cannot adequately be distinguished from pores made by sponges and filled with sponge tissue (Schönberg 2001). Another important aspect of sponge bioerosion is the study of its microstructures, especially the scars that result from the etching and removal of the so-called sponge chips. These patterns may be useful for sponge taxonomy (Calcinai et al. 2003, 2004; Borchiellini et al.

2004). To date only SEM was known to provide a resolution sufficient to recognise such microstructures.

Recently another technology has been adapted for ichnology: high resolution X-ray micro-computed tomography. It was developed in the mid-1980s and became available as a desktop scanning unit about ten years ago. Micro-CT is a technique that non-destructively images external and internal structures of living or non-living samples at an increased resolution to previously available methods such as plain film X-radiography and CAT scans (Van Geet et al. 2001; Farber et al. 2003; Itoh et al. 2004). Micro-CT captures a series of X-radiographs at regular increments as the sample is rotated between a fixed polychromic X-ray source and a detector. The resolution (or degree of magnification) is determined by the distance between the source and the sample, i.e. resolution is increasing with decreasing distance. However, this correspondingly decreases the total area of the sample scanned. The Skyscan 1172 scanning unit used in the present study achieves a theoretical resolution of up to 1 µm, however, real resolving power is effectively limited to 4 µm by the source spot size (<5 µm) and the size of the detector camera (Lin and Miller 1996).

The series of X-radiographs acquired during scanning provide many angular views of the sample taken through 180 or 360 degrees of rotation. These are synthesised by cone-beam reconstruction software (using the Feldkamp algorithm) to produce a stack of 1000 virtual cross-sections (or 'slices') through the object (Feldkamp et al. 1984). The pixel size in these images is equal to the scanning resolution. The resulting slice thickness corresponds to the pixel size of the cross-sections, so that the resolution is the same in XY, ZX and ZY planes.

During the present study the cross-sections were processed with commercial volume rendering software that produces three-dimensional images and animations that can be viewed from all angles, digitally 'clipped' to reveal internal structures and analysed for volume measurements (e.g., Tapanila this volume; Bromley et al. this volume). The present publication will test the use of a Skyscan 1172 in the evaluation of sponge erosion traces, focusing on the comparison of traces of two different species and on the resolution of endolithic microstructures in situ, and will consider this in contrast to known results possible with other microscopic techniques.

Material and methods

Optical or Iceland spar rhombohedra from Naica, Mexico of about 1 x 1 x 2 cm and larger were obtained through gemstone trading (Mineralien-Online.Net, Schlechting, Germany). Spar was chosen as the main experimental substrate as its transparency allows an initial viewing of endolithic structures under the dissecting microscope to monitor the infection by a given sponge. Optical spar is also a very homogeneous, dense material. Therefore, erosion structures can extend evenly without encountering barriers or hollow spaces. As a consequence, allospecific erosion traces can be compared as they developed under the same conditions. The lack of pores in spar furthermore allows evaluation of the erosion traces with radiographic techniques and does not pose the problem that existing pores and sponge cavities filled with tissue cannot be sufficiently distinguished from each other (Schönberg 2001). Cleavage planes of the spar enabled us to split off one slim layer after initial observations, removing the outer surface corroded and obscured by endolithic algae, and facilitating a clear view for more detailed studies and photographs.

Spar rhombohedra were attached to naturally occurring bioeroding sponges in Little Pioneer Bay, Orpheus Island, Palm Island Group, central Great Barrier Reef, Australia (Fig. 1). They were fixed to sponge-infested materials on the shore-side behind the reef crest by nailing cable ties across each rhombohedron (Fig. 2). Spar rhombohedra were attached to the tops and papillae of live colonies of *Cliona orientalis* Thiele, 1900, *Cliona celata* Grant, 1826 (sensu Schönberg 2000), *Cliona caesia* (Schönberg, 2000), *Cliona vermifera* Hancock, 1867, *Pione* cf. *vastifica* (Hancock, 1849), *Cliothosa hancocki* (Topsent, 1888) and *Cliothosa* cf. *aurivillii* (Lindgren, 1897). Rhombohedra stayed in place for 3.5 or 11.5 months. Several rhombohedra were invaded by the encrusting growth form of *C. orientalis*, one by the papillate growth form of *C. celata* (sensu Schönberg 2000), but unfortunately none of the other species infected the spar. Species identities were confirmed by spicule preparations after the icro-CT scans. Rhombohedra were dried in an oven at 40°C for several days and kept dry for preliminary scans.

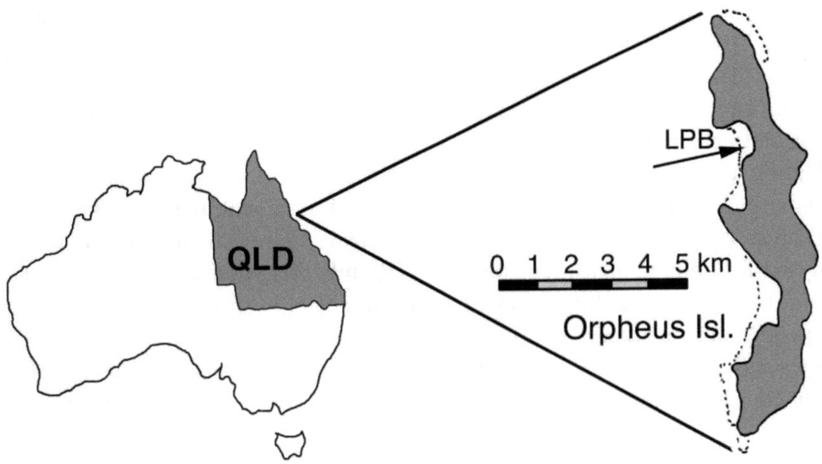

Fig. 1 Sample site and experimental environment. Orpheus Island, Palm Islands, central Great Barrier Reef, Australia. Rhombohedra were fixed in the field on the leeward side of the island, on the island side of reef flat behind the reef crest between 1 and 3 m water depth (arrow). QLD = Queensland. LPB = Little Pioneer Bay

Later investigations involved some light scraping of the outer surfaces of the rhombohedra with a scalpel to remove epibionts. Rhombohedra were viewed under a dissecting microscope (Wild). The view was nevertheless somewhat obscured by endolithic algae. After scanning internal structures with micro-CT, in some cases a layer of spar was cleaved off to allow a much clearer view, and photographs were taken (ColorView I digital camera for microscopy in combination with AnalySIS 3.2 software). As long as the light came from above, the quality of the views was excellent.

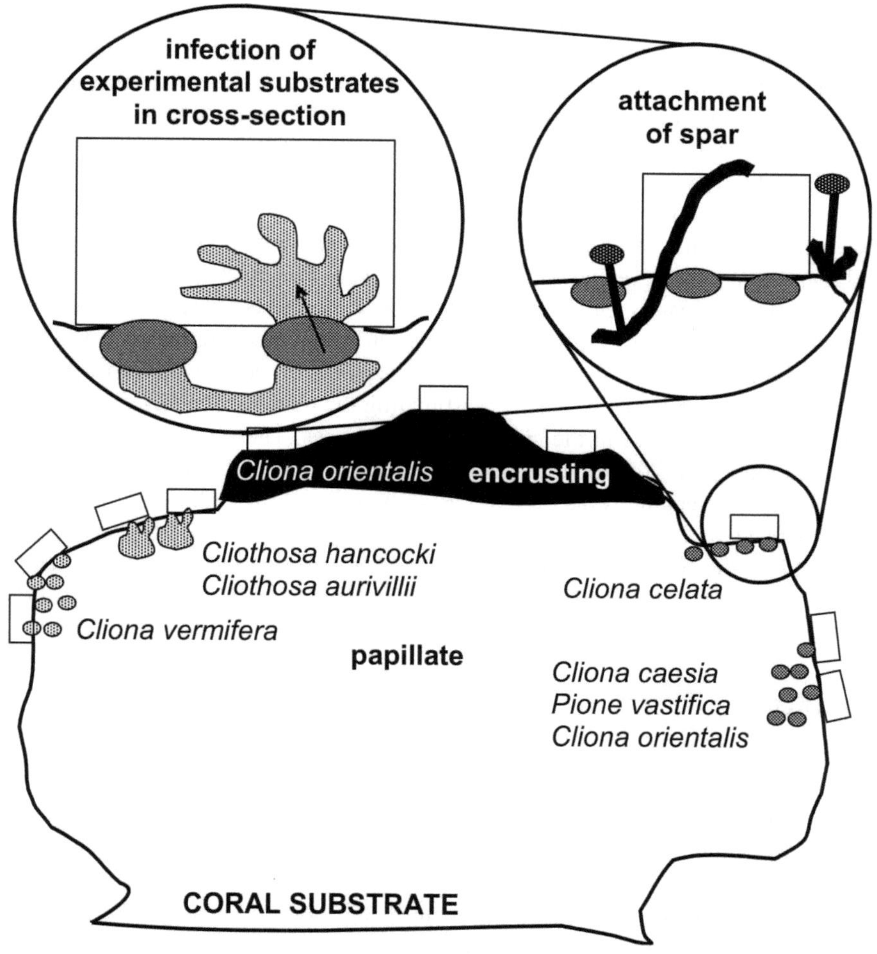

Fig. 2 Principle of subjecting spar rhombohedra to bioeroding sponge erosion. More than 35 rhombohedra were placed onto the surfaces and papillae of various colonies of the most common bioeroding sponge species occurring at the field site by nailing cable ties across each spar. Any sponge that infected a rhombohedron grew upwards into the clean substrate and spread within

Sponge-spar samples were micro-CT-scanned at the Australian Key Centre for Microscopy and Microanalysis at The University of Sydney, using a Skyscan 1172 equipped with a 10 megapixel CCD camera. During scanning the sample was rotated in 0.2° angular increments over 360° total rotation, and an X-ray absorption radiograph was digitally recorded at each step, producing a series of 1800 images. The X-ray source was set at 100 kV and 98 μA. An aluminium filter was placed between the sample and the camera to minimise beam-hardening effects. Beam-hardening occurs because the sample differentially absorbs lower energies from the X-ray spectrum, thereby transmitting a higher distribution of energies and producing artificially bright edges in the final cross-sections (Ramakrishna et al. 2006). Scans were performed at resolutions of5 μm and 17 μm for both *C. celata* and *C. orientalis*. This corresponds to total volumes of 0.125 cm^3 and 4.913 cm^3 of the samples that fit in the field of scanning. A 2 μm scan of *C. orientalis* was also performed in order to resolve further microstructural erosion patterns. This necessitated cutting of one sample down to a 4 x 4 x 10 mm subsection to be able to place it accurately within the narrow field of view visible at this level of magnification.

These projection images were used with cone-beam reconstruction software (Nrecon, Skyscan) resulting in a series of 1000 axial cross-sections with Z-dimensional spacing equal to the XY resolution. The axial cross-sections were imported as an image stack into VGStudiosMax (Volume Graphics), a commercial volume rendering software package. A Gaussian smoothing filter using a 3 x 3 x 3 voxel kernel was applied to reduce image noise. Erosion pores were differentiated from the surrounding spar by visually estimating the range of greyscale values that represent the pore space in the cross-sectional images. These values were used as threshold values in the three-dimensional image histogram to allow erosion pores to be visualised separately, as digital 'casts'. A series of three-dimensional images was produced, and a short animation of the rotating erosion trace of *C. celata* was deposited on the website of Pangaea®, a site for geoscientific and environmental data available to the public (see reference list). Erosion pore volumes were measured in VGStudiosMax using the same threshold values.

Results

Checking the rhombohedra under a dissecting microscope, it became apparent that only *Cliona orientalis* in the encrusting growth form (5 specimens) and *Cliona celata* in the papillate growth form (1 specimen) had infected the optical spar. Therefore, neither statistics nor meaningful taxonomic comparison was possible. The single occurrence of *C. celata* represented only a small invasion, smaller than a 7th of the rhombohedron (Fig. 3; growth of 3.5 months), whereas *C. orientalis* usually evenly infected about ¼ to >½ of a rhombohedron (Fig. 4; growth of 11.5 months). Microscopic observations thus proved to be a good means to obtain a general overview of bioerosion patterns in optical spar. It also facilitated detailed examination of a number of other characters.

Erosion morphology of *C. celata* resembled a bush thinning out toward the margins. Erosion traces ended in slightly juvenating ducts with blunt, rounded tips (Fig. 3A-C) that opened into papillar pores where reaching the surface (Fig. 3D-E). Erosion was structured as short, robust branches that were anastomosing (Fig. 3A-C), rather than in form of rounded erosion chambers and narrow connecting ducts. Overall, the structure resembled what is known for *Entobia megastoma* Fischer, 1868 (rotating image at doi:10.1594/PANGAEA.615181). However, pioneering threads were rare. Slim threads (Fig. 3D) were the exception, and more conical explorative erosion ducts were the common form of pioneering structures (Fig. 3E).

Fig. 3 Spar rhombohedron infected with *Cliona celata* viewed under the dissecting microscope. **A** Entire colony viewed from the edge of the rhombohedron. **B** Close-up of left

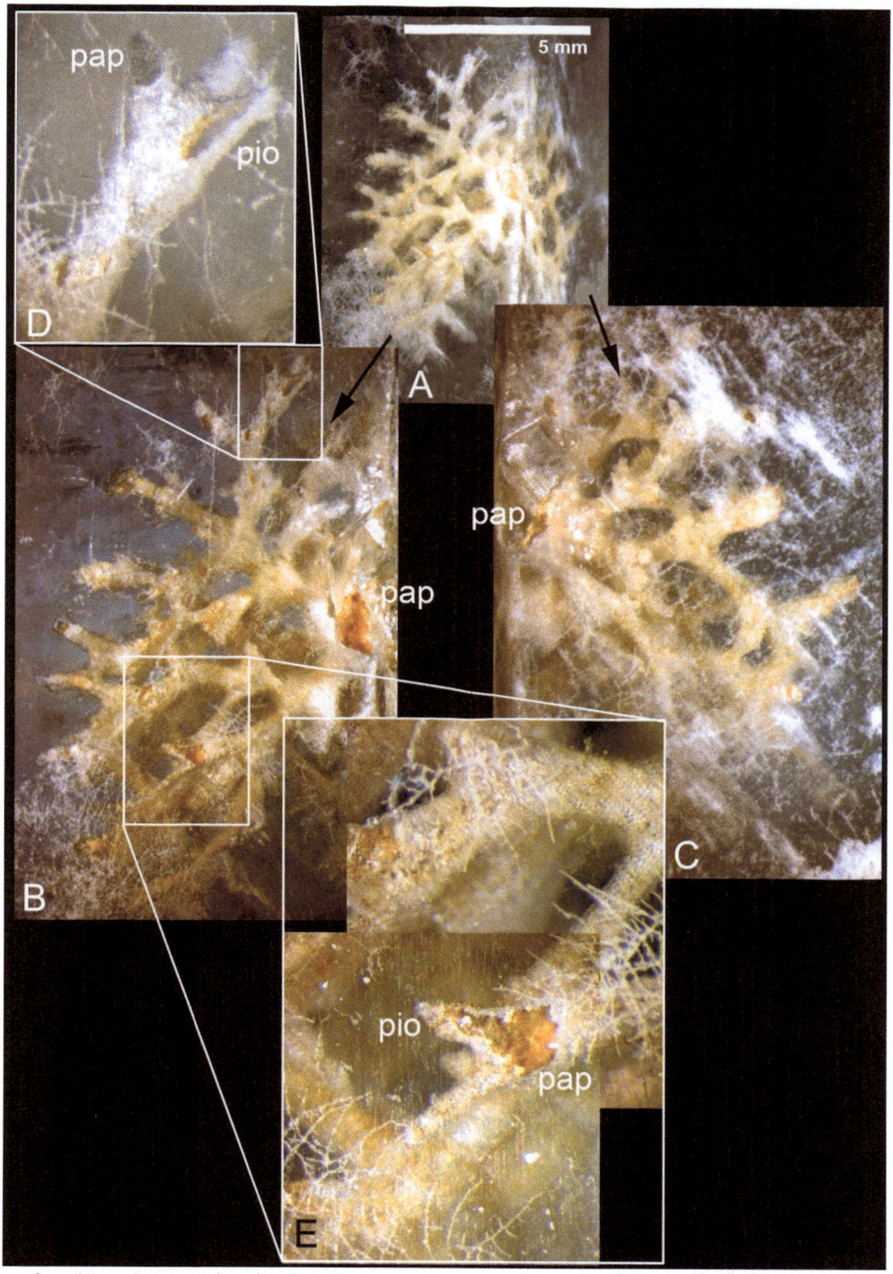

surface (arrow) presenting the clearer view and main papillar opening on the right (pap). Smaller, secondary papillar openings on branch ends on the left. **C** Close-up of right surface (arrow), stronger obscured by microendoliths, with main papillar opening on the left (pap). **D** Close-up with marginal branches, one thread-like pioneering channel (pio), one papillar opening (pap) and shagreen structure caused by sponge scars. **E** Close-up showing shagreen-like patterns, one pioneering element reaching toward the surface (pio) and a papillar opening (pap)

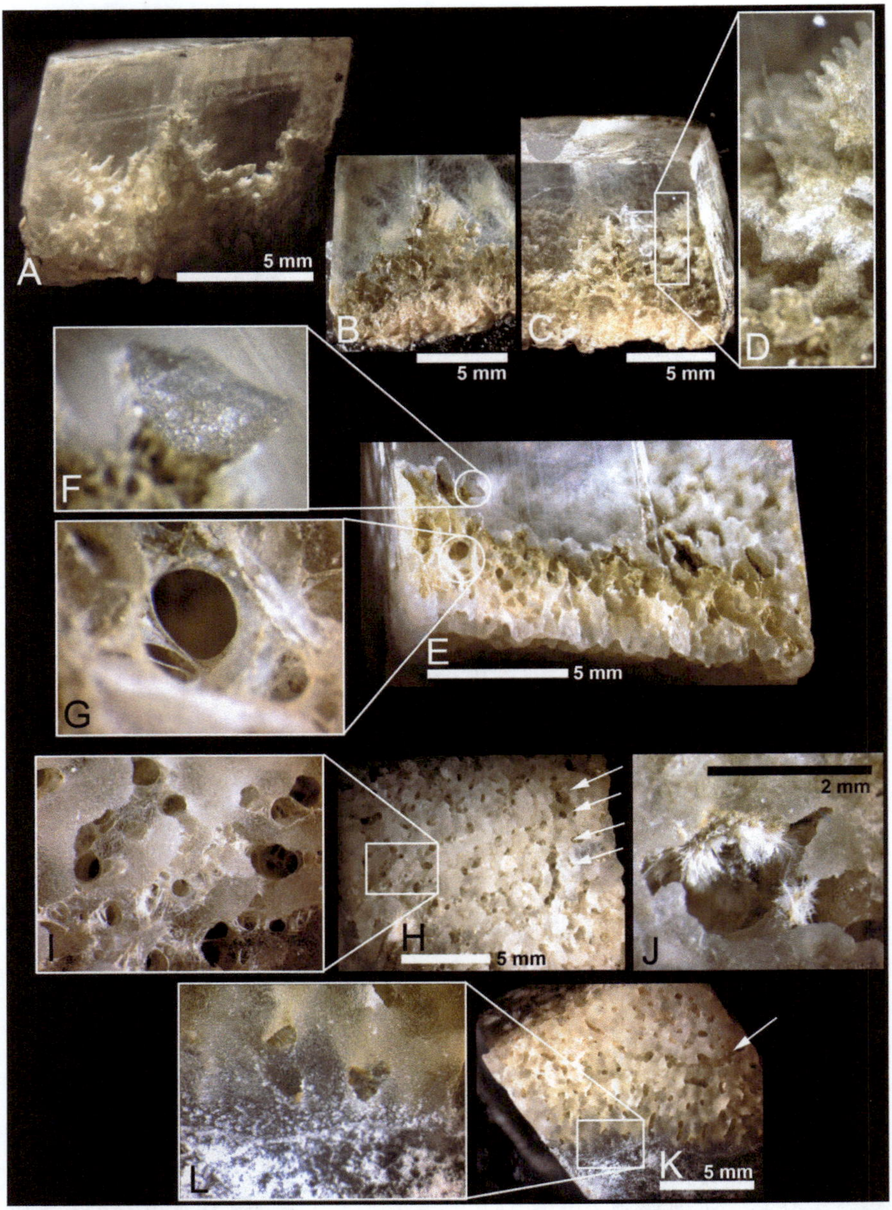

Fig. 4 Spar rhombohedra infected with *Cliona orientalis* viewed under the dissecting microscope. Photographs were taken after cleaving off one layer of spar to remove superficial contaminations by microendoliths, and by shining the light onto the object rather than through it (A-C and E). Infection proceeded either from the bottom up (A-G) or from the surface into the Figure (I-K). **A** Spar 1 with dense erosion traces and short pioneering channels in the marginal regions. **B-D** Spar 2 with dense erosion traces that break the surface in **B** and show different marginal forms in C, enlarged in D. **E** Cross-section through spar 3 (slightly obscured by double refraction in the right half). **F** Close-up of erosion chamber with sponge

Erosion traces of *C. orientalis* were very crowded in central areas, forming a bulky meshwork that did not thin out as strongly as in *C. celata*, but terminated in short, finger-like, slightly conical ducts that were occasionally arranged in bouquets (Fig. 4A-D). Again, no clear separation into chambers and connecting ducts was possible. No pioneering threads were found. The trace appears to be undescribed to date and has not been named. Microscopic erosion structures could be observed through the spar where a layer was cleaved off (Fig. 4E). Scars, chips (Fig. 4F) and even tissue morphology (Fig. 4G) could be studied in clear detail. The dissecting microscope did not provide sufficient magnification to make out the micropatterns within the scars. When rhombohedra were removed from the field, the areas of contact with mother colonies of *C. orientalis* came away without adhering tissue and allowed the investigation of external erosion structures. The entire surface displayed the 'shagreen' pitting caused by the erosion scars and, on a larger scale, a 'pillar-and-well landscape' (Fig. 4H-L). Pores were often arranged in rows radial to the centre of infection, whereas pronounced rings were formed in increasing distance from the infection centre (Fig. 4H, K). Occasionally, openings allowed insights into the arrangement of spicules, being cemented into place by the dried tissue (Fig. 4J). Overall, *C. orientalis* erosion most distinctly differed from that of *C. celata* by the coherently eroded and remodelled surface (Fig. 4K), whereas in *C. celata* only papillae break through. *C. orientalis* removed a surface layer in such a way that a shallow trough was formed with a step against the still uneroded surface area (Fig. 4L).

Occasionally, the effect of the double refraction of the optical spar was confusing, wrongly suggesting sponge erosion patterns to be arranged in a repetitive array of galleries (Figs. 3E; 4D-E). Apart from sponge tissue, the spar rhombohedra were otherwise mainly colonised by green algae (Figs. 3A-E; 4B; M. Gektidis pers. comm.).

Micro-CT scans revealed clear differences between *C. celata* and *C. orientalis* erosion, either as representation of three-dimensional casts (Fig. 5) or as the slices showing the distribution of erosion ducts (Fig. 6). Whereas the single sample of *C. celata* appeared to originate from one centre of infection, likely from one papilla (Figs. 5-6), *C. orientalis* traces suggested that the infection of the new substrate occurred over a larger surface area or spread faster (Figs. 5-6). On average, *C. celata* removed less substrate material than *C. orientalis* (Figs. 5-6). In none of the experimental substrates was a state reached in which the entire rhombohedron was penetrated and a phase of equilibrium was reached. Bioerosion traces of both investigated species could be matched to results obtained from using the dissecting

scars and chips. **G** Close-up of chamber in which tissue has dried into delicate strands and membranes framing pores. **H** Spar 3; contact surface of spar with live sponge in the field, showing a pronounced pillar-and-well structure entirely shagreened by sponge scars (enlarged in **I**) and radial erosion traces with regular erosion pores aligned (arrows). Infection probably started in the vicinity of the inset. **J** Spar 3; erosion chamber with dried bundles of tylostyles in palisade. **K** Spar 4; contact surface of spar with live sponge in the field, with slight step between the original and eroded area. Infection proceeded from uppermost corner and formed concentric grooves (arrow). **L** Close-up with sponge chips accumulated along the step

Fig. 5 Micro-CT images of *Cliona celata* (full rotation; an animation can be viewed at doi:10.1594/PANGAEA.615181) and *Cliona orientalis* (part of spar 3 viewed in different angles). Respective portions of spar rhombohedra were added for bearings and are oriented as in Figure 4. Pioneering structures from Figure 3D and E could not be visualised (boxes and respective enlargements), which is, however, not due to resolution problems.

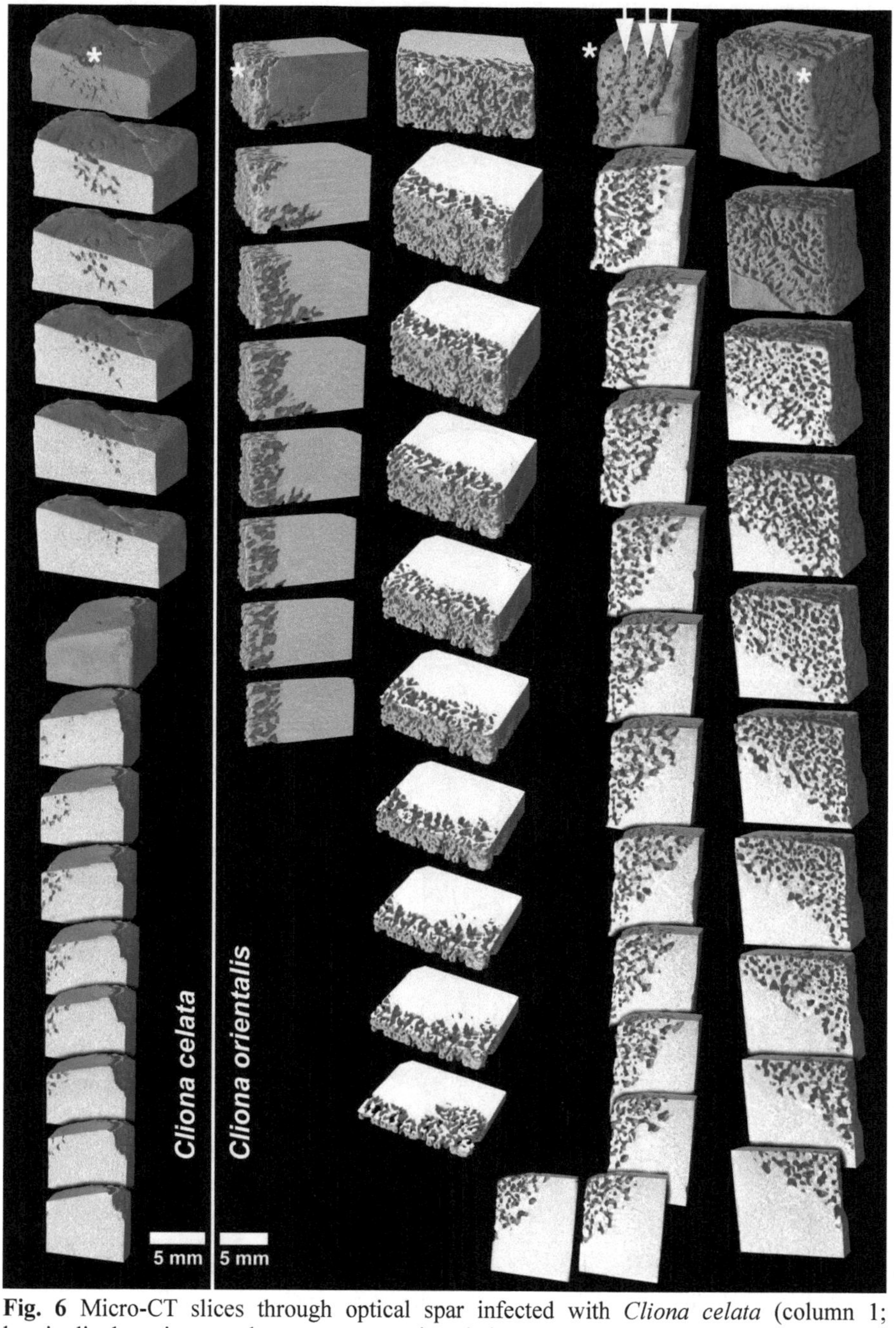

Fig. 6 Micro-CT slices through optical spar infected with *Cliona celata* (column 1; longitudinal sections on the top, cross-sections below) and *Cliona orientalis* (column 2+3: spar 3 with 90° rotation; column 4+5: spar 4 with 90° rotation). Asterisks indicate main origin of sponge infection. Spar 4 clearly displays ring-like grooves spreading on the spar surfaces (arrows, top). Please also note the step on the spar surface in column 5 top.

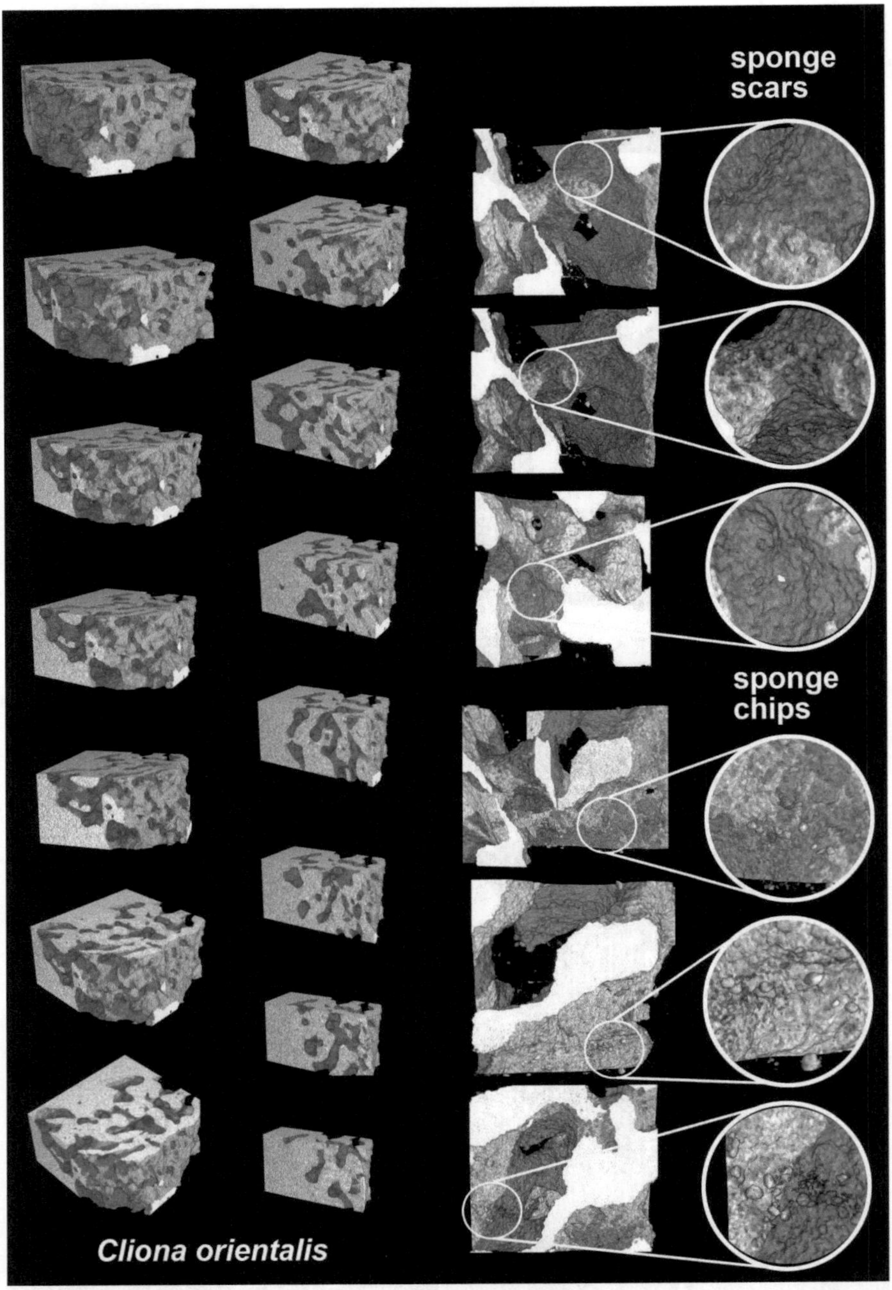

Fig. 7 High-resolution micro-CT scans of erosion patterns of *C. orientalis*. Subsample views on the left in two columns: chamber structure and their connections can be seen, but no pioneering threads. Subsample views on right: top three scans show sponge erosion scars, lower three sponge chips. Both structures are unmistakable proof of sponge erosion, but could not be resolved to a clarity that is possible with SEM

microscope. However, due to sample processing superficial structures could be cut off in micro-CT images (Fig. 5 for *C. celata*), but allowed a clearer view on deeper parts of the rhombohedra (Fig. 5 for both species). *C. orientalis* dense, reticulate erosion meshworks were difficult to visualise in three dimensions (Fig. 5). Micro-CT showed clearly that in contrast to *C. celata, C. orientalis* erosion structure or distribution was comparatively even with only very little difference between central and peripheral regions: *C. celata* removed 9.3% substrate from central vs. 20.9% from marginal regions (10.6 mm^3 total volume), and *C. orientalis* 48.4% vs. 42.3%, respectively. *C. orientalis* thus removed more material per unit area and produced much denser erosion patterns than *C. celata* (Fig. 6). Diameters of ducts also varied slightly between species and between central and marginal parts of the colony: *C. celata* had branch diameters of 0.17 mm vs. 0.14 mm, and *C. orientalis* 0.25 mm vs. 0.23 mm, in central and marginal regions, respectively. Micro-CT scans for *C. orientalis* confirmed the above observation of radial and circular structures of spreading erosion in *C. orientalis* (Fig. 6).

During the present study microstructures could not be resolved to adequate levels. Sponge scars were well visible, however, the microstructural patterns of the scar cups in the images was not clear enough to allow assessment of putative rings or grooves (Fig. 7). Remnant chips were observed in the tissue and erosion chambers (Fig. 7), but again, the quality of resolution remained far lower than what is possible with SEM. Due to their very small size, Skyscan 1172 could not visualise sponge spicules either. Other tomographs may provide better results (e.g. Synchrotron; anonymous reviewer pers. comm.).

Discussion

According to our results micro-CT is a promising tool for investigations of bioerosion on small macro- to microscopic scale (e.g., Tapanila this volume; Bromley et al. this volume). Some of its most important benefits are that it is non-destructive if the sample is small, can be conducted on live organisms (but high X-ray doses make multiple scans detrimental), and allows three-dimensional and animated visualisation of endolithic structures (e.g., Beuck et al. 2007). It is thus very useful in the study of sponge bioerosion traces (*Entobia*). Differences between traces of various sponge species can be easily compared by three-dimensional visualisation in slices or as casts, all of which can be viewed from different angles and in a series of layers (Figs. 5-6). In animations, 'sliding' through the sample and viewing of rotating samples is possible (e.g., doi:10.1594/PANGAEA.615181), allowing a much more detailed understanding of the complex structures of *Entobia* than previously possible. An enormous benefit of micro-CT is the opportunity to measure pore volumes, enabling an exact estimate of sponge bioerosion, which was previously impossible and usually included remnant substrate (e.g., Schönberg 2001, 2003, 2006). However, as micro-CT is here used in combination with radiography, this is best performed on dense substrates such as spar, marble or mollusc shells. Porous substrates such as many coral skeletons may still present problems in not permitting a reliable distinction of existing coral-derived pores from sponge-generated, tissue-filled pores, as the difference in density may be too small in comparison to the substrate (Schönberg 2001). To be able to visualise soft parts of the sample with the used tomograph, voltage would have to be reduced, using lower energy X-rays, necessitating in a smaller sample size. Staining techniques and imaging based on phase contrast may make simultaneous visualisation of denser and softer parts possible when using high energies; however, previous pilot studies with phase contrast radiography were unsatisfactory (Schönberg 2001).

Presently used micro-CT technology did not allow non-destructive viewing of endolithic microstructural patterns in sponge erosion scars (Fig. 7; Calcinai et al. 2003, 2004; Borchiellini et al. 2004). A maximum scanner resolution of 2 µm was not good enough to recognise minute linear structures such as circular grooves or radial ridges in the erosion scars of diameters of 20-80 µm in *C. celata* and 20-60 µm in *C. orientalis* (Schönberg 2000). Similarly, facets on sponge chips of average diameters of 25-35 µm in *C. celata* and 30 µm in *C. orientalis* (Schönberg 2000) could not be resolved. Even though sponge spicules were considerably longer than scars and chips (270 x 7 µm on average for both species; Fig. 4J and Schönberg 2000) and their mineral properties and density should be sufficiently different to the spar, we could not visualise them, probably because they are too thin or mostly aligned in parallel to the substrate, adhering to the walls. Minute traces of algal erosion seen under the dissecting microscope were not detected either. Again, it may be due to the very slim form of the erosion canals algae produce, and other tomographs may reach a better detail. Pioneering canals are relatively large with ca. 500 x 50 µm (Hoeksema 1983 for European *C. celata*) and were found in our sample of *C. celata* (Fig. 5), but were by far not as pronounced as in samples depicted by Wesche et al. (1997). All our results indicate that micro-CT performed with Skyscan 1172 is an excellent tool for comparative and quantitative studies in small volumes, but for qualitative studies on microstructural level SEM is still the better method. Other drawbacks of micro-CT include comparatively high usage costs and a presently restricted access to the technique.

Overall, our data obtained by micro-CT and by using a dissecting microscope to study the clear spar rhombohedra resulted in similar insights. Our trace for *C. celata* is quite typical for this species (i.e., it resembles *Entobia megastoma*), but is slightly more organised and appears more simple, stouter and crowded than what has been previously published from the Netherlands, Italy and Denmark (Bromley and Tendal 1973; Hoeksema 1983; Bromley and D'Alessandro 1989). This may either be due to slight differences caused by substrate properties, to the fact that our *C. celata* was only in initial stages of colonisation, or to the possibility that *C. celata* is composed of several, very similar but different species. Latter appears to be a likely hypothesis, as our specimen is from the Australian Great Barrier Reef, but the other studies mentioned here have been conducted with *C. celata* from Europe. Also, it has been repeatedly discussed that *C. celata* contains more than one species and that specimens with tylostyles and raphides (as in our case) should be separated from the classical *C. celata* sensu Hancock (1849) that only has tylostyles (as for the Dutch, Italian and Danish samples see Rosell and Uriz 2002; Schönberg et al. 2006). This is a controversy, which can probably only be solved by involving molecular analyses (P. Rachello-Dolmen and J. Xavier pers. comm.). Nevertheless, more detailed micro-CT studies could provide additional data.

The *Entobia* produced by *C. orientalis* did not resemble any traces previously known for sponges and is probably new to ichnoscience (R. Bromley pers. comm.). The structure of the dense network of the observed erosion does not match known traces. Most clionaid erosion is multicamerate, the sub-globular chambers being connected with slim canals (e.g., Bromley and D'Alessandro 1989). Traces for *C. orientalis* resemble a dense meshwork of widening and slimming canals and are more similar to condensed erosion traces of *C. celata* than to other clionaids. A formal ichnotaxonomic description of this trace is, however, beyond the scope of the present publication.

The most interesting result of our study is the opportunity micro-CT provides to exactly estimate the volume of the sponge erosion as long as this is conducted on dense substrates. The values obtained for erosion intensity are in very good agreement with results derived from non-visual methods (Rützler 1975; Bergman 1983). When comparing *C. celata* and *C. orientalis* it becomes very obvious that the latter sponge is the more destructive. *C. celata* on the Great Barrier Reef can cause erosion seemingly as intense as that of *C. orientalis* and can form large colonies of several tens of cm^2 surface area. However, the area of *C. orientalis* colonies can reach several m^2, and during the period of our experiment *C. celata* did not nearly reach the effect as *C. orientalis* did. We have only one colony of *C. celata* for comparison, but the fact that *C. orientalis* infected almost every spar placed onto established, encrusting *C. orientalis* colonies in the field, whereas the colonisation success of all other species was almost nil, shows that *C. orientalis* is the better space competitor. This may be due to its symbiotic dinoflagellates of the genus *Symbiodinium* Freudenthal, 1962 that have been shown to affect the speed of tissue growth and erosion (Schönberg and Loh 2005; Schönberg 2006).

In conclusion, we recognise large potential in micro-CT technology for sponge bioerosion studies. Even though results are presently mainly restricted to small macroscopic features, these can be used to advantage when combined with traditional methods such as microscopy, thin sectioning and cast studies.

Acknowledgements

CS would like to thank everybody involved for enabling her to conduct studies at the Centre for Marine Studies (director O. Hoegh-Guldberg, support staff and members of the department), The University of Queensland and field work at Orpheus Island Research Station (OIRS). At OIRS she received very competent assistance from the station staff and dive buddies H. Cambourn, E. Cebrian, W. Loh and R. Olbrich. Very much appreciated was the help for transporting scientific equipment and materials provided by Palm Island Barges and the Heron Island Resort. Significant gratitude goes to funding sources. Funding included a Feodor-Lynen-Fellowship from the Alexander von Humboldt Foundation and the University of Queensland

Chancellor's Fund to CS and support from the Centre for Marine Studies and the Australian Research Council made available by O. Hoegh-Guldberg. GS acknowledges his supervisor A. S. Jones at the Australian Key Centre for Microscopy and Microanalysis and his assistance at the Electron Microscopy Unit of The University of Sydney. R. Sieger helped with getting the animation into the Internet and explained everything about doi-s. Both authors thank the organisers of the 5th International Bioerosion Workshop for allowing related data to be presented as a talk in Erlangen. Last but not least, L. Beuck and two anonymous referees are kindly acknowledged for their valuable comments and suggestions.

References

Becker LC, Reaka-Kudla ML (1997) The use of tomography in assessing bioerosion in corals. Proc 8th Int Coral Reef Symp, Smithson Trop Res Inst, Panama 1996, 2:1819-1824

Bergman KM (1983) The distribution and ecological significance of the boring sponge *Cliona viridis* on the Great Barrier Reef, Australia. MSc Thesis, Geol Dept McMaster Univ, Hamilton / Canada, 69 pp

Beuck L, Vertino A, Stepina E, Karolczak M, Pfannkuche O (2007) Skeletal response of *Lophelia pertusa* (Scleractinia) to bioeroding sponge infestation visualised with micro-computed tomography. Facies 53:157-176

Borchiellini C, Alivon E, Vacelet J (2004) The systematic position of *Alectona* (Porifera, Demospongiae): a tetractinellid sponge. Boll Mus Ist Univ Genova 68:209-217

Bromley RG, D'Alessandro A (1989) Ichnological study of shallow marine endolithic sponges from the Italian coast. Riv Ital Paleont Stratigr 95:279-314

Bromley RG, Tendal OS (1973) Example of substrate competition and phobotropism between two clionid sponges. J Zool London 169:151-155

Bromley RG, Beuck L, Taddei Ruggiero E (this volume) Endolithic sponge versus terebratulid brachiopod, Pleistocene, Italy: accidental symbiosis, bioclaustration, deformity and death. In: Wisshak M, Tapanila L (eds) Current developments in bioerosion. Springer, Berlin, pp 361-368

Bronn HG (1837) Lethaea Geognostica oder Abbildungen und Beschreibungen der für die Gebirgs-Formationen bezeichnendsten Versteinerungen. Schweizerbart, Stuttgart, 1:1-544 and plates

Calcinai B, Arillo A, Cerrano C, Bavestrello G (2003) Taxonomy-related differences in the excavating micro-patterns of boring sponges. J Mar Biol Assoc UK 83:37-39

Calcinai B, Bavestrello G, Cerrano C (2004) Bioerosion micro-patterns as diagnostic characteristics in boring sponges. Boll Mus Ist Univ Genova 68:229-238

Farber L, Tardos G, Michaels J (2003) The use of X-ray tomography to study the porosity and morphology of granules. Powder Technol 132:57-63

Feldkamp LA, Davis LC, Kress JW (1984) Practical cone-beam algorithm. J Optic Soc Amer, Ser A 1:612-619

Fischer MP (1868) Recherches sur les éponges perforantes fossiles. Nuov Arch Mus Hist Nat Paris 4:117-173

Freudenthal GH (1962) Symbiodinium gen. nov. and *Symbiodinium microadriaticum* sp. nov., a zooxanthella: taxonomy, life cycle, and morphology. J Protozool 9:45-52

Grant RE (1826) Notice of a new zoophyte (*Cliona celata*, Gr.) from the Firth of Forth. Edinburgh N Phil J 1:78-81

Hancock A (1849) On the excavating powers of certain sponges belonging to the genus *Cliona*; with descriptions of several new species, and an allied generic form. Ann Mag Nat Hist, Ser 2, 3:321-348

Hancock A (1867) Note on the excavating sponges; with descriptions of four new species. Ann Mag Nat Hist, Ser 3, 19:229-242

Hoeksema BW (1983) Excavation patterns and spiculae dimensions of the boring sponge *Cliona celata* from the SW Netherlands. Senckenbergiana Marit 15:55-85

Itoh M, Shimazu A, Hirata I, Yoshida Y, Shintani H, Okazaki M (2004) Characterization of CO_3Ap-collagen sponges using X-ray high-resolution microtomography. Biomaterials 25:2577-2583

Johnson JY (1899) Notes on some sponges belonging to the Clionidae obtained at Madeira. J Roy Microscop Soc 9:461-463

Lin C, Miller J (1996) Cone beam X-ray microtomography for three-dimensional liberation analysis in the 21st century. Int J Miner Process 47:61-73

Lindgren NG (1897) Beitrag zur Kenntnis der Spongienfauna des Malayischen Archipels und der chinesischen Meere. Zool Anz 20:486-487

PANGAEA (accessed June 25th, 2007) Publishing network for geoscientific and environmental data. http://www.pangaea.de/

Ramakrishna K, Muralidhar K, Munshi P (2006) Beam-hardening in simulated X-ray tomography. Nondestruct Test Eval Int 39:449-457

Rosell D, Uriz M-J (2002) Excavating and endolithic sponge species (Porifera) from the Mediterranean: species descriptions and identification key. Org Diver Evol 2:55-86

Rützler K (1974) The burrowing sponges of Bermuda. Smithson Contr Zool 165:1-32

Rützler K (1975) The role of burrowing sponges in bioerosion. Oecologia 19:203-216

Schönberg CHL (2000) Bioeroding sponges common to the central Australian Great Barrier Reef: description of three new species, two new records, and additions to two previously described species. Senckenbergiana Marit 30:161-221

Schönberg CHL (2001) Estimating the extent of endolithic tissue of a Great Barrier Reef clionid sponge. Senckenbergiana Marit 31:29-39

Schönberg CHL (2003) Substrate effects on the bioeroding demosponge *Cliona orientalis*. 2. Substrate colonisation and tissue growth. Publ Stn Zool Napoli Mar Ecol 24:59-74

Schönberg CHL (2006) Growth and erosion of the zooxanthellate Australian bioeroding sponge *Cliona orientalis* are enhanced in light. In: Suzuki Y, Nakamori T, Hidaka M, Kayanne H, Casareto BE, Nadao K, Yamano H, Tsuchiya M (eds) Proc 10th Int Coral Reef Symp, Okinawa, Japan, pp 166-174

Schönberg CHL, Loh WK (2005) Molecular identity of the unique symbiotic dinoflagellates found in the bioeroding demosponge *Cliona orientalis*. Mar Ecol Prog Ser 299:157-166

Schönberg CHL, Grass S, Heiermann AT (2006) *Cliona minuscula*, sp. nov. (Hadromerida: Clionaidae) and other bioeroding sponges that only contain tylostyles. Zootaxa 1312:1-24

Tapanila L (this volume) The medium is the message: imaging a complex microboring (*Pyrodendrina cupra* igen. n., isp. n.) from the early Paleozoic of Anticosti Island, Canada. In: Wisshak M, Tapanila L (eds) Current developments in bioerosion. Springer, Berlin, pp 123-146

Thiele J (1900) Kieselschwämme von Ternate. 1. Abh Senckenb Natf Ges 25:19-80

Topsent E (1888) Contribution à l'étude des Clionides. Mém Soc Zool France 2:1-165

Topsent E (1905) *Cliothosa seurati*, Clionide nouvelle des Îles Gambier. Bull Mus Hist Nat Paris 2:94-96

Van Geet M, Swennen R, Wevers M (2001) Towards 3-D petrography: application of microfocus computer tomography in geological science. Computer Geosci 27:1091-1099

Wesche SJ, Adlard RD, Hooper JNA (1997) The first incidence of clionid sponges (Porifera) from the Sydney rock oyster *Saccostrea commercialis* (Iredale and Roughley, 1933). Aquaculture 157:173-180

A history of sponge erosion: from past myths and hypotheses to recent approaches

Christine H. L. Schönberg[1]

[1] Institut für Bio- und Umweltwissenschaften, Carl von Ossietzky Universität Oldenburg, 26111 Oldenburg, Germany, (christine.schoenberg@uni-oldenburg.de)

Abstract. Bioeroding sponges have historically been mystical beasts of the sea. Originally, they were classified as half cnidarian, half sponge. It required much time and scientific convincing to confirm their status as active bioeroders. The scientists Hancock and Nasonov were pioneers of their time. They recognised and defended the main concepts: the endolithic organisms are sponges, they produce the cavities they inhabit, and their activities are likely to involve chemical and mechanical processes. However, this viewpoint was often scathingly challenged, and the notion of actively bioeroding sponges was hotly disputed. Once the concept was firmly established in the mid 1900s, related research experienced a significant leap. Notably studies by Pomponi and Hatch on the ultrastructure of etching cells and their associated biochemical properties left no room for doubt: The enzymes carbonic anhydrase and acid phosphatase are associated with sponge bioerosion, providing means for mineral dissolution and digestion of organic components, thus enabling the removal of the so-called sponge chips. However, the exact etching agent remained undetected, and since 1980 research on this phenomenon significantly slowed down. Further studies predominantly focused on environmental control of bioerosion and the taxonomic value of sponge erosion traces. A recent study with electrochemical liquid ion exchange microsensors revisited the question of how sponge bioerosion is achieved and whether acid is involved. This study is presented here to conclude the summary of current knowledge in this context. Microgradients of pH and calcium ions in the tissues of *Cliona celata* and the non-eroding sponge *Halichondria panicea* from the North Sea were compared. The pH slightly decreased with distance into the tissue of *C. celata*, whereas after an initial drop it remained stable in *H. panicea*. Calcium concentrations in *C. celata* increased slightly more with tissue depth than in *H. panicea*. *C. celata* bioerosion may be periodic (few hours cycle) as evidenced by oscillating pH values at the sponge-substrate interface, which may explain micro-terracing in sponge scars. Nevertheless, measured pH changes were too weak to prove beyond doubt that sponge bioerosion employs acid, and further studies will be necessary to confirm present preliminary findings.

Keywords. Bioerosion, mechanism, history, mechanical, chemical, microsensors, *Cliona celata*

M. Wisshak, L. Tapanila (eds.), *Current Developments in Bioerosion*. Erlangen Earth Conference Series, doi: 10.1007/978-3-540-77598-0_9,

Introduction

Historically, sponges have thoroughly puzzled the observer, e.g., with the question whether they are plants or animals (e.g., Dujardin 1838a; Priest 1881a). Within the Porifera, the bioeroding sponges posed an even larger riddle: Were they cnidarians or sponges (e.g., Grant 1826), squatting in vacated worm borings (e.g., Bowerbank 1866) or did they make their own 'digs' (e.g., Hancock 1849a)? If they were actively eroding, did they use chemicals or minute tools facilitating mechanical erosion? If assuming a combination of chemical and mechanical erosion (e.g., Nasonov 1883), what is the compound they employ for substrate dissolution and is it secreted over the entire surface of the sponge or only very locally? All these questions created a multitude of hypotheses that were repeatedly contested, supported, contradicted, defended and re-erected, yet some issues still remain unsolved.

Parts of the available knowledge were previously summarised by Bianconi (1841: 455-463), Topsent (1888: 3-18), Vosmaer (1933: 309-315), Hartman (1958: 89-90), Warburton (1958: 555-556), Cobb (1969: 783-784), Rützler and Rieger (1973: 144-145) and Pomponi (1980: 302-315). Of these, Vosmaer's account is the most detailed and complete. After 1980 the interest in the mechanisms of sponge bioerosion markedly decreased and only recently received renewed attention (e.g., Zundelevich et al. 2007).

As the historical development of the ideas is enlightening and the ensuing debates often enjoyable, a review on contrasting viewpoints on the mechanisms of sponge bioerosion was compiled for the present publication, providing some quotes and original illustrations, and in consequence giving a significantly more complete account than previous authors. This review on sponge bioerosion will conclude with a recent study that employed a modern, frontier technique: pH and calcium liquid ion exchange microsensors. Following one of the historical and yet unsolved questions, the microsensors were used to test whether chemical dissolution in bioeroding sponges involves acid secretion. *Cliona celata* Grant, 1826 was compared to the non-eroding *Halichondria panicea* (Pallas, 1766) in the effort to detect increased acid concentration and calcium carbonate dissolution when approaching the sponge-substrate interface in bioeroding sponges.

Material and methods

The first part of this publication is a review of available knowledge on the mechanisms of sponge bioerosion that is based on literature reaching back to 1802. In this context the Systema Porifera (Hooper and van Soest 2002) and two bibliographies on sponges and marine bioeroders (Vosmaer 1928; Clapp and Kenk 1963) were of significant help for accessing full citation records of historic references. Passages will be cited, which were translated by the present author if they were not in English. Figures of publications that are older than 100 years were adapted and used for the present purpose. Photographs of historic authors were provided by museums or galleries. Missing information on biographies of historic scientists was obtained at the websites of the Oxford Dictionary of National Biography and Wikipedia (see references). New images of sponge bioerosion traces were obtained at The Carl von Ossietzky University Oldenburg with a Leica phase-contrast microscope together with a ColorView digital camera and the software AnalySIS 3.2, viewing slide preparations of sponges obtained 1995-1997 at Orpheus Island, Palm Island Group, central Great Barrier Reef, Australia (Schönberg 2000) and from Topsent material obtained from the Paris National Museum of Natural History.

This publication additionally contains results of one original, previously unpublished study that was conducted at the Max Planck Institute for Marine Microbiology in Bremen, Germany. Calcium and pH microprofiles were measured within the tissue of *Cliona celata* Grant, 1826 and *Halichondria panicea* (Pallas, 1766) to test whether (i) calcium and pH environments differ between bioeroding and non-eroding sponges, (ii) sponges use acid for bioerosion, (iii) increased calcium concentrations resulting from sponge bioerosion can be detected in the tissue, and (iv) whether changes in pH and calcium concentrations are gradual or sudden (latter suggesting localised application of etching agent). A recent technology was employed to pursue these questions: electrochemical liquid ion exchange microsensors (LIX; Kühl and Revsbech 2001). The idea behind this study was that microsensor tips with a diameter of 3-10 μm might be fine enough to reach the area of bioerosion without significant damage to the sponge.

Live samples of *C. celata* and *H. panicea* were taken at Helgoland Island, North Sea by dredging (material service of the Research Station, M. Krüß). Contained in buckets they were transported to Bremen together with water from the sample site. At the Bremen Max Planck Institute for Marine Microbiology, sample material was transferred into a 200 l holding tank and supplemented with artificial seawater mixed to the salinity at the sample site (Meersalz HW Professional, Wiegandt GmbH, Krefeld, Germany). The holding tank was cooled down and maintained at sample site ambient temperatures, the pH was automatically regulated with a CO_2 reactor. Taxonomic identity of the sponges was confirmed by viewing spicule preparations under a Leica phase-contrast microscope with ocular graticule.

LIX microsensors for pH and calcium were hand-made and used within 1-3 days of adding the membranes (Kühl and Revsbech 2001). Each sensor was calibrated and checked with standard solutions before and after each series of measurements and each time data irregularities occurred that might have signified a broken sensor tip. Data acquisition and experimental setup followed the description given in Schönberg et al. (2004), i.e., data were monitored with a voltmeter and recorded with a strip chart recorder and on computer. Sponges were taken from their holding tank one at a time and placed into a 12 x 12 x 22 cm flow chamber for measurements. The flow chamber was connected to a reservoir of 10 l of water for circulation by pump. Flow rates were kept constant. The whole system was temperature-controlled and aerated with a bubble-stone (see fig. 2 in Schönberg et al. 2004).

Microsensors were placed and moved with the aid of a manually-adjusted micromanipulator. Sensors were driven into *C. celata* through papillae emerging from the substrate (Fig. 1) and near oscula in *H. panicea*. Position and progress were monitored with a modified microscope. Microsensors were positioned on the surface of a given sponge, brought to 2 mm above the sponge and then driven towards the sponge and into the tissue, usually in 100 μm increments. At each step three voltammetric measurements were taken at steady state, of which the means were used. Measurements for *H. panicea* were terminated after 2 mm, for *C. celata* when the background was reached, i.e., the calcium carbonate substrate (Fig. 1; as evidenced either by feeling the resistance when using the micromanipulator or by a jump in data signals monitored on a strip chart recorder, see also Schönberg et al. 2004). Finally, voltammetric values were converted to pH and calcium concentrations using the calibrations for each individual sensor.

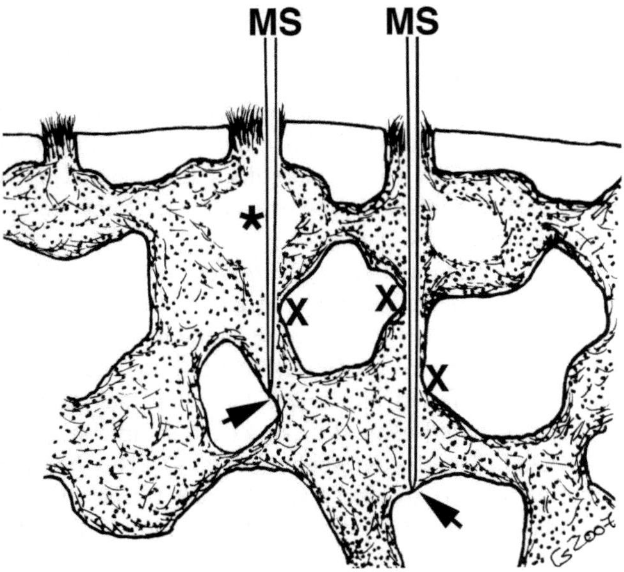

Fig. 1 Application of microsensors in the tissue of *Cliona celata* (in schematic cross-section). Microsensors (MS) were driven into the sponge through the papillae and lowered until they met with resistance from the calcium carbonate substrate (arrows). The measured profiles were probably influenced by heterogeneities in the sponge morphology (* = water in the canal system, likely to reflect external conditions) and the occasional proximity to remaining substrate material (X = conditions may resemble those at the ends of the profiles)

These measurements were repeated three times in a given sponge and conducted with three sponge individuals for *C. celata* and *H. panicea* each. As the initial sample series of *C. celata* did not provide the expected strong evidence in support of the present hypotheses, measurements were repeated with three fresh specimens of *C. celata*, again sampled at Helgoland Island. For a small number of *C. celata* individuals, measurements over longer periods were attempted, i.e., sensors were carefully driven into the sponges until the background (substrate) was reached and remained in place for at least 20 h to ascertain whether sponge erosion followed temporal patterns. Of these data series, four for pH and one for calcium are available.

Literature review

Early 1800s: What is this endolithic organism?

Bosc (1802a: 147-148) was probably the first scientist to describe a bioeroding sponge. In contrast to some of the later authors he immediately recognised it as a sponge. The entire description reads as follows: *'Éponge pézize, Spongia peziza. Jaune; les rameaux sortant des cavités des pierres, sous la forme d'un petit champignon. Voyez la fig. 8 pl. 30 qui la représente de grandeur naturelle. Cette espèce ne se trouve que dans les cavités des pierres et des bois, qui sont dans le mer; elle en remplit l'intérieur, et sort, par leurs orifices, sous la forme d'une petite pézize de couleur jaune. Les graves, que l'où jette dans le mer, à Charleston, observe Bosc, à qui on doit la connaissance de cette espèce, en sont quelquefois si couverts qu'ils ont l'air d'un lichen tuberculeux.'* (Translation: 'Cup fungus sponge [cup fungus gen. *Peziza*], *Spongia peziza*. Yellow; the branches exit from cavities in stones in form of a small mushroom. See fig. 8, pl. 30 that represents the natural size. This species can only be found in cavities of stones and timber that are in the sea [timber: erroneous observation]; it fills the interior and exits through pores in the form of a small cup fungus of yellow colour. The gravel that extends into the sea at Charleston is sometimes so covered [in it] that it has the appearance of a tuberculate lichen, observes Bosc to whom the recognition of this species is due.') Bosc did not try to explain how this sponge came to live in a stone. Judging from the sample site, the colour and the form of the erosion (resembling *Entobia megastoma* (Fischer, 1868); see Fig. 2), we can assume that he described a member of the *Cliona celata* species complex. He should thus receive part of the credit for discovering bioeroding sponges. However, as the description was published in a little-known journal and is very short and vague, it has all but been forgotten.

Fig. 2 *Spongia peziza* (Bosc, 1802:pl. 30, fig. 8) (= *Cliona celata*?). The first known drawing of a bioeroding sponge, slightly modified to remove spots

Only four years later fossil traces of sponge erosion were noted by Parkinson (1808: 76 and pl. 8), who assumed organisms similar to cnidarians to be the tracemakers: *'That the formation of these bodies has been the work of some animal, of a nature similar to the polypes, by which the known zoophytes are formed, cannot, I think, be doubted. But in what genus in the order of zoophytes can they be admitted?'*

Fig. 3 Some of the players in the history of sponge erosion research. **A** The British anatomist Robert Edmont Grant (1793-1874). Photograph provided by C. Valentine, courtesy Porifera Section, Natural History Museum, London. **B** The British zoologist Albany Hancock (1806-1873). Photograph taken by R. B. Bowman, provided by J. Holmes, courtesy of the Natural History Society of Northumbria, Hancock Museum (archive no. NEWHM:1996.H54). **C** The British naturalist John Hogg (1800-1869). **D** The British naturalist and palaeontologist James Scott Bowerbank (1797-1877), well-known for his sponge monographs. C and D © National Portrait Gallery, London [archive nos. P120(50) and P120(1), respectively], provided by H. Trompeteler. **E** American naturalist Prof. Joseph Leidy (1823-1891). Photograph taken 1863. Courtesy of E.

Despite these earlier records, the credit for discovering and naming bioeroding sponges has traditionally been given to Grant (1826: 79; Fig. 3A), who found some *'pulpy matter'* and *'fleshy mass'* that inhabited excavations within a common oyster. He created the new genus *Cliona* in reference to its *'high degree of contractile power'*. Similar to Parkinson (1808), he believed *C. celata* to be *'zoophyte'*, and to be an intermediate form between cnidarians and sponges. In the same year Osler (1826: 364) mentioned *'a fibrous yellow pulp, filling a number of irregular cells, which open freely into each other, and eventually occupy and destroy the whole shell.'* He recognised it as *'a kind of sponge'*. This view was shared by Dujardin (1838a and b) and Michelin (1846), but other authors understood *Cliona* as an alcyonarian, even though they knew sponges (e.g., de Blainville 1834). The latter was puzzling over sponge erosion traces and obtained advice from Nardo (1839) before he formed his opinion. Bianconi (1841: 458) summarised this by stating: '... *e per conseguenza che non trattavasi qui di un alcionario, ma di un vero spongiale.'* (Translation: '...and consequently this was not a matter of an alcyonarian, but of a true sponge.'). Hancock (1848, 1849a and b, 1867) firmly established that the genus *Cliona* should be understood as a sponge.

He elaborated (Hancock 1849a: 322; Fig. 3B): *'I have examined with much care the papillae of* Cliona *when just removed from the sea, but have not succeeded in detecting any polypes. The propriety, nevertheless, of retaining it as a distinct genus would appear evident; for though it undoubtedly possesses many characters in common with* Halichondria, *yet* Cliona *differs widely from it in its habits, and particularly in its contractile power, a quality surely of great importance...'*. Hogg (1851: 192-193; Fig. 3C) again disputed that the organisms in question are really sponges: *'...these perforating 'sponges' do not seem to be true sponges – merely species of* Cliona *– a genus, according to the accurate accounts of Dr. Grant, Dr. Johnston, De Blainville, &c., belonging to the class Zoophytes, and which is described by them as a polype furnished with about eight short tentaculata.'* But due to the minute studies of Hancock (1849a; see above), Hogg failed to find acceptance.

Grant (1826) remains the accepted founder of the group despite Bosc's (1802a) earlier account of *Spongia peziza* and Nardo's (1839) and Bianconi's (1841) insistence it should be called *Vioa*, because Grant did not recognise his *Cliona* as

Mathias, © Academy of Natural Sciences, Ewell Sale Stewart Library (archive details: coll. 9, Leidy seated at desk). **F** The Russian-Soviet researcher Prof. Nicolai Victorovich Nasonov (1855-1939) at the age of 80 years. Photograph taken 1934, provided by C. Eckert, courtesy of the Historical Image and Script Collections at the Museum für Naturkunde, Humboldt University Berlin (MfN HUB, HBSB, Zool. Mus., B I/1787). **G** French spongiologist Prof. Emile Topsent (1862-1951) who contributed significantly to sponge taxonomy and systematics. Photograph provided by C. Valentine, courtesy Porifera Section, Natural History Museum, London. **H** Dutch spongiologist and taxonomist Prof. Gualtherus Carel Jacob Vosmaer (1854-1916), best known for his Bay of Naples monograph (Vosmaer 1933). Photograph by Vanderstok, Leiden, provided by C. Eckert, courtesy of the Historical Image and Script Collections at the Museum für Naturkunde, Humboldt University Berlin (MfN HUB, HBSB, Zool. Mus., B I/2078). In the background an oyster shell from Houlgate, Côte Fleurie, Normandie, France, riddled with bioerosion traces produced by *Cliona celata*. Scale in cm

a sponge. Since the description of *Cliona celata* a wave of nominations of new species of bioeroding sponges was triggered and this curious group of poriferans attracted the interest of many scientists. This quickly led to the next question: if the sponges are found inside substrates harder than their own tissue, do they actively erode?

Mid to late 1800s: Does the sponge make the hole it occupies or is it a squatter?

In the mid 1800s the intense discussion erupted: does the sponge itself produce cavities or does it invade spaces created by other organisms (e.g., Ehrenberg 1834: 62): *'Ostreas perforans. An perforatas occupans?'* (Translation: Perforating in oysters. Or occupying perforations?) Several authors assumed at an early stage that the sponges were able to erode, e.g., Osler (1826), Nardo (1839), Bianconi (1841) and Lereboullet (1841: 131): *'Il est probable que les galeries tapissées par l'Éponge ont été creusées par cette substance animale elle-même'.* (Translation: 'It is likely that the chambers that are coated with the sponge were eroded by the animal substance itself.')

Others claimed that the sponges invaded spaces created by various other eroding organisms, especially by worms. Grant (1826: 78 and 81) set precedence by reasoning that the cavities were *'perforated by some marine worms'* and only inhabited by the sponge as a nestler. But he allowed that perhaps *'the sharp siliceous specula, and constant currents of its papillae'* in the organism provided some means *'in forming or enlarging the inhabitation of this zoophyte'.* Duvernoy (1839: 685) could not imagine how the sponges should be able to erode by themselves: *'Ici les moyens mécanique sont évidemment nuls. Il n'y a que les moyens chimiques qui puissent être mis en jeu par un organisme privé de toute espèce de force motrice apparente.'* (Translation: 'Here, the mechanic means are obviously nil. Only chemical means can be employed by an organism entirely withdrawn from any species by an obvious driving force.') Dujardin (1838b: 5) thought the cavities were made by sabellids: *'...occupant dans les pierres calcaires, des trous qu'elle n'a pas creusés, mais qui sont dus à une Sabelle trouvée souvent dans les mêmes pierres...'* (Translation: '... occupies calcareous stones and holes that it did not erode, but which are due to a sabellid often found in the same stones...').

Bowerbank (1866: 215-217; Fig. 3D) passionately held the view that the sponges nestled in excavations left by annelids and explained in detail: '*...lithodomus Annelids which probably excavated the tortuous passages which have subsequently been taken possession of by the sponge.* [...] *There is no British sponge regarding which there has existed a greater diversity of opinion than the present subject of investigation, and this is, perhaps, in a great measure due to the singularity of its habit, in selection the perforations of lithodomus annelids, and other marine animals as its habitation, and very few oyster or other shells in which such perforations exist, are free from this parasitical sponge, but it does not confine itself to the sinuous canals thus formed; if they happen to open into the bases of large parasitic Balani attached to the shell, the whole of the interiour of the Balani become coated with the sponge, and in the excavated stones of Tenby it frequently entirely fills the smaller* *cavities, or completely coats the larger one made by the lithodomus molluscs so abundant in the surfaces of the limestone rocks between high and low-water marks in those districts.'*

Waller (1871: 269) was utterly mystified how people could believe in bioeroding sponges: *'The errors and the dreams of science have been numerous. We have had the flints on the upper chalk formation attempted to be accounted for as the coprolites of whales; the* 16th *and* 17th *century gives us the wonderful story of the goose which developed from barnacles, and now we have a 'burrowing or boring sponge.'* [...] *It is extraordinary how completely it has been accepted, and how widely disseminated.'* Waller (1871, 1881) agreed with Bowerbank and insisted that bioeroding sponges are nestlers. He pointed out encrusting sponges that live in a similar situation, but do not erode, claiming that it is extremely unlikely that only *Cliona* should be able to dissolve carbonate materials. Regrettably he overlooked that he had already minutely studied more than one species of bioeroding sponges, not noticing the difference between the spicules: *C. celata* with subterminal tylostyles (Waller 1871) and an *Aka* with smooth oxeas (Waller 1881). He observed the surprising accuracy of the scalloped, excavated walls, marvelled at the precision of the erosion scars and noted that *'a thin membrane* [of sponge tissue] *overlies the work described, having upon it its spicules'* and accurately portrayed the papillar spicule palisades (Waller 1881: 257; Fig 4).

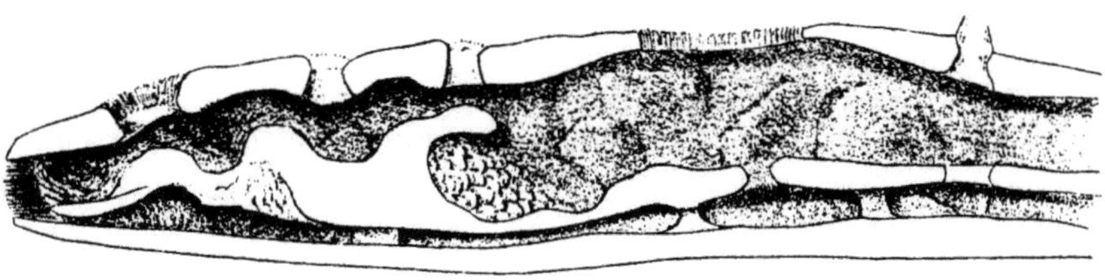

Fig. 4 Sponge traces *Entobia* of *Cliona celata* as observed by Waller (1871: pl. 20, fig. 1). Despite accurately representing the morphology of the erosion and the arrangement of the sponge tissue in the cavities, Waller believed the chambers to be the product of worms

Nevertheless, he was unable to see spicules arranged in a way that would suggest mechanical erosion. Waller considered and rejected three possible pathways of erosion:

1. Acid – he regarded the necessary self-protection of the sponge as unlikely.
2. Wear by soft tissue – the observed sharp edges of the sponge scars did not quite agree with this hypothesis.
3. Wear by hard tissue – he found that the spicules were arranged in parallel to the substrate, and he could not find linear or angular traces of scraping that would reflect the grinding of hard materials.

Waller also thought this ineffective (1871: 272): *'To my mind this process would be as effectual as mining the tunnel of Mont Cenis with darning needles.'* In the same publication (1871: 269) he lectured: *'There is nothing more commonly witnessed in historical literature, or in the records of science than the persistence of error. This is especially the case if it contains something of the romantic or of the marvellous. A writer of credit puts forward a statement; he may, or he may not, give an authority for it: let it be accepted without dispute, it gets copied by one writer after another and passes as an established feat. At length an inquisitive eye, by chance perhaps, happens to refer to the authorities, and it is found, either that they are inconclusive, or, as it has often happened, absolutely disprove the statement which has long been accredited. In fact, though it is not flattering, man has much of a sheep in his composition, and likes to follow a bell-weather.'* Overall, Waller blamed annelid larvae and rejected the concept of bioerosion by sponges (1881: 265): *'It is my hope that the subject may now be definitely settled, for, in friendly warfare, both victor and vanquished are gainers equally in the contest, when the struggle is not for victory but for truth.'*

Hancock (1848, 1849a, b, 1867) was just as unsure about the exact means of erosion as were the above authors, but presented observations that clearly supported that the sponges were able to actively erode calcium carbonate. He hotly and sometimes scathingly defended their excavating powers against the views of other scientists. His arguments were (1849a, 1867):

1. Healthy sponges always fill the entire excavation.
2. Their traces were observed in different materials and geological periods.
3. Observations involved different growth stages and a variety of sponge species that share common, but complicated, partly species-specific erosion traces (sponge scars = 'shagreen', globular and connected cavities = 'lobes' (Fig. 5), a specific number of papillar openings).
4. Sponge erosion clearly differs from worm traces, latter are tubular and have smooth surfaces.
5. Why should sponges not be able to erode hard substrate if the ability has been accepted for other groups?
6. Sponges have chips in their tissues that they have obviously removed from the substrate.

Hancock (1867: 232) challenged: *'But here, again, we are unfortunately at issue with Dr. Bowerbank* [1866], *who asserts that these burrows are made by 'lithodomous Annelids'* [...] *the sponge is lodged, being moulded, in fact, in worm-burrows.* [...] *It may* [...] *be asked how it is that, while* C. celata *is found in vast abundance on our coasts* [...] *the worm or annelid assumed to have made the cavities has never yet been determined.'* And (1867: 234): *'...or that the worm made the openings purposely, in strict accordance with the requirements of the sponge that in some future day might take up its abode in the deserted excavation?'* He observed that it was more likely to find bioeroding sponges in larger oyster shells than in smaller ones, the sponge sometimes conquering the oyster with gregarious growth and simultaneously weakening its own dwelling (Hancock 1849a: 323): *'...the whole system of elaborately wrought chambers becoming exposed soon gives way, and* Cliona, *Sampson-like perishes amidst the ruin produced by its own energy.'* He resolutely stated (1849a: 327): *'...it would seem impossible to arrive at any other conclusion.'*

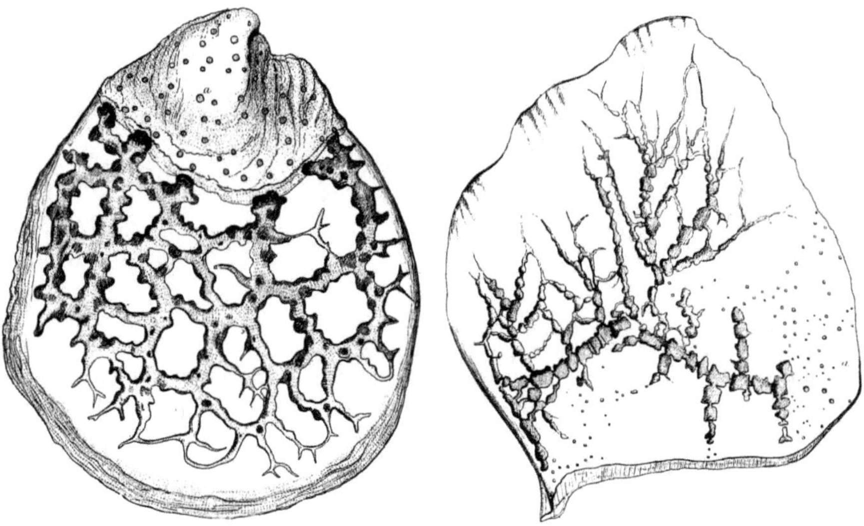

Fig. 5 Sponge traces *Entobia* of *Cliona celata* (left) and *Cliona corallinoides* (= *Pione vastifica*; right) according to Hancock (1849a: pl. 13, fig. 3 and pl. 15, fig. 1). He recognised that even though sponge traces differ between species, the underlying principles are the same

Schmidt (1862: 77) shared Hancock's opinion but reserved the possibility that the sponges may also benefit from existing pores: *'Daß die Vioen sich vorzugsweise selbst ihre Wohnhöhlen bilden, ist bei manchen Arten schon aus der regelmäßigen Stellung der Auströmungslöcher in Reihen ersichtlich. Jedoch scheinen manche Arten auch schon vorhandene Bohrlöcher zu benutzen,...'*. (Translation: 'That the vioas preferentially make their own dwelling cavities is visible in some species that have a regular arrangement of the exhalant pores in rows. However, some species also appear to utilise already existing erosion cavities,...')

After first supporting Bowerbank's view, Priest (1881a, b: 270-271) convinced himself of the sponges' activities: *'Even though Dr. Bowerbank himself tells us that very few oyster, or other shells in which the perforations exist, are free from this Sponge, and yet he is trying hard to prove that it is only as a parasite that it exists in them. It seems strange to me that, if such is the case, the same species or class of Sponge should almost always be found to occupy those cavities.* [...] *As I said before, I certainly agreed with Dr. Bowerbank until I came across shells where the cavities ramified right and left into such fine processes, and those cavities being filled with the Sponge I could not at all make the burrowing of Annelids agree with them'*. He may also be the first to provide evidence for variability in the diameters of sponge scars (Fig. 6). Hyatt's (1882: 83) views were very similar. He explicitly pointed out the differences between sponge and worm erosion traces and argued: '[...] *large annelids* [...] *are, from the character of their jaws, incapable of cutting such depressions, and moreover, from their size, unable even to enter the burrows.'*

Leidy (1856; Fig. 3E) also believed that the sponges were capable of bioerosion. He summarised some of the opposing views, but did not make clear whose opinion he shared (Leidy 1889). In the end, Hancock won through and was supported by

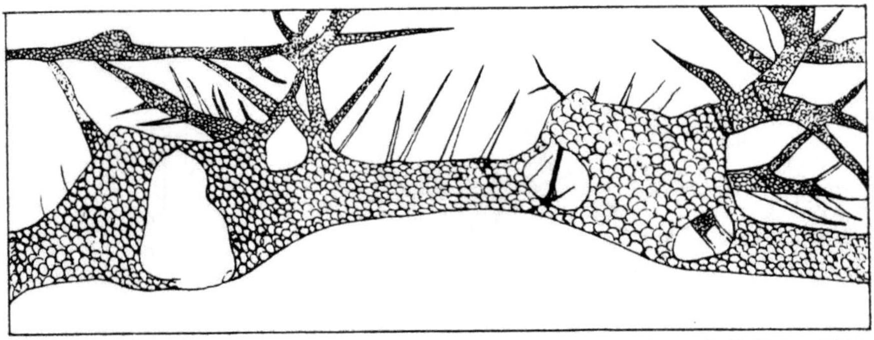

Fig. 6 Erosion traces of an unknown sponge (trace: *Entobia*) in a *Haliotis* shell (Priest 1881a: pl. 17, fig. 6). The author fairly accurately represented main chambers with pitting by sponge scars, connecting canals and very fine pioneering canals. It is interesting to note that even at this early stage, Priest recognised the different diameters of sponge scars in central areas of the erosion (larger) and in pioneering areas (smaller)

later studies. Nasonov's (1881, 1883; Fig. 3F) outstanding, detailed and accurate observations on the settling of clionaid larvae described very clearly that cell extensions were in contact with circular fissures and that they removed ellipsoidal bodies from the substrate that were expelled by the sponge (Fig. 7). Topsent's (Fig. 3G) excellent PhD thesis (1888) added to this evidence and pictured sponge chips more accurately than Nasonov. Topsent recognised that chips removed from surfaces already scalloped by sponge erosion have concave facettes on their upper surfaces (Fig. 8A). He also noted the difference in the sponge scars of the main erosion chambers and pioneer areas (Fig. 8B).

By the end of the 19th century it was largely established that the sponges were actively eroding calcium carbonate substrates (e.g., Hinde 1883; Nasonov 1883; Topsent 1888; Keller 1891; Topsent 1900). However, this decision increasingly

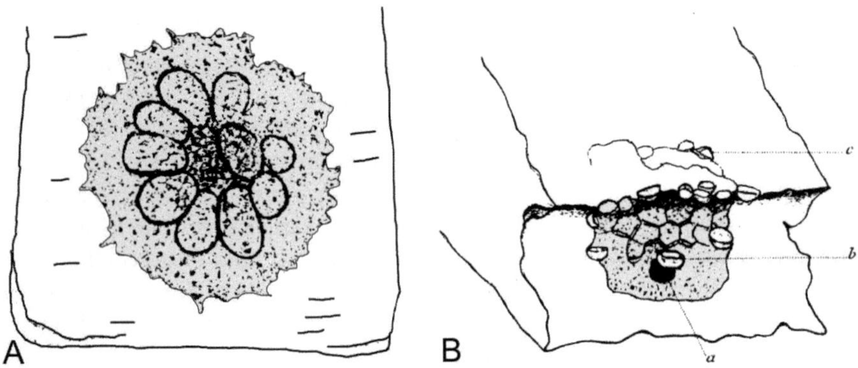

Fig. 7 Initial erosion traces of *Clione stationis* (Nasonov, 1883: pl. 18, figs. 4-5) (= *Pione vastifica*), the figure was manually enhanced. **A** A larva shortly after settling on a transparent calcium carbonate fragment. It can be clearly seen that at this early stage of development fissures for the removal of sponge chips are already cut. **B** Similar situation with a sponge scar (a) from which a sponge chip is being removed (b). On the upper surface sponge tissue can be seen through the transparent material, showing filipods and threads (c) that reach into freshly cut fissures. At this stage the sponge does not yet have spicules

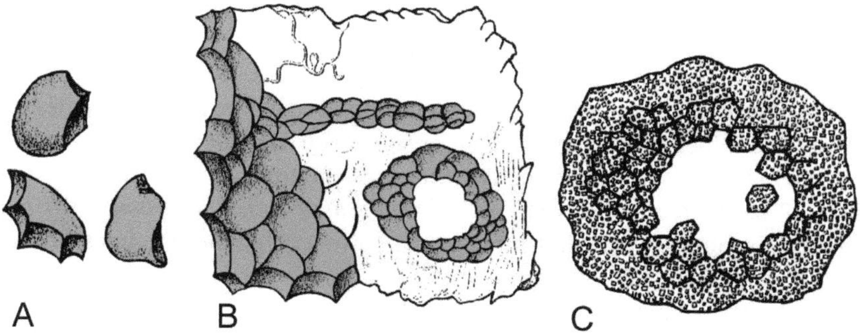

Fig. 8 Sponge erosion as illustrated by Topsent (1888: pl. 3, fig. 13 and pl. 4, figs. 1, 3), the figure was manually enhanced. **A** Sponge chips (grey) figured with smooth, convex basal sides and concave-facetted sides nearer the sponge tissue. **B-C** Sponge scars (grey) in calcium carbonate substrate (B) and conchiolin (C). B Note the difference in diameter in established areas of erosion (left) and pioneering regions (extending towards right and circular area)

provoked the question: how? Nasonov (1887: 362) suggested that different species may have different ways to erode and noted with sorrow: *'Malgré l'intérêt que présente cette particularité au point de vue scientifique comme au point de vue pratique biologique, elle n'a pas étudié avec tous le soin désirable'*. (Translation: 'Despite the interest of this peculiarity, both in scientific and practical biological context, it has not been studied with the desired care.') But more research was about to be published.

Mid 1800s to early 2000s: How does the sponge make the hole?

Accepting that sponges are able to erode, it still could not be explained how they achieve this. Several possibilities were hotly discussed: mechanical erosion by scraping of spicules or other particles, by wear with the soft tissue, including the removal of fragments by traction, erosion with water jets created by the pumping, chemical erosion through the action of acid or other agents and any combination of the above. Early opinions tended to favour mechanical erosion and occasionally mused over possibilities of additional means. Bate (1849: 74) proposed: *'Cliona first obtains a footing in some crevice, where it developes itself so as to penetrate the whole fabric, destroying the shell or pebble by simply fulfilling the condition of its existence, which is by pouring its currents in a given direction, until a passage be broken through by the corroding power of the carbonic acid in those currents.'* Hancock (1848: 243) supported mechanical erosion: *'...there can be no doubt that* [the cavities] *are the work of this creature, most probably aided by its siliceous spicula, which penetrating the surface* [have] *the character of rasping-paper.* [...] *The excavations* [...] *can only be effected by the surface of the sponge, aided either by minute mechanical instruments in connexion with it, or by a solvent'*. He discounted the possibility that water jets are used to dig into the substrate. If a chemical was used, he reasoned it should be exuded from the entire surface of *'this humble animal'* with its *'extreme simplicity of the organic structure'* (Hancock 1849a: 328). But he did not think it could be acid, doubting that the sharp-edged, replicate structures of the sponge scars could have been the product of etching. He reflected that the occurrence of particles (sponge chips) in the tissue was evidence for mechanical means of erosion, the tool being either the sponge spicules or large acid-resistant crystals he noticed on preparations (Hancock 1849a, 1867). However, Nasonov (1881, 1883) found sponge erosion to proceed before sponge spicules were formed. The hypothesis of mechanical erosion with crystals was also supported by Fischer (1868), but the structures represented an artefact: According to Waller (1871: 273), Hancock observed *'only disintegrated cells of carbonate of lime'* (remains of sponge chips?). It is more likely, however, that the structures he found were the yellow crystals that sometimes form in saturated nitric acid digestions – Leidy (1889) had thus difficulties to find them.

Letellier (1894) presented a new possibility for mechanical erosion and claimed that the sponge created traction with tissue contractions to break off small particles. Topsent (1894: 11) discussed the feasibility of this idea: *'La question est évidemment de savoir si les fragments de calcaire ou de conchyoline que détache l'Èponge pour s'enfoncer dans son support sont extirpés par simple traction ou, au contraire, découpés et façonnés.'* (Translation: 'The question is obviously to know if the fragments of limestone or conchiolin that are detached as the sponge subsides into its support are torn out by simple traction or, in contrast, cut out and formed.') He further pointed out politely that the observed minute fragments are unlikely the outcome of tearing by the entire body of tissue. Then, becoming less polite, he suggested that Letellier (1894) had based his idea on a hypothesis without any direct

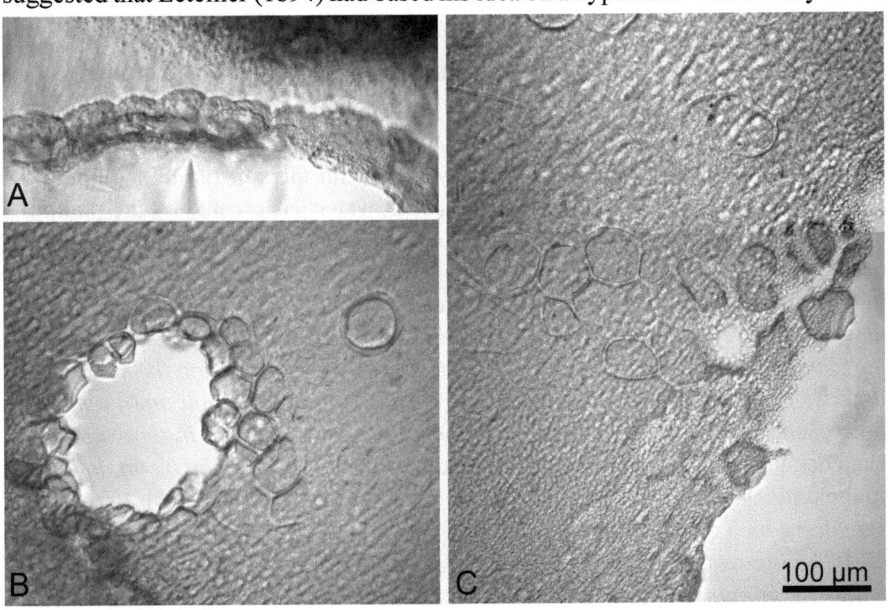

Fig. 9 Bioerosion traces in the conchiolin of a bivalve, produced by an unknown *Pione* species (Topsent sample from the Paris Museum). **A** Edge of an area, where a thick layer of conchiolin was entirely penetrated. **B** Small hole in conchiolin with partially freed chips and beginning etching grooves. Small chip diameters suggest that this is a marginal, pioneering erosion area. **C** Partial etching grooves with larger diameters in a more central area of the sample. Scale bar applies for A-C

observations supporting it. Topsent was unable to offer a better explanation, but he argued that the force needed to tear fragments out of nacre should be too large for tissue acting in parallel to its surface. Finally he firmly declared that the very specifically shaped sponge chips could not be the product of a general tearing action that would usually result in larger, more irregularly formed particles corresponding to the layering characteristics of a given shell. Also, sponge chips always have the same form, regardless of the material they stem from, calcareous material or conchiolin (Figs. 8-9).

The use of acid was repeatedly proposed, sometimes as the only means of erosion (e.g., Parfitt 1871). Several scientists tried to detect acid secretion in bioeroding sponges, but failed (Table 1). Some of them reflected that the reason might be very local application of the etching agent (Cotte 1902; Hatch 1980). Warburton (1958: 561-562) mused: '*... an acid, or some other solvent, is secreted in minute quantities by the threads of cytoplasm* [...] *carving out blocks with acid dispensed from a pipette.*' Other scientists suggested that instead of acid other chemicals might be used (Pomponi 1979a; Hatch 1980; Pomponi 1980). Nasonov (1883: 300) reasoned that a combined chemical-mechanical erosion was most likely: '*Auf diese Weise übt in diesem Falle der Schwamm zugleich eine chemische und mechanische Zerstörungswirkung aus, wodurch er bedeutend viel weniger Kraft aufwendet. Anstatt jedes einzelne Kalkpartikelchen gänzlich aufzulösen, löst er eine dünne Kalkschicht auf, die der konvexen Oberfläche des Partikelchens entspricht.*' (Translation: 'In this manner the sponge simultaneously affects chemical and mechanic destruction and in consequence needs to spend significantly less energy. Instead of entirely dissolving each separate little chalk particle it etches a thin layer of chalk that matches the convex surface of the little particle.') He thought that the etching agent is probably acid.

In 1924 Nasonov described how larvae of *Clione stationis* (Nasonov, 1883) (= *Pione vastifica*) settle on oyster shells and actively penetrate the surface of the carbonate, eventually proceeding inwards, the scars then covering all internal walls of the endolithic cavities. He again concluded that sponge erosion is partially chemical as it can dissolve calcium carbonate and conchiolin. Other authors agreed (e.g., Vosmaer 1933; Fig. 3H; Old 1942), but the agent still remained unknown (e.g., Otter 1937).

Table 1 Previous attempts to ascertain the existence of acid as etching agent in bioeroding sponges

Experimental approach	Author	Year
failed to detect acid when applying sponge extracts to litmus paper	Hancock	1849a
failed to detect acid, but believed the sponge obtains it from the seawater	Parfitt	1871
failed to detect a lowered pH in the sponge culture water using indicator solutions	Cotte	1902
detected an increase in dissolved calcium carbonate	Old	1942
failed to detect a lowered pH, could not demonstrate increase in dissolved calcium carbonate, argumented that it may be applied very locally	Warburton	1958
observed flocculent material of unknown nature	Rützler and Rieger	1973

Roughly between 1970 and 1980 the most significant experimental results with regards to sponge bioerosion were generated. In 1969, Cobb provided excellent summaries and new data and urged not to forget the earlier findings of particles cut out by the sponges that indicated that part of the erosion must be mechanical by removal of these fragments (Nasonov 1881, 1883; Cotte 1902; Warburton 1958). He experimentally confirmed Nasonov's (1883, 1924) results by using explants of *Cliona celata*, a method established by Warburton (1958). Cobb (1969: 785) detected about 1 µm wide, cup-shaped fissures around unfinished sponge chips. After staining the tissue next to them and finding it to reach into the fissures, he proposed amoebocytes with *'prominent nucleus and nucleolus and numerous basophilic granules'* to be responsible for the etchings. Following Pomponi 1979b, Cobb probably chose the wrong cell type, i.e., osmiophilic cells that are closely associated with etching cells; see also Cotte (1902) for *Pione vastifica* (Hancock, 1849) and Rützler and Rieger (1973) for *Pione lampa* (de Laubenfels, 1950). Cobb (1969) concluded that one cell produced one scar. Several cells together remove the resulting chip of calcium carbonate, a chip with the characteristic shape of a smooth convex lower half and a facetted upper side. He noted that calcite and aragonite substrates with different organic content as well as conchiolin layers in bivalve shells and *Mytilus* Linnaeus, 1758 periostracum were eroded in the same manner (Figs. 8-9), but that possibly the chip-size varied with substrate. Agreeing with Cotte (1902), Warburton (1958) and Vosmaer (1933), Cobb further reasoned that the ability to dissolve mineral and organic materials suggests the secretion of acid and an additional enzyme, but he could not be more precise. Cobb (1969, 1975) was the first to coin the expression 'sponge chip' and to describe the terraced microstructure of concentric rings within *Cliona* erosion scars, which he ascribed to intermittent activity of the etching cells and which differed between substrate types.

Cobb's work was supported by the excellent studies of Rützler and Rieger (1973) on *Pione lampa*. However, whereas Cobb's (1969) data suggested that in *Cliona celata* one chip is excavated by a single cell etching along its edges, *Pione lampa* chips are made by several cells contributing with a web of filopods that form a basket to lift out the chip. This is an observation, which was already made by Warburton (1958: 560): *'network of thread-like interconnections and pseudopodia, often 50 µ or more long'*. Etching cells undergo cytoplasmolysis during very localised etching and the removal of chips and produce a *'flocculent secretory product'* that was not identified (Rützler and Rieger 1973: 158). However, Rützler and Rieger allowed (1973: 159): *'it is quite possible that the cell types, organelles, and secretory products involved vary among species'*, which appears to be the case when their observations from *P. lampa* are compared to those from *C. celata* (Cobb 1969).

Despite ongoing disagreements over the exact mechanisms, research came full cycle since Nasonov (1881, 1883) and the basic concept was finally sustained by the late 1970s: *'The most tenable mechanism is the combination of a localized chemical dissolution coupled with mechanical dislodging of fragments of the substratum and their subsequent transport out of the sponge galleries* [...] *the mechanism of excavation involves a localized modification of the calcium carbonate solubility equilibrium'* (Hatch 1980: 135-136). Pomponi (1977, 1979a-c) conducted excellent research on the ultrastructure of the etching cells of 11 different species of bioeroding sponges. The etching cells of these species have a diameter of 10 x 5 μm and processes of 0.25 x 3-20 μm with a high degree of coordination (Pomponi 1977, 1979b). They contain *'an anucleolate nucleolus, Golgi, mitochondria, well-developed rough endoplasmatic reticulum, phagosomes, glycogen granules, and numerous vacuoles'* (Pomponi 1977: 485) and are capable of *'protein synthesis, secretion, absorption and intracellular digestion'* (Pomponi 1979b: 777). Pomponi's (1979a, c) enzyme essays supplemented Hatch's results (1980; see below) that were available to Pomponi in the form of his 1975 PhD thesis. She found carbonic anhydrase activity was associated with etching cell bodies, their processes and spaces between the processes, whereas acid phosphatase activity was most intense on the outer surfaces of the cell processes, but also detectable in cell organelles. She reasoned that phosphatase was involved in the extra- and intracellular digestion of organic components of the substrate, carbonic anhydrase in the dissolution of mineral materials. In her 1980 publication, Pomponi summarised all of the above and put it into a larger context, also pointing out the resemblance between sponge bioerosion and osteoclast bone resorption.

Hatch (1980) first provided biochemical evidence of the enzyme carbonic anhydrase being involved in the shifting of the carbonate equilibrium. Papillate sponges showed slightly higher enzyme activity than encrusting sponges of the *Cliona celata* species complex. Respective free-living specimens exhibited strong enzyme activity in the cortex, but a significantly reduced rate of activity in inner parts. Enzyme inhibition caused a decrease in erosion activity, as evidenced by chip production. He discussed the possibilities that carbonic anhydrase may be responsible for:

1. transporting hydrogen ions across membranes; or
2. exchanging hydrogen ions for bicarbonate ions, lowering the pH and consequently substrate dissolution; or
3. providing a pH optimum for other chemical processes possibly involving chelators or other enzymes.

When studying secondary metabolites, Sullivan and Faulkner (1990) were able to demonstrate the occurrence of calcium chelators in *Aka coralliphaga* (Rützler, 1971). Various siphonodictyals of this sponge were able to bind calcium ions and removed them from the test solution. The authors displayed the reaction as a cycle in which the chelator molecule releases H^+ and receives calcium ions at the site of dissolution and releases calcium into the water column in exchange for a new H^+ ion (Sullivan and Faulkner 1990: fig. 4). Therefore, in this pathway of erosion a lowering of pH would be involved at the sponge-substrate interface. This exciting discovery has since largely been overlooked and has not been shown for any other bioeroding sponge.

Considering the proportion of chemical etching of the crevices around the chips compared to the mechanical removal of the chips, several authors presented different estimates. Warburton (1958) failed to detect an increase in the calcium carbonate content of his experimental dishes containing *Cliona celata* (Table 1) and deducted that the share of chemical dissolution should be less than 10%. Judging from the width of the etched crevice and the size of the freed sponge chips, Rützler and Rieger (1973) reasoned that 2-3% of the eroded material is chemically dissolved and 97-98% mechanically removed. Very recently, Zundelevich et al. (2007) experimentally evaluated changes in total alkalinity and the amount of expelled chips of *Pione* cf. *vastifica* and concluded that volumes removed by chemical dissolution were three times that of volumes removed as chips. This is an enormous difference compared to earlier estimates and may be related to the fact that Zundelevich et al. (2007) did not consider that a certain amount of chips at least temporarily remains within the tissue of bioeroding sponges (e.g., Warburton 1958; Fig. 10A). However, the erosion traces of *P.* cf. *vastifica* markedly differed from almost all other observations, except for one case of *Aka* sp. (Calcinai et al. 2003: fig. 1E; Calcinai et al. 2004: fig. 4C). Therefore, *P.* cf. *vastifica* and perhaps Calcinai's *Aka* sp. rely much stronger on chemical erosion than most other known bioeroding sponges. Moreover, the role of chemical etching may shift with changing environments (e.g., with ocean acidification).

The above account demonstrates how much effort has been invested into resolving the question of the etching agent, with research on the mechanisms of sponge erosion spanning a period over 180 years (from Grant 1826 and Osler 1826 to Zundelevich et al. 2007). Nevertheless, we still do not know which exact chemical agent sponges use.

Late 1900s to early 2000s: The larger picture – what does sponge bioerosion mean within the system?

Bioeroding sponges attack biogenic calcium carbonate substrates such as corals, mollusc and barnacle shells, and coralline algae (e.g., Hartman 1958; Hoeksema 1983; Mao Che et al. 1996), composite materials containing calcium carbonate and organic materials (e.g., Hoeksema 1983; Schönberg 2002a), natural limestone (e.g., Hoeksema 1983), and even man-made substrates (Scott et al. 1988; Brusco et al. 2005). They cannot penetrate stylolites (Hoeksema 1983), and are slowed down by organic layers such as conchiolin (Nasonov 1924; Old 1942; Warburton 1958; Cobb 1969, 1975; Hoeksema 1983). Encrusting species of bioeroding sponges can overpower live organisms such as neighbouring corals (Schönberg and Wilkinson 2001; Rützler 2002; López-Victoria and Zea 2004; López-Victoria et al. 2006). Avoidance reactions have been described for *Cliona celata* encountering *Polydora* Bosc, 1802 tunnels (Hoeksema 1983) and another individual of *C. celata* (Bromley and Tendal 1973), creating stunted patterns in bioerosion. However, *Cliona mucronata* Sollas, 1878 often appears to mingle much more closely with other species of *Cliona* (Sollas 1878; and *Cliona ecaudis* Topsent, 1932; Schönberg pers. obs. in several preparations made by Topsent, at the Paris Museum).

Fig. 10 Sponge chips of *Aka mucosa* (Bergquist, 1965) from Orpheus Island, Palm Island Group, central Great Barrier Reef, Australia. **A** A considerable amount of sponge chips can be found in the endosome of the sponge. **B** Chips of relatively uniform size, but variable shape. **C** Chips of different size. Scale bar applies for A-C

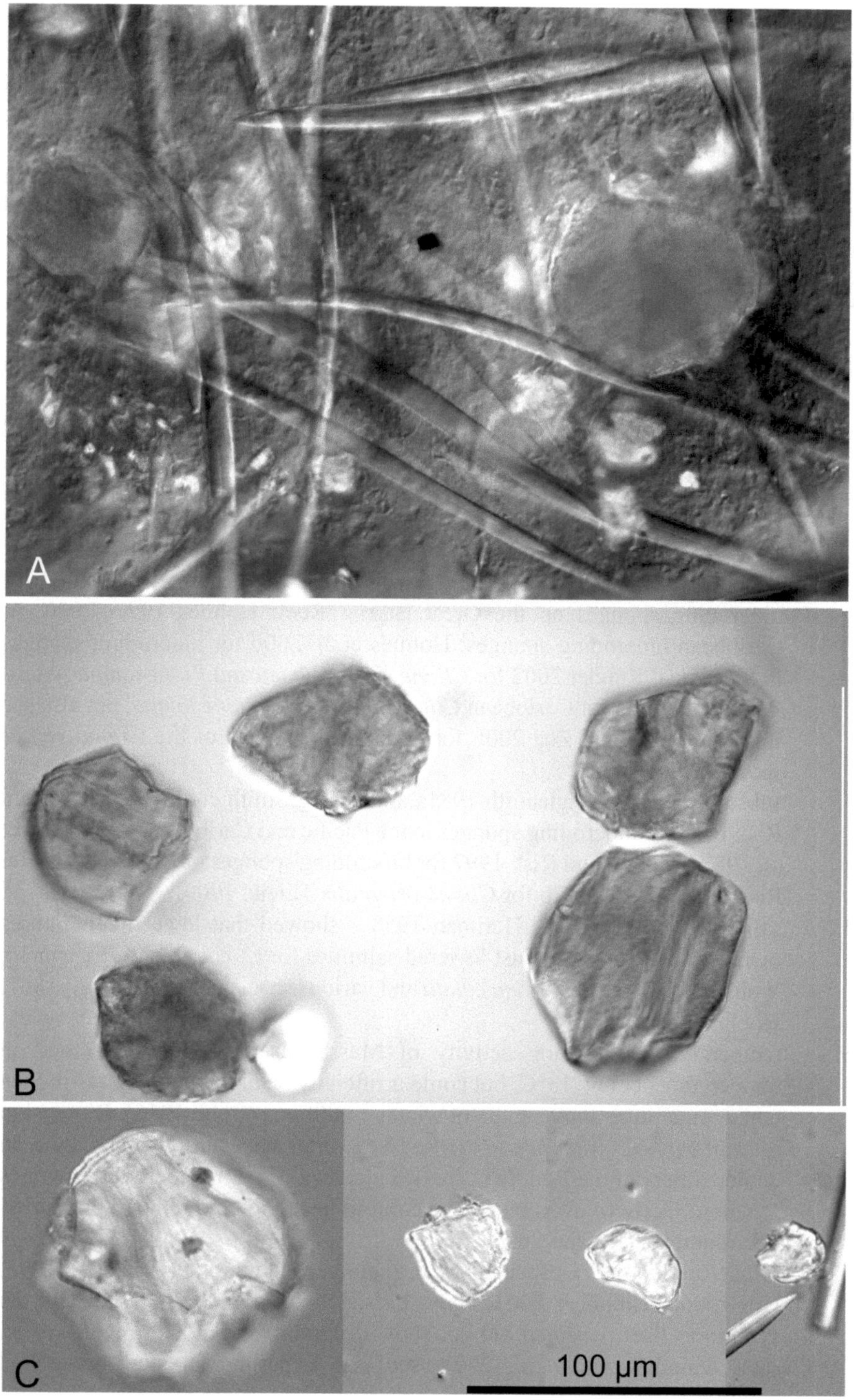
A
B
C
100 µm

A variety of field observations and experimental results became available. They largely ignored the 'how', but focused on estimating rates (erosion: Neumann 1966; Hein and Risk 1975; Rützler 1975; Bak 1976; Hudson 1977; Moore and Shedd 1977; Scoffin et al. 1980; Acker and Risk 1985; Kiene and Hutchings 1992, 1994; Peyrot-Clausade et al. 1999; Pari et al. 2002; Schönberg 2002b; Hutchings et al. 2005; Osnorno et al. 2005; Zundelevich et al. 2007; sediment production: Goreau and Hartman 1963; Neumann 1966; Fütterer 1974; Scoffin et al. 1980; Acker and Risk 1985; Rose and Risk 1985; Young and Nelson 1985; Adjas et al. 1990; Siegrist et al. 1991; Peyrot-Clausade et al. 1999) and pointing out controls and consequences. Sponge bioerosion and the abundance of bioeroding sponges appear to be related to:

1. water flow (de Laubenfels 1950b and Rützler 1975 for *Pione lampa*; Hatch 1980 for *Cliona celata*; Carballo et al. 1996 for *C. celata* and *Cliona viridis* (Schmidt, 1862); López-Victoria and Zea 2005 for *Cliona tenuis* Zea and Weil, 2003 and *Cliona caribbaea* Carter, 1882);
2. nutrient or sewage concentration (Rose and Risk 1985 for *Cliona delitrix* Pang, 1973; Carballo et al. 1996 for *C. celata* and *Cliona viridis*; Muricy 1991 for *C. celata*; Sammarco and Risk 1990; Edinger and Risk 1997; Schönberg et al. 1997; Hutchings et al. 2005 and Osnorno et al. 2005 for bioeroding sponges on the Great Barrier Reef; Holmes 1997, 2000 for Caribbean bioeroding sponges; Holmes et al. 2000 for bioeroding sponges in Indonesia; Rützler 2002 for *Cliona* cf. *caribbaea* and *Pione lampa*; Ward-Paige et al. 2005 for Caribbean *Cliona delitrix* and *Pione lampa*; but also see López-Victoria and Zea 2005 for Caribbean sponges of the *Cliona viridis* species complex);
3. substrate density (Highsmith 1981a and b; Highsmith et al. 1983; Rose and Risk 1985 for bioeroding sponges in the Pacific and Caribbean; Sammarco et al. 1987; Edinger and Risk 1997 for bioeroding sponges on the Great Barrier Reef; Schönberg 2002b for *Cliona orientalis* Thiele, 1900);
4. salinity (Hopkins 1956; Hartman 1958 – showed that lower temperatures increased tolerances against lowered salinities for *C. celata* and *P. vastifica*; Wells 1959, 1961 for *Cliona celata* and various species of *Pione* Gray, 1867; Hopkins 1962);
5. temperature: bioerosion activity of Massachusetts *C. celata* ceased in temperatures below 13°C, but could artificially be maintained by raising the tank temperature above ambient winter conditions (Cobb 1975). Bioerosion rates of *Pione* cf. *vastifica* in the Red Sea varies with season and is lower in cooler waters of the Red Sea (Zundelevich et al. 2007). Bioeroding sponges are more likely to overpower live coral in heat and cold temperature-stress environments (Rützler 2002);
6. light (Rützler 1975), especially if the sponge is symbiotic (Uriz et al. 1992; Hill 1996; Schönberg 2001, 2006; López-Victoria and Zea 2005). But the symbiotic *P.* cf. *vastifica* did not show difference in day and night erosion rates (Zundelevich et al. 2007; see also experimental study described below);

7. water depth (Goreau and Hartman 1963; Acker and Risk 1985; Kobluk and Kozelj 1985; Uriz et al. 1992; Reed and Pomponi 1997; Hill 1999; Schönberg 2001; López-Victoria and Zea 2005);
8. trauma (Rützler 1975; Thomas 1997);
9. general stress situations acting on reef organisms that are more weakened in comparison to bioeroding sponges (Rützler 2002; Márquez et al. 2006; Schönberg and Suwa in press); and
10. age / size of the sponge, i.e., whether a larva or fragment is freshly settling or attaching (fast growth and erosion) or whether it is an established colony (slow growth, less erosion) (Neumann 1966; Rützler 1975).

2000 to today: Sponge bioerosion traces – can they be used in taxonomy and how do they relate to the etching process?

Macroscopic patterns of sponge bioerosion traces (*Entobia* Bronn, 1837) have traditionally been used for taxonomic purposes in detailed descriptions (e.g., Rützler 1974) or for morphometric approaches (e.g., Hoeksema 1983; Rose and Risk 1985; Rosell 1994). However, a considerable range of overlap exists, i.e., one sponge species can produce more than one kind of trace and one trace can be the result of the action of different sponge species (Bromley and D'Alessandro 1984, 1989; see also Calcinai et al. this volume, in press).

More recent approaches focused on the microscopic traces: sponge chips and scars. Originally Rützler and Rieger (1973: table 2) noted considerable variation in the size range of these structures, but thought this may be more strongly related to circumstances than to taxonomy. Their chip diameters ranged from 15 to 71 µm for clionaids, with respective scar diameters of 18-94 µm. Schönberg (2000) agreed that mean sponge chip diameters provide insufficient information for species distinction within the clionaids, but allowed restricted usefulness to distinguish genera, with mean chip diameters from *Aka mucosa* (Bergquist, 1965) being noticeably larger (45 µm) than those from *Zyzzya criceta* Schönberg, 2000 (30 µm) and the 5 studied clionaids (30-35 µm). However, sponge chip and sponge scar sizes are subject to location within the cavity (larger diameters in central, established regions, smaller diameters in pioneer regions; Figs. 6, 8, 10; see also Rützler and Rieger 1973; Ward and Risk 1977; Calcinai et al. 2004). They may also vary with substrate type (Rützler and Rieger 1973; Pomponi 1976). Pomponi (1976) was the first to note differences in the microstructure of sponge scars between genera. Italian and French scientists proposed the use of these patterns for taxonomic purposes (Omnes 1991; Bavestrello et al. 1996; Calcinai et al. 2003, 2004; Borchiellini et al. 2004). Clionaid scars are much smoother, less size-variable and more concentric in outline than scars produced by *Aka* (Pomponi 1976; Calcinai et al. 2003, 2004). *Aka* de Laubenfels, 1936, *Holoxea* Topsent, 1892, *Spiroxya* Topsent, 1896, *Alectona* Carter, 1879, *Delectona* de Laubenfels, 1936 and *Thoosa* Hancock, 1849 produce scars with concentric rings, i.e., microterracing (Pomponi 1976; Calcinai et al. 2003, 2004; Borchiellini et al. 2004). *Alectona* exhibits radial lines overlying the circular grooves (Omnes 1991; Vacelet 1999; Calcinai et al. 2004; Borchiellini et al. 2004). Despite these consistent reports above, comparisons should only be conducted using traces from the same kind of substrate. Even though clionaids are said to erode smooth scar-walls, concentric groves have been observed in clionaid scars in bivalve shells, presumably caused by the alternating layers of calcium carbonate and conchiolin (Cobb 1975), but also in the very homogeneous substrate of Iceland spar (Cobb 1969 for *Cliona celata*; Rützler and Rieger 1973 for *Pione lampa*). Similar variations to the basic patterns may be observed in a wider range of substrate types and environmental situations (Pomponi 1976). Recently Zundelevich et al. (2007) found sponge scars in which a central knob of material remained standing. This microtrace is presently only known for the Red Sea *Pione* they used and which appears to be closely related to *Pione vastifica*, and for *Aka* sp. from Indonesia (Calcinai et al. 2003, 2004).

Very recently, Calcinai et al. (this volume, in press) studied erosion traces of a single species (*Cliona albimarginata* Calcinai, Bavestrello and Cerrano, 2005) in different types of substrata. They came to the conclusion that while microsculpturing of scars remained the same in different substrates, other microscopic traces (diameter of sponge scars) and macroscopic traces (*Entobia*) can vary between substrates. This is in part caused by the substrate microtexture (orientation of the calcium carbonate crystals) and in part by its mineralogy (Calcinai et al. this volume).

One alleged consequence of sponge erosion is the phenomenon of the putative growth 'phases' of bioeroding sponges: initial alpha-papillate, later beta-encrusting and last gamma-massive, free-living morphology (Topsent 1888; Topsent 1900; Vosmaer 1933; Hartman 1958). This principle has been studied in most detail in *Cliona celata* in the Atlantic, where the sponge can be found in all three forms. It has also been postulated for species of the *Cliona viridis* group (e.g., *C. viridis* being the papillate growth stage of *Cliona nigricans* (Schmidt, 1862) in massive growth, e.g., Rosell and Uriz 1991). It is important to note, however, that

1. no sponge has ever been monitored from larval settlement to free-living sponge, and the principle was never experimentally confirmed (see Hartman 1958),
2. not all growth forms can be found in all habitats (no gamma form bioeroding sponges observed on the Great Barrier Reef, C. Schönberg pers. obs.; no beta and gamma form of *C. viridis* group sponges found in Japan, Y. Ise pers. comm.; Mediterranean gamma form *C. celata* only in deep water, T. Perez pers. comm.),
3. some growth 'phases' are lacking or do not very commonly occur in distinctive morphology (e.g., the fully encrusting beta form for *Cliona viridis / nigricans* in the Mediterranean), and
4. most known species of bioeroding sponges retain the alpha growth form, even when occurring in small, restrictive substrates (e.g., Hartman 1958).

There are also physiological observations that may indicate that different growth forms may represent different species. Gamma form sponges of *C. celata* have been shown to be capable of bioerosion, but with lesser efficiency than the respective alpha form (Hartman 1958). Moreover, its gamma form contains unusually high concentrations of carbonic anhydrase in the cortex when compared to alpha and beta form, with carbonic anhydrase being involved in the process of bioerosion (Hatch 1980). Alpha and gamma form sponges of the Mediterranean *C. viridis / nigricans* group showed different adaptation to hetero- and autotrophy via their dinoflagellate symbionts under the same conditions (Schönberg et al. 2005). If (alpha, beta, gamma) growth form is species-specific, rather than related to age or growth stage, then this morphological consequence of sponge bioerosion can be used for taxonomic purposes as well.

1802 to 2007: A summary of knowledge on sponge bioerosion

After two centuries of research, we know that the organisms in question are endolithic sponges that produce the cavities they inhabit. Sponge erosion proceeds chemically (etching of cup-shaped fissures) and mechanically (removal of sponge chips). Proportions of the two components of sponge erosion were estimated between 2% chemical to 98% mechanical and 75% chemical and 25% mechanical activity. The sponges are able to attack calcium carbonates and organic materials, which involves specialised etching cells with filipodia providing a very localised application of an etching agent and the enzymes carbonic anhydrase and phosphatase. Sponge bioerosion does not apparently follow diurnal rhythms. It produces characteristic traces: the chambers and tunnels in the substrate that often end in minute pioneering ducts (*Entobia*), the pitting of the chamber walls (sponge scars) and the loosened silt-size particles with a smooth convex lower side and a concavely-facetted upper side (sponge chips). These traces may be of limited help in taxonomy. A large variety of environmental parameters can influence bioerosion rates and bioeroding sponge abundances, of which flow, nutrient concentration and substrate density appear to be the main magnifiers in common species. Despite over 200 years of research some questions still remain unsolved: Which is the exact chemical agent used in sponge erosion? Can sponge erosion proceed from larval settlement over papillate and encrusting growth forms to the entire removal of substrate and free-living sponges?

Present results

Microprofiles of pH and calcium concentrations in the bioeroding sponge *Cliona celata* and the free-living sponge *Halichondria panicea* did not reveal immediately clear patterns (Fig. 11). Overall, data taken from *H. panicea* were more stable than those from *C. celata* and were more uniform along the microprofiles into the sponges (Fig. 11C). However, some data series of *C. celata* were comparable to those of *H. panicea* (e.g., Fig. 11A-B; CEL 1.1 and CEL 3.3), and some of the *H. panicea* profiles were as erratic as were some of those for *C. celata* (Fig. 11C; HAL 1.2 and HAL 2.3). Calcium concentrations in *C. celata* did not uniformly increase with distance into the sponge, even though eight of the 18 profiles showed an increase of 0.5 to 1 mM Ca^{2+} ions when considering the entire profile (Fig. 11B; CEL 1.1 and 1.3, CEL 3.1 and 3.2, CEL 5.1 and 5.3, CEL 6.1 and 6.3). Of these eight profiles, two were so erratic that they should be discounted (Fig. 11A-B; CEL 1.2 and CEL 5.3). There were also eight profiles of the 18 for *C. celata* for which the calcium concentration decreased

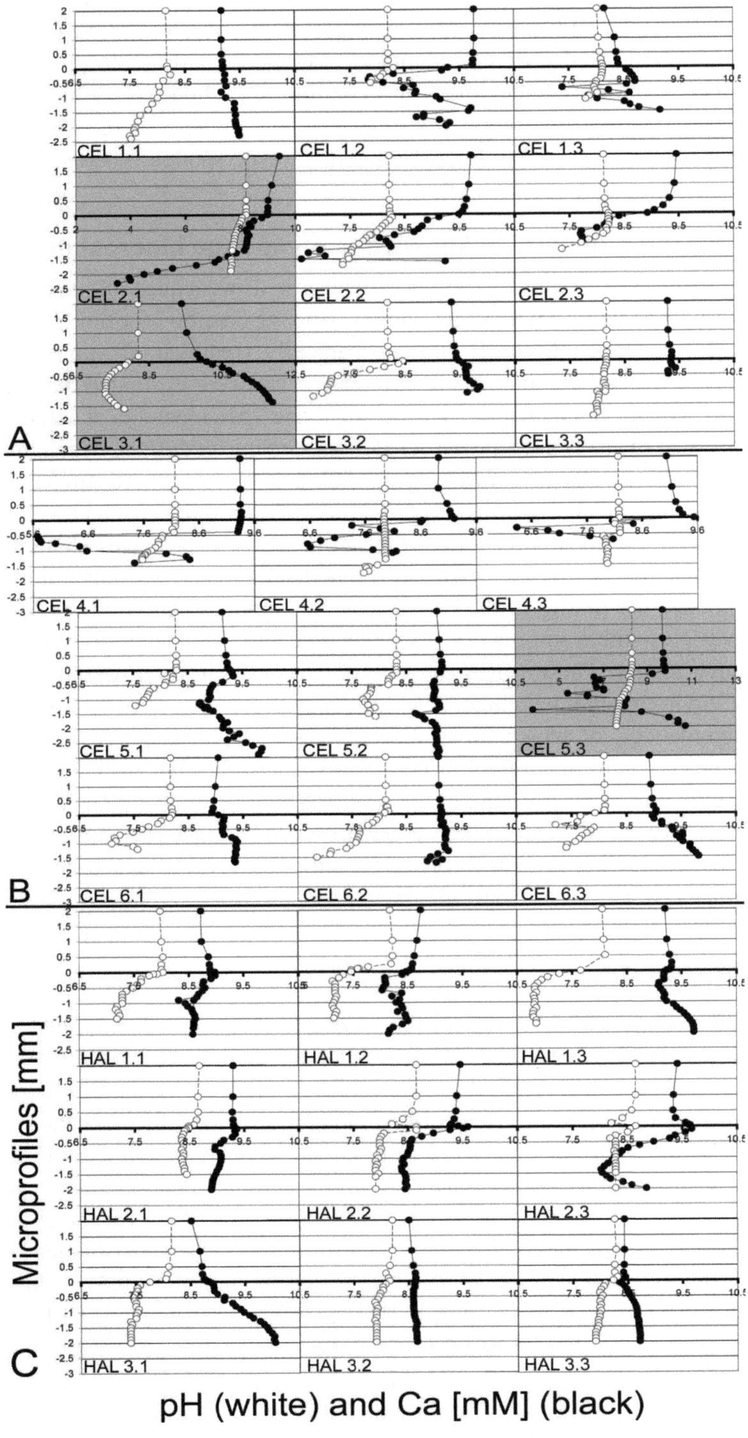

A
B
C
CEL 1.1
CEL 1.2
CEL 1.3
CEL 2.1
CEL 2.2
CEL 2.3
CEL 3.1
CEL 3.2
CEL 3.3
CEL 4.1
CEL 4.2
CEL 4.3
CEL 5.1
CEL 5.2
CEL 5.3
CEL 6.1
CEL 6.2
CEL 6.3
HAL 1.1
HAL 1.2
HAL 1.3
HAL 2.1
HAL 2.2
HAL 2.3
HAL 3.1
HAL 3.2
HAL 3.3
Microprofiles [mm]
pH (white) and Ca [mM] (black)

with distance into the sponge (Fig. 11A-B; CEL 1.2, CEL 2.1-2.3, CEL 4.1-4.3, CEL 6.2), but of these, four profiles were so erratic, that they should be discounted (Fig. 11A-B; CEL 1.2, CEL 4.1-4.3). Three of the eight decreasing calcium profiles were fairly regular, but still displayed strong negative trends that cannot be explained (Fig. 11A; CEL 2.1-2.3). In two profiles calcium concentrations appeared to remain more or less constant with tissue depth (Fig. 11A-B; CEL 3.3 and CEL 5.2). Nevertheless and overall, calcium concentrations in *C. celata* increased with tissue depth (0.5 mM in the first series and 0.3 mM in the second series on average; see Table 2). But there were a few *H. panicea* profiles that also displayed calcium increases with distance into the sponge (Fig. 11C; HAL 1.3, HAL 3.2 and 3.3). The overall mean for *H. panicea* was an increase by 0.03 mM Ca^{2+} (Table 2), that was mainly influenced by the strong increase in the first measurement of the third specimen (Fig. 11; HAL 3.1).

The pH measurements yielded additional evidence for sponge erosion, as 15 of 18 cases of *C. celata* revealed a decrease of pH with distance into the sponge tissue (Fig. 11A-B). In the remaining profiles, pH values rose slightly with depth but remained lower than in the superficial layers (Fig. 11A-B; CEL 3.1, CEL 5.2, CEL 6.1). In *H. panicea* the tissue pH was lower than that of the ambient water, but once the sensor penetrated the sponge's superficial layers, the values remained fairly constant (Fig. 11C).

Table 2 Means of pH and calcium concentrations in the culture water (at 2 mm above the sponges) and each individual of *Cliona celata* (CEL) and *Halichondria panicea* (HAL). Erratic and atypical microprofiles for calcium were omitted when calculating the values below (CEL 1.2-1.3, CEL 2.1-2.3, CEL 4.1-4.3, CEL 5.3). Data are given for the following steps in the microprofiles: in the water above the sponge (2 mm), on the surface of the sponge (0 mm), in superficial layers of the sponge, but below the level of the papillae (-0.3 mm) and at the endpoint of the measurements (varying depths). An overall difference (Δ) of the measured parameters was calculated between the values on the surface and the endpoints

	Sponge #	pH					Calcium [mM]				
		(2 mm)	(0 mm)	(-0.3 mm)	endpoint	Δ pH	(2 mm)	(0 mm)	(-0.3 mm)	endpoint	Δ Ca^{2+}
Series 1	CEL 1	8.1	8.2	8.1	7.7	-0.5	9.2	9.2	9.2	9.5	+0.3
									(CEL 1.1)		
	CEL 2	8.2	8.2	8.1	7.4	-0.8	-	-	-	-	-
	CEL 3	8.2	8.2	7.8	7.5	-0.7	9.3	9.6	9.9	10.3	+0.7
Series 2	CEL 4	8.1	8.1	8.1	7.7	-0.4	-	-	-	-	-
	CEL 5	8.3	8.3	8.2	7.7	-0.6	9.1	9.2	9.1	9.4	+0.2
									(CEL 5.1-5.2)		
	CEL 6	8.1	8.0	7.8	7.3	-0.7	9.0	9.1	9.1	9.4	+0.3
	HAL 1	8.1	7.7	7.2	7.1	-0.6	8.9	8.9	8.7	8.8	+0.1
	HAL 2	8.7	8.6	8.2	8.2	-0.4	9.4	9.5	9.1	8.7	-0.4
	HAL 3	8.2	8.0	7.9	7.7	-0.3	8.5	8.6	8.7	9.1	+0.4

Fig. 11 Microsensor microprofiles for pH (white) and calcium concentration (black) for the sponges *Cliona celata* (**A** and **B**) and *Halichondria panicea* (**C**). Each series was conducted on three sponge specimens (CEL 1-3, CEL 4-6 and HAL 1-3; graphs vertically distributed), measured in three places (graphs horizontally distributed). All graphs with white backgrounds were made to scale; grey background indicates differing scale. Standard deviations not displayed, as they were smaller than the data points

Patterns became clearer after taking means of the replicate measurements per sponge and then means of the sponges per series (Fig. 12). Whereas pH and calcium concentration in *H. panicea* remained fairly uniform from -500 µm downwards (Fig. 12C), in *C. celata* pH continually decreased and calcium concentration slightly increased (Fig. 12A-B). However, as these means were obtained by omitting erratic data sets, results are not entirely objective and should be treated with care.

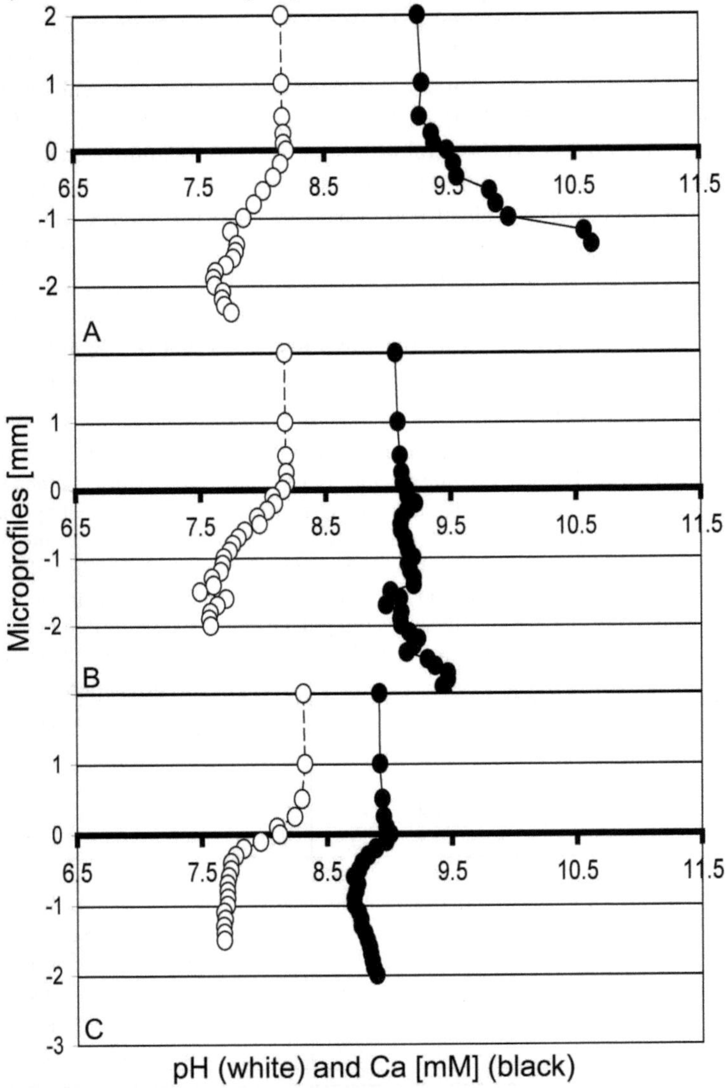

Fig. 12 Gradients of mean pH (white) and calcium concentrations (black) in *Cliona celata* (**A** and **B**) and *Halichondria panicea* (**C**). Very erratic calcium profiles of *C. celata* and the extreme negative trend of calcium in CEL 2 were omitted before calculations: CEL 1.2-1.3, 2.1-2.3, 4.1-4.3 and 5.3. Of all replicate measurements per sponge means were taken, and then the means of the sponges per series. Standard deviations not displayed, as they were smaller than the data points in the graphic

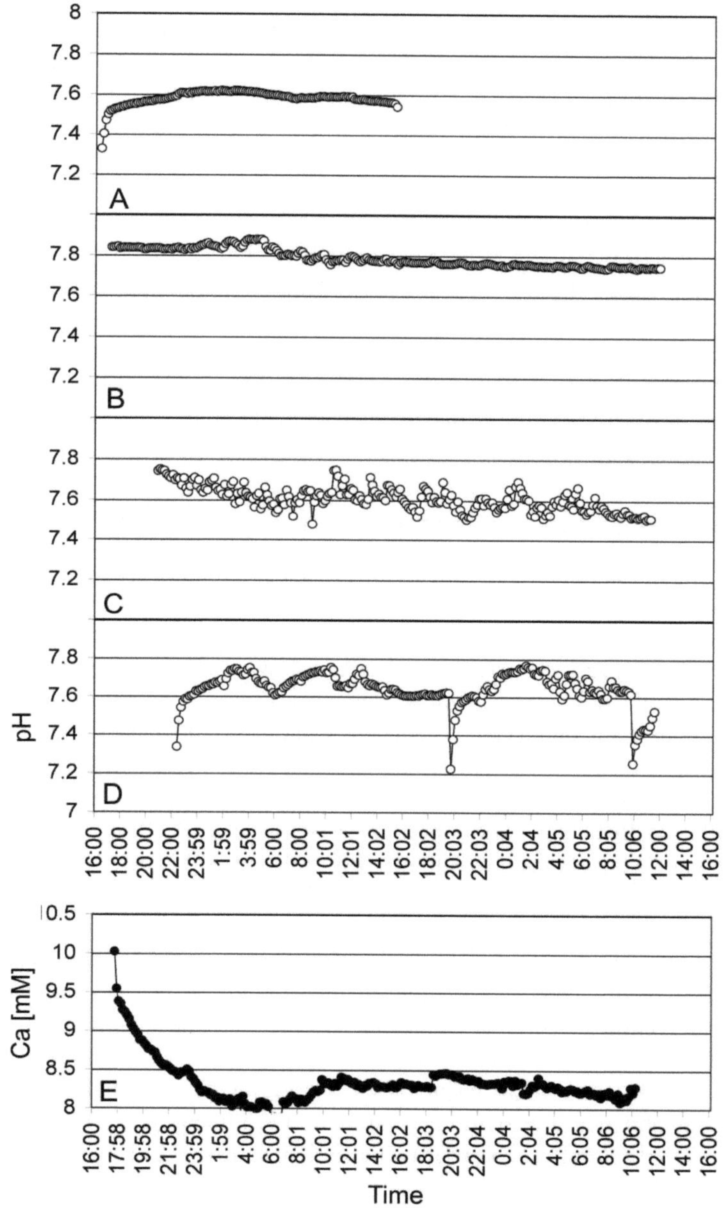

Fig. 13 Time series of microsensor measurements for pH (**A-D**) and calcium concentration (**E**) at the tissue-substrate interface of *Cliona celata*. The sensors were positioned in the sponge tissue against the substrate and not disturbed over the period of the measurements. Extreme negative peaks in A and D are unlikely to represent real decreases in pH, but were caused by the initial disturbance of applying the sensor (A) and at least in one case by a slamming door (D). The steadily decreasing values in the first third of E may not represent the true calcium concentration, but may have been caused by a sensor drift. Please note that curves for B and C display some areas with small, periodic oscillations of the pH, which may also be the case for the last third of D

Moreover, absolute values of pH and calcium concentration were weaker evidence of the acid-etching theory than expected. Mean pH and calcium concentrations at the tissue-substrate interface in *C. celata* were 7.55 and 9.65 mM, respectively, compared to ambient conditions of pH 8.2 and 9.2 mM Ca^{2+}. Especially the pH values at the endpoints of the profiles were thus much more subtle than expected, and only went below neutral in two cases: CEL 3.2 and CEL 6.2, both with a pH of 6.8 at the endpoint of measurements (Fig. 11A-B).

Time series of pH and calcium concentration at the tissue-substrate interface of *C. celata* showed no diurnal rhythm (Fig. 13). Initial high values for calcium were probably due to a sensor drift (Fig. 13E). It is interesting to note, however, that a shorter-term, slightly irregular periodicity may exist in some individuals, with small pH peaks occurring in intervals of about 1 to 3 hours (Fig. 13B-C). Extremely negative peaks were very likely due to physical disturbance of the work place (e.g., door slamming; Fig. 13A, D).

Discussion of present results: the mystery continues

Microsensor measurements revealed that pH and calcium properties in the tissue of the bioeroding sponge *Cliona celata* slightly differed from those of the free-living *Halichondria panicea.* Regrettably, differences were not as clear as could be wished, but in contrast to *H. panicea, C. celata* showed a continually decreasing pH with tissue depth, and in the first measurement series for *C. celata* calcium strongly increased with tissue depth (Fig. 12). These values were opposite in trend and very similar in dimension to previously found gradients for calcification processes in corals and the calcifying alga *Halimeda* measured with LIX microsensors (de Beer et al. 2000; de Beer and Larkum 2001; Al-Horani et al. 2003). However, as the present means were obtained by subjectively omitting erratic data sets, results have to be regarded with care.

Much variation between and within individuals existed, masking the observations. Within a given individual of *C. celata* variation was probably influenced by the heterogeneity of the sponge tissue and the distribution of the remaining substrate (Fig. 1). *C. celata* consists of branching erosion chambers (see Schönberg and Shields this volume) in which its tissue is located, which in turn surrounds the canal system. The tip of the microsensor may occasionally pass through lumina of the canal system and measure conditions similar to the ambient water (Fig. 11; e.g., pH for CEL 3.1), or come close to a protruding shelf of substrate and measure conditions similar to the endpoints of the profiles (Fig. 11; e.g., pH for CEL 6.3). Measured values in the first 250 µm of the profiles could occasionally be slightly different compared to the rest of the profiles and represented conditions in the sponge's papillae and superficial layers that contain less cellular and much more skeletal material than the rest of the sponge. Additionally confounding were situations in which equipment problems may have played a role (Fig. 11; e.g., calcium values for CEL 4.1 to CEL 4.3 and CEL 5.3). Unexpected results from *H. panicea* cannot easily be explained (occasional increases in calcium, initial decrease in calcium upon insertion of microsensor; Fig. 11). Calcium plays an important role in marine sponges for cell motility and tissue contractions (e.g., Prosser 1967; Pavans de Ceccatty 1971; Lorenz et al. 1996). Sponges have no nervous system, and cell-to-cell communication is used to affect coordination. Intracellular calcium concentrations may be elevated by stress or by tissue contractions (Leys and Meech 2006). As *H. panicea* presented more resistance when trying to insert the sensors into the tissue than *C. celata* (higher spongin content), changed calcium concentrations may be a consequence of squeezed cells before the sensors pierced the tissue and possibly of a resulting if unobserved contraction.

If we assume that the gradual decrease in pH in *C. celata* is related to its etching activities, the present study is the first example of experimentally showing that acid production may play a role in sponge bioerosion. Previous attempts to detect acid in bioeroding sponges failed, because employed tools were not sensitive enough and not applied locally enough (Table 1). It may be reasoned, however, that the overall mean pH of 7.55 at the tissue-sponge interface of *C. celata* may not be low enough to signify active bioerosion, especially when considering that human stomach contents have a pH between 1 and 2 and the oral pH commonly reaches a value of 5.5 during sugar intake (Cleffmann 1979). But one has to keep in mind that this value is maintained against an external pH of about 8.2 and that the positioning of the microsensors was conducted 'blindly' (Fig. 1), i.e., due to the opacity of the tissue and substrate a sensor would only by chance come to rest directly on a erosion crevice between sponge chip and scar to detect the very local application of the etching agent. This may have been more or less achieved for CEL 3.2 and 6.2 (Fig. 11A-B). If it were possible to visually guide a microsensor to fit directly over or into the etched crevice of a future chip and the remaining substrate, we could probably expect more striking results. Moreover, if *C. celata* employs chelators as has been found for *Aka coralliphaga* (see Sullivan and Faulkner 1990), then the pH values may not have to be as low as for purely acid-achieved etching. Nevertheless, very similar values of pH in the deeper tissue of *H. panicea* make the above reasoning perilous and present data should be understood as the product of a pilot study that provides preliminary evidence only. A repetition of the approach would probably help to confirm the observations. Calcium gradients may be the safer indicators in this case and definitively point towards calcium dissolution in *C. celata*, at least for the first series of measurements (Fig. 12A-B).

One intriguing outcome is the possible short-term periodicity in sponge etching as evidenced in oscillating pH values at the tissue-substrate interface of some individuals of *C. celata* (Fig. 13B-C). Sponge scar microstructures contain concentric grooves that may be caused by substrate properties (e.g., Cobb 1969; Rützler and Rieger 1973), but were previously also explained by a putative periodic etching activity of the sponges (Cobb 1969). If the latter theory is true and the present changes in pH are directly related to the bioerosion activity, then one etching cycle takes 1 to 3 hours and the production of one chip that leaves a scar with about 5 distinct concentric grooves may take 5-15 hours (deepest grooves counted in figures from, e.g., Calcinai et al. 2003). In this context it is also important to consider that there may be individual differences in the etching process, as one of the present individuals of *C. celata* did not show any periodic changes in pH measured at the tissue-substrate interface (Fig. 13A), when another individual displayed pronounced oscillations of pH (Fig. 13C). Therefore, the clear occurrence, the depth and the number of concentric grooves in sponge scars may be not as good a taxonomic aid as previously thought (Omnes 1991; Bavestrello et al. 1996; Calcinai et al. 2003, 2004; Borchiellini et al. 2004). Earlier findings that sponge bioerosion does not display a diurnal rhythm (Zundelevich et al. 2007) are here supported.

Overall, it still remains somewhat unclear whether observed gradients of pH and calcium in the tissue of *C. celata* were related to sponge bioerosion, as most of them were gradual and subtle, and the means were not distinctly different compared to the free-living sponge (Table 2). However, as changes in pH were uniformly represented in different individuals of *C. celata* and because changes in calcium concentrations were so pronounced in the first series, it is quite likely that the findings demonstrate chemical etching involving acid production. Hopefully this study can be repeated with more replicates and time, making it possible to obtain better means. In this case it would be interesting to include different growth forms of bioeroding sponges or a symbiotic and an asymbiotic species in comparison.

Acknowledgements

During times of increasingly restrictive rules concerning literature that is older than 100 years, I was very grateful for generous assistance by colleagues who sent me photocopies of rare references or gave me full access to ancient books in special collections (e.g., C. Lueter, Zoologic Museum Berlin; R. W. M. van Soest, Zoologic Museum Amsterdam; C. Valentine, British Museum of Natural History). Interlibrary loan services of The Carl von Ossietzky University were also of much assistance, repeatedly showing patience with the requests for strange and ancient publications and the complaints about missing plates. Some colleagues provided access to good images of historic researchers: C. Eckert, Berlin Museum for Natural Science; J. Holmes, Hancock Museum; C. Valentine, Natural History Museum London. The experimental study was conducted at The Max Planck Institute for Marine Microbiology in Bremen, Germany, supported by a scholarship from the Max Planck Society. That this was possible is due to D. de Beer and B. Jørgensen. During the experiment I was allowed to use equipment of and received much appreciated instructions from F. Al-Horani, G. Eickert, A. Gieseke, H. Jonker, S. Köhler-Rink, H. Ploug, I. Schröder and E. Walpersdorf, and infrastructure support from H. Bartholomäus, M. Fitze, A. Fouquet, O. Gundermann, J. Kallmeyer and C. Pranz. Samples were provided by M. Krüß from the Helgoland Marine Station, and transport involved the kind assistance by the Helgoland-Cuxhaven ferry staff. Using the opportunity I would also like to sincerely acknowledge the continuous and faithful support of my German professor H. K. Schminke who recently retired and without whom this and many other publications would not have come into existence. Last but not least, E. Walpersdorf and an anonymous referee are kindly acknowledged for their valuable comments and suggestions.

References

Acker KL, Risk MJ (1985) Substrate destruction and sediment production by the boring sponge *Cliona caribbaea* on Grand Cayman Island. J Sediment Petrol 55:705-711

Adjas A, Masse J-P, Montaggioni LF (1990) Fine-grained carbonates in nearly closed reef environments: Mataiva and Takapoto Atols, central Pacific Ocean. Sediment Geol 67:115-132

Al-Horani FA, Al-Moghrabi SM, de Beer D (2003). The mechanism of calcification and its relation to photosynthesis and respiration in the scleractinian coral *Galaxea fascicularis*. Mar Biol 142:419-426

Bak RPM (1976) The growth of coral colonies and the importance of crustose coralline algae and burrowing sponges in relation with carbonate accumulation. Netherlands J Sea Res 10:285-337

Bate CS (1849) Notes on the boring of marine animals. Rep Meet Brit Assoc Adv Sci 19:73-75

Bavestrello G, Calcinai B, Cerrano C, Pansini M, Sarà M (1996) The taxonomic status of some Mediterranean clionids (Porifera: Demospongiae) according to morphological and genetic characters. Bull Inst Roy Sci Nat Belgique, Biol 66 Suppl:85-195

Bergquist PR (1965) The sponges of the Palau Islands. Part 1. Pac Sci 19:123-204

Bianconi GG (1841) Sopra alcuni zoofiti descritti sotto i nomi di *Cliona calata* (Grant), *Vioa* (Nardo) e *Spongia terebrans* (Duvernoy). N Ann Sci Nat Rend Lav Accad Bologna 6:455-469

Borchiellini C, Alivon E, Vacelet J (2004) The systematic position of *Alectona* (Porifera, Demospongiae): a tetractinellid sponge. Boll Mus Ist Biol Univ Genova 68:209-217

Bosc LAG (1802a) Histoire naturelle des éponges. Éponge pézize, *Spongia peziza*. Hist Nat Vers 3:147-148

Bosc LAG (1802b) Histoire naturelle des vers, contienant leur description et leurs moeurs, avec figures dessinées d'après nature. Déterville, Paris, 1:1-324

Bowerbank JS (1866) A monograph of the British sponges. Ray Soc, Hardwicke, London, 2, 388 pp

Bromley RG, D'Alessandro A (1984) The ichnogenus *Entobia* from the Miocene, Pliocene and Pleistocene of southern Italy. Riv Ital Paleont Stratigr 90:227-296

Bromley RG, D'Alessandro A (1989) Ichnological studies of shallow marine endolithic sponges from the Italian coast. Riv Ital Paleont Stratigr 95:279-314

Bromley RG, Tendal OS (1973) Example of substrate competition and phobotropism between two clionid sponges. J Zool London 169:151-155

Bronn HG (1837) Lethaea Geognostica oder Abbildungen und Beschreibungen der für die Gebirgs-Formationen bezeichnendsten Versteinerungen. Schweizerbart, Stuttgart, 1:1-544 and plates

Brusco F, Calcinai B, Bavestrello G, Cerrano C (2005) Attività di erosione di *Cliona celata* Grant, 1826 e *Pione vastifica* (Hancock, 1849) (Porofera, Demospongiae) perforanti substrati artifiali. Biol Mar Medit 12:240-243

Calcinai B, Arillo A, Cerrano C, Bavestrello G (2003) Taxonomy-related differences in the excavating micro-patterns of boring sponges. J Mar Biol Assoc UK 83:37-39

Calcinai B, Bavestrello G, Cerrano C (2004) Bioerosion micro-patterns as diagnostic characters in boring sponges. Boll Mus Ist Biol Univ Genova 68:229-238

Calcinai B, Bavestrello G, Cerrano C (2005) Excavating sponge species from the Indo-Pacific Ocean. Zool Stud 44:5-18

Calcinai B, Azzini F, Bavestrello G, Gaggero L, Cerrano C (in press) Excavating rates and boring pattern of *Cliona albimarginata* (Clionaidae, Demospongiae, Porifera) in different substrata. In: Custódio MR, Hajdu E, Lôbo-Hajdu G, Muricy G (eds) Porifera research: biodiversion, innovation and sustainability. Proc 7th Int Sponge Symp Búsioz, Brazil

Calcinai B, Bavestrello G, Cerrano C, Gaggero L (this volume) Substratum microtexture affects the boring pattern of *Cliona albimarginata* (Clionaidea, Demospongiae). In: Wisshak M, Tapanila L (eds) Current developments in bioerosion. Springer, Berlin, pp 203-212

Carballo JL, Naranjo SA, García-Gómez JC (1996) Use of marine sponges as stress indicators in marine ecosystems at Algeciras Bay (southern Iberian Peninsula). Mar Ecol Prog Ser 135:109-122

Carter HJ (1879) On a new species of excavating sponge (*Alectona millari*); and on a new species of *Rhaphidotheca* (*R. affinis*). J Roy Microscop Soc 2:493-499

Carter HJ (1882) Some sponges from the West Indies and Acapulco in the Liverpool Free Museum described, with general and classificatory remarks. Ann Mag Nat Hist, Ser 5, 9(52):266-301 and 346-368

Clapp WF, Kenk R (1963) Marine borers. An annotated bibliography. Off Nav Res, Dept Navy, Washington DC, 1136 pp

Cleffmann G (1979) Stoffwechselphysiologie der Tiere. Stoff- und Energieumsetzungen als Regelprozesse. Ulmer, Stuttgart, 296 pp

Cobb WR (1969) Penetration of calcium carbonate substrates by the boring sponge, *Cliona*. Amer Zoologist 9:785-790

Cobb WR (1975) Fine structural features of destruction of calcareous substrata by the burrowing sponge *Cliona celata*. Trans Amer Microscop Soc 94:197-202

Cotte MJ (1902) Note sur le mode de parforation des Cliones. C R Séance Soc Biol Paris 54:636-637

De Beer D, Larkum AWD (2001) Photosynthesis and calcification in the calcifying alga *Halimeda discoidea* studies with microsensors. Plant Cell Environ 24:1209-1217

De Beer D, Kühl M, Stambler N, Vaki L (2000) A microsensor study of light enhanced Ca^{2+} uptake and photosynthesis in the reef-building hermatypic coral *Favia* sp. Mar Ecol Prog Ser 194:75-85

De Blainville HMD (1834) Manuel d'actinologie ou de zoophytologie. Levrault, Strasbourg, 694 pp [plates in extra volume]

De Laubenfels MW (1936) A discussion of the sponge fauna of the Dry Tortugas in particular and the West Indies in general with material for a revision of the families and orders of the Porifera. Carnegie Inst Wash Publ, Pap Tortugas Lab 467, 30:1-225

De Laubenfels MW (1950a) The Porifera of the Bermuda Archipelago. Trans Zool Soc London 27:1-154

De Laubenfels MW (1950b) An ecological discussion of the sponges of Bermuda. Trans Zool Soc London 27:155-201

Dujardin MF (1838a) Observations sur les éponges. C R Hebd Séance Acad Sci Paris 6:676

Dujardin MF (1838b) Observations sur les éponges et un particulier sur la Spongille ou éponge d'eau douce. Ann Sci Nat, Ser 2, 10:5-13 and 34

Duvernoy M (1839) Note sur une espèce d'éponge qui se loge dans la coquille de l'huître à pied de chaval (*Ostrea hippopus*, Lamarck), en creusant des canaux dans l'épaisse des valves de cette coquille. C R Hebd Séances Acad Sci Paris 9:683-686

Edinger EN, Risk MJ (1997) Sponge borehole size as a relative measure of bioerosion and palaeoproductivity. Lethaia 29:275-286

Ehrenberg CG (1934) Die Corallenthiere des rothen Meeres physiologisch untersucht und systematisch verzeichnet. Roy Acad Sci, Berlin, 156 pp

Fischer MP (1868) Recherches sur les éponges perforantes fossiles. Nouv Arch Mus Hist Nat Paris 4:117-173

Fütterer DK (1974) Significance of the boring sponge *Cliona* for the origin of fine grained material of carbonate sediments. J Sediment Petrol 44:79-84

Goreau TF, Hartman WD (1963) Boring sponges as controlling factors in the formation and maintenance of coral reefs. In: Sognnaes RF (ed) Mechanisms of hard tissue destruction. Publ Amer Assoc Adv Sci 75:25-54

Grant RE (1826) Notice of a new zoophyte (*Cliona celata*, Gr.) from the Firth of Forth. Edinburgh New Phil J 1:78-81

Gray JE (1867) Notes on the arrangement of sponges, with the description of some new genera. Proc Zool Soc London 1867:492-558

Hancock A (1848) On the boring of the Mollusca into rocks etc. and on the removal of portions of their shells. Ann Mag Nat Hist, Ser 2, 2:225-248

Hancock A (1849a) On the excavating powers of certain sponges belonging to the genus *Cliona*; with descriptions of several new species, and an allied generic form. Ann Mag Nat Hist, Ser 2, 3:321-348

Hancock A (1849b) Observations on Mr. Morris's paper on the excavating sponges. Ann Mag Nat Hist, Ser 2, 4:355-357

Hancock A (1867) Note on the excavating sponges; with descriptions of four new species. Ann Mag Nat Hist, Ser 3, 13:229-242

Hartman WD (1958) Natural history of the marine sponges of southern New England. Bull Peabody Mus Nat Hist 12:1-155

Hatch WI (1979) The implication of carbonic anhydrase in the physiological mechanism of penetration of substrata by the burrowing sponge *Cliona celata*. PhD Thesis, Boston Univ, 171 pp

Hatch WI (1980) The implication of carbonic anhydrase in the physiological mechanism of penetration of carbonate substrata by the marine burrowing sponge *Cliona celata*. Biol Bull 159:135-147

Hein FJ, Risk MJ (1975) Bioerosion of coral heads: inner patch reefs, Florida reef tract. Bull Mar Sci 25:133-138

Highsmith RC (1981a) Coral bioerosion: damage relative to skeletal density. Amer Nat 117:193-198

Highsmith RC (1981b) Coral bioerosion at Enewetak: agents and dynamics. Int Rev Ges Hydrobiol 66:335-375

Highsmith RC, Lueptow RL, Schonberg SC (1983) Growth and bioerosion of three massive corals on the Belize barrier reefs. Mar Ecol Prog Ser 13:261-271

Hill MS (1996) Symbiotic zooxanthellae enhance boring and growth rates of the tropical sponge *Anthosigmella varians* forma *varians*. Mar Biol 125:649-654

Hill MS (1999) Morphological and genetic examination of phenotypic variability in the tropical sponge *Anthosigmella varians*. Mem Queensland Mus 44:239-247

Hinde GJ (1883) Catalogue of the fossil sponges in the geological department of the British Museum (Natural History). Trustees British Mus, London, 248 pp

Hoeksema BW (1983) Excavation patterns and spicule dimensions of the boring sponge *Cliona celata* from the SW Netherlands. Senckenbergiana Marit 15:55-85

Hogg J (1851) On Dr. Nardo's classification of the spongiae. Ann Mag Nat Hist, Ser 2, 7:190-193

Holmes KE (1997) Eutrophication and its effects on bioeroding sponge communities. In: Lessios HA, Macintyre IG (eds) Proc 8th Int Coral Reef Symp Panama, 2:1411-1416

Holmes KE (2000) Effects of eutrophication on bioeroding sponge communities with the description of new West Indian sponges, *Cliona* spp. (Porifera: Hadromerida; Clionidae). Invertebr Biol 119:125-138

Holmes KE, Edinger EN, Hariyadi, Limmon GV, Risk MJ (2000) Bioerosion of live massive corals and branching coral rubble on Indonesian coral reefs. Mar Poll Bull 40:606-617

Hooper JNA, van Soest RWM (2002) Systema Porifera. A guide to the classification of sponges. Kluwer Acad / Plenum Publ, New York, 1, 1149 pp

Hopkins SH (1956) Notes on the boring sponges in Gulf Coast estuaries and their relation to salinity. Bull Mar Sci Gulf Caribbean 6:44-58

Hopkins SH (1962) Distribution of species of *Cliona* (boring sponge) on the eastern shore of Virginia in relation to salinity. Chesapeake Sci 3:121-124

Hudson JH (1977) Long-term bioerosion rates on a Florida reef: a new method. In: Taylor DL (ed) Proc 3rd Int Coral Reef Symp Miami, Florida, 2:491-497

Hutchings P, Peyrot-Clausade M, Osnorno A (2005) Influence of land runoff on rates and agents of bioerosion of coral substrates. Mar Poll Bull 51:438-447

Hyatt JD (1882) The boring sponge – does it excavate the burrows in which it is found? Amer Month Microscop J 3(5):81-84

Keller C (1891) Die Spongienfauna des rothen Meeres. 2. Hälfte. Z Wiss Zool 52:294-368

Kiene WE, Hutchings PA (1992) Long-term bioerosion of experimental coral substrates from Lizard Island, Great Barrier Reef. In: Richmond RH (ed) Proc 7th Int Coral Reef Symp Guam, 1:397-403

Kiene WE, Hutchings PA (1994) Bioerosion experiments at Lizard Island, Great Barrier Reef. Coral Reefs 13:91-98

Kobluk DR, Kozelj M (1985) Recognition of a relationship between depth and macroboring distribution in growth framework reef cavities, Bonaire, Netherland Antilles. Bull Canad Petroleum Geol 33:462-470

Kühl M, Revsbech NP (2001) Biogeochemical microsensors for boundary layer studies. In: Boudreau BP, Jørgensen BB (eds) The benthic boundary layer. Transport processes and biogeochemistry. Oxford Univ Press, Oxford, pp 180-210

Leidy J (1856) On a boring sponge. Amer J Sci Arts, Ser 2, 22(66):281-282

Leidy J (1889) The boring sponge *Cliona*. Proc Acad Nat Sci Philadelphia 41:70-75

Lereboullet A (1841) Sur une espèce d'éponge perforante qui occupe l'épaisseur des valves de l'huître commestible (*Ostrea edulis*). Institut 9:131-132

Letellier A (1894) Une action purement mécanique suffit aux Cliones pour creuser les galeries dans les valves des huîtres. C R Hebd Séance Acad Sci Paris 118:986-989

Leys SP, Meech RW (2006) Physiology of coordination in sponges. Canad J Zool 84:288-306

Linnaeus C (1758) Systema naturae per regna tria naturae, secundum classes, ordines, genera, species, cum characteribus, differentiis, synonymis, locis. Tomus I. Editio decima, reformata. Laurentii Salvii, Holmiae, 824 pp

López-Victoria M, Zea S (2004) Storm-mediated coral colonization by an excavating Caribbean sponge. Climate Res 26:251-256

López-Victoria M, Zea S (2005) Current trends of space occupation by encrusting excavating sponges on Colombian coral reefs. Mar Ecol 26:33-41

López-Victoria M, Zea S, Weil E (2006) Competition for space between encrusting excavating Caribbean sponges and other coral reef organisms. Mar Ecol Prog Ser 312:113-121

Lorenz B, Bohnensack R, Gamulin V, Steffen R, Müller WEG (1996) Regulation of motility of cells from marine sponges by calcium ions. Cell Signal 8:517-524

Mao Che L, Le Campion-Alsumard T, Boury-Esnault N, Payri C, Golubic S, Bézac C (1996) Biodegradation of shells of the black pearl oyster, *Pinctada margaritifera* var. *cumingii*, by microborers and sponges of French Polynesia. Mar Biol 126:509-519

Márquez JC, Zea S, López-Victoria M (2006) Is competition for space between the encrusting excavating sponge *Cliona tenuis* and corals influenced by higher-than-normal temperatures? Bol Invest Mar Cost 35:259-265

Michelin H (1846) Note sur différentes espèces du genre *Vioa* (famille des Spongiaires). Rev Zool Ann 9:56-61

Moore Jr CH, Shedd WH (1977) Effective rates of sponge bioerosion as a function of carbonate production. In: Taylor DL (ed) Proc 3rd Int Coral Reef Symp Miami, Florida, 2:499-505

Muricy G (1991) Structure des peuplements des spongiaires autour de l'égout de Cortiou (Marseille, France). Vie Milieu 41:205-221

Nardo DG (1839) Sopra un nuovo genere di spugne, le quali perforano le pietri ed i gusci marini. Letta al veneto Ateneo nel giorno 29 Aprile 1839, ed all'Assemblea de' Naturalisti tenutasi in Pisa il giorno 7 Ottobre dell' anno stesso. Ann Sci Regno Lombardo-Veneto 9:221-226

Nasonov N (1881) Über die aushöhlende Kraft und zum feineren Bau der Clione. Vorläufige Mitteilungen. Zool Anz 4:459-460

Nasonov N (1883) Zur Biologie und Anatomie der Clione. Z Wiss Zool 39:295-308

Nasonov N (1887) Les éponges perforantes de la famille de clionides. Bull Soc Amis Sci Nat Moscou TL 236:362-366

Nasonov N (1924) Sur l'éponge perforante *Clione stationis* Nason. et le procédé du creusement des galeries dans les valves des huîtres. C R Acad Sci Russia 1924:113-115

Neumann AC (1966) Observations on coastal erosion in Bermuda and measurements of the boring rate of the sponge, *Cliona lampa*. Limnol Oceanogr 11:92-108

Old MC (1942) The boring sponges and their effect on shellfish culture. Convention Pap, Natl Shellfish Assoc Philadelphia, Mimeo, pp 1-3

Omnes P (1991) Micromorphologie des galleries de démosponges perforantes. Rep Univ Aix Marseille II, 20 pp

Osler E (1826) On burrowing and boring animals. Phil Trans Roy Soc London 116:342-371

Osnorno A, Peyrot-Clausade M, Hutchings PA (2005) Patterns and rates of erosion in dead *Porites* across the Great Barrier Reef (Australia) after 2 years and 4 years of exposure. Coral Reefs 24:292-303

Otter GW (1937) Rock-destroying organisms in relation to reef corals. Brit Mus Great Barrier Reef Exp 1928-29 Sci Rep 1(12):323-351

Oxford Dictionary of National Biography (as of March 14th, 2007) http://www.oxforddnb.com

Pang RK (1973) The systematics of some Jamaican excavating sponges (Porifera). Postilla Peabody Mus Yale Univ 161:1-75

Pallas PS (1766) Elenchus Zoophytorum sistens generum adumbrations generaliores et specierum cognitarum succinctas descriptions cum selectis auctorum synonymis. Van Cleef, The Hague, 451 pp

Parfitt E (1871) On the boring of molluscs, annelids and sponges into rocks, wood and shells. Rep Trans Devonshire Assoc Adv Sci 4:456-466

Pari N, Peyrot-Clausade M, Hutchings PA (2002) Bioerosion of experimental substrates on high islands and atoll lagoons (French Polynesia) during five years of exposure. J Exp Mar Biol Ecol 276:109-127

Parkinson J (1808) Organic remains of a former world. An examination of the mineralized remains of the vegetables and animals of the Antediluvian world generally termed extraneous fossils. Vol 2. The fossil zoophytes. Whittingham, London, 286 pp

Pavans de Ceccatty M (1971) Effects of drugs and ions on a primitive system of spontaneous contractions in a sponge (*Euspongia officinalis*). Experientia 27:57-59

Peyrot-Clausade M, Chazottes V, Pari N (1999) Bioerosion in the carbonate budget of two Indo-Pacific reefs: La Réunion (Indian Ocean) and Mooréa (Pacific Ocean). Bull Geol Soc Denmark 45:151-155

Pomponi SA (1976) An ultrastructural study of boring sponge cells and excavated substrata. Scan Electron Microscop 1976:569-574

Pomponi SA (1977) Etching cells of boring sponges: an ultrastructural analysis. In: Taylor DL (ed) Proc 3rd Int Coral Reef Symp, Miami, Florida, 2:485-490

Pomponi SA (1979a) Cytochemical studies of acid phosphatase in etching cells of boring sponges. J Mar Biol Assoc UK 59:785-789

Pomponi SA (1979b) Ultrastructure of cells associated with excavation of calcium carbonate substrates by boring sponges. J Mar Biol Assoc UK 59:777-784

Pomponi SA (1979c) Ultrastructure and cytochemistry of the etching area of boring sponges. In: Lévi C, Boury-Esnault N (eds) Biologie des Spongiaires. Coll Int CNRS Paris 291:317-323

Pomponi SA (1980) Cytological mechanisms of calcium carbonate excavation by boring sponges. Int Rev Cytol 65:301-319

Priest BW (1881a) On the natural history and histology of sponges. J Quekett Microscop Club 6:229-238

Priest BW (1881b) Further remarks on the histology of sponges. J Quekett Microscop Club 6:269-271

Prosser CL (1967) Ionic analyses and effects of ions on concentractions of sponge tissue. Z Vgl Physiol 54:109-120

Reed JK, Pomponi SA (1997) Biodiversity and distribution of deep and shallow water sponges in the Bahamas. In: Lessios HA, Macintyre IG (eds) Proc 8th Int Coral Reef Symp Panama, 2:1387-1392

Rose C, Risk MJ (1985) Increase in *Cliona delitrix* infestation of *Montastrea cavernosa* heads on an organically polluted portion of the Grand Cayman fringing reef. Publ Stn Zool Napoli Mar Ecol 6:345-362

Rosell D (1994) Morphological and ecological relationships of two clionid sponges. Ophelia 40:37-50

Rosell D, Uriz M-J (1991) *Cliona viridis* (Schmidt, 1962) and *Cliona nigricans* (Schmidt, 1862) (Porifera, Hadromerida): evidence which shows they are the same species. Ophelia 33:45-53

Rützler K (1971) Bredin-Archbold-Smithsonian Biological Survey of Dominica: burrowing sponges, genus *Siphonodictyon* Bergquist, from the Caribbean. Smisthon Contr Zool 77:1-37

Rützler K (1974) The burrowing sponges of Bermuda. Smithson Contr Zool 165:1-32

Rützler K (1975) The role of burrowing sponges in bioerosion. Oecologia 19:203-216

Rützler K (2002) Impact of crustose clionid sponges on Caribbean reef corals. Acta Geol Hisp 37:61-72

Rützler K, Rieger G (1973) Sponge burrowing: fine structure of *Cliona lampa* penetrating calcareous substrata. Mar Biol 21:144-162

Sammarco PW, Risk MJ (1990) Large-scale patterns in internal bioerosion of *Porites*: cross-continental shelf trends on the Great Barrier Reef. Mar Ecol Prog Ser 59:145-156

Sammarco PW, Risk MJ, Rose C (1987) Effects of grazing and damselfish territoriality on internal bioerosion of dead corals: indirect effects. J Exp Mar Biol Ecol 112:185-199

Schmidt O (1862) Die Spongien des adriatischen Meeres. Engelmann, Leipzig, 88 pp

Schönberg CHL (2000) Bioeroding sponges common to the central Great Barrier Reef: descriptions of three new species, two new records, and additions to two previously described species. Senckenbergiana Marit 30:161-221

Schönberg CHL (2001) Small-scale distribution of Great Barrier Reef bioeroding sponges in shallow water. Ophelia 55:39-54

Schönberg CHL (2002a) *Pione lampa*, a bioeroding sponge in a worm reef. Hydrobiologia 482:49-68

Schönberg CHL (2002b) Substrate effects on bioeroding demosponge *Cliona orientalis*. 1. Bioerosion rates. Publ Stn Zool Napoli Mar Ecol 23:313-326

Schönberg CHL (2006) Growth and erosion in the zooxanthellate Australian bioeroding sponge *Cliona orientalis* are enhanced in light. In: Suzuki Y, Nakamori T, Hidaka M, Kayanne H, Casareto BE, Nadao K, Yamano H, Tsuchiya M (eds) Proc 10th Int Coral Reef Symp Okinawa, Japan, pp 166-174

Schönberg CHL, Shields G (this volume) Micro-computed tomography for studies on *Entobia*: transparent substrate versus modern technology. In: Wisshak M, Tapanila L (eds) Current developments in bioerosion. Springer, Berlin, pp 147-164

Schönberg CHL, Suwa R (in press) Why bioeroding sponges are better hosts for symbiotic dinoflagellates than many corals. In: Custódio MR, Hajdu E, Lôbo-Hajdu G, Muricy G (eds) Porifera research: biodiversion, innovation and sustainability. Proc 7th Int Sponge Symp Búsioz, Brazil

Schönberg CHL, Wilkinson CR (2001) Induced colonization of corals by a clionid bioeroding sponge. Coral Reefs 20:69-76

Schönberg CHL, Wilkinson CR, Hoppe KN (1997) Coral skeletal destruction – boring sponges of the Great Barrier Reef. In: Campbell J, Dalliston C (eds) The Great Barrier Reef. Science, use and management. Proc Nat Conf Townsville, Australia, 2:67-71

Schönberg CHL, Hoffmann F, Gatti S (2004) Using microsensors to measure sponge physiology. Boll Mus Ist Biol Univ Genova 68:593-604

Schönberg CHL, de Beer D, Lawton A (2005) Oxygen microsensor studies on zooxanthellate clionaid sponges from the Costa Brava, Mediterranean Sea. J Phycol 41:774-779

Scoffin TP, Stearn CW, Boucher D, Frydl P, Hawkins CM, Hunter IG, MacGeachy JK (1980) Calcium carbonate budget of a fringing reef on the west coast of Barbados. Bull Mar Sci 30:475-508

Scott PJB, Moser KA, Risk MJ (1988) Bioerosion of concrete and limestone by marine organisms – a 13 year experiment from Jamaica. Mar Poll Bull 19:219-222

Siegrist Jr HG, Randall RH, Edwards CA (1991) Shallow reef-front detrital sediments from the northern Mariana Islands. Micronesia 24:231-248

Sollas WJ (1878) On two new and remarkable species of *Cliona*. Ann Mag Nat Hist, Ser 5, 1:54-67

Sullivan BW, Faulkner DJ (1990) Chemical studies of the burrowing sponge *Siphonodictyon coralliphagum*. In: Rützler K (ed) New perspectives in sponge biology. Proc 3rd Sponge Conf Washington, pp 45-50

Thiele J (1900) Kieselschwämme von Ternate 1. Abh Senckenberg Natf Ges 25:19-80

Thomas PA (1997) Destruction of coral reef by boring sponges. In: Hoon V (ed) Proc Regional Workshop Conservation Sustainable Management Coral Reefs. Proc 22 CRSARD, Chennai, Madras, pp C-117 - C-122

Topsent E (1888) Contribution à l'étude des Clionides. Arch Zool Exper Gén, Ser 2, 5:1-165

Topsent (1892) Diagnoses d'éponges nouvelles de la Méditerranée et plus particulièrement de Banyuls. Arch Zool Exper Gén, Ser 2, 10:17-28 [Notes et Revue 6]

Topsent E (1894) Sur le Mécanisme de la Perforation des Cliones. Arch Zool Exper Gén, Ser 3, 2:10-13

Topsent E (1896) Matériaux pour servir à l'étude de la faune des spongiaires de France. Mém Soc Zool France 9:113-133

Topsent E (1900) Étude monographique des spongiaires de France. 3. Monaxonida (Hadromerina). Arch Zool Exper Gén, Ser 3, 8:1-331

Topsent E (1932) Notes sur des Clionides. Arch Zool Exper Gén 74(28):549-579

Uriz MJ, Rosell D, Martín D (1992) The sponge population of the Cabrera Archipelago (Balearic Islands): characteristics, distribution, and abundance of the most representative species. Publ Stn Zool Napoli Mar Ecol 13:101-117

Vacelet J (1999) Planctonic armoured propagules of the excavating sponge *Alectona* (Porifera: Demospongiae) are larvae: evidence from *Alectona wallichii* and *A. mesatlantica* sp. nov. Mem Queensland Mus 44:627-642

Vosmaer GCJ (1928) Bibliography of sponges 1551-1913. Cambridge Univ Press, London, 234 pp [Edited and published after Vosmaer's death by Bidder GP and Vosmaer-Röell CS]

Vosmaer GCJ (1933) The sponges of the Bay of Naples; Porifera incalcaria. Nijhoff, The Hague, 1, 456 pp

Waller JG (1871) On the so-called boring or burrowing sponge. *Cliona celata* (Grant). *Halichondria celata* (Johnston). *Hymeniacidon celata* (Bowerbank). J Quekett Microscop Club 2:269-277

Waller JG (1881) On *Cliona celata* (Grant). *Hymeniacidon celata* (Bowerbank). Does the sponge make the hole? J Quekett Microscop Club 6:251-268

Warburton FE (1958) The manner in which the sponge *Cliona* bores into calcareous objects. Canad J Zool 36:555-562

Ward PW, Risk MJ (1977) Boring pattern of the sponge *Cliona vermifera* in the coral *Montastrea annularis*. J Paleont 51:520-526

Ward-Paige CA, Risk MJ, Sherwood OA, Jaap WC (2005) Clionid sponge surveys on the Florida Reef Tract suggest land-based nutrient inputs. Mar Poll Bull 51:570-579

Wells HW (1959) Boring sponges (Clionidae) of Newport River, North Carolina. J Elisha Mitchell Sci Soc 75:168-173

Wells HW (1961) The fauna of oyster beds, with special reference to the salinity factor. Ecol Monogr 31:139-266

Wikipedia, the free encyclopedia (accessed March 14th, 2007) http://en.wikipedia.org/wiki

Young HR, Nelson CS (1985) Biodegradation of temperate-water skeletal carbonates by boring sponges on the Scott Shelf, British Columbia, Canada. Mar Geol 65:33-45

Zea S, Weil E (2003) Taxonomy of the Caribbean excavating sponge species complex *Cliona caribbaea – C. aprica – C. langae* (Porifera, Hadromerida, Clionaidae). Caribbean J Sci 39:348-370

Zundelevich A, Lazar B, Ilan M (2007) Chemical versus mechanical bioerosion of coral reefs by boring sponges – lessons from *Pione* cf. *vastifica*. J Exp Biol 210:91-96

Substratum microtexture affects the boring pattern of *Cliona albimarginata* (Clionaidae, Demospongiae)

Barbara Calcinai[1], Giorgio Bavestrello[1], Carlo Cerrano[2], Laura Gaggero[2]

[1] Dipartimento di Scienze del Mare, Università Politecnica delle Marche, 60131 Ancona, Italy, (b.calcinai@univpm.it)
[2] Dipartimento per lo studio del Territorio e delle sue Risorse, Università di Genova, 16132 Genova, Italy

Abstract. This work focuses on the erosion pattern morphology of the Indonesian excavating sponge *Cliona albimarginata* in three different calcareous substrates (Carrara marble, branches of the coral *Acropora*, and the umbo of the bivalve *Hippopus*). The experimental data demonstrate that the sponge produced galleries that followed the preferred orientation of fibrous aragonite crystals (*Acropora*), or the parallel lamination within the *Hippopus* shell. Conversely, the homogeneous granoblastic, mosaic texture of Carrara marble was etched without preferred directions. SEM analysis showed that pit size also varies between the three substrates in relation to the respective substrate porosity.

Keywords. *Cliona albimarginata,* bioerosion, excavating sponge, crystal texture, petrography, calcium carbonate, Indonesia

Introduction

Excavating sponges produce a series of connected galleries within calcareous substrata in which they live. The sponges excavate by chemical and mechanical means employing etching cells (Rützler and Rieger 1973; Pomponi 1980) and producing typical microscopic scars (pits) on the walls of the excavations. Recently Calcinai et al. (2003, 2004) have demonstrated that pits of different taxa of boring sponges have different morphological characteristics, such as concentric grooves or ornamentations that may be used as a supplementary character in the diagnosis of super-specific taxa.

It is well known that sponges are able to excavate both inorganic and biogenic calcium carbonates (e.g., Neumann 1966; Rützler and Rieger 1973) but very little is known about the influence of different forms of carbonate on the erosion process. Several authors (Neumann 1966; Highsmith 1982; Rose and Risk 1985; Schönberg 2002) demonstrated that the density of the substrate affects the erosion rate while no evidence was obtained about the influence of the substrate mineralogy on the process.

M. Wisshak, L. Tapanila (eds.), *Current Developments in Bioerosion.* Erlangen Earth Conference Series, doi: 10.1007/978-3-540-77598-0_10,

At a microscopic level the comparison of the size of the erosion scars produced by the same species in different substrata (Calcinai et al. 2004) demonstrated that the pits of *Cliona vermifera* Hancock, 1867 and *Cliona rhodensis* Rützler and Bromley, 1981 are larger when the sponges bore calcareous rock rather than coralline algae. Pit size is also thought to vary in relation to the micro-topography of the substratum: in pioneer tunnels the chips are smaller than in large boring chambers (Rützler and Rieger 1973; Calcinai et al. 2004).

In a recent experiment (Calcinai et al. in press) we explored the importance of different mineralogical features on the erosion rate of the Indopacific species *Cliona albimarginata* Calcinai, Bavestrello and Cerrano, 2005. In that case we recorded differences of one order of magnitude of eroded material (kg/m^2/y) in relation to the kind of substrate the sponge is growing in.

Here we study at different scales the differences in the erosion pattern of *C. albimarginata* in three different calcareous substrata (marble, coral and mollusc shell) characterised by a different microtexture.

Materials and methods

Our experimental sponge, *Cliona albimarginata*, is a very destructive, symbiotic species occurring in the Bunaken Marine Park (North Sulawesi, Indonesia) and is known also from East Africa (F. Azzini pers. comm.). It was observed exclusively in beta growth form covering wide portions of coralline substrate. To evaluate whether the substrate microtexture affects the boring pattern we have tied blocks of different mineral composition, texture and porosity on a single specimen of *C. albimarginata* (about 15 m^2 wide). The blocks (6 x 4 x 2 cm) were directly tied onto the sponge surface using wire and nails. In this way, avoiding sponge grafts commonly used in this kind of experiment (Neumann 1966; Rützler 1975; Schönberg and Wilkinson 2001), we limited the stress to the sponge due to the handling procedure. The blocks used were: 15 blocks obtained from the skeleton of the coral *Acropora* Oken, 1815; 15 blocks obtained from the umbo of the shell of *Hippopus* Lamarck, 1799; 15 blocks of Carrara marble. After 200 days the blocks were removed and dried. The bored post-experimental blocks were embedded and filled with epoxy resin, then processed to produce 30 μm thick sections glued onto slides by conventional petrographic techniques. The optical mineralogical and petrographic analyses were carried out by stereoscopic and transmitted light microscopy on the biogenic or inorganic substrata, identifying the mineralogical components, their shape and overall texture. The mineral phases were confirmed using a Siemens diffractometer with Ni filtered CuK radiation X-ray.

To evaluate the size of the pits produced by the sponge inside the different substrata, some fragments of *Acropora*, *Hippopus* and marble were cleaned in sodium hypochlorite solution, washed in water and air dried. The fragments were put onto stubs for scanning electron microscopy and sputtered with gold. SEM was conducted using a Philips EM 515. Pit maximal dimensions were measured on SEM images with the freeware program ImageJ 1.37; diameters of 120 pits per each kind of substrate were measured from central region of the main erosion chambers avoiding pioneer areas.

Results

The observation of the sections of Carrara marble revealed a granoblastic texture (Fig. 1A-D) due to calcite grains homogeneous in size (0.3-0.5 mm). Boundaries between calcite grains were generally straight and tended to display triple point junctions. Lobate and indentate boundaries were scarce, as were subgrains. Erosions, visible at low magnification, were represented by isodiametric, oval, homogeneously distributed galleries (Fig. 1A-B). Pronounced etching frequently resulted in coalescent-lobated galleries (Fig. 1C). The average diameter of discrete cavities was 1.5-2 mm increasing in the coalescent galleries. Single calcium

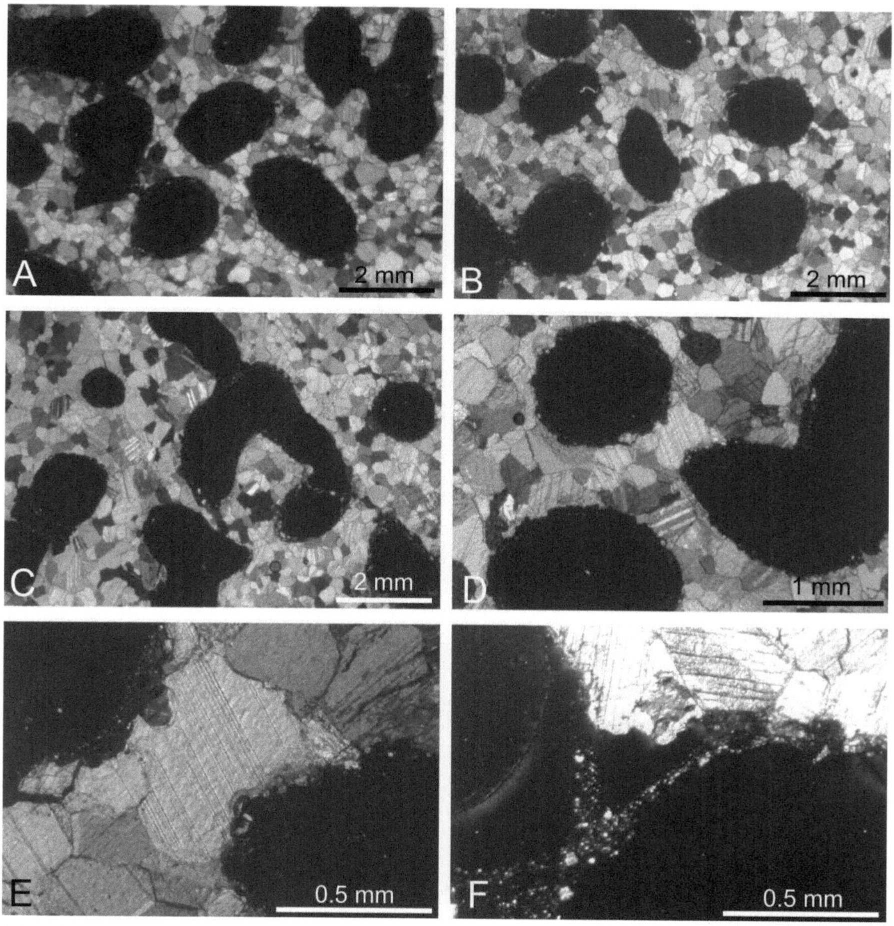

Fig. 1 Transmitted light microphotographs of Carrara marble perforated by *Cliona albimarginata.* **A-B** Erosion galleries quite homogeneously distributed inside the homogeneous granoblastic texture of the marble. **C** Fusion of different galleries to bigger, lobate ones. **D** Rounded galleries with lobate boundaries due to the pitted sponge erosion. The calcite grains are attacked independent of grain orientation. **E** Detail of the wall between two galleries showing the pitted erosion. **F** Detail of joint galleries. A detrital fraction derived from erosion is preserved by the impregnation in epoxy of the substratum

carbonate crystals were etched, resulting in size reduction of the crystals showing the typical pitting surface (Fig. 1D-E). At the end of the process, detrital calcite from the eroded walls between adjacent chambers was preserved by the impregnation in epoxy of the substratum (Fig. 1F).

The sections showed that texture of the coral skeleton of *Acropora* was composed of acicular (i.e., needle shaped) to fibre-shaped aragonite crystals (average diameter <0.05 mm) arranged in a gently bent to wavy pattern (Fig. 2A-B). Chambers produced by sponge erosion were elongated, following the orientation of the aragonite crystal bundles (Fig. 2A-D). The texture at microscale (Fig. 2C-F) provided evidence that the pitting caused by sponge erosion developed perpendicularly to the direction of

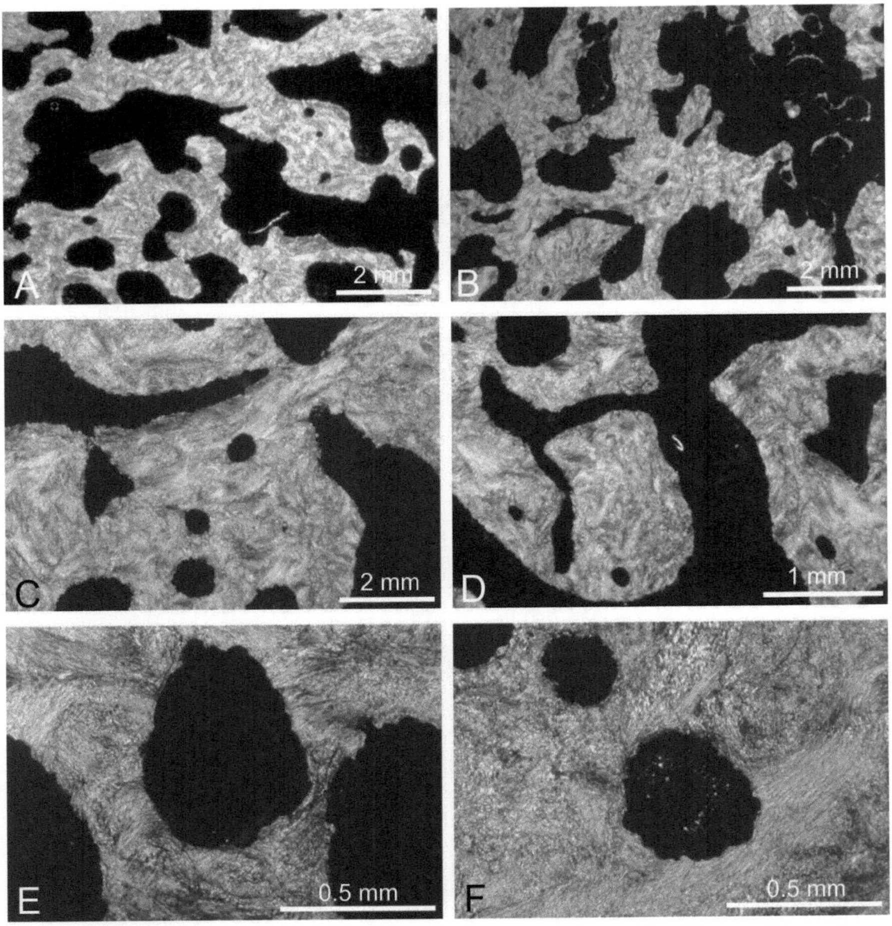

Fig. 2 Transmitted light microphotographs of *Acropora* perforated by *Cliona albimarginata*. **A-B** The inner fabric of *Acropora* skeleton is characterised by sinuous aragonite fibres. The erosions follow the sinuous orientations of crystals. **C-D** Similar erosion patterns at different magnifications. Holes are generally aligned with the crystal orientation. **E-F** Details of galleries with the pitted boundaries, where chips of substratum were removed. In F, a fine-grained detrital fraction derived from erosion is preserved by the impregnation in epoxy of the substratum

the aragonite fibre. A ring of detrital aragonite was sometimes preserved inside the excavation (Fig. 2F).

The shell structure of *Hippopus* displayed a gently bent stratified geometry (Fig. 3A-B). Aragonite crystals <0.05 mm in diameter with chevron kinks perpendicular to the surface of the shell formed the layers (Fig. 3A-D). The etching followed the crystal orientation (Fig. 3A). In the inner part, the etching developed tangential galleries infesting the entire layer of crystals (Fig. 3B-C). The walls of the eroded galleries had a pitting geometry in particular correspondence of the kink band suggesting that the highest excavation rate occurred where lattice defects were localised (Fig. 3F). In this case the resin embedding preserved detrital remains (Fig. 3F).

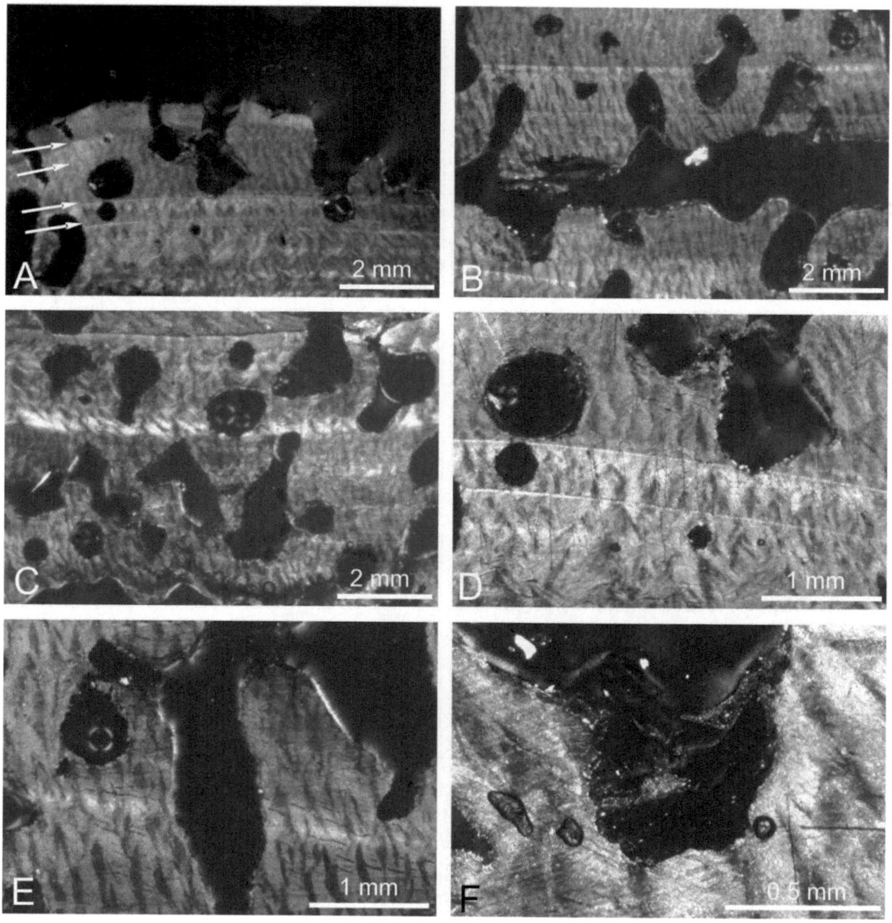

Fig. 3 Transmitted light microphotographs of *Hippopus* perforated by *Cliona albimarginata*. **A-B** The texture of the shell is characterised by aragonite crystals with chevron kinks perpendicular to the shell surface, organised in layers of variable thickness parallel to the surface (arrows indicate boundaries). The sponge penetrates from the surface and follows the direction of aragonite crystals. In the inner part of the shell, the etching developed as tangential galleries in the entire layer of a given crystal. **C-F** Details of the vertical penetration that follows the crystal orientation. In D, the pitting pattern and in F the remains of detrital calcite preserved by the resin are visible

The SEM study showed that the microscopic pattern of the erosion scars produced by *C. albimarginata* was similar in the three different substrata. In accordance with previous observations (Calcinai et al. 2003, 2004) conducted on several species of *Cliona*, the pits of *C. albimarginata* were polygonal, elongated, with a smooth surface. Very slight microtopographic differences of the surface of the pits were probably due to the crystal texture of the substratum (Fig. 4).

The diameters of the pits produced by *C. albimarginata* were 49.8 ± 10 µm (28-72 µm), 38.7 ± 7.9 (19-62 µm), 50.6 ± 12.4 µm (23-77 µm), respectively, in the marble, in *Acropora* and in the *Hippopus* (Fig. 5).

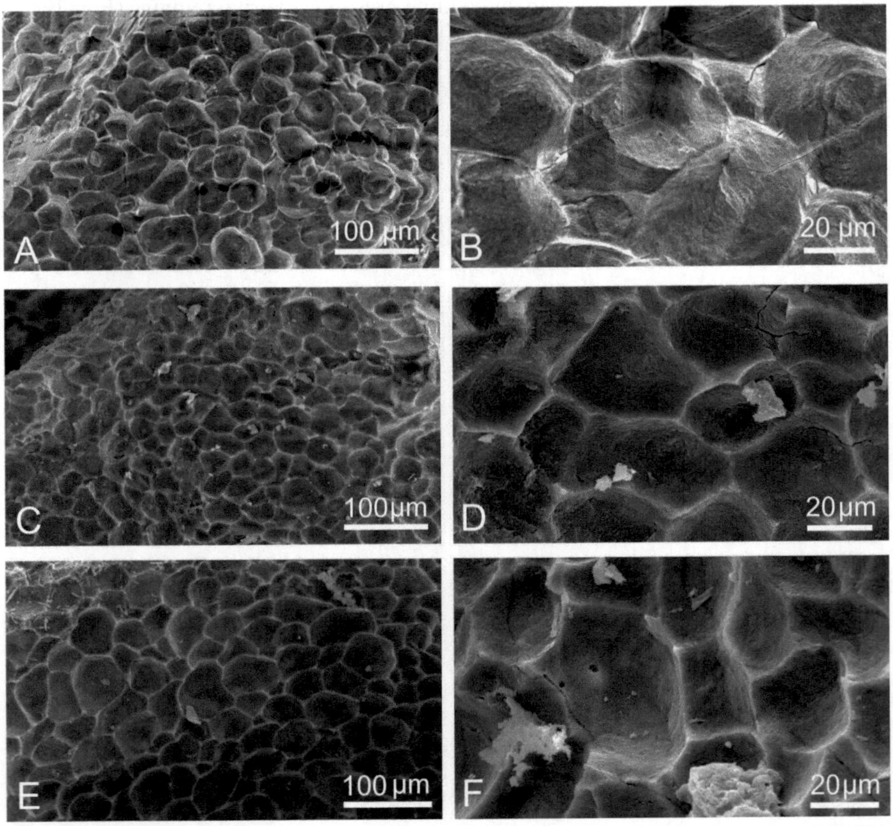

Fig. 4 SEM of the microscopic erosion traces (pits) produced by *C. albimarginata*. **A-B** Pits in marble. **C-D** Pits in *Acropora*. **E-F** Pits in *Hippopus*

Discussion

The data showed that microtexture of calcareous substrata affected the erosion pattern of *Cliona albimarginata*, evidencing that the same sponge can produce different excavation morphologies in different substrata. The sponge produced galleries that followed the preferred orientation of fibrous aragonite crystals (*Acropora*), or the parallel lamination in the *Hippopus* shell. Conversely, the homogeneous granoblastic texture of Carrara marble was etched without preferred directions.

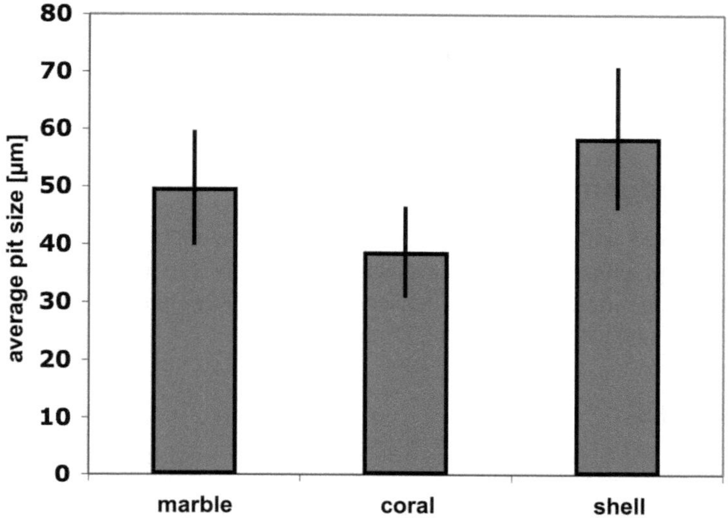

Fig. 5 Average size (±sd) of the pits made by *C. albimarginata* in the tested substrates

It is assumed that the boring mechanism of sponges is both chemical and mechanical, although the respective importance of the two steps is controversial (Rützler and Rieger 1973; Zundelevich et al. 2007). Our data suggest that the chemical etching drives the erosion morphology and enables the dissolution of the carbonate along preferential directions aligned with the crystal orientation.

This evidence provides a basis for a more consistent discussion about the specificity of the erosion pattern in boring sponges. Some authors assumed that the shape and size of the chambers and cavities are to be considered species-specific and therefore are suitable for the identification of boring sponges (e.g., Bromley and Tendal 1973; Rützler 1974; De Groot 1977; Risk and MacGeachy 1978). On the other hand, other authors have suggested that the excavating patterns are related to some substratum characteristics or to the developing stage of the specimen (Volz 1939; Cobb 1975; Hoeksema 1983; Bromley and D'Alessandro 1989, 1990).

We suggest that the morphology of the excavation is the result of the interaction between a genetically determined pattern of growth of the sponge inside the substratum corresponding to the general morphology of the epilithic sponges and the textural features of the substratum that is able to drive the direction and shape of the excavations. For example *C. albimarginata* completely pervades its natural substratum occupying its natural pores and producing irregular, spherical or elliptic erosion chambers and galleries that extend deep in the coral skeleton (Calcinai et al. 2005). This erosion pattern is preserved in the three experimental substrata but in the section the erosion traces are sub-rounded in the marble and elongated in the biogenic substrata due to the different shape of the crystals.

In Calcinai et al. (2004), we already demonstrated that pit size varies within a single species excavating in algae compared to rock. The present data agree with these previous results confirming that in porous substrata (coralline algae, coral) the pit size is reduced in comparison to denser substrata (marble, shell). It is unlikely that these differences are due to the substratum microtexture; in fact the fibrous crystals of the coral are more similar to those of the shell than to those of marble. Future studies will be necessary to clarify this point.

Acknowledgements

We are indebted with Dr. Christine Schönberg for the comments and the help in the manuscript preparation. The project was partially funded by the Executive program of scientific co-operation between the Italian Republic and the Republic of Indonesia in 2004 to 2007 (project STE-16).

References

Bromley RG, D'Alessandro A (1989) Ichnological study of shallow marine endolithic sponges from the Italian coast. Riv Ital Paleont Stratigr 95:279-314

Bromley RG, D'Alessandro A (1990) Comparative analysis of bioerosion in deep and shallow water, Pliocene to recent, Mediterranean Sea. Ichnos 1:43-49

Bromley RG, Tendal OS (1973) Example of substrate competition and phobotropism between two clionid sponges. J Zool 169:151-155

Calcinai B, Arillo A, Cerrano C, Bavestrello G (2003) Taxonomy-related differences in the excavating micro-patterns of boring sponges. J Mar Biol Assoc UK 83:37-39

Calcinai B, Bavestrello G, Cerrano C (2004) Bioerosion micro-patterns as diagnostic characteristics in boring sponges. In: Pansini M, Pronzato R, Bavestrello G, Manconi R (eds) Sponge sciences in the new millennium. Boll Mus Ist Biol Univ Genova 68:239-252

Calcinai B, Bavestrello G, Cerrano C (2005) Excavating sponge species from the Indo-Pacific Ocean. Zool Stud 44:5-18

Calcinai B, Azzini F, Bavestrello G, Gaggero L, Cerrano C (in press) Excavating rates and boring pattern of *Cliona albimarginata* (Clionaidae, Demospongiae, Porifera) in different substrata. In: Custódio MR, Hajdu E, Lôbo-Hajdu G, Muricy G (eds) Porifera research: biodiversity, innovation & sustainability. Proc 7th Int Sponge Symp Búsioz, Brazil

Cobb WR (1975) Fine structural features of destruction of calcareous substrata by the burrowing sponge *Cliona celata*. Trans Amer Microscop Soc 94:197-202

De Groot RA (1977) Boring sponges (Clionidae) and their trace fossils from the coast near Rovinj (Yugoslavia). Geol Mijnb 56:168-181

Hancock A (1867) Note on the excavating sponges; with descriptions of four new species. Ann Mag Nat Hist, Ser 3, 19:229-242

Highsmith RC (1982) Reproduction by fragmentation in corals. Mar Ecol Prog Ser 7:207-226

Hoeksema BW (1983) Excavation patterns and spiculae dimensions of the boring sponge *Cliona celata* from the SW Netherlands. Senckenbergiana Marit 15:55-85

Lamarck JBPA de Monet de (1799) Prodromo d'une nouvelle classification de coquilles, comprenent une réaction appropiée de charactères génériques, et l'établissement d'un grand nombre de genres nouveaux. Mém Soc Hist Nat Paris 1:63-91

Neumann AC (1966) Observations on coastal erosion in Bermuda and measurements of the boring rate of the sponge *Cliona lampa*. Limnol Oceanogr 11:92-108

Oken L (1815) Okens Lehrbuch der Naturgeschichte. Dritter Theil. Zoologie. Mit vierzig Kupfertafeln. Erste Abtheilung. Fleischlose Thiere. Reclam, Leipzig, 850 pp

Pomponi SA (1980) Cytological mechanisms of calcium carbonate excavation by boring sponges. Int Rev Cytol 65:301-319

Risk MJ, McGeachy JK (1978) Aspects of bioerosion of modern Caribbean reefs. Rev Biol Trop 26:85-105

Rose CS, Risk MJ (1985) Increase in *Cliona delitrix* infestation of *Montastrea cavernosa* heads on an organically polluted portion of the Grand Cayman. PSZN Mar Ecol 6:345-363

Rützler K (1974) The burrowing sponges of Bermuda. Smithson Contr Zool 165:1-32

Rützler K (1975) The role of burrowing sponge in bioerosion. Oecologia 19:203-216

Rützler K, Bromley RG (1981) *Cliona rhodensis*, new species (Porifera: Hadromerida) from the Mediterranean. Proc Biol Soc 94:1219-1225

Rützler K, Rieger G (1973) Sponge burrowing: Fine structure of *Cliona lampa* penetrating calcareous substrata. Mar Biol 21:144-162

Schönberg CHL (2002) Substrate effects on the bioeroding demosponge *Cliona orientalis*. 1. Bioerosion rates. PSZN Mar Ecol 23:313-326

Schönberg CHL, Wilkinson CR (2001) Induced colonization of corals by a clionid bioeroding sponge. Coral Reefs 20: 69-76

Volz P (1939) Die Bohrschwämme (Clioniden) der Adria. Thalassia 3:1-64

Zundelevich A, Lazar B, Ilan M (2007) Chemical versus mechanical bioerosion of coral reefs by boring sponges - lessons from *Pione* cf. *vastifica*. J Exper Biol 210:91-96

Two new dwarf *Entobia* ichnospecies in a diverse aphotic ichnocoenosis (Pleistocene / Rhodes, Greece)

Max Wisshak[1]

[1] Institute of Palaeontology, Erlangen University, 91054 Erlangen, Germany, (wisshak@pal.uni-erlangen.de)

Abstract. This study explores and extends the lower size range of the sponge boring ichnotaxon *Entobia*, thereby introducing two new ichnospecies – *E. mikra* and *E. nana* – both reaching only a fraction of a millimetre in size. They are components of a diverse aphotic ichnocoenosis, comprising a total of 18 ichnotaxa, recorded in skeletons of the cold-water scleractinian *Lophelia pertusa* from Lower Pleistocene deposits of Rhodes. Recent analogs are studied and morphometrically analysed from various aphotic, cold-temperate settings in the NE Atlantic.

The first new ichnospecies, *E. mikra*, is characterised by solitary or clustered, irregular cavities ranging from 60 to 335 µm in size with a mean of only 171 µm, and comprises elongate, branched and anastomosing forms. The traces posses a characteristic botryoidal microsculpture with etching scars measuring 3.6 to 8.1 µm at a mean of only 5.6 µm. The scars are locally encountered in various growth stages resembling initial rims, semi- and nearly-closed etching cells. The other new ichnotaxon, *E. nana*, is slightly larger with solitary or clustered cavities measuring 102 to 756 µm in diameter at a mean of 298 µm, and show a bulged to weakly botryoidal surface texture. The chambers posses varying numbers of apophyses and exploratory threads which terminate bluntly or lead to neighbouring chambers.

Both new ichnospecies are assumed to represent the work of unknown boring micro-sponges, given the lack of alternatives among potential heterotroph bioerosion agents in aphotic environments, and considering the compelling resemblance in the botryoidal microsculpture of *E. mikra* (recording the activity of archaeocyte etching cells) and the presence of interconnected chambers in the case of *E. nana*. The possibility that they are merely juvenile stages of larger entobians is recejcted because of missing intermediate growth stages and the Gaussian size class frequency distribution, mirroring normal interspecies variability and population structure.

The ultimate proof can only be drawn by histological analyses or by finding siliceous sponge spicules – if present at all in the case of the responsible sponges – allowing for a taxonomical evaluation of the trace maker and shedding new light on the actual size range and diversity of boring sponges.

Keywords. Bioerosion, ichnotaxonomy, *Entobia mikra*, *Entobia nana*, boring sponge, etching scars, *Lophelia pertusa*

M. Wisshak, L. Tapanila (eds.), *Current Developments in Bioerosion*. Erlangen Earth Conference Series, doi: 10.1007/978-3-540-77598-0_11, © Springer-Verlag Berlin Heidelberg 2008

Introduction

The complex trace fossil *Entobia*, introduced by Bronn as early as 1837, is one of the most renowned ichnotaxa and its numerous ichnospecies display a wide range of morphologies reflecting the high diversity of their corresponding trace makers, mainly clionaid boring sponges (see Bromley and D'Alessandro 1984 for a review). The extent of *Entobia* systems is usually in the magnitude of a few centimetres but may reach more than a decimetre when large calcareous substrates are infested, then forming chambers of up to 500 cm^3 in volume (Rützler 1971). As this study will show, the lower limit in their size range extends to surprisingly small forms with a size of only a fraction of a millimetre, some of which possessing typical botryoidal etching scars of only around 5 µm in dimension. Two of these – formerly only dealt with in open nomenclature (e.g., Günther 1990; Beuck and Freiwald 2005; Wisshak 2006) or as morphological variants of other ichnotaxa (e.g., Radtke 1991; Försterra et al. 2005) – are now sufficiently understood to warrant ichnotaxonomical treatment. They are herein formally established as new *Entobia* ichnospecies and their implications for the size range and diversity of boring sponges are discussed.

The fossil holotype material comes from skeletons of the cold-water scleractinian coral *Lophelia pertusa* (Linnaeus, 1758) from Lower Pleistocene deposits of Rhodes (Greece). The traces are constituents of a diverse aphotic boring ichnocoenosis which was investigated in more detail by Bromley (2005) and is only briefly touched upon in this paper with the addition of some new findings.

Materials and methods

Sample material

The holotype material of both new *Entobia* ichnospecies originates from *Lophelia* skeletons from the Early Pleistocene Lindos Bay Clay (Rhodes Formation) from Rhodes. The type locality is located one km southwest of Lardos on the southwestern flank of a prominent isolated hill. General descriptions of the Rhodes Formation, which forms an extensive transgressive-regressive cycle, were given by Hanken et al. (1996) and Cornée et al. (2006), microfaunistic investigations and their palaeoenvironmental implications are presented in a recent special volume edited by Thomsen (2005), and preliminary results on the aphotic trace fossil assemblage recorded in *Lophelia* skeletons were given by Bromley (2005). *Lophelia pertusa* represents a relatively homogenous and sufficiently sized substrate composed of very fine needle aragonite, allowing for an idiomorphic development of small boring traces, hence, providing an ideal type substrate.

In addition, the traces were encountered in several Recent settings in the NE Atlantic – most of which also occur in cold-water coral skeletons and / or associated bivalves. A detailed list of locations, coordinates, sampling gear applied and substrates investigated is given in Tables 3 and 5.

The holotypes (MB.W 2020 and MB.W 2021) and selected Recent material (MB.W 2022 to 2027) are housed at the Museum für Naturkunde, Humboldt University, Berlin (MfNB).

Vacuum cast embedding technique

The calcareous skeletons were processed with the approved modified cast-embedding technique based on Golubic et al. (1970, 1983). The selected samples were placed in a vacuum chamber purpose-designed for this method (Struers, Epovac). This allows the infiltration of several samples with a low viscosity epoxy resin (Araldite BY158 + Aradur 21) while the samples are held under vacuum. This way, a complete infiltration of the borings is assured and trapping of air is avoided. After curing, the samples were formatted and decalcified by a treatment of diluted HCl (~5%) and consecutive rinsing in distilled water and drying at 70°C. The samples – now exhibiting the positive casts of the borings – were then sputter-coated with gold and finally analysed and photographed using SEM. The encountered traces were semiquantitatively grouped in four abundance classes (very common, common, rare and very rare). In the descriptions, morphological terms (e.g., bothryoidal) refer to the cast.

Morphometry

For exact measurements of the material, the SEM images were imported in image analysis software (Soft Imaging Systems, AnalySIS) which allows calibration of the scale and subsequent measurement of diameters, outlines and areas. The data of >70 specimens of each new ichnotaxon were then compiled (Microsoft, EXCEL) for basic statistical analyses (mean, standard deviation SD, min., max., n, %). The parameters considered were maximum diameter, width perpendicular to maximum diameter, outline, area (excluding exploratory threads), size of etching scars, presence of incomplete scars, number of apophyses + exploratory threads, and number of traces with close (<100 μm) proximity to each other. The summarised numerical data are given in Tables 2 and 4, and are graphically displayed in Figures 7 and 8.

Results

The bioerosion trace fossil assemblage

The trace fossil assemblage (Table 1) includes 18 different ichnotaxa and forms, and is dominated by the two key ichnospecies *Saccomorpha clava* (Fig. 1A) and *Orthogonum lineare* (Fig. 1B) which form the typical diagnostic aphotic microboring ichnocoenosis. In addition, various unspecified large *Entobia* ichnospecies (Fig. 1C) are the main bioerosion traces encountered. Other common taxa are the presumed foraminiferal boring *Semidendrina pulchra* (Fig. 1D) and one of the new ichnospecies *Entobia nana* (Figs. 5-6). Rare traces are represented by the brachiopod holdfast etchings *Podichnus centrifugalis* (Fig. 1E), polychaete worm borings *Caulostrepsis cretacea* (Fig. 1F), *Maeandropolydora* (Fig. 1G) and *Palaeosabella* (Fig. 1H), foraminiferal etchings (Fig. 2A) and potential foraminiferal pits *Oichnus simplex* (Fig. 2C). In addition, previously unrecorded possible bacterial microborings *Scolecia serrata* (Fig. 2B) and the new ichnospecies *Entobia mikra* (Figs. 3-4) were encountered. The phoronid worm boring *Talpina* (Fig. 2E) and the attachment scars *Centrichnus eccentricus* of anomiid bivalves (Fig. 2H) are very rare constituents of the present ichnocoenosis. They are complemented by the new findings of the fungal boring *Polyactina araneola* (Fig. 2D), the worm boring *Trypanites* (Fig. 2F) and predatory borings produced by the foraminifer *Hyrrokkin sarcophaga* (Fig. 2G), established as new ichnotaxon *Kardopomorphos polydioryx* by Beuck et al. (this volume).

Table 1 Ichnotaxa inventory in order of relative abundance as recorded in epoxy resin casts taken from fragments of the cold-water scleractinian *Lophelia pertusa*, with their known or inferred modern trace maker and relative abundance (+ + = very common; + = common, - = rare; - - = very rare)

Ichnotaxon	Known or assumed modern producer	Abundance
Saccomorpha clava Radtke, 1991	*Dodgella priscus* Zebrowski, 1937 (fungus)	+ +
Orthogonum lineare Glaub, 1994	? ? (fungus)	+ +
Entobia Bronn, 1837	e.g., *Cliona* Grant, 1826 (sponges)	+ +
Entobia nana isp. n.	? (micro-sponge)	+
Semidendrina pulchra Bromley et al., 2007	cf. *Globodendrina* Plewes, Palmer, Haynes, 1993 ? (foraminiferan)	+
Podichnus centrifugalis Bromley and Surlyk, 1973	? (brachiopod)	-
Caulostrepsis cretacea (Voigt, 1971)	e.g., *Polydora* Bosc, 1802 and other spionids (polychaete worm)	-
Maeandropolydora Voigt, 1965	e.g., *Polydora* Bosc, 1802 and other spionids (polychaete worm)	-
Palaeosabella Clarke, 1921	? (polychaete worm)	-
'larger dish-shaped etchings' (see Bromley 2005)	? (foraminiferan)	-
Scolecia serrata Radtke, 1991	? ? (bacterium)	-
Entobia mikra isp. n.	? (micro-sponge)	-
Oichnus simplex Bromley, 1981	? ? (foraminiferan)	-
Polyactina araneola Radtke, 1991	*Conchyliastrum* Zebrowski, 1937 (fungus)	- -
Talpina Hagenow, 1840	? (phoronid worm)	- -
Trypanites solitarius (Hagenow, 1840)	? (polychaete or sipunculan worm)	- -
Kardopomorphos polydioryx Beuck, López Correa and Freiwald, this volume	*Hyrrokkin sarcophaga* Cedhagen, 1994 (foraminiferan)	- -
Centrichnus eccentricus Bromley and Martinell, 1991	e.g., *Anomia aculeata* Gmelin, 1791 (anomiid bivalve)	- -

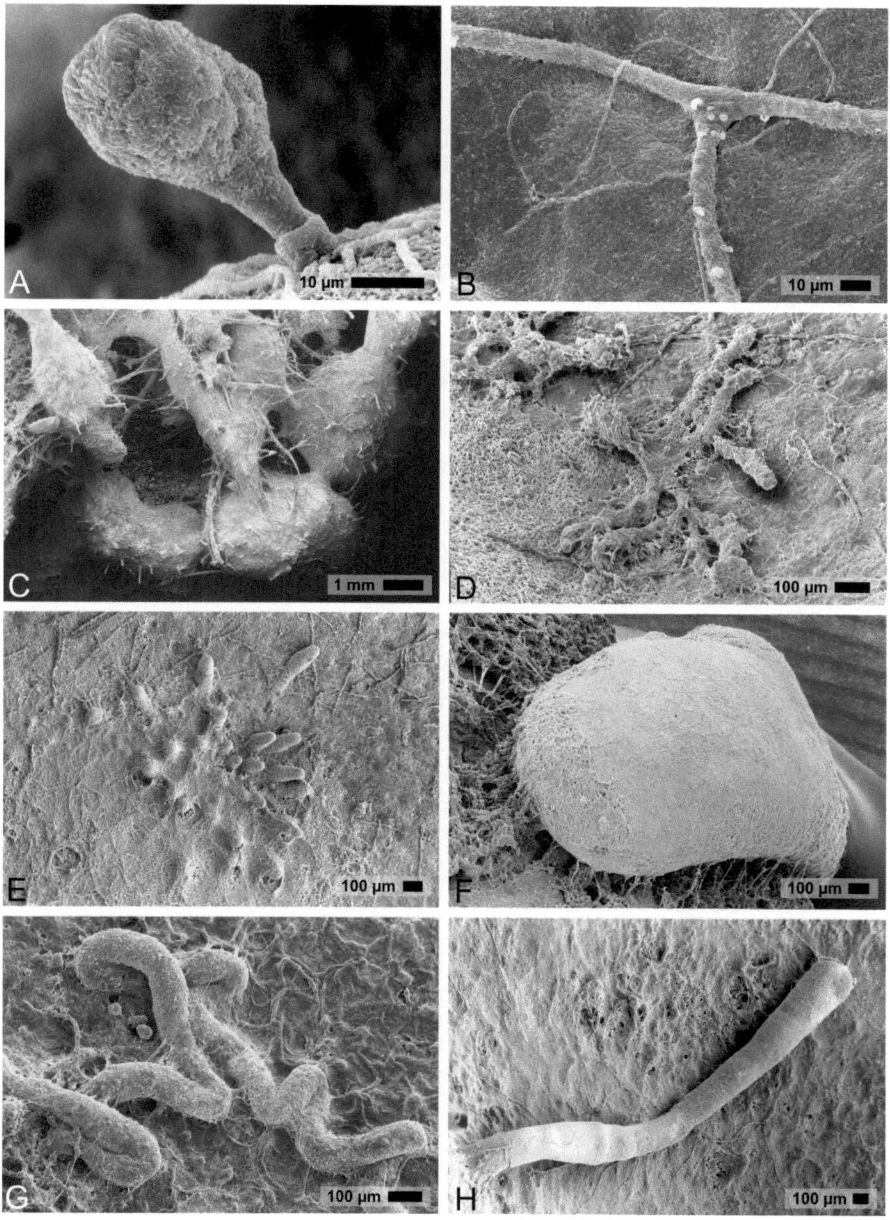

Fig. 1 SEM images of trace fossils recorded in epoxy resin casts of *Lophelia pertusa*, Lower Pleistocene, Rhodes, Greece. **A** Sack-shaped sporangial cavity of the fungal boring *Saccomorpha clava*. **B** Perpendicular branching of fungal galleries *Orthogonum lineare*. **C** Part of a camerate sponge boring, an unspecified ichnospecies of *Entobia*. **D** Two presumable foraminiferal borings *Semidendrina pulchra*. **E** Brachiopod attachment scar *Podichnus centrifugalis*. **F** Polychaete worm boring *Caulostrepsis cretacea*. **G** Substrate-parallel meanders of *Maeandropolydora*. **H** The distally enlarging blunt tunnel *Palaeosabella*

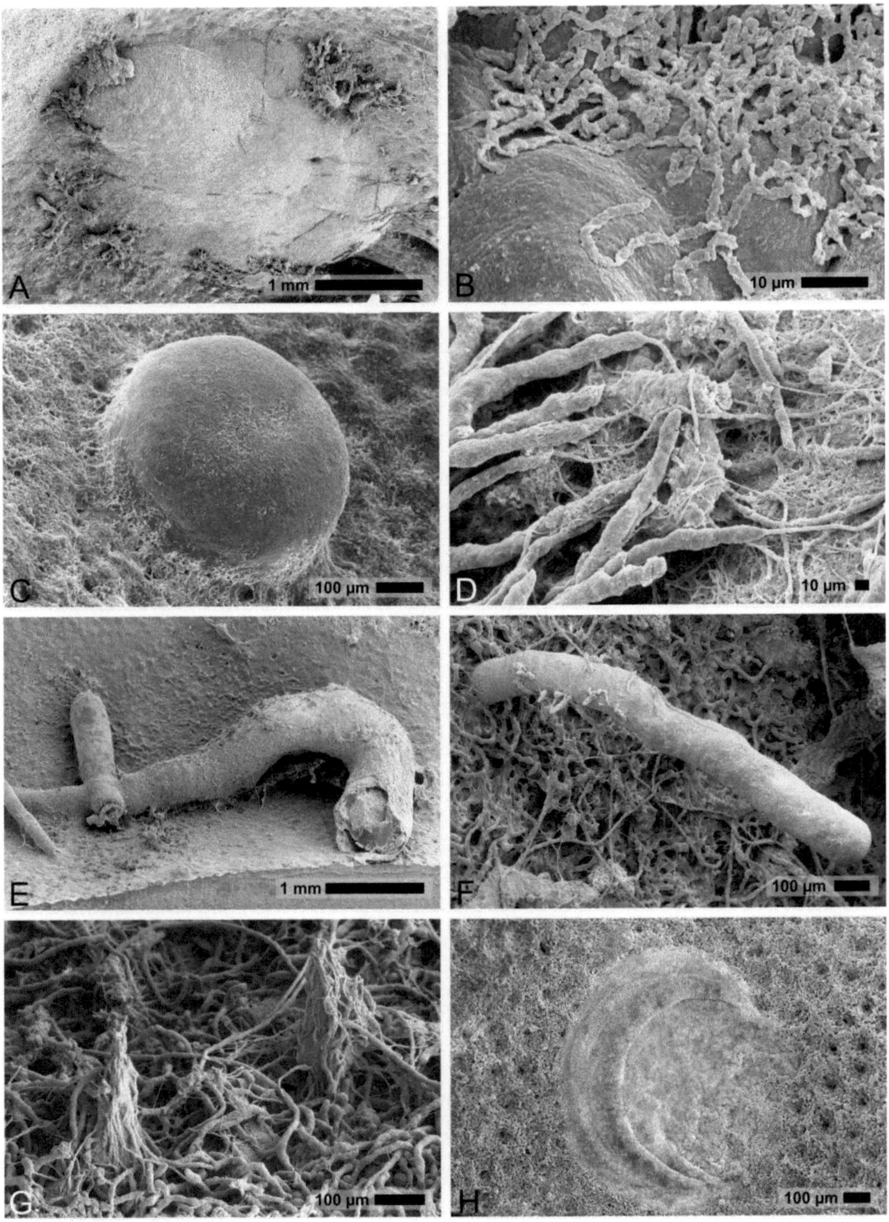

Fig. 2 SEM images of trace fossils recorded in epoxy resin casts of *Lophelia pertusa*, Lower Pleistocene, Rhodes, Greece: **A** 'Larger dish-shaped etching' of a foraminiferan surrounded by several *Semidendrina pulchra*. **B** The minute meandering microboring *Scolecia serrata*, possibly the work of bacteria. **C** Potential foraminiferan pit *Oichnus simplex*. **D** Abruptly thinning fungal microborings of the ichnospecies *Polyactina araneola*. **E** Small part of the phoronid worm boring *Talpina*. **F** The constant-diameter blunt boring *Trypanites solitarius*. **G** Predatory borings produced by the foraminiferan *Hyrrokkin sarcophaga*. **H** The ichnospecies *Centrichnus eccentricus*, an attachment scar of an anomiid bivalve

Systematic ichnology

Ichnogenus *Entobia* Bronn, 1837

Ichnogenus diagnosis: Boring in carbonate substrates comprising a single chamber or networks or boxworks of galleries connected to the surface by several or numerous apertures. Morphology changes markedly with ontogeny. The galleries show progressive increase in diameter during growth; in some forms, inflation at more or less regular distances produces a system of closely interconnected chambers; in other forms, chamber development is restricted to only a brief ontogenetic stage; in still other forms, no cameration is developed. The surface of the boring bears a cuspate microsculpture that may be lost in gerontic specimens. Fine apophyses arise from all or most surfaces of the system (emended diagnosis, Bromley and D'Alessandro 1984).

Entobia mikra isp. n.

Figs. 3-4

2005 'Sponge form I'.- Wisshak et al., 991, fig. 7A
2005 Foraminiferan Trace.- Försterra et al., 957-959, fig. 9C; [partim]
2005 microboring Sponge Type 1.- Beuck and Freiwald, 928-929, fig. 6A; [partim]
2006 'Microsponge-form 1'.- Wisshak, 83-84, figs. 23H, 28C

Diagnosis: Very small ichnospecies with an irregular-shaped, elongate and in places branched or anastomosing, substrate-parallel chamber or array of chambers, connected to the surface by one or several apertures. The boring bears a cuspate (in casts botryoidal) microsculpture with small, globular to irregular-shaped scars, rarely in an initial or semi-closed state.

Differential diagnosis: This ichnospecies is distinguished from all other *Entobia* ichnospecies by its much smaller dimension, both with respect to the whole colony as well as the size of the botryoidal microsculpture. Being aware that size is a problematic ichnotaxobase, in the present case the size differences are so pronounced (more than an order in magnitude) that it warrants consideration. In addition, the camerate architecture is less pronounced than in other entobians.

Etymology: *mikros*, Greek = small

Type material, locality and horizon: Epoxy resin cast (MfNB: MB.W 2020) of a *Lophelia pertusa* skeleton, exhibiting plenty of specimens including the holotype (Fig. 3A-C). SW flank of isolated hill one km SW of Lardos, Rhodes, Greece. Lindos Bay Clay, Rhodes Formation, Early Pleistocene.

Description: This very small *Entobia* ichnospecies (Figs. 3-4) is characterised by irregularly shaped cavities ranging in maximum length from 60 to 335 µm at a mean of 171 µm and a maximum width from 33 to 208 µm at a mean of 99 µm (see Table 2 for detailed numerical data). The traces have an elongate, slightly meandering, branching and anastomosing pattern. The cavities may be separated in little-distinct, swollen chambers and interconnecting canals forming camerate aggregates. The traces are connected to the surface by at least one thick apertural canal, whereas other connections to the surface are mostly obscured by the substrate parallel orientation of the borings. The traces posses a characteristic botryoidal microsculpture of variably-shaped etching scars with maximum diameters in the range of 3.6 to 8.1 µm at a mean of 5.6 µm. In 33% of the traces, incomplete scars in various stages are observed, resembling initial rims (Figs. 3C, 4B-C), semi- and nearly-closed etching cells (Figs. 3B, 4B). Well preserved etching scars may feature faint growth lines (Fig. 4B-C). The traces appear solitary or clustered in large numbers (Fig. 4D).

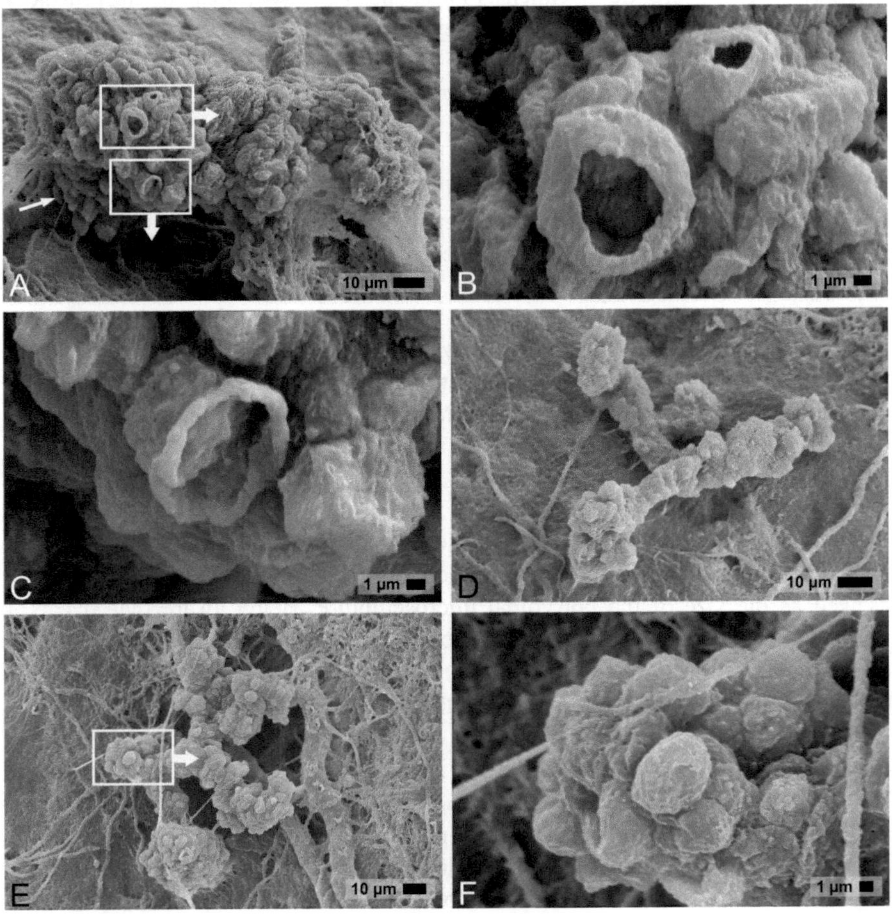

Fig. 3 SEM images of epoxy resin casts with *Entobia mikra* isp. n. taken from *Lophelia pertusa* skeletons, Lower Pleistocene, Rhodes, Greece. **A** The holotype with a large apertural canal (arrow) and characteristic botryoidal surface microsculpture. **B** Close-up of semi- and nearly-closed etching scars. **C** Close-up of initial etching cell rim. **D** Juvenile branched specimen with swollen chambers. **E** Another irregularly branched specimen oriented parallel to the substrate surface towering fungal microborings. **F** Close-up of botryoidal microsculpture

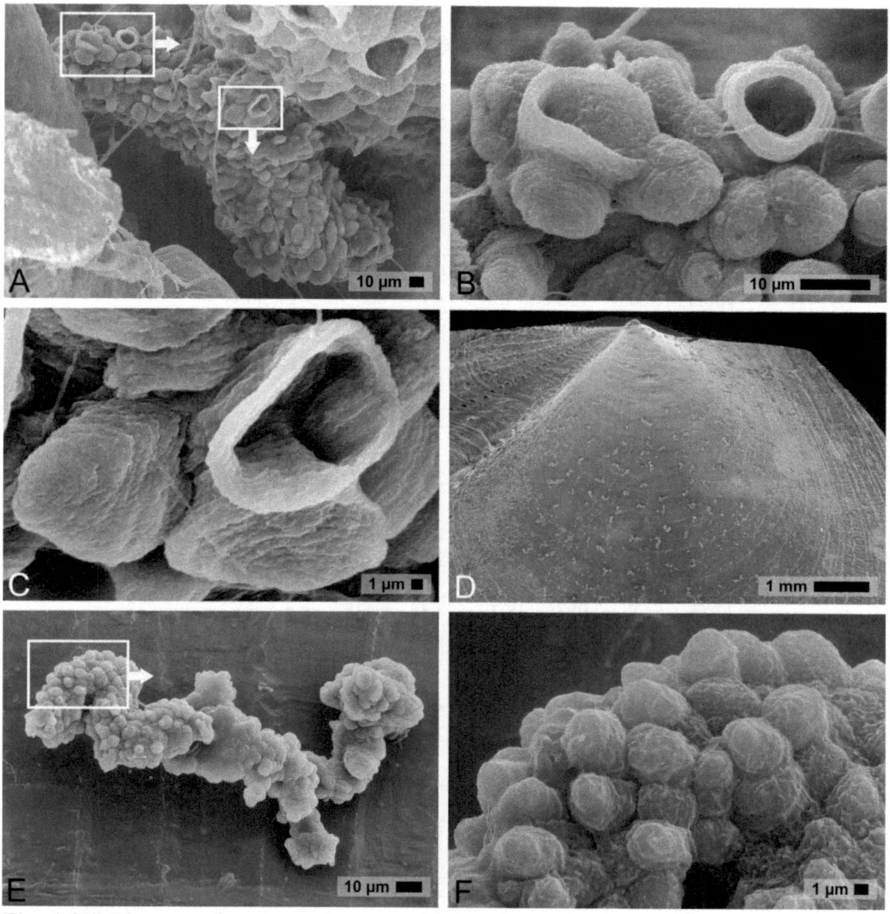

Fig. 4 SEM images of epoxy resin casts with *Entobia mikra* isp. n. taken from a *Lophelia pertusa* skeleton, Recent, Kosterfjord, Sweden (A-C) and a *Delectopecten vitreus* shell, Recent, Sula Ridge, Norway (D-F). **A** Irregularly shaped cavity featuring typical botryoidal surface microsculpture; note the pronounced size difference from the scars exhibited by a common *Entobia* boring in the upper right. **B** Close-up of incomplete etching scars in various stages. **C** Close-up of initial boring scar; note the faint growth increments on the scars. **D** Inner side of bivalve shell bearing a cluster of several hundred specimens. **E** Typical irregular branched cavity, oriented parallel to the substrate suface. **F** Close-up of diagnostic botryoidal microsculpture

Provenance and stratigraphic range: The trace was encountered in Early Pleistocene strata and Recent environments. A detailed list of the studied and previously published material, their provenance and stratigraphic position as well as the infested substrates is given in Table 3.

Remarks: Corresponding traces were repeatedly reported in the literature (see synonymy and Table 3 for details) under open nomenclature. The material reported by Wisshak et al. (2005) and Wisshak (2006) was also included in the present study.

Beuck and Freiwald (2005) and Försterra et al. (2005) encountered corresponding traces and considered an unknown microsponge as likely trace maker.

Table 2 Numerical data for *Entobia mikra* isp. n.

site / n	mean max. diameter + SD / min. + max. [µm]	mean diameter perpendicular to max. + SD / min. + max. [µm]	mean surface + SD [µm²] / mean outline + SD [µm]	mean scar diameter + SD / min. + max. [µm]	percentage of traces with incomplete scars
Kosterfjord n = 4	230.2 ± 65.4 180.3 to 326.5	147.8 ± 10.9 132.6 to 158.1	20510 ± 8415 777 ± 68	8.4 ± 2.2 5.5 to 11.7	25
Sula Ridge n = 34	149.7 ± 48.0 60.1 to 335.4	82.7 ± 27.0 38.2 to 159.3	6043 ± 3504 519 ± 180	5.3 ± 1.3 3.4 to 7.3	32
Jan Mayen n = 1	142.2	105.1	6593 559	5.8	0
Svalbard n = 2	230.9 ± 38.1 203.9 to 257.8	123.8 ± 22.6 107.8 to 139.7	10906 ± 2490 889 ± 119	5.5 ± 1.2 2.8 to 8.2	100
Stjernsund n = 22	181.4 ± 52.8 85.6 to 287.4	114.2 ± 40.9 49.9 to 207.7	9372 ± 4717 718 ± 271	5.6 ± 1.3 3.6 to 8.4	18
Rhodes n = 16	179.3 ± 55.2 89.1 to 263.5	96.4 ± 27.6 33.0 to 131.9	9058 ± 4855 604 ± 168	5.4 ± 1.2 3.5 to 8.1	50
Total n = 79	170.5 ± 54.9 60.1 to 335.4	98.9 ± 35.2 33.0 to 207.7	8443 ± 5387 615 ± 223	5.6 ± 1.3 3.6 to 8.1	33

Table 3 Recent and fossil occurrences of *Entobia mikra* isp. n.

Locality	Coordinates	Water depth [m] / gear	Stratigraphical unit	Substrate	Reference
Kosterfjord (off Sweden)	59°03'06"N 11°09'49"E	7 m (experimental substrate), 85 m ROV	Recent	*Callista chione, Lophelia pertusa*	Wisshak et al. (2005), Wisshak (2006), this study
Jan Mayen, Straumsflaket (off Greenland)	70°44'48"N 8°53'37"W	78 m Boxcorer	Recent	*Chlamys islandica*	this study
Sørkappbanken (off Svalbard)	76°22'N 15°57'E	85-75 m Agassiz-Trawl	Recent	*Chlamys islandica*	this study
Stjernsund (off Norway)	70°15'99"N 22°28'35"E	236 m Boxcorer	Recent	*Lophelia pertusa*	this study
Rhodes (Greece)	36°05'22"N 28°00'30"E		Early Pleistocene / Lindos Bay Clay	*Lophelia pertusa*	this study (including holotype material)
Propeller Mound (off SW Ireland)	52°08'42"N 12°46'19"W	780 m Dredge	Recent	*Lophelia pertusa*	Beuck and Freiwald (2005)
Reñihue Fjord, Patagonia (Chile)	42°32'47"S 72°37'00"W	28 m Scuba diving	Recent	*Desmophyllum dianthus*	Försterra et al. (2005)
Sula Ridge (off Norway)	64°05'N 8°00'E	~270 m Boxcorer	Recent	*Delectopecten vitreus*	this study

Entobia nana isp. n.

Figs. 5-6

1990 Boring Sponge, Form 3.- Günther, 238, plate 57, figs. 1-2
1991 *Orthogonum tubulare*.- Radtke, 60-64, plate 5, figs. 5-6; [partim]
1996 'Fungoid-Form C'.- Hofmann, 57-59, plate 3, fig. 4; [partim]
2004 *Orthogonum tubulare*.- Glaub, 70, fig. 4b
2005 'Sponge form III'.- Wisshak et al., 991, fig. 7C
2005 *Polyactina araneola*.- Beuck and Freiwald, 925, figs. 4C, 7C-D
2005 microboring sponge.- Beuck and Freiwald, 927, fig. 6C
2006 'Microsponge-form 3'.- Wisshak, 84-85, fig. 28E

Diagnosis: Very small ichnospecies with irregular, potato-sack-shaped cavities oriented parallel to the substrate surface, often bearing few to many apophyses and exploratory threads that terminate bluntly or lead to neighbouring chambers. The traces show a bulged to weakly cuspate (in casts weakly botryoidal) microsculpture.

Differential diagnosis: This ichnospecies is distinguished from all other *Entobia* ichnospecies by its much smaller dimension and the lack of a pronounced botryoidal surface texture.

Etymology: *nanos*, Greek = dwarf

Type material, locality and horizon: Epoxy resin cast (MfNB: MB.W 2021) of a *Lophelia pertusa* skeleton, exhibiting plenty of specimens including the holotype (Fig. 5A-E). SW flank of isolated hill one km SW of Lardos, Rhodes, Greece. Lindos Bay Clay, Rhodes Formation, Early Pleistocene.

Table 4 Numerical data for *Entobia nana* isp. n.

site / n	mean max. diameter + SD / min. + max. [μm]	mean diameter perpendicular to max. + SD / min. + max. [μm]	mean surface + SD [μm²] / mean outline + SD [μm]	mean number of exploratory threads + SD	percentage of traces in close proximity (<100 μm) to each other
Kosterfjord n = 5	205.1 ± 82.2 121.3 to 336.3	154.4 ± 77.8 103.9 to 292.0	29340 ± 29171 648 ± 301	0.6 ± 0.9 0 to 2	0
Stjernsund n = 22	195.6 ± 59.2 101.8 to 319.1	154.4 ± 47.1 89.8 to 256.4	22035 ± 13838 579 ± 174	8.5 ± 21 0 to 75	32
New Zealand n = 3	261.9 ± 62.5 224.5 to 336.3	219.0 ± 49.6 177.1 to 292.0	39537 ± 14791 814 ± 208	0.3 ± 0.6 0 to 2	0
Palamos n = 1	162.5	93.6	10548 408	11	0
Rhodes n = 43	366.5 ± 139.2 153.3 to 756.2	244.8 ± 90.8 88.3 to 448.2	64329 ± 48018 1092 ± 414	3.5 ± 2.5 0 to 10	65
Total n = 74	297.7 ± 139.5 101.8 to 756.2	208.7 ± 88.4 88.3 to 448.2	47659 ± 42904 889 ± 417	4.7 ± 11.8 0 to 75	47

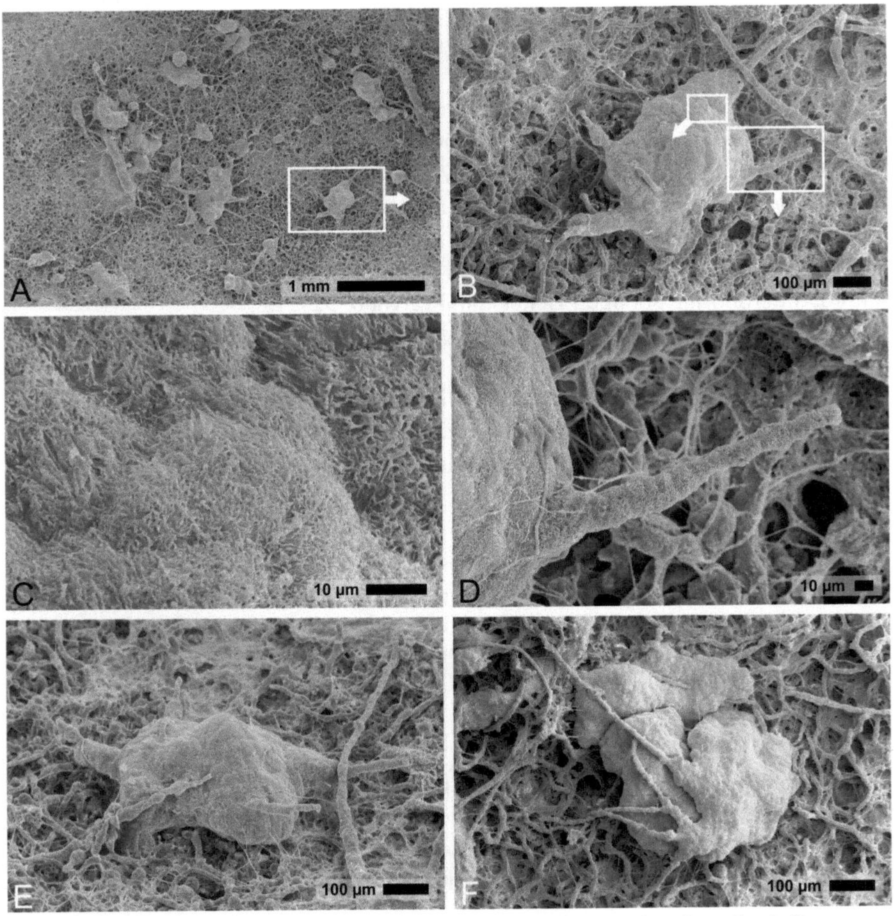

Fig. 5 SEM images of epoxy resin casts with *Entobia nana* isp. n. taken from *Lophelia pertusa* skeletons, Lower Pleistocene, Rhodes, Greece. **A** Cluster of numerous, partly interconnected specimens including the holotype in the lower right, superimposing a dense microboring carpet. **B** The holotype possessing several exploratory threads. **C** Close-up of bulged surface texture. **D** Close-up of blunt exploratory thread. **E** Lateral view of the holotype. **F** Closely spaced chambers with exploratory threads

Description: This very small *Entobia* ichnospecies (Figs. 5-6) is characterised by solitary or clustered, irregularly shaped cavities with the appearance of a sack of potatoes with a bulged to weakly botryoidal surface texture (Fig. 5C). Individual cavities measure 102 to 756 µm in maximum diameter at a mean of 298 µm, and a perpendicular width in the range of 88 to 448 µm at a mean of 209 µm (see Table 4 for detailed numerical data). The chambers posses either no (Fig. 6B), some (Figs. 5B-F, 6C-E), and in a few cases many (Fig. 6F) apophyses and exploratory threads. Their number ranges from 0 to 75 at a mean of 5 per specimen. Where not terminating bluntly, these canals rarely lead to neighbouring chambers, then forming a camerate colony pattern. In other cases, neighbouring chambers are in close contact to each other, only separated by a thin veneer of substrate (Fig. 5F). In total, 47% of the cavities were found at the close distance of less than 100 µm from each other.

Provenance and stratigraphic range: The trace was encountered in Middle Jurassic, Upper Cretaceous, Eocene and Upper Miocene strata as well as in Recent environments. A detailed list of the studied and previously published material, its provenance and stratigraphic position as well as the infested substrates is given in Table 5.

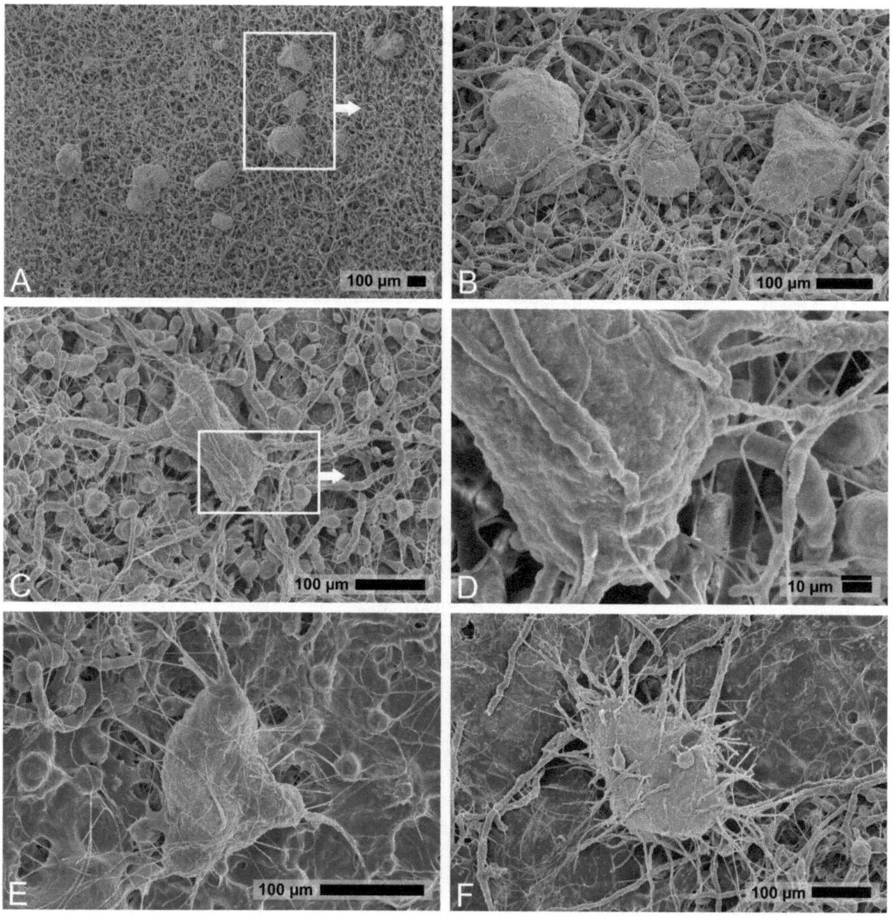

Fig. 6 SEM images of epoxy resin casts with *Entobia nana* isp. n. taken from *Lophelia pertusa* skeletons, Recent, Stjernsund, Norway. **A** Cluster of numerous specimens, superimposing a dense microboring carpet. **B** Close up of three chambers lacking exploratory threads or apophyses. **C** Typical specimen with several exploratory threads. **D** Close-up showing details of partly collapsed apophyses. **E** Specimen with several exploratory threads, partly overlain by fungal microborings. **F** Specimen featuring an unusualy large number of apophyses and exploratory threads

Table 5 Recent and fossil occurrences of *Entobia nana* isp. n.

Locality	Coordinates	Water depth [m] / gear	Stratigraphical unit	Substrate	Reference
Kosterfjord (off Sweden)	59°03'06"N 11°09'49"E	85 m ROV	Recent	*Lophelia pertusa*	Wisshak et al. (2005), Wisshak (2006), this study
Hinakura (New Zealand)	41°16'03"S 175°40'11"E		Upper Miocene / Tongaporutuan Stage	*Lophelia pertusa*	this study
Stjernsund (off Norway)	70°15'99"N 22°28'35"E	236 m Boxcorer	Recent	*Lophelia pertusa*	this study
Rhodes (Greece)	36°05'22"N 28°00'30"E		Early Pleistocene / Lindos Bay Clay	*Lophelia pertusa*	this study (including holotype material)
Propeller Mound (off SW Ireland)	52°08'42"N 12°46'19"W	780 m Dredge	Recent	*Lophelia pertusa*	Beuck and Freiwald (2005)
Shelf off Mauritania (off W Africa)	19°40'N 16°40'W	41-156 m Dredge, grab sampler	Recent	Molluscan shells	Glaub (2004)
Maastricht (Netherlands)			Upper Cretaceous / Lower Maastrichtian	*Ostrea* sp.	Hofmann (1996)
Lägerdorf (Germany)			Upper Cretaceous / Lower Campanian	*Belemnitella mucronata*	Hofmann (1996)
Hervest / Dorsten (Germany)			Upper Cretaceous / Santonian	*Ostrea semiplana*	Hofmann (1996)
Mülheim (Germany)			Upper Cretaceous	*Ostrea* sp.	Hofmann (1996)
Possagno (Italy)			Upper Eocene / Priabonien	*Turritella* sp.	Radtke (1991)
Cozumel, Caribbean Sea (off Mexico)		1.5 and 28 m Scuba diving	Recent	Bivalve shells	Günther (1990)
off Palamos, Costa Brava (off Spain)	41°53'N 2°15'E	Dredge	Late Pleistocene / Last Glacial	*Chlamys islandica*	this study

Remarks: Corresponding traces were repeatedly reported in the literature (see synonymy and Table 5 for details) under open nomenclature. The material reported by Wisshak et al. (2005) and Wisshak (2006) was also included in the present study. Beuck and Freiwald (2005) considered some of the specimens as mature *Polyactina araneola*, produced by the fungus *Conchyliastrum*. Radtke (1991) tentatively assigned similar traces also to the work of boring fungi and considered them as variants of her ichnotaxon *Orthogonum tubulare* (mistakenly labelled as holotype in her original publication; Radtke pers. comm). This identification was later also applied by Glaub (2004) for corresponding traces. Hofmann (1996) found resembling traces also assigning them to a fungal producer, whereas Günther (1990) and Beuck and Freiwald (2005) considered a boring micro-sponge as likely producer.

Morphometry

Some of the basic morphometrical data were already stated in the descriptions above and the numerical data are summarised in Tables 2 and 4. Figures 7 and 8 graphically display selected morphometrical parameters as a basis for evaluating ichnospecies variability, ontogenetic questions and in order to relate different sites. Figures 7A and 8A show a class-frequency distribution of the size range (max. diameter) of the species and give the percentage of each size class for *Entobia mikra* isp. n. and *Entobia nana* isp. n., respectively. Both histograms show a positively skewed Gaussian distribution. The average etching scar diameter of *Entobia mikra* isp. n. exhibits a nearly linear increase with increasing maximum trace diameter and the maximum percentage of traces with initial or incomplete scars is found with 53% in the 200 to 250 μm size class (Fig. 7B). The number of exploratory threads and apophyses in *Entobia nana* isp. n. is not linked to the size classes (Fig. 8B). When plotting the mean maximum diameter + standard deviation against mean surface area + standard deviation (Fig. 7C) or the mean scar width + standard deviation (Fig. 7D) of the various *Entobia mikra* isp. n. occurrences investigated, the fossil Rhodes material lies close to the total average. This also applies for *Entobia nana* isp. n. when considering the number of exploratory threads (Fig. 8D), whereas the Rhodes material is in the upper end of the size range of this ichnospecies (Fig. 8C).

Discussion

The immediate question the present study raises is whether these dwarf entobians are also produced by boring sponges like their larger cousins and what lines of evidence lead to a positive answer in this issue. Most of the material stems from fossil and modern aphotic (palaeo)environments, ruling out all but heterotrophic organisms in the pool of known bioerosion agents. Among these, there is no convincing alternative potential trace maker. Bacteria are much too small and a fungal producer is also unlikely because of the same reason. Bryozoans form very different traces and all known macroborers are much too large in size. Hence, from our current knowledge on bioerosion agents, boring sponges are indeed the most likely trace makers of the new dwarf entobians. Especially when considering the compelling resemblance in the botryoidal microsculpture, demonstrating the activity of specialised archaeocyte etching cells (e.g., Rützler and Rieger 1973; see Schönberg this volume for a review of the boring mechanism) in the case of *Entobia mikra* isp. n., and the presence of partly interconnected chambers in the case of *Entobia nana* isp. n.

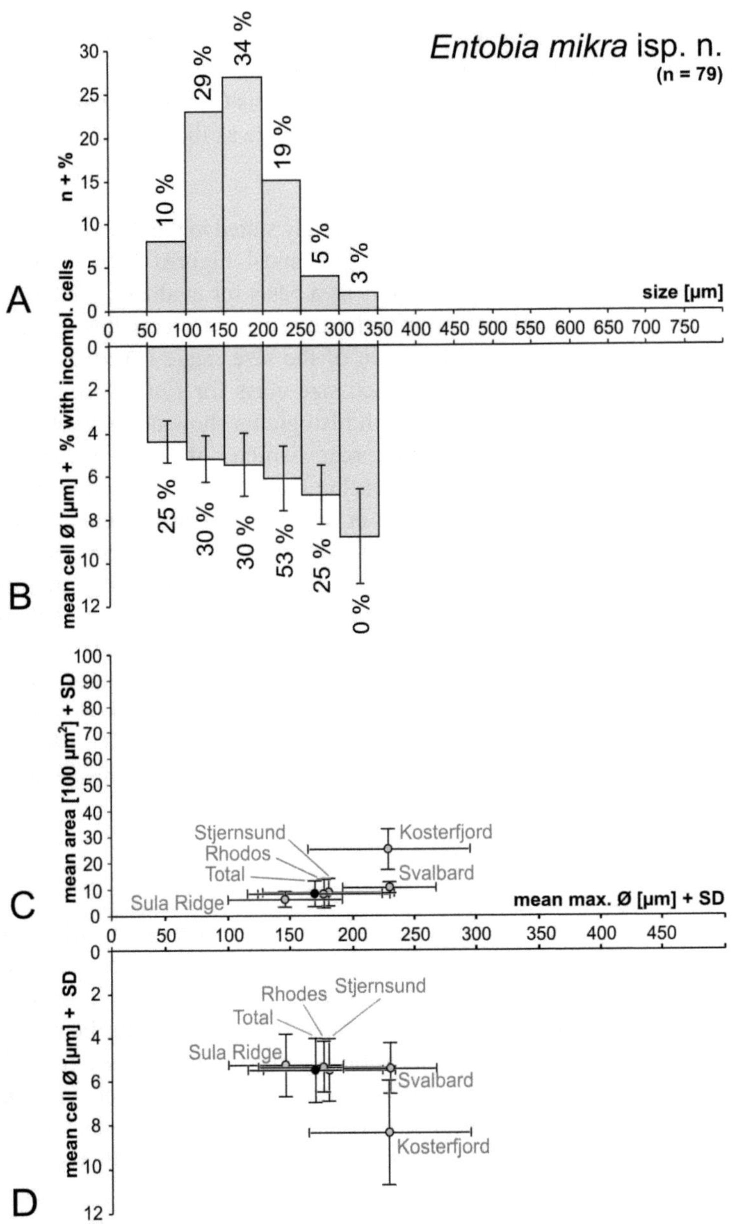

Fig. 7 Graphic display of selected morphometric parameters of *Entobia mikra* isp. n. **A** Histogram of the size range (max. diameter) and the percentage of specimens in each size class. **B** The average scar size and percentage of traces with incomplete etching scars in each size class. **C** Comparison of the different occurrences investigated in terms of area versus size. **D** Comparison in terms of scar size versus size

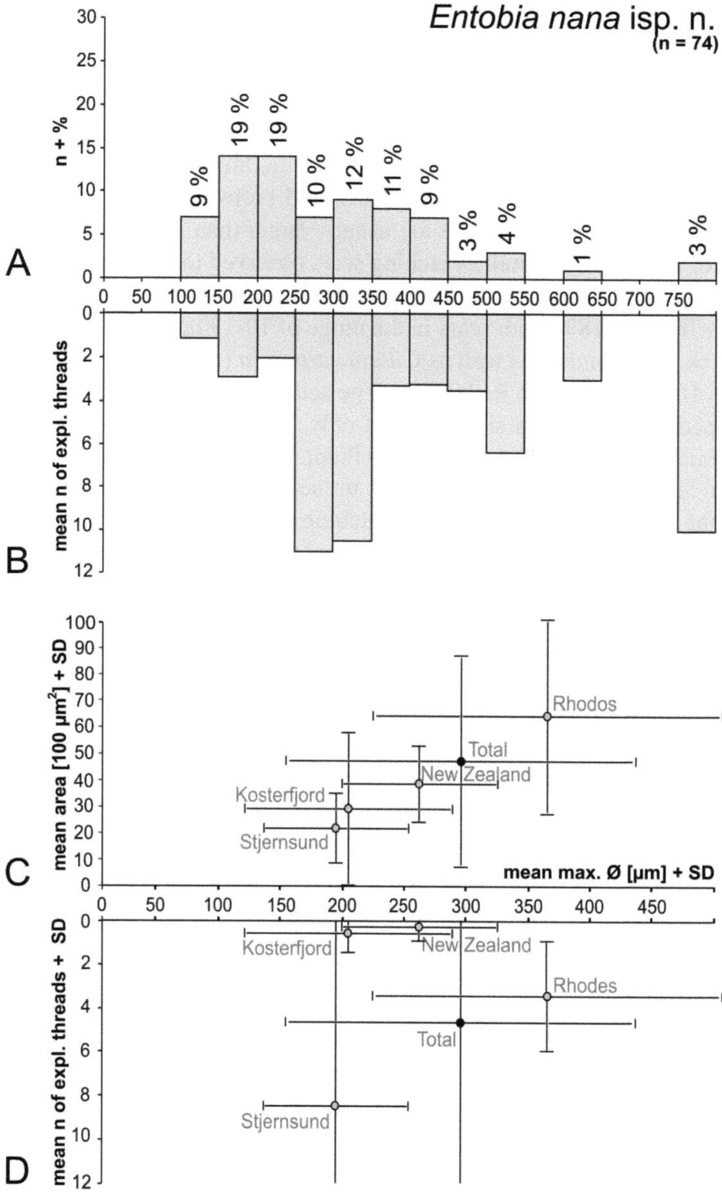

Fig. 8 Graphic display of selected morphometric parameters of *Entobia nana* isp. n. **A** Histogram of the size range (max. diameter) and the percentage of specimens in each size class. **B** The average number of exploratory threads and apophyses in each size class. **C** Comparison of the different occurrences investigated in terms of area versus size. **D** Comparison in terms of number of exploratory threads versus size

But if these traces were indeed produced by boring sponges, is it possible that they represent juvenile stages (as suggested for instance by Beuck and Freiwald 2005) rather than distinctive populations of boring sponges? The data suggest 'no', based on two observations. Firstly, no traces intermediate in size between the present traces and larger entobians (which are also very common in most of the studied material) were observed with the exception of raspberry-shaped initial entobians formed for instance by *Alectona millari* Carter, 1879 which do, however, posses much larger etching scars right from the start (e.g., Beuck and Freiwald 2005; Wisshak 2006). The etching scars of boring sponges are usually larger than 20 µm and may exceed diameters of 100 µm. The smallest etching scars observed to date are found only on exploratory threads of *Alectona millari* with scar diameters of 5-11 µm and *Spiroxya heteroclita* Topsent, 1896 with scars in the range of 10-15 µm (Beuck and Freiwald 2005; Beuck pers. comm.) as well as *Cliona vermifera* (Hancock, 1867) with scars larger than 10 µm (Ward and Risk 1977). The actual etching cells on the other hand in some species may have a size range of only 3-20 µm (Pomponi 1977) and an average dimension of only 9.6 by 5.1 µm (Pomponi 1979), as revealed by detailed cytological investigations of boring sponge tissue. However, there is not necessarily always a one to one relationship between etching cell and etching scar (Cobb 1969) but there are also strategies were several cells produce the scars in a combined effort (Warburton 1958). The second line of reasoning against juvenile stages is grounded on the size class-frequency distribution of both new ichnospecies (Figs. 7A, 8A), which show a Gaussian distribution supporting the hypothesis that the investigated specimens reflect the size range and population structure of a specific tracemaking bioerosion agent for each one of the new ichnospecies.

The ultimate proof can only be drawn when finding and isolating siliceous sponge spicules, paving the road to a proper taxonomical evaluation of the trace maker. This is, however, a difficult task considering the minute size of the traces in question, requiring preparation work well within the range of light microscopy and expected spicule size in the range of SEM visualisation. On the other hand, it is also well conceivable that these very small forms – like juvenile stages of larger boring sponges – do not contain spicules since they are physiologically not essential. In such a case, the etching scar morphology may provide an important taxonomical character, such as it has only recently been revived as an additional taxonomic feature for larger boring sponges (e.g., Calcinai et al. 2003, 2004; Beuck and Freiwald 2005). In any case, sponge workers are encouraged to find and identify the actual trace maker, and to expect and explore species in a considerably lower size and larger size range, respectively, than previously thought.

Acknowledgements

I would like to express my sincere thanks to Jürgen Titschack and André Freiwald (both Erlangen, Germany) for making available the fossil *Lophelia* material and the Recent samples, respectively. Nadine Sutalo (Erlangen, Germany) kindly provided SEM images of the Stjernsund material. Furthermore I would like to thank Lydia Beuck (Erlangen, Germany), Christine Schönberg (Oldenburg, Germany) and

Leif Tapanila (Pocatello, USA) for fruitful discussions. Last but not least, the manuscript benefited from the valuable referee comments provided by Klaus Rützler (Washington, USA) and Richard G. Bromley (Copenhagen, Denmark).

References

Beuck L, Freiwald A (2005) Bioerosion patterns in a deep-water *Lophelia pertusa* (Scleractinia) thicket (Propeller Mound, northern Porcupine Seabight). In: Freiwald A, Roberts JM (eds) Cold-water corals and ecosystems. Springer, Berlin Heidelberg, pp 915-936

Beuck L, López Correa M, Freiwald A (this volume) Biogeographical distribution of *Hyrrokkin* (Rosalinidae, Foraminifera) and its host-specific morphological and textural trace variability. In: Wisshak M, Tapanila L (eds) Current developments in bioerosion. Springer, Berlin, pp 329-360

Bosc LAG (1802) Histoire naturelle des vers, contenant leur description et leurs mœurs, avec figures dessinées d'après nature. Deterville, Paris, pp 1-258

Bromley RG (1981) Concepts in ichnotaxonomy illustrated by small round holes in shells. Acta Geol Hisp 16:55-64

Bromley RG (2005) Preliminary study of bioerosion in the deep-water coral *Lophelia*, Pleistocene, Rhodes, Greece. In: Freiwald A, Roberts JM (eds) Cold-water corals and ecosystems. Springer, Berlin Heidelberg, pp 895-914

Bromley RG, D'Alessandro A (1984) The ichnogenus *Entobia* from the Miocene, Pliocene and Pleistocene of southern Italy. Riv Ital Paleont Stratigr 90:227-296

Bromley RG, Martinell J (1991) *Centrichnus*, new ichnogenus for centrically patterned attachment scars on skeletal substrates. Bull Geol Soc Denmark 38:243-252

Bromley RG, Surlyk F (1973) Borings produced by brachiopod pedicles fossil and recent. Lethaia 6:349-365

Bromley RG, Wisshak M, Glaub I, Botquelen A (2007) Ichnotaxonomic review of dendriniform borings attributed to foraminifera: *Semidendrina* igen. nov. In: Miller III W (ed) Trace fossils: concepts, problems, prospects. Elsevier, Amsterdam, pp 518-530

Bronn HG (1837) Lethaea Geognostica oder Abbildungen und Beschreibungen der für die Gebirgs-Formationen bezeichnendsten Versteinerungen. Schweizerbart, Stuttgart, 1:1-544 and plates

Calcinai B, Arillo A, Cerrano C, Bavestrello G (2003) Taxonomy-related differences in the excavating micro-patterns of boring sponges. J Mar Biol Assoc UK 83:37-39

Calcinai B, Bavestrello G, Cerrano C (2004) Bioerosion micro-patterns as diagnostic characteristics in boring sponges. Boll Mus Ist Biol Univ Genova 68:229-238

Carter HJ (1879) On a new species of excavating sponge (*Alectona millari*); and on a new species of *Rhaphidotheca* (*R. affinis*). J Roy Microscop Soc 2:493-499

Cedhagen T (1994) Taxonomy and biology of *Hyrrokkin sarcophaga* gen. et sp. n., a parasitic foraminiferan (Rosalinidae). Sarsia 79:65-82

Clarke JM (1921) Organic dependence and disease, their origin and significance. New York State Mus and Sci Serv Bull 221-222, 113 pp

Cobb WR (1969) Penetration of calcium carbonate substrates by the boring sponge, *Cliona*. Amer Zoologist 9:785-790

Cornèe J-J, Moissette P, Joannin S, Suc J-P, Quillévéré F, Krijgsman W, Hilgen F, Koskeridou E, Münch P, Lécuyer C, Desvignes P (2006) Tectonic and climatic controls on coastal sedimentation: The Late Pliocene-Middle Pleistocene of northeastern Rhodes, Greece. Sed Geol 187:159-181

Försterra G, Beuck L, Häussermann V, Freiwald A (2005) Shallow-water *Desmophyllum dianthus* (Scleractinia) from Chile: characteristics of the bioeroding community, heterotrophic interactions and (palaeo)-bathymetrical implications. In: Freiwald A, Roberts JM (eds) Cold-water corals and ecosystems. Springer, Berlin Heidelberg, pp 937-977

Glaub I (1994) Mikrobohrspuren in ausgewählten Ablagerungsräumen des europäischen Jura und der Unterkreide (Klassifikation und Palökologie). Courier Forschinst Senckenberg 174:1-324

Glaub I (2004) Recent and sub-recent microborings from the upwelling area off Mauritania (West Africa) and their implications for palaeoecology. In: McIlroy D (ed) The application of ichnology to palaeoenvironmental and stratigraphic analysis. Geol Soc London, Spec Publ 228:63-77

Gmelin JF (1791) Caroli a Linné, systema naturae per regna tria naturae. Editio decima tertia. Beer, Lipsiae, Vol 1, Pars 6, pp 3021-3910

Golubic S, Brent G, LeCampion T (1970) Scanning electron microscopy of endolithic algae and fungi using a multipurpose casting-embedding technique. Lethaia 3:203-209

Golubic S, Campbell S, Spaeth C (1983) Kunstharzausgüsse fossiler Mikroben-Bohrgänge. Präparator 29:197-200

Grant RE (1826) Notice of a new zoophyte (*Cliona celata* gr.) from the Firth of Forth. Edinburgh New Phil J 1:78-81

Günther A (1990) Distribution and bathymetric zonation of shell-boring endoliths in recent reef and shelf environments: Cozumel, Yucatan (Mexico). Facies 22:233-262

Hagenow KF von (1840) Monografie der Rügenschen Kreideversteinerungen II. Abth. Radiarien u. Annulaten. N Jb Mineral Geogn Geol Petrefaktenkd 1840:631-672

Hancock A (1867) Note on the excavating sponges; with descriptions of four new species. Ann Mag Nat Hist 19:229-242

Hanken N-M, Bromley RG, Miller J (1996) Plio-Pleistocene sedimentation in coastal grabens, north-east Rhodes, Greece. Geol J 31:393-418

Hofmann K (1996) Die mikro-endolithischen Spurenfossilien der borealen Oberkreide Nordwest-Europas und ihre Faziesbeziehungen. Geol Jb, A 136, 151 pp

Linnaeus C (1758) Systema naturae per regna tria naturae, secundum classes, ordines, genera, species, cum characteribus, differentiis, synonymis, locis. Tomus I. Editio decima, reformata. Laurentii Salvii, Holmiae, 824 pp

Plewes CR, Palmer TJ, Haynes JR (1993) A boring foraminiferan from the Upper Jurassic of England and Northern France. J Micropalaeont 12:83-89

Pomponi SA (1977) Etching cells of boring sponges: an ultrastructure analysis. Proc 3rd Coral Reef Symp, Miami, pp 485-491

Pomponi SA (1979) Ultrastructure of cells associated with excavation of calcium carbonate substrates by boring sponges. J Mar Biol Assoc UK 59:777-784

Radtke G (1991) Die mikroendolithischen Spurenfossilien im Alt-Tertiär West-Europas und ihre palökologische Bedeutung. Courier Forschinst Senckenberg 138:1-185

Rützler K (1971) Bredin-Archbold-Smithsonian biological survey of Dominica: Burrowing sponges, genus *Siphonodictyon* Bergquist, from the Caribbean. Smithson Contr Zool 77:1-37

Rützler K, Rieger G (1973) Sponge burrowing: fine structure of *Cliona lampa* penetrating calcareous substrata. Mar Biol 21:144-162

Schönberg CHL (this volume) A history of sponge erosion: from past myths and hypotheses to recent approaches. In: Wisshak M, Tapanila L (eds) Current developments in bioerosion. Springer, Berlin, pp 165-202

Thomsen (eds, 2005) Lagoon to deep-water foraminifera and ostracods from the Plio-Pleistocene Kallithea Bay Section, Rhodes, Greece. Cushman Foundation for Foraminiferal Research, Spec Publ 39:1-290

Topsent E (1896) Matériaux pour servir à l'étude de la faune des spongiaires de France. Mém Soc Zool France 9:113-133

Voigt E (1965) Über parasitische Polychaeten in Kreide-Austern sowie einige andere in Muschelschalen bohrende Würmer. Paläont Z 39:193-211

Voigt E (1971) Fremdskulpturen an Steinkernen von Polychaeten-Bohrgängen aus der Maastrichter Tuffkreide. Paläont Z 45:144-153

Warburton FE (1958) The manner in which the sponge *Cliona* bores into calcareous objects. Canad J Zool 36:555-562

Ward P, Risk MJ (1977) Boring pattern of the sponge *Cliona vermifera* in the coral *Montastrea annularis*. J Paleont 51:520-526

Wisshak M (2006) High-latitude bioerosion: the Kosterfjord experiment. Lect Notes Earth Sci 109:1-202

Wisshak M, Freiwald A, Lundälv T, Gektidis M (2005) The physical niche of bathyal *Lophelia pertusa* in a non-bathyal setting: environmental controls and palaeoecological implications. In: Freiwald A, Roberts JM (eds) Cold-water corals and ecosystems. Springer, Berlin Heidelberg, pp 979-1001

Zebrowski G (1937) New genera of Cladochytriaceae. Ann Missouri Bot Gard 23:553-564

Borings, bodies and ghosts: spicules of the endolithic sponge *Aka akis* sp. nov. within the boring *Entobia cretacea*, Cretaceous, England

Richard G. Bromley[1], Christine H. L. Schönberg[2]

[1] Department of Geography and Geology, University of Copenhagen, 1350 Copenhagen K, Denmark, (rullard@geol.ku.dk)
[2] Institut für Bio- und Umweltwissenschaften, Carl von Ossietzky Universität Oldenburg, 26111 Oldenburg, Germany

Abstract. The fossilised borings of endolithic sponges are generally abundant in marine sediments since the Jurassic, and sparse occurrences date back to the Lower Palaeozoic. However, the zoological identity of the boring sponge is not revealed by the morphology of the boring alone. The preservation potential of the boring is far greater than that of the spicular skeleton of the endolithic sponge. The spicules are opaline silica, which is readily soluble in seawater. Thus, while the borings are abundant and diverse, we have almost no knowledge of the bioeroding sponges that produced them.

In situations where early diagenetic cementation of the sediment occurs before the dissolution of the silica, the spicules may be preserved as casts. Such a case exists in association with hardground formation. A hardground in the Upper Cretaceous (Turonian) Chalk Rock of Buckinghamshire, SE England, yielded an intraclast containing a sponge boring referable to *Entobia cretacea* Portlock, 1843. Many of the chambers contain sparite-filled casts of sponge spicules. Only megascleres are present, comprising smooth oxeas about 2 mm in length and 74 µm in width, indicating an extinct species of the phloeodictyid genus *Aka*. No part of the original skeleton of the sponge is preserved, only very accurate external moulds of the spicules, representing little more than the ghost of the animal. But this is not unusual in palaeontology and, as it appears that this body fossil is new to science, it is herein named *Aka akis* sp. nov. This is only the fifth account of fossilised *Aka* spp. where spicule morphology and erosion traces can be observed. It is hoped that further search for spicules preserved within *Entobia* will allow an investigation of the endolithic sponge communities of the Mesozoic and Tertiary seas, about which almost nothing is known at present.

Keywords. *Entobia cretacea*, chalk hardgrounds, Chalk Rock, sponge body fossil, oxeas, *Aka akis* sp. nov.

M. Wisshak, L. Tapanila (eds.), *Current Developments in Bioerosion*. Erlangen Earth Conference Series, doi: 10.1007/978-3-540-77598-0_12,

Introduction

Borings attributable to endolithic sponge bioerosion are uncommon in the Palaeozoic (e.g., Mikuláš 1994; Schönberg and Tapanila 2006; Tapanila 2006), but are generally abundant since the Jurassic. However, we still have little understanding of which biological genera they represent (e.g., Bromley 2004), because the preservation potential of the boring is far greater than that of the spicular skeleton of the tracemaker. Spicules of endolithic sponges are composed of hydrated, amorphous, opaline silica, which is soluble in seawater (Siever 1962; Lang and Steiger 1985; Uriz et al. 2003). Owing to the common lack of spicular remains, many palaeontologists simplistically consider as synonyms 'boring sponge' and 'clionid sponge' (or, after the revision by Rützler [2002] of the family name, 'clionaid sponge'). However, although the family Clionaidae may be the dominant and most diverse family of endolithic sponges today, it is not the only one, and the families represented in, for example, the Cretaceous are hardly understood at all, despite the great abundance of their borings. It would also appear that the clionaids have a shorter history than *Aka* de Laubenfels, 1936, at least in their boring habit.

Spicule preservation

Spiculation is the basic means of sponge taxonomy (e.g., Uriz et al. 2003), but the opaline spicules are rarely preserved in fossil material. After the death of an endolithic sponge, however, it is unlikely that all the spicules would be physically washed out of the intricately chambered forms of boring. Instead, the most usual cause of loss is apparently through early diagenetic dissolution of the silica.

Preservation of the spicules requires special taphonomic conditions. An early diagenetic cementation is required that will cement the matrix of the spicules before they are dissolved. Their subsequent dissolution will create a void that may later be filled with cement. Such a diagenetic environment is found in hardgrounds, in which calcitic cement is precipitated from the marine pore water of the seafloor (e.g., Bathurst 1971; Bromley 1975). It might be pointed out that the use of 'preservation' in this scenario is somewhat misleading. The spicules themselves are not preserved; only their external shape is preserved, like a ghost of the original sponge. But this is a common phenomenon in palaeontological material, and the ghosts are herein regarded as body fossils.

Material and methods

In the Upper Cretaceous Chalk Group of northern Europe, the pure, white, micritic chalk is weakly lithified and generally preserves sponges poorly, and at many levels, not at all. Their original presence is divulged, however, by borings of endolithic sponges in skeletal fossils such as oysters and belemnites (e.g., Bromley 1970). Spicules have not been found in these borings, except where these are embedded within flint concretions, but here visibility of the spicules is generally poor.

However, in the Turonian Chalk Rock Formation of southern England (Bromley and Gale 1982), sometimes regarded as a diachronous member of the Lewis Nodular Chalk Formation (Mortimore et al. 2001; Woods and Aldis 2004; Hopson et al. 2006), the topmost of a group of hardgrounds, the Hitch Wood Hardground,

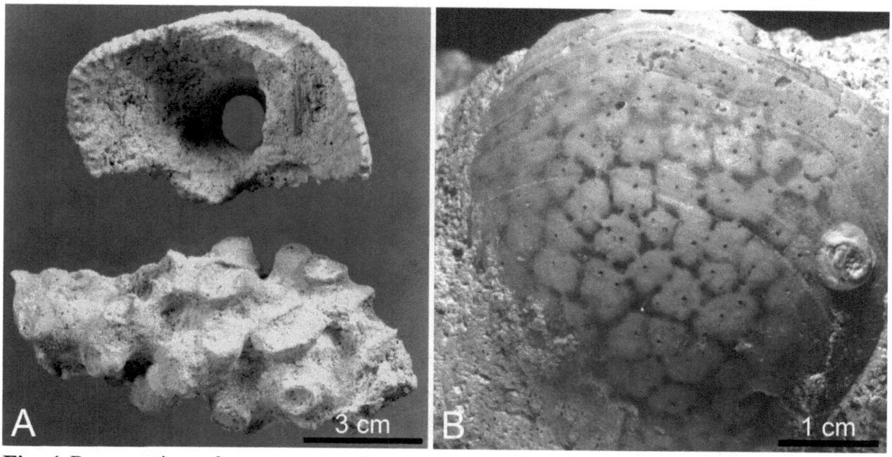

Fig. 1 Preservation of sponges and their borings in the Hitch Wood Hardground at Hitch Wood quarry. **A** Two hexactinellid sponges preserved as phosphatic body fossils, the spicules as empty or cement-filled cavities. Extracted from the matrix using acetic acid. **B** The transparent shell of a terebratulid brachiopod, *Gibbithyris* sp, containing the post mortem boring *Entobia cretacea*

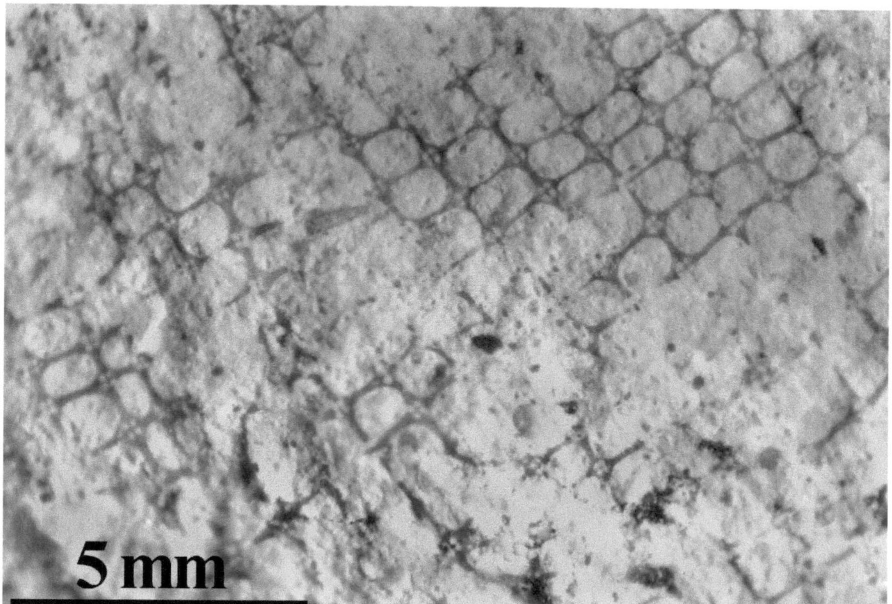

Fig. 2 The skeleton of a lychniscosan hexactinellid sponge, the silica replaced by glauconite. Hitch Wood Hardground, Chalk Rock, Hitch Wood quarry

contains abundant non-boring sponge body fossils as well as *Entobia* Bronn, 1837 (Fig. 1). The hardground is well mineralised with phosphates and glauconite. The spicular skeleton of the non-boring sponges is well preserved by early cementation of the sediment followed by phosphatisation and glauconitisation at or close to the seafloor. In some cases, the spicules themselves are replaced by glauconite (Fig. 2). Intraclastic pebbles and mollusc shells commonly contain *Entobia*.

At an old quarry at The Frogmore, High Wycombe, Buckinghamshire, England (Bromley and Gale 1982), an intraclast, containing *Entobia cretacea*, was collected from the Hitch Wood Hardground. Using a rock saw, the intraclast was sectioned (Fig. 3), and was found to contain spicules in several of the chambers of the boring (Figs. 4-6). Parallel serial surfaces were ground and polished, about 30-50 µm apart, and an acetate peel of each was taken (17 peels labelled A to Q). The specimen was illustrated by Bromley (1970: pl. 5a) as it was before the peels were taken.

Measurements of spicules were taken from enlarged images obtained from peels (Figs. 4-6). From consecutive images, the widest diameters were used for the erosion traces. Spicule widths were defined as the smallest diameters of bright circular areas representing spicule cross-sections seen in Figures 4-6, because slightly diagonal sections would artificially increase diameters. Where spicules were sectioned longitudinally, the widest diameters were used, as the opposite effect took place. Spicule lengths were measured in a straight line from tip to tip, always assuming to underestimate, as the spicule was unlikely to be sectioned in a plane that allowed viewing of the full length. Photographs of acetate peels and spicules of recent sponges were taken with a ColorView I digital camera for microscopy, using

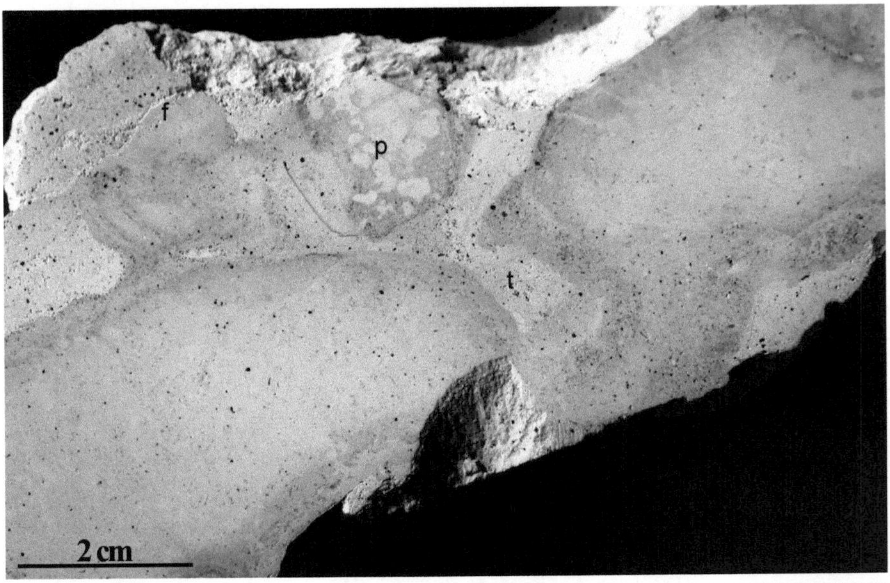

Fig. 3 Sample of the Hitch Wood Hardground from High Wycombe quarry, sawn and polished surface. The framework of the hardground, some distance below seafloor (f) shows no bioerosion. Omission-suite, pre-cementation *Thalassinoides* penetrate the cemented chalkstone (t). The bioeroded pebble (p) has fallen into the *Thalassinoides* network with the filling sediment

the software AnalySIS 3.2, a Heerbrugg dissecting microscope and a Leica phase contrast microscope at Oldenburg University.

Type material is housed in the Geological Museum of the University of Copenhagen, Denmark. Photographs of recent sponge spicules were taken from preparations in the Schönberg collection.

Results and discussion

The intraclast revealed sponge erosion traces filled with heterogeneous, cemented chalk sediment (Fig. 4). The morphology of the boring corresponded to that of *Entobia cretacea* Portlock, 1843 in having rather small, well-inflated chambers 2-6.5 mm in diameter that maintained a fairly thick interchamber wall of about 360 µm (Figs. 4, 5A). The chambers were connected by rather few intercameral canals of widths around 350 µm, and the surface apertures also were slender but not accurately measurable. The wall of *Entobia cretacea* has well-developed chip-ornament (Bromley 1970; Ekdale et al. 1984: fig. 10-6) and this was clearly visible in sections in the studied intraclast (Fig. 5B-D, arrows). Sponge chips are unlikely to survive, as they lie unattached within the soft endosomal tissues and presumably become dislodged when the tissue disintegrates. If chips were preserved in the present sample, they would be in the darker colour of the eroded calcareous substrate. It may be possible that we located sponge chips in two instances (Fig. 5E, circles). Bright areas of similar diameters and shapes did not represent sponge chips, but cross sections through sponge spicules (Fig. 5F).

The spicules were very uniform and appeared to represent only one type (Figs. 4-6). They tended to lie geopetally on the floors of the larger chambers closer to the substrate surface (Fig. 4), but the sections were not taken in the original vertical plane. In some cases the spicules had a more ordered distribution, e.g., in parallel alignment or as fan-shaped structures, almost suggesting their position as in life (Fig. 4B-F, i.e., in chambers deeper within the substrate). We can assume that spongin originally held the spicules in place and allowed them to form bands and bundles typical for *Aka* (see Rützler 1971). Spongin does not as easily break down in water as opal will and thus may have held some of the spicules in their former position, especially in the chambers more remote from the surface. There can be no doubt that the spicules originated within the boring and were not washed in allochthonously. They are the skeleton of the borer.

The spicules pictured here represent casts of dissolution cavities after silica, filled with sparry calcite cement. It is important, therefore, to remember that we are looking at random sections of them, not at the spicules themselves. The shapes seen in the peels are not those of the whole spicules. In most cases, what appears to be the termination of the spicule is spurious, and merely represents the place where the plane of section passed out of the spicule. In only very few cases is anything resembling the full length revealed (Figs. 4F-G, 6E-F). The true morphology of the terminations is seen only where the plane of section passes longitudinally through the termination of the spicule (Fig. 6E-G). For these reasons, lengths given in earlier studies may be underestimates (Reitner and Keupp 1991).

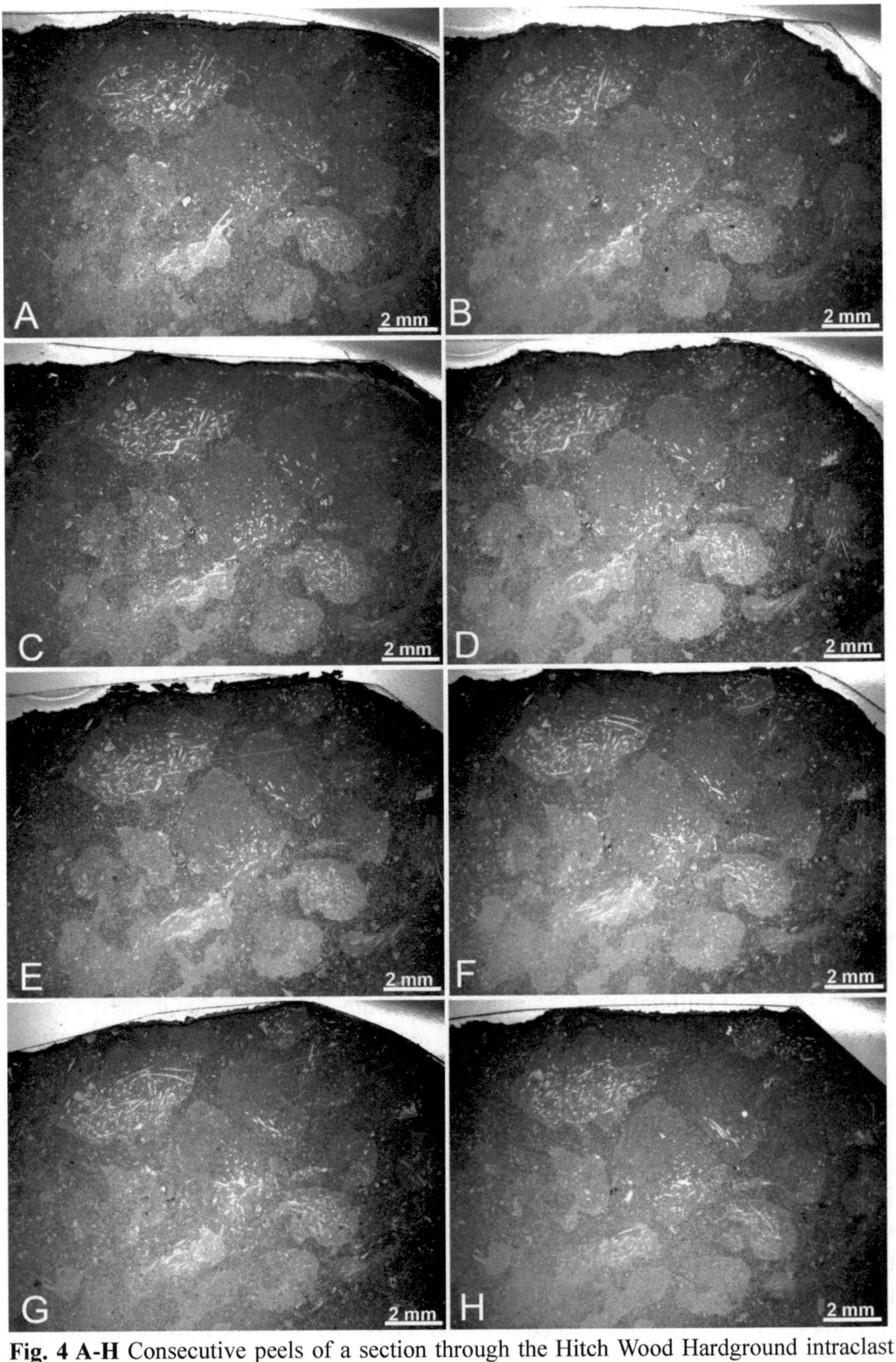

Fig. 4 A-H Consecutive peels of a section through the Hitch Wood Hardground intraclast seen in Figure 3 (MGUH 28734), allowing some understanding of the three-dimensional arrangement of the chambers and spicules. Peels show the pale fill of the *Entobia cretacea*, itself containing sparry cement-filled dissolution casts of spicules of the endolithic sponge

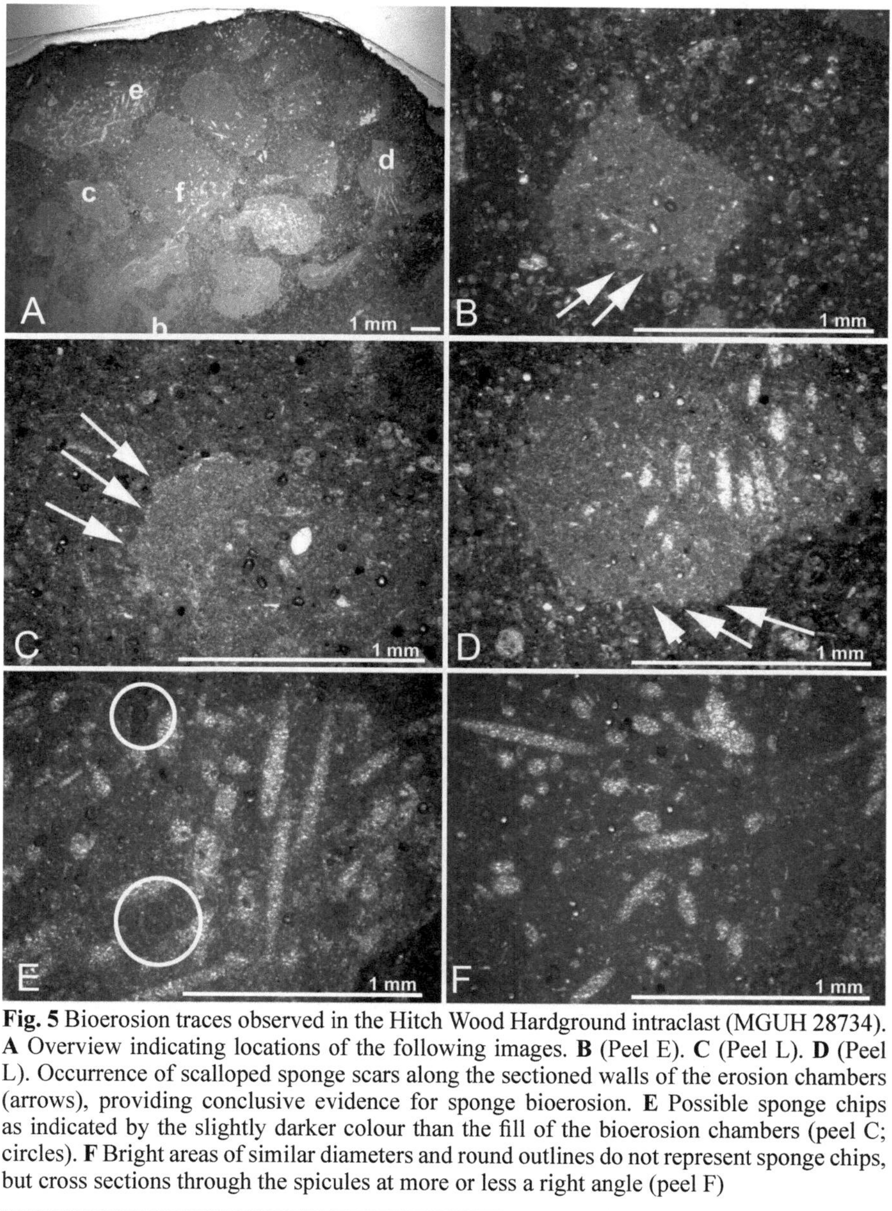

Fig. 5 Bioerosion traces observed in the Hitch Wood Hardground intraclast (MGUH 28734). **A** Overview indicating locations of the following images. **B** (Peel E). **C** (Peel L). **D** (Peel L). Occurrence of scalloped sponge scars along the sectioned walls of the erosion chambers (arrows), providing conclusive evidence for sponge bioerosion. **E** Possible sponge chips as indicated by the slightly darker colour than the fill of the bioerosion chambers (peel C; circles). **F** Bright areas of similar diameters and round outlines do not represent sponge chips, but cross sections through the spicules at more or less a right angle (peel F)

Aka, in white. Chambers of the multicamerate boring were more or less oval, but irregular in outline, connected to each other by slim, short ducts. The largest chamber in the upper left contained most spicule casts. In the figured peels they had the appearance of having fallen to the bottom without apparent order. However, in the largest chamber in the middle, spicules were all arranged at right angles to the section, and in the chamber underneath to the left, they are parallel with the chamber walls, being partly lodged in the connecting duct between the two chambers (A). **D** Fan-shaped arrangements of spicules can be seen on the right of the image. **F-G** Peels F and G, top left chamber, the longest spicule sections about 2 mm long

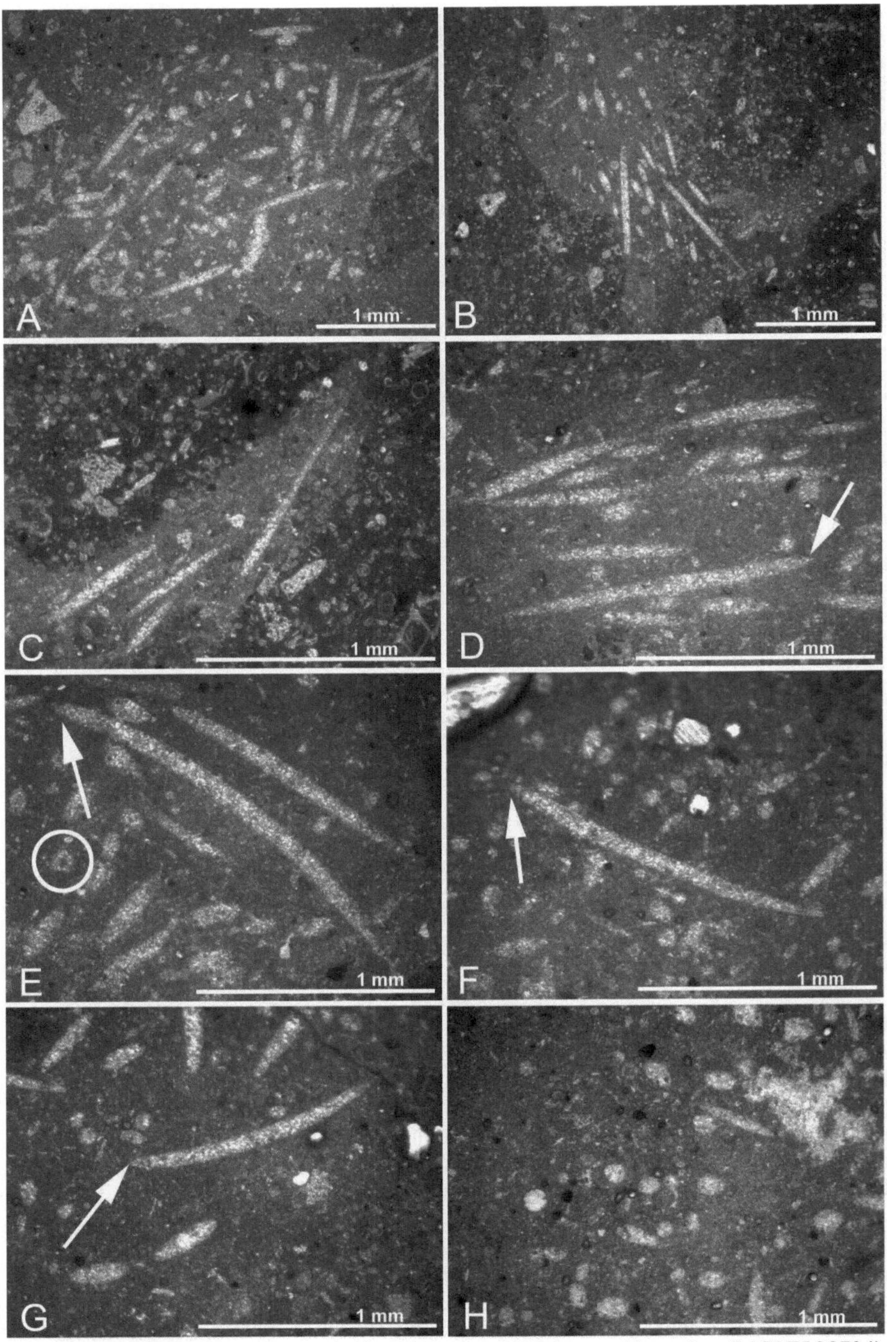

Fig. 6 Cement-filled spicule casts in the Hitch Wood Hardground intraclast (MGUH 28734). **A** Distribution of spicules in the largest chamber near the surface appeared without order in the uppermost 8 peels. **B-C** However, fan-shaped arrangement indicated that some spicules might have been preserved in their original position (peel D). **D-E** Spicule arrangement in

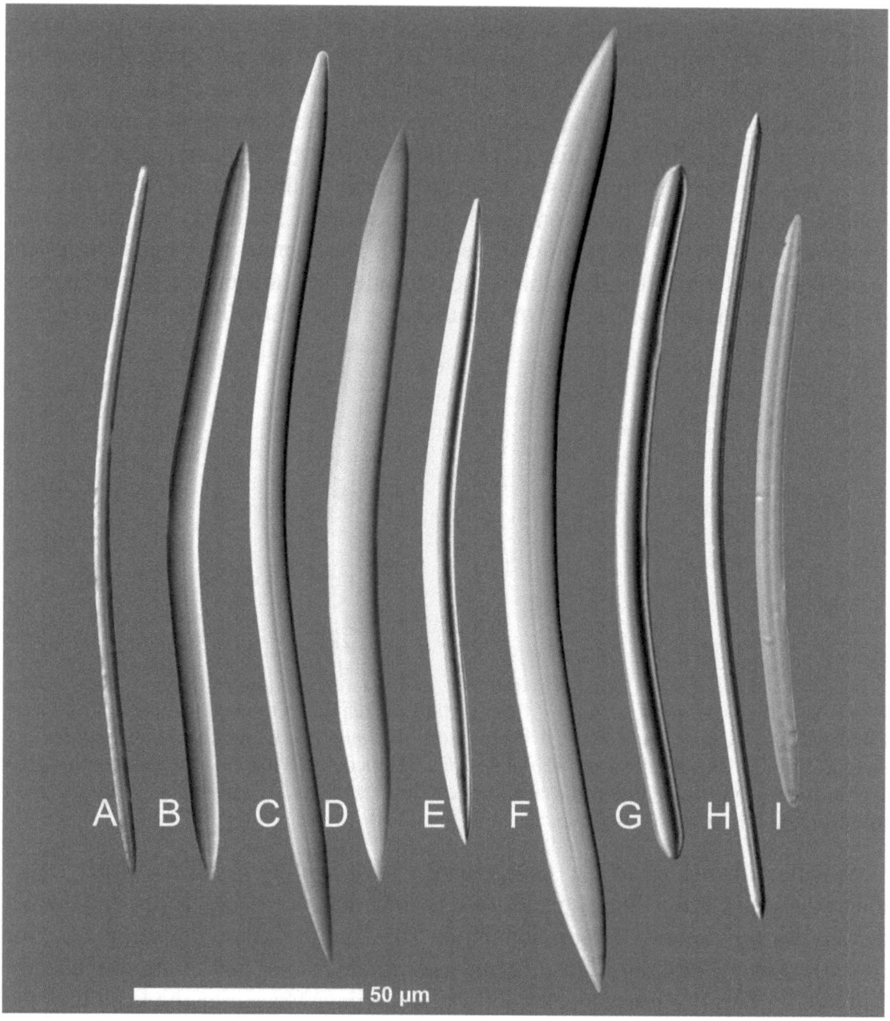

Fig. 7 Oxeas of living *Aka* spp. **A** *Aka brevitublata* (Pang, 1973) (Caribbean, from Calcinai sample Clio 75). **B** *Aka cachacrouense* (Rützler, 1971) (Dominica, from holotype USNM 24094). **C** *Aka coralliphaga* forma *typica* (Rützler, 1971) (Jamaica, from holotype USNM 24095). **D** *Aka infesta* (Johnson, 1899) (Adriatic, Schönberg collection CR-73-1). **E** *Aka minuta* Thomas, 1972 (central Great Barrier Reef, QM G322595). **F** *Aka mucosa* (Bergquist, 1965) (Palau, from holotype USNM 23697). **G** *Aka paratypica* Fromont, 1993 (central Great Barrier Reef, Schönberg collection LPB TS-1.3). **H** *Aka siphona* de Laubenfels, 1949 (Bahamas, from USNM 24107). **I** *Aka xamaycaense* (Pulitzer-Finali, 1986) (Jamaica, from holotype MSNG 47907)

parallel suggested preservation of original orientation (peel F). One cross section in E may display an open axial canal of the spicule (circle). **F-G** Curvature of spicules was similar to oxeas of living *Aka* (peel O). **H** Cross sections of spicules more or less at right angles, from the central large cavity (peel L). The best longitudinal sections were observed in peels F and G (E), where the spicule may be displayed more or less at maximum length (peel F).

The spicules appeared to be smooth, curved oxeas, suggesting the phloeodictyid genus *Aka*, the only endolithic genus in this family (Desqueyroux-Faúndez and Valentine 2002). Microscleres were not seen, but microscleres are not present in living species of *Aka*. Comparison with the often short-tipped, acerate shape of oxeas of living species was close (Figs. 7, 8A), showing little tapering of the shaft. There was a superficial likeness to the acanthoxeas of the clionaid *Pione vastifica* (Hancock, 1849). These, however, have long, tapering (i.e., hastate) terminations, and are covered with small spines (Fig. 8B, left inset). These spines appeared to be absent in the fossil material. Also absent were the tylostyles that categorise clionaid sponges (Fig. 8B), and the microrhabds of the genus *Pione* Gray, 1867 (Fig. 8B, right inset).

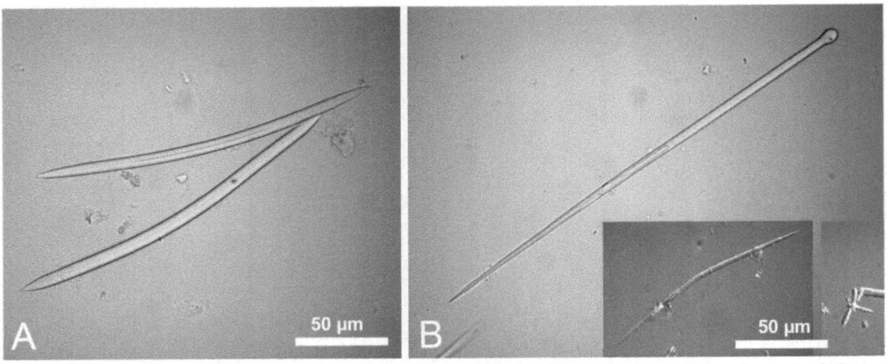

Fig. 8 Spicules from living sponges, from Little Pioneer Bay, Orpheus Island, central Great Barrier Reef. All spicules to scale. **A** Oxeas from *Aka mucosa* (Bergquist, 1965), Schönberg sample LPB-14. **B** Spicules from *Pione vastifica* (Hancock, 1849), Schönberg sample LPB-1.130, with a tylostyle (large), an acanthoxea (left inset) and a cluster of 3 cigar-shaped microspined microrhabds (right inset)

Spicule dimensions of the sponge in the intraclast are about ten times larger than those known for living species of *Aka* (e.g., Rützler 1971). Mean oxea width is 73.6 µm (minimum 47.6 µm, maximum 103.2 µm, standard deviation 13.0, for n = 50). Spicule lengths were difficult to estimate, as we had to assume that we had only incomplete spans, due to oblique sectioning. Measuring the lengths of the ten longest spicule sections, we obtained a mean length of 1327.2 µm, an obvious underestimate owing to the oblique section planes of the spicules. A length / width ratio of 18 is typical for an oxea as displayed in Figure 7D. Dividing the longest and probably most complete spicule length (of the spicule in Fig. 6E) by the mean width gives a ratio of 25.9, which is more typical of *Aka* spp. (Schönberg and Beuck, in press). Common ratios in *Aka* are approximately 23, therefore we can assume that the spicule in Figure 3E shows a section in more or less full length and that the putative mean spicule length in this fossil *Aka* was likely around 2 mm. Of the measured oxeas, about a tenth may represent immature spicules that were still in the process of formation when the sponge died and was buried, because a histogram of spicule widths showed a slightly bimodal distribution on the side of the slimmer spicules (Fig. 9).

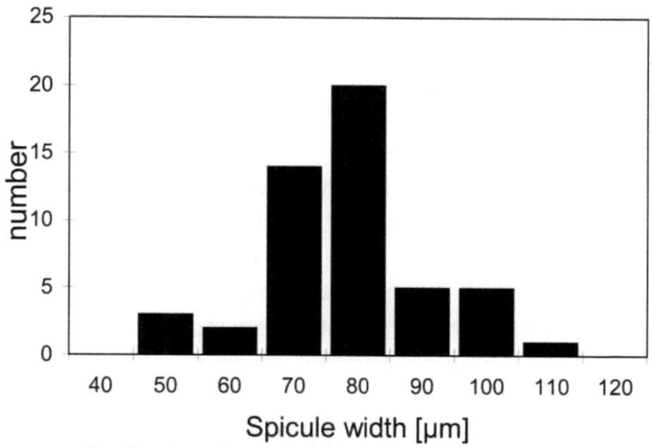

Fig. 9 Size frequency distribution of oxea width from spicules in the Hitch Wood Hardground intraclast (MGUH 28734). The slight bimodality with a secondary, smaller peak on the left indicates the occurrence of a few immature oxeas that were unusually slim compared to the other oxeas and may have been unfinished when the sponge died

Systematics

Phylum Porifera Grant, 1826
Class Demospongiae Sollas, 1885
Family Phloeodictyidae Carter, 1882
Genus *Aka* de Laubenfels, 1936

Diagnosis: Boring sponges, excavating calcareous substrates (molluscs, coral rocks); externally visible by their fistulose tubes. Skeleton formed only of spicule-bundles in sinuous bands, not in a network, lining walls of fistules. Spicules short, slender, curved oxeas. Without microscleres. (From Desqueyroux-Faúndez and Valentine 2002).

***Aka akis* sp. nov.**

Diagnosis: *Aka* having very large, smooth oxeas, mean width ca. 74 µm, approximate length 2000 µm. Other spicule types absent. Tips of the oxeas acerate, shafts with only little curvature, soft, not angular.

Differential diagnosis: The herein described *Aka* has oxeas that are about ten times the size of those of extant *Aka* spp. It has similar spicule characters compared to *Aka cassianensis* Reitner and Keupp, 1991 from the Italian Triassic and *Aka boltanaensis* Reitner and Keupp, 1991 from the Spanish Tertiary. This new sponge is from a different period and location, and differs from the above species in the larger size of its spicules. We therefore describe it as *Aka akis* sp. nov.

Etymology: *akis*, Greek, needle.

Type material, locality and horizon: The holotype is a bioeroded hardground intraclast containing sparry calcite cement-filled casts of spicules, MGUH 28734, together with a series of 17 peels made from successive sections of the same specimen, respectively MGUH 28734 A-Q. Repository: the Geological Museum, University of Copenhagen, Denmark. Quarry at The Frogmore, High Wycombe, Buckingham, SE England (Bromley and Gale 1982), no longer exposed. Hitch Wood Hardround, Chalk Rock Formation, Upper Chalk Group, Turonian, Cretaceous.

Description: Spicules of holotype occurring within sponge boring *Entobia cretacea*. The only spicule present, a smooth oxea, mean width 73.6 µm, maximum width recorded 103.2 µm. Spicules observed in section only, leaving measurement of length uncertain, but 10 measurements resulting in mean of 1327 µm. Considering length : width ratios typical for *Aka*, a spicule length of 2000 µm appears more likely. Most spicules observed straight, some with subtle, soft, evenly formed curve. Shaft only with little tapering. Short, acerate terminations where points appeared complete. Skeletal arrangement partly preserved, showing fan-shaped and band-like formations. No epilithic structures preserved.

Conclusions

The present account of *Aka akis* sp. nov. as a tracemaking species producing *Entobia cretacea* is the first positive identification of an *Entobia*-maker in the geological record that preserves the trace fossil and the (ghosts of) spicules in combination. Earlier accounts of fossil *Aka* described the sponges without naming the trace fossils (Reitner and Keupp 1991) or did not have spicules at their disposal, preserved in the bioerosion chambers (Schönberg and Tapanila 2006). The Hitch Wood Hardground contains other (unnamed) forms of *Entobia*. If further examples can be found containing autochthonous spicules, an idea at last may be developed of the ecology, evolution and systematics of endolithic sponges through time. However, there are important questions that still require answering. What was the diversity of endolithic sponges in the Mesozoic? When did the Clionaidae arise? Hardgrounds may supply the answers.

Acknowledgements

We thank S.B. Andersen (Aarhus) for kindly making the peels and L. Beuck (Erlangen), B. Calcinai (Ancona) and K. Rützler (Washington) for providing spicule slides or tissue fragments for taxonomic studies on living *Aka* spp. C. Plagge (Oldenburg) took the photographs of living-sponge spicules in Figure 8.

References

Bathurst RGC (1971) Carbonate sediments and their diagenesis. Dev Sedimentol 12, 620 pp

Bergquist PR (1965) The sponges of the Palau Islands. Part 1. Pac Sci 19:123-204

Bromley RG (1970) Borings as trace fossils and *Entobia cretacea* Portlock, as an example. In: Crimes TP, Harper JC (eds) Trace fossils. Geol J Spec Issue 3:49-90
Bromley RG (1975) Trace fossils at omission surfaces. In: Frey RW (ed) The study of trace fossils. Springer, New York, pp 399-428
Bromley RG (2004) A stratigraphy of marine bioerosion. In: McIlroy D (ed) The application of ichnology to palaeoenvironmental and stratigraphic analysis. Geol Soc London Spec Publ 228:455-479
Bromley RG, Gale AS (1982) The lithostratigraphy of the English Chalk Rock. Cret Res 3:273-306
Bronn HG (1837) Lethaea Geognostica oder Abbildungen und Beschreibungen der für die Gebirgs-Formationen bezeichnendsten Versteinerungen. Schweizerbart, Stuttgart, 1:1-544 and plates
Carter HJ (1882) New sponges, observations on old ones, and a proposed new group. Ann Mag Nat Hist, Ser 5, 10:106-125
De Laubenfels MW (1936) A discussion of the sponge fauna of the Dry Tortugas in particular and the West Indies in general with material for a revision of the families and orders of the Porifera. Carnegie Inst Wash Publ, Pap Tortugas Lab 467, 30:1-225
De Laubenfels MW (1949) Sponges of the Western Bahamas. Amer Mus Novitates 1431:1-25
Desqueyroux-Faúndez R, Valentine C (2002) Family Phloeodictyidae Carter, 1882. In: Hooper JNA, Van Soest RWM (eds) Systema Porifera: a guide to the classification of sponges. Kluwer Acad/Plenum, New York, pp 893-905
Ekdale AA, Bromley RG, Pemberton SG (1984) Ichnology – trace fossils in sedimentology and stratigraphy. SEPM Short Course 15, 317 pp
Fromont J (1993) Descriptions of species of the Haplosclerida (Porifera: Demospongiae) occurring in tropical waters of the Great Barrier Reef. Beagle Rec North Territ Mus Arts Sci 10:7-40
Grant RE (1826) Animal kingdom. In: Todd RB (ed) The encyclopedia of anatomy and physiology. Vol 1, Sherwood, Gilbert and Piper, London, pp 107-118
Gray JE (1867) Notes on the arrangement of sponges, with the description of some new genera. Proc Zool Soc London 1867:492-558
Hancock A (1849) On the excavating powers of certain sponges belonging to the genus *Cliona*; with descriptions of several new species, and an allied generic form. Ann Mag Nat Hist, Ser 2, 3:321-348
Hopson PM, Woods MA, Aldiss DT, Ellison RA, Farrant AR, Booth KA, Wilkinson IP (2006) Invited comment on Wray & Gale's 'The palaeoenvironment and stratigraphy of the Late Cretaceous Chalks'. Proc Geol Assoc 117:145-162
Johnson JY (1899) Notes on some sponges belonging to the Clionidae obtained at Madeira. J Roy Microscop Soc Trans 9:461-463
Lang B, Steiger T (1985) Paleontology and diagenesis of Upper Jurassic siliceous sponges from the Mazagan Escarpment. Oceanol Acta 5:93-100
Mikuláš R (1994) Sponge borings in stromatoporoids and tabulate corals from the Devonian of Moravia (Czech Republic). Věstn Česk Geol Ústavu 69:69-74
Mortimore RN, Wood CJ, Gallois RW (2001) British Upper Cretaceous stratigraphy. Geol Conserv Rev Ser 23, Joint Nat Conserv Comm, Peterborough, 558 pp
Pang RK (1973) The systematics of some Jamaican excavating sponges (Porifera). Postilla Peabody Mus Yale Univ 161:1-75
Portlock JE (1843) Report on the geology of the County of Londonderry, and Parts of Tyrone and Fermanagh. Her Majesty Stationary Office, Dublin, 748 pp

Pulitzer-Finali G (1986) A collection of West-Indian Demospongiae (Porifera), in appendix, a list of the Demospongiae hitherto recorded from the West Indies. Estr Ann Mus Civ Storia Nat Giacomo Doria Genova 86:65-216

Reitner J, Keupp H (1991) The fossil record of the haplosclerid excavating sponge *Aka* de Laubenfels. In: Reitner J, Keupp H (eds) Fossil and recent sponges. Springer, Berlin, pp 102-120

Rützler K (1971) Bredin-Archbold-Smithsonian Biological Survey of Dominica: burrowing sponges, genus *Siphonodictyon* Bergquist, from the Caribbean. Smithsonian Contr Zool 77:1-37

Rützler K (2002) Impact of crustose clionid sponges on Caribbean reef corals. Acta Geol Hisp 37:61-72

Schönberg CHL, Beuck L (in press) Where Topsent went wrong: *Aka infesta* a.k.a. *A. labyrinthica* (Demospongiae: Phloeodictyidae). J Mar Biol Assoc UK

Schönberg CHL, Tapanila L (2006) The bioeroding sponge *Aka paratypica*, a modern tracemaking analogue for the Paleozoic ichnogenus *Entobia devonica*. Ichnos 13:147-157

Siever R (1962) Silica solubility, 0°-200° C, and the diagenesis of siliceous sediments. J Geol 70:127-150

Sollas WJ (1885) A classification of the sponges. Ann Mag Nat Hist, Ser 5, 16(95):395

Tapanila L (2006) Devonian *Entobia* borings from Nevada, with a revision of *Topsentopsis*. J Paleont 80:760-767

Thomas PA (1972) Boring sponges of the reefs of Gulf of Mannar and Palk Bay. J Mar Biol Assoc India, Spec Vol Proc Symp Corals Coral Reefs 14:333-362

Uriz M-J, Turon X, Becerro MA, Agell G (2003) Siliceous spicules and skeleton frameworks in sponges: origin, diversity, ultrastructural patterns and biological functions. Microscop Res Techn 62:279-299

Woods MA, Aldis DT (2004) The stratigraphy of the Chalk Group of the Berkshire Downs. Proc Geol Assoc 115:249-265

Role of polychaetes in bioerosion of coral substrates

Pat Hutchings[1]

[1] The Australian Museum, Sydney NSW 2010, Australia, (path@austmus.gov.au)

Abstract. Polychaetes are important in the early stages of bioerosion of newly available coral substrate often as a result of coral death by disease, bleaching, Crown-of-Thorns starfish attack etc., and physical destruction during storms. A succession of polychaetes are recruited to the substrate, which it is hypothesised, facilitates subsequent recruitment by sponges, sipunculans and molluscs that are dominant in 'mature' boring communities. Recruitment of boring polychaetes varies according to the type of substrate available, season and geographical location of substrate, and environmental factors such as light, water quality, depth and wave exposure. The mechanisms by which polychaetes bore still require further investigations.

Keywords. Bioerosion, polychaetes, early stages, succession

Introduction

Present day coral reefs are dynamic systems as were fossil reefs (Vogel 1993), resulting from the balance between reef growth and erosion both physical and biological. Biological erosion consists of losses of calcium carbonate by organisms boring into the substrate. It involves a wide variety of borers such as microalgae, fungi, sponges, molluscs, polychaetes and sipunculans (Hutchings 1986), which often results in a honeycombed structure. Then when endolithic algae and fungi are embedded in the substrate (Le Campion-Alsumard et al. 1995), it makes it highly susceptible to grazing by herbivores such as echinoderms, molluscs and certain species of fish (Bellwood and Choat 1990; Conand et al. 1998), which physically remove the substrate containing the algae. This porous structure also makes the substrate more susceptible to physical and chemical erosion, as water is pumped into the borings and buffeted by waves and currents that may cause dislodgement of parts of the coral substrate, especially upright dead coral colonies (Glynn 1997). Water movement facilitates the chemical dissolution of calcium carbonate substrates and while no detailed measurements have been made on the dissolution of reef substrate, a study by Fulton and Bellwood (2005) on rates of dissolution of gypsum (calcium sulfate) in reefal environments found that rates varied significantly with wave motion. While gypsum dissolves faster than calcium carbonate it does provide an indication of the relative rates of dissolution in a variety of reefal environments and that areas of high water movement experience the highest rates of dissolution,

M. Wisshak, L. Tapanila (eds.), *Current Developments in Bioerosion*. Erlangen Earth Conference Series, doi: 10.1007/978-3-540-77598-0_13, © Springer-Verlag Berlin Heidelberg 2008

although relative to biological erosion rates are very small. It is important to stress that these agents of reef destruction, biological, chemical and physical, interact all together and vary between habitats and geographical areas in space and time (Le Campion-Alsumard et al. 1993). All rates of bioerosion will of necessity include a component of chemical and physical erosion, as dissecting out the individual components is not possible.

Once a coral colony dies as a result of various events such as predation, disease or being physically dislodged by storms, it is rapidly colonised by micro-organisms including endolithic algae, viruses (Tribollet et al. 2005) and presumably bacteria, which are attracted to the dying or dead coral tissue. The surface of this dead colony rapidly becomes pale green as algae colonises both the surface and the substrate immediately below the surface (Chazottes et al. 1995).

Recruitment of boring organisms is believed to be entirely via pelagic or benthic larvae which settle on the substrate, metamorphose and rapidly bore into the substrate (Hutchings 1986). As live corals are active carnivores, recruitment onto live coral colonies is difficult or restricted to areas where polyps are damaged or very scarce. In contrast, dead corals or dead parts of live coral colonies and coral substrate are actively bored. Traditionally, only mature bored colonies have been studied, where conspicuous borers include extensive sponge colonies, bivalves and sipunculans (Glynn 1997). However, it is also important to understand the initial stages of bioerosion, and to determine the substrate susceptibility, and to assess if these initial stages have an influence on the development of mature boring communities. In situ studies on massive colonies of *Porites lobata* Dana, 1846 ranging from living colonies to 100% dead colonies in French Polynesia showed that boring rates increased with the time elapsed since the death of the colony, and that the dominant agents of boring also changed with increasing age of the coral structure (Peyrot-Clausade et al. 1992). As coral reefs are becoming increasingly degraded by bleaching events, diseases, Crown-of-Thorns starfish attacks etc. (Wilkinson 2004), it is becoming more and more critical to understand how the balance between reef construction and reef destruction is being altered, and which can lead to loss of reef framework and all the ensuing consequences of this (Glynn 1997; Holmes et al. 2000; Salvat et al. 2001).

To date virtually all the experimental studies have been carried out on massive species of *Porites* Link, 1807, with only one study (Musso 1994) using colonies of *Acropora* Oken, 1815. Preliminary studies on the complete infauna (borers + nestlers) on a range of coral substrates composed of a variety of morphological types of coral at One Tree Island, Great Barrier Reef, found considerable differences between branching, massive and columnar type corals in terms of the biomass and diversity of infauna (Hutchings 1974). The highest biomass and diversity occurred in the substrates composed of the massive species and the least in the samples of dead organ pipe corals. This suggests that borers favour substrates composed of large colonies of dead massive species as these habitats are less vulnerable to physical destruction and provide sufficient space for borers to either become large as in the case of some polychaetes, sipunculans or bivalves, or form massive colonies such as sponges (Hutchings 1974). Branching thickets of *Acropora* have limited numbers of borers as the physical dimensions of the branches limits large borings from forming (Hutchings et al. unpublished data).

Traditionally, boring sponges, molluscs and sipunculans have been regarded as major bioeroders and, certainly, they are responsible for substantial losses of coral substrate in mature boring communities (Hutchings 1986; Perry 1998). The mechanisms by which these animals bore are reasonably well documented and include both chemical dissolution as well as mechanical abrasion or a combination of both (Lazar and Loya 1991; Mao Che et al. 1996; Schönberg and Wilkinson 2001). These organisms are conspicuous when coral colonies are broken open and as well, they are easily recognised in fossil coral reefs. During the past 20 years using experimental techniques, the initial stages of bioerosion of recently killed coral has been investigated in the Pacific both in relatively pristine and disturbed environments. This review discusses the findings of these studies dealing with polychaete borers. To assist in this review line drawings of the common polychaete borers and the structures which they use to bore are provided in Figures 1 to 3.

Study techniques

To study the recruitment of the early borers to recently killed coral substrates, colonies of live coral are collected, typically one of the massive species of *Porites*. The layer of live polyps are trimmed off and regular sized blocks are cut using a band saw. Any blocks with signs of boring are discarded. The resultant blocks are then washed, dried and measured and so when the samples are returned to the reef and firmly attached, one is certain that no living organisms are present in them. The blocks (Fig. 2A) are then collected at intervals to document the distribution and abundance of the infauna, over time, allowing the succession of borers to be determined. Early studies used to place the blocks in an acid mixture, to dissolve the calcium carbonate substrate and to extract the animals, which are often small and impossible to extract in any other way (Kiene and Hutchings 1994). Following dissolution, the resultant residue consists of a matrix of endolithic algae and a suite of animals, and it is relatively easy to extract them intact and to sort them to species. More recently, the collected blocks are cut in half, one to be treated as above, the other to be used to determine boring, accretion and grazing rates, and allowing net rates of gains or losses to be calculated (Fig. 2B; Tribollet et al. 2002).

Bioeroders

Distinguishing between nestlers and borers

The polychaetes extracted after acid dissolution include both borers and species that enter the coral substrate after borings have been established to settle inside. The latter are referred to as ‘nestlers’ or ‘opportunistic species’ (Hutchings 1983) and separating them from borers can be problematic. However, knowing the morphology and ecology of the groups, as well as examining the species in situ

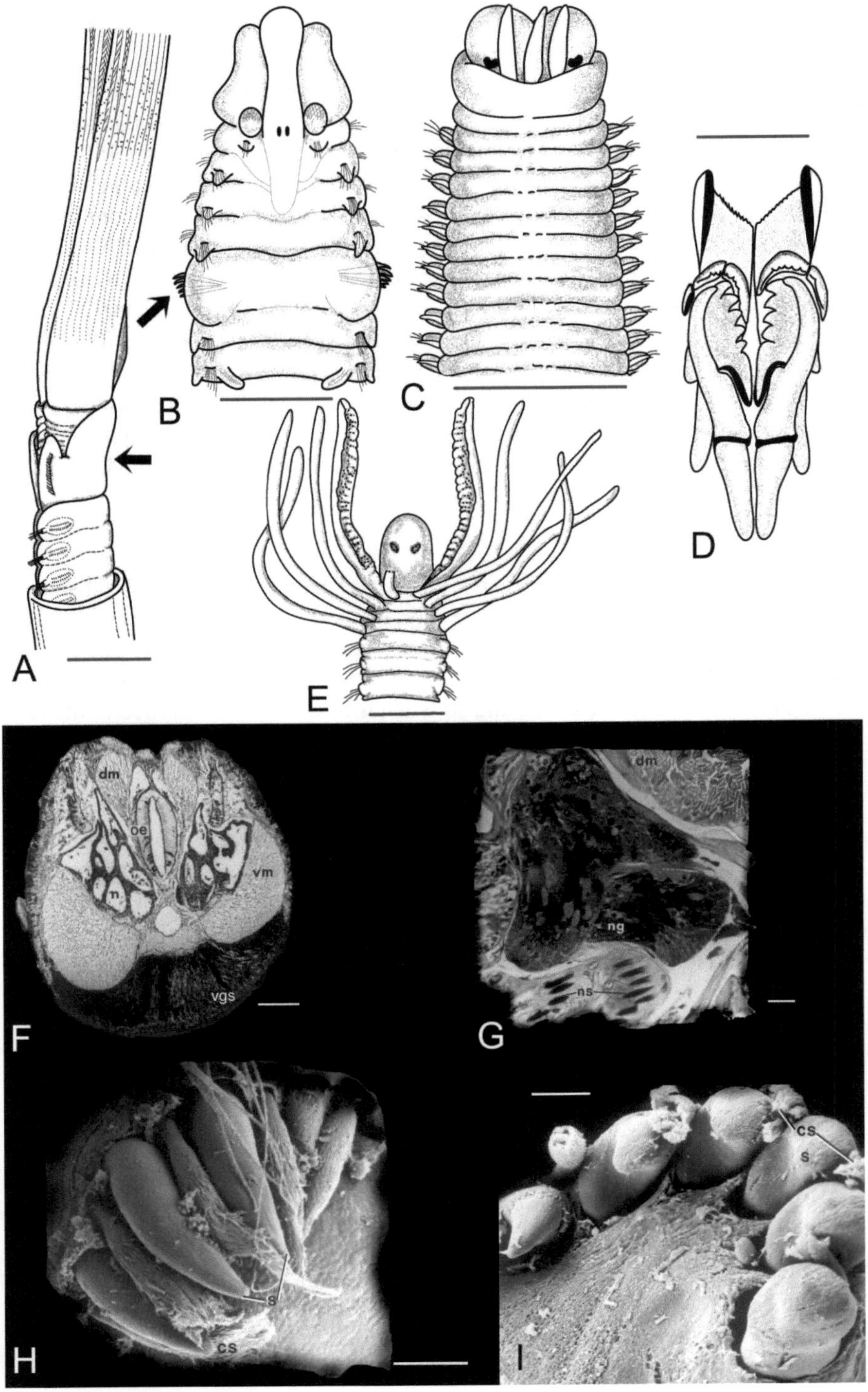
A
B
C
D
E
F
G
H
I
dm
oe
vm
n
vgs
ng
ns
s
cs

in coral substrate, allows the identification of the real borers, and to characterise their borings. Some of these polychaete borers are also found in other calcareous substrates such as limestone and mollusc shells, especially species of spionids and sabellids. One genus of the polychaete family Sabellidae, *Caobangia* Giard, 1893, are found boring into mollusc shells. They occur in *'blind-ending tear drop-shaped burrows'* according to Giard (1893). Later, Jones (1969, 1974) was able to expand these descriptions based on additional material becoming available and moved the genus into a new family Caobangiidae. The genus represented by several species was found in 22 species of molluscs from fast flowing streams in South East Asia and the Philippines at elevations from 800 to 1300 m. This genus appears to be restricted to freshwater and only bores into mollusc shells belonging to several genera. Recent phylogenetic studies have suggested that the family should be placed back within the Sabellidae (Beesley et al. 2000).

Which polychaetes bore?

Polychaetes which dominate the initial stages of colonisation of experimental substrates belong to a wide range of families and species. They have been observed to occupy tight borings, similar to its body width, although sometimes more than twice its body length (Chughtai and Knight-Jones 1988). Those belonging to sedentary species (such that they cannot be extracted from the borings easily) are certainly borers, while those occupying cavernous borings and moving freely are nestlers.

Basically, there are only two options for providing an explanation of how boring species can actually bore into hard substrates. First they can secrete substances to dissolve the substrate or, secondly they can use hard structures such as teeth or spines (Fig. 1D, H-I) to physically erode the substrate. Perhaps a combination of both mechanisms is also possible. Among polychaetes, one of the best known groups of borers are the polydorids, which include several species of Spionidae (e.g., *Polydora* Bosc, 1802 and *Dipolydora* Augener, 1914) having modified chaetae on the 5th chaetiger; Fig. 1B, H-I). Polydorids are known to bore into a wide variety of calcium carbonate substrates including mollusc shells and limestone as well as coral substrates (Blake and Evans 1973; Zottoli and Carriker 1974; Simon et al. 2006). While some polydorids are found in benthic muddy substrates (Wilson 2000), all species occurring in coral substrates are regarded as borers and are always found in tight fitting borings. The exact method of boring is unknown in corals but has been extensively studied in commercial mollusc species (Evans 1969). Thus, it is not surprising that the boring mechanism by these polychaetes has been the

Fig. 1 **A** Anterior end of *Notaulux* sp.; arrow indicates glandular collar; scale 1 mm. **B** Anterior end of *Polydora* sp.; arrow indicates position of modified chaetiger 5; scale 1.5 mm. **C** Anterior end of *Lysidice* Savigny, 1818; scale 2 mm. **D** Dissected jaws of *Lysidice* sp.; scale 0.6 mm. **E** Anterior end of *Dodecaceria* sp.; scale 1 mm. **F** Transverse section of *Notaulux* in thoracic region; Alcian blue / fast red stain; scale 150 µm. **G** Transverse section of thoracic segment of *Eunice* Cuvier, 1817 showing notopodial gland. **H-I** SEM of spines of *Polydora* sp. from *Porites* spp. at Lizard Island; scale 10 µm. (dm = dorsal musculature, n = nephridium, ng = notopodial gland, ns = notochaetae, oe = oesophagus, vgs = ventral glandular shield, vm = ventral musculature, s = chaeta, cs = capillary chaeta; Figures A-E: redrawn from Day 1967)

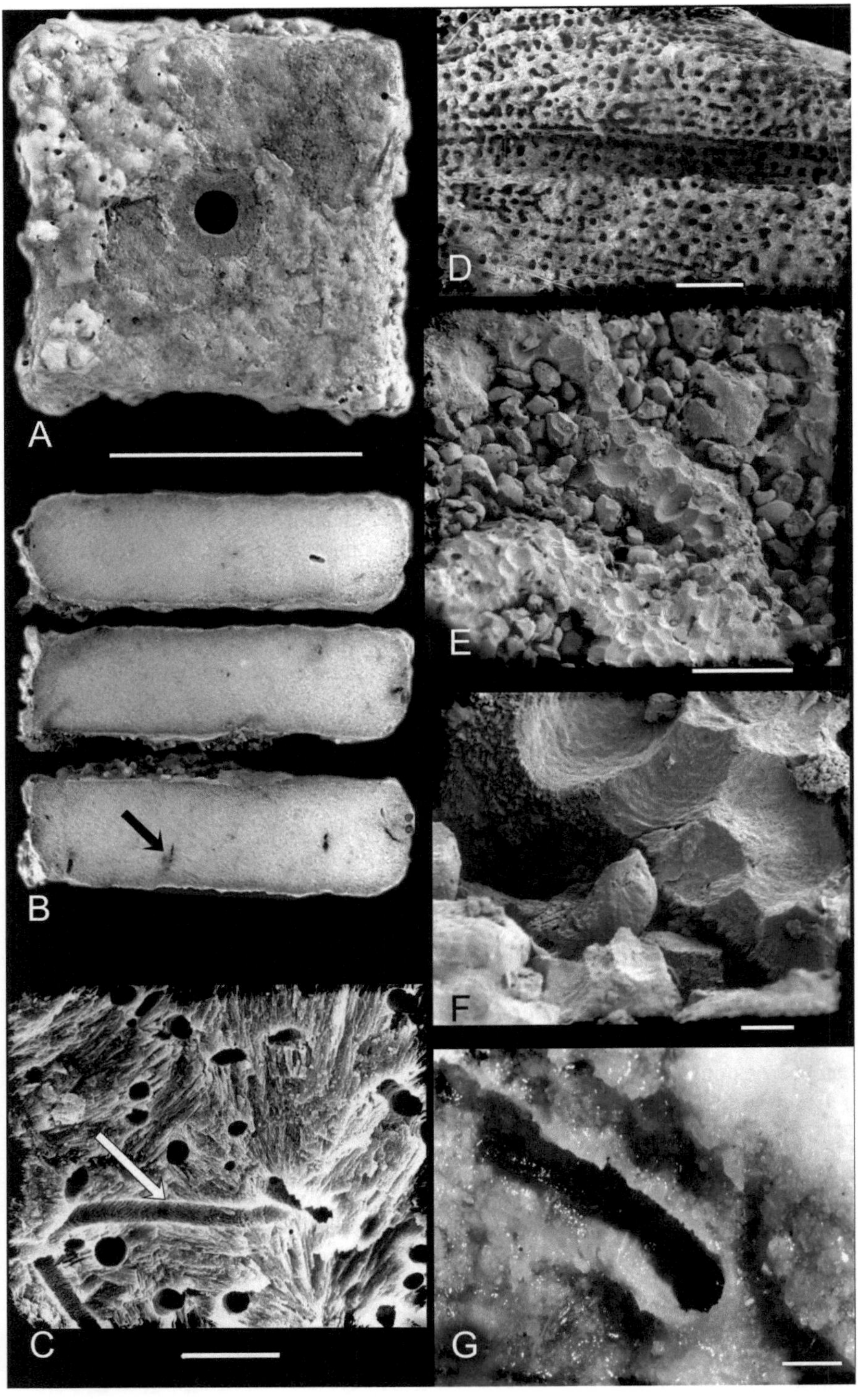
A
B
C
D
E
F
G

best studied, as their extensive borings reduce the value of the infested shells and, if heavily infested, may be lethal to the molluscs. It was suggested that boring was assisted by the modified chaetae on chaetiger 5 (Fig. 1B, H-I) and that the ventral epithelium and segmental mucus glands produced acid mucopolysaccharides (Dorsett 1961), which may bring about decalcification by acting as chelating agents (Simkiss and Tyler 1957). Nevertheless, *Polydora websteri* Hartman, 1943, could still bore after removing the modified chaetae (Haigler 1969). The same species boring into *Crassostrea* Sacco, 1897 and *Mytilus* Linnaeus, 1758, secretes a viscous fluid, which dissolves interprismatic and interlamellar organic matrices and then dissolves the exposed crystals (Zottoli and Carricker 1974). However, the actual composition of the fluid was not determined. Although this substance has not been further identified, it was suggested that it is secreted along the body of the worm and that it is able to dissolve mollusc shells (Sato-Okoshi and Okoshi 1993). It seems likely that similar processes occur when polydorids bore into coral substrate. These polychaetes are among the earliest colonisers of the experimental substrates and after just 3 months, they may already form dense colonies. Species of *Polydora* may spawn in either summer or spring, while other polydorids deposit egg capsules in winter, which release pelagic larvae in spring (Blake 1969). Recruitment varied between sites, depending on the local composition of the polydorid populations and environmental conditions, which suggests that competent larvae may be available for recruitment throughout much of the year (Hutchings et al. 1992).

Another well represented group are species of *Dodecaceria* Örsted, 1843 (Cirratulidae). After successful recruitment by pelagic larvae, and establishment, these worms can then rapidly build up large populations within a chamber (Figs. 1E, 3E). They use asexual reproduction where adult worms split up into individual segments, and each developing into miniature worms, using the process of fragmentation (Gibson 1977). However, these asexually produced recruits appear unable to produce their own borings (Gibson and Clark 1976). This group of polychaetes is also commonly found boring into calcareous substrates in both temperate and tropical areas. These animals must secrete an acid like secretion, as they have no hard structures capable of eroding the substrate. The flask-shaped chambers are connected to the exterior by a narrow passage and these surface deposit feeding animals protrude their long palps from the chamber to collect fine sediments which are washed into or onto the substrate (Fig. 3E). *Dodecaceria concharum* Örsted, 1843 produces a straight boring in bivalve shells, which tend to expand towards the blind end, but the method by which it bores is not known (Evans 1969). *Dodecaceria* species are present within 3-6 months of exposure in experimental coral blocks and they are abundant in areas where the substrates are subjected to high sediment loads from river run off or from nearby inter reefal areas where regular trawling occurs (Hutchings et al. 2005; Osorno et al. 2005).

Fig. 2 A Experimental block of *Porites* collected after 12 months of exposure. **B** Block cut into slices revealing internal borings; arrow indicates worm boring. **C** Longitudinal section along *Notaulux* boring under the SEM; scale 30 µm. **D** Longitudinal section of *Notaulux* under SEM; scale 30 µm. **E** *Lysidice* sp. boring in *Porites* under SEM; scale 30 µm. **F** *Lysidice* sp. boring in *Porites* under SEM; scale 30 µm. **G** *Eunice* sp. in *Porites*; scale 30 µm

Another family which has several boring representatives are the Sabellidae (Chughtai and Knight-Jones 1988), which include species of *Notaulux* Tauber, 1879, *Megalomma* Johansson, 1926, *Pseudopotamilla* Bush, 1904, *Perkinsiana* Knight-Jones, 1983 and *Terebrasabella* Fitzhugh and Rouse, 1999. Common in all experimental coral blocks is a species of *Notaulux,* which has been referred to as *Hypsicomus* Grube, 1870 (see Peyrot-Clausade et al. 1992 and subsequent papers; Fig. 1A). Most species of *Hypsicomus* were transferred to *Notaulux* (see Perkins 1984) and subsequently an additional species of *Notaulux* including an illustration of a longitudinal section of the tube buried deep within *Porites* was described (Nishi and Nishihira 1999). *Notaulux* seems to be very common throughout the Indo-Pacific and typically includes boring species (J. Gil pers. comm.). However, it is still unclear, whether it represents one widely distributed species or several. This group of fan worms are found in tight fitting flask-shaped tubes lined with chitin and, in both experimental and 'natural' habitats, they are found penetrating deep into almost pristine coral, but still maintaining an opening allowing them to expand the branchial crown for filter feeding and respiration (Figs. 2C-D, 3A-B, D). Like *Dodecaceria* they have no hard structures capable of eroding the substrate. Thus, boring by an acid-like secretion must be assumed. Sabellids have well developed glands at the base of the branchial crown, although attempts to identify the nature of the substances extruded by these glands were unsuccessful (Hutchings unpub. data; Fig. 1F). However, this is the most likely secretory area, given that the rest of the body is far less glandular. Studies on species boring into limestone on the Welsh coast strongly indicate that the substrate was removed by chemical dissolution aided by the boring irrigation of the cavity by muscular action and faecal removal by cilia (Chughtai and Knight-Jones 1988). These authors also found that the chaetae rarely exhibited any sign of damage or wear which could have resulted from mechanical boring into hard rock. Another group of sabellids, the fabriciniids, have also been identified as borers (Peyrot-Clausade et al. 1995) but the data supporting this is weak as the animals are very small and about the size of the pores in the coral matrix. While present in the earliest experimental blocks, fabriciniid borings are impossible to distinguish from the coral structure.

The other group of polychaete borers are the eunicids and several species are involved (Fig. 1C). They typically do not occur until several months after the experimental substrates have been exposed and they become more abundant as the boring communities mature. Eunicids (including species of Eunicidae and Dorvilleidae) have well developed, complex jaws (Fig. 1C-D) and occur in well developed borings (Figs. 2G, 3C, F). Detailed examinations of the walls of these borings using SEM reveal gouges resembling tooth marks, and fragments of coral substrate are often found between the numerous teeth which constitute these complex jaws (Hutchings unpubl. data; Fig. 2E-F). Acid secretion may also occur although this has not been documented (Fig. 1G).

In summary the identification of boring species of polychaetes depends on the shape of the boring relative to the animal, the distribution of borings in the substrate both in the experimental substrates as well as in in situ substrates. Polychaete borings

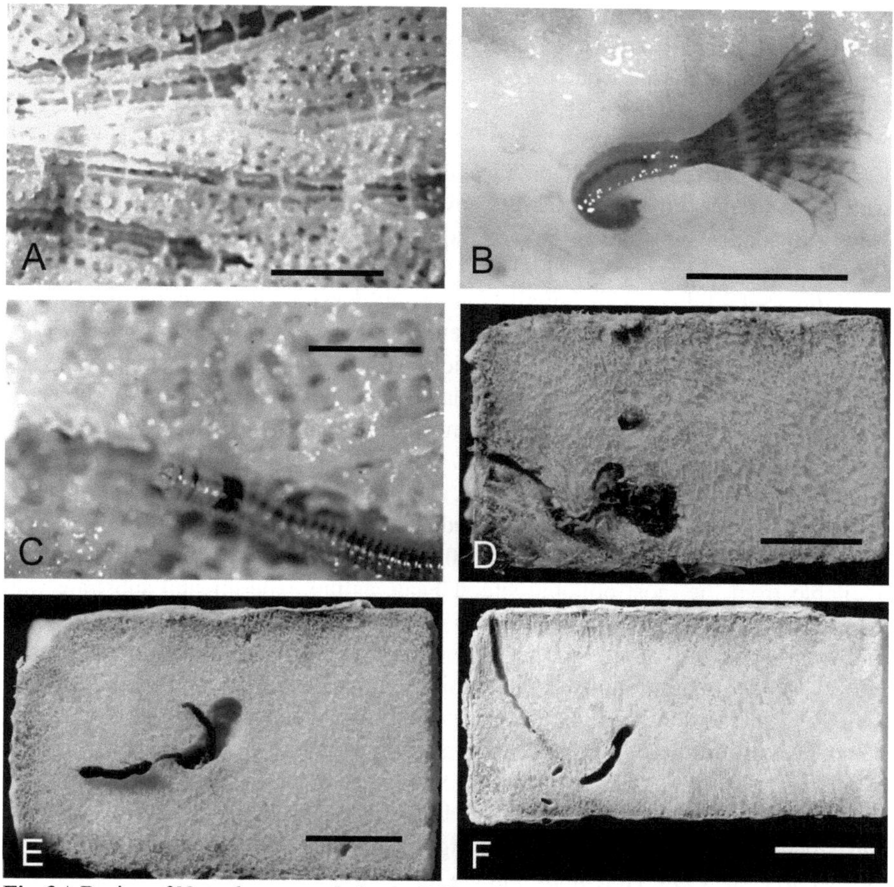

Fig. 3 **A** Boring of *Notaulux* sp.; scale 1 mm. **B** *Notaulux* sp. emerging from boring; scale 1 mm. **C** *Eunice* sp. in boring; scale 1 mm. **D** Sabellid boring in section of experimental block; scale 10 mm. **E** *Dodecaceria* sp. in section of experimental block; scale 10 mm. **F** Eunicid boring in section of experimental block together with a sipunculan tube; scale 10 mm

are characteristic in shape so that examining thin sections of substrate in order to measure rates of loss of calcium carbonate, will allow identification of borings as belonging to polychaete worms rather than to other borers (Hutchings et al. 2005; Fig. 2B-C). More detailed information on the exact mechanism of boring is needed but it is difficult to study these small worms. However, it should be possible now to actually measure the microenvironments within individual borings to detect acid secretion and characterise its composition, although to date this has not been done. The only detailed investigation was undertaken by Zottoli and Carriker (1974) who studied boring in *Polydora* by examining live worms living in artificial preparations of polished shell covered with plastic film. They found that the worms can settle in crevices on the shell surface and they slowly penetrate the shell forming a U- or flask-shaped cavity by secreting a viscous fluid which dissolves the interprismatic and interlamellar organic matrices and then the exposed crystals. SEM of the borings reveal a complex pattern of internal dissolution. They found no evidence of chaetal abrasion on the shell walls even under high magnification. It seems likely that a similar process of boring would occur in coral substrates.

Succession of borers

Experiments carried out both on the Great Barrier Reef and the Coral Sea as well as in French Polynesia clearly show a succession in the polychaete borers. Newly available substrate is heavily colonised by *Polydora* spp., (Spionidae), *Schistomeringos* Jumars, 1974 (Dorvilleidae belonging to the eunicids) and *Notaulux* (Sabellidae), with initially increasing densities that decline after 12-15 months. With increasing exposure, other eunicids arrive as well as *Dodecaceria* spp. In mature boring communities, large eunicids are common, often reaching many cm in length (Hutchings et al. 1992). The relative densities of the different borers vary according to habitat and location. In the northern Great Barrier Reef, *Polydora* sp., favoured more oceanic sites, whereas *Dodecaceria* sp. preferred more inshore sites with higher levels of turbidity (Osorno et al. 2005). In contrast, in French Polynesia *Polydora* spp. favoured the more polluted sites where turbidity run off levels were high (Hutchings and Peyrot-Clausade 2002). Unfortunately, no long term data were available from the most polluted site at Faaa, as the experimental substrate was completely eroded after only two years of exposure (Pari et al. 1998). At Faaa, the densities of polychaete borers were determined after 6 months of exposure. However, the grazing rate by the echinoid *Echinometra mathaei* Blainville, 1825 which was present in high numbers, was such that the surface of the blocks was being heavily eroded, so that the borers present in the peripheral areas of the experimental substrate were constantly being removed, so the recorded densities of the borers were seriously underestimated. Nevertheless, the presence of *Polydora* and *Dodecaceria* was reported at this highly polluted site (Peyrot-Clausade et al. 1995; Pari et al. 1998; Hutchings and Peyrot-Clausade 2002).

Due to the nature of the boring communities, the substrate must always be destroyed to extract the borers. Thus, while one can document the densities and species composition at a particular point and time, and then compare them with the next sampling time, this may not reflect the real dynamics of the boring community during the period between sampling. Some polychaetes have life cycles ranging from a few weeks to several months, so that it is possible that generations have come and gone between the sampling periods. All studies on early stages of bioerosion have found considerable variation between replicate blocks collected at the same sampling period. However, these differences are less than those found between sites, suggesting that recruitment patterns at the local scale are variable. Certainly, larvae recently settled on the block are extremely vulnerable to predation or they may be easily washed off by local currents (Sebens and Johnson 1991) until they actually bore into the substrate. In fact, the highest density of larvae of boring sabellids occurred around surface concavities, which provides shelter from wave action, but also harbor many predators (Chughtai and Knight Jones 1988). They suggest that, at this stage, larvae are most vulnerable so that a high mortality must be presumed.

Seasonality of recruitment

A distinct seasonality of recruitment has been demonstrated, reflecting that of breeding for many of these polychaete borers (Hutchings and Murray 1982; Hutchings et al. 1992; Hutchings and Peyrot-Clausade 2002). Maximum recruitment occurs during the summer months, with windward and reef flat sites being preferred. Recruitment varies over time at a particular site and patterns of recruitment vary between sites over time (Hutchings and Peyrot-Clausade 2002). However, any newly available substrate is colonised rapidly regardless of the time of year (Hutchings et al. 1992). In addition, at Lizard Island (Great Barrier Reef), significant variations between years occurred at individual sites, likely reflecting variations in larval supply, and / or recruitment success, which in turn will be modified by variations in weather patterns between years (Hutchings et al. 1992). In addition, significant differences occurred between sites over a four-year period at Lizard Island, with a windward slope site at 10 m and the reef flat being the preferred recruitment sites for borers (Kiene and Hutchings 1994). Similar patterns were found over a five-year period, in French Polynesia (Pari et al. 2002).

Rates of boring

The few published rates of worm borings include both polychaetes and sipunculans (Peyrot-Clausade et al. 1992; Kiene and Hutchings 1994; Osorno et al. 2005). However, they are not directly comparable as Osorno et al. (2005) determined rates of loss from examining thin sections and estimating the volume of the boring volumes, whereas the other two studies estimated rates of loss of calcium carbonate by measuring the volume of the worms. The highest rates recorded at Lizard Island (Great Barrier Reef) were 0.104 kg/m^2 (Kiene and Hutchings 1994), whereas those recorded in French Polynesia were 0.1979 kg/m^2 (Peyrot-Clausade et al. 1992). But rates which included both polychaetes and sipunculans across the Great Barrier Reef varied from 0.113 ± 0.066 to 0.356 ± 0.243kg/m^2 (Osorno et al. 2005). It must be taken into account that all these net rates are almost certainly underestimates, as worm borings in the outer layers of the substrate are habitually being removed by grazers, and the borings of small worms are difficult to identify in thin sections.

Discussion and future areas of research

While polychaetes have been shown to be important colonisers of newly available coral substrate, their importance declines as the boring community ages so that other boring agents tend to become dominant. It has been suggested that the small borings created by the polychaetes and their high rate of turnover facilitates the recruitment of other borers such as sponges, bivalves and sipunculans which are never found in substrates made recently available. However, it seems difficult to actually test this hypothesis, taking into account our knowledge of the actual settlement and larval boring process. An additional problem is to actually measure the polychaete boring rates, as those calculated from thin sectioning of bored substrates by Pari et al. (2002) and Osorno et al. (2005) are almost certainly underestimates. Moreover, although sipunculans tended to be less numerous than polychaetes, these authors grouped polychaetes and sipunculans together as worm borings. Measuring rates may also be problematic as these early colonising borers occur in the surface layers of the substrate, and this is the region more susceptible to being removed by grazers unless it is protected by heavy layers of silt, which inhibit the development of endolithic algae (Osorno et al. 2005).

Another area requiring urgent attention is the resolution of the species-level taxonomy of the taxa involved. This will reveal whether some boring species are really widely distributed Indo-Pacific species, as currently suggested, or a suite of cryptic species. However, this will require major taxonomic revisions of the groups.

Finally, while this review has by default concentrated on experimental studies in French Polynesia and the Great Barrier Reef (Australia), similar data are required from the eastern Pacific and the Caribbean. Some bioerosion studies have been carried out in Indonesia (e.g., Holmes et al. 2000) but did not look at the polychaetes. In the tropical eastern Pacific, the bioerosion by polychaetes in *Pocillopora* Lamarck, 1816 was investigated, revealing that, although present at all the study sites, were the least important of the bioeroders (Londoño-Cruz et al. 2003).

Most of the experimental work referred to above used blocks of *Porites*. However, one study on *Acropora* has been carried out (Musso 1994). In this case, polychaetes were found to be important early colonisers, suggesting that similar processes would occur in both coral species. As the amount of dead coral continues to increase with increasing reef degradation (Wilkinson 2004), we are witnessing increasing bioerosion rates (Pari et al. 2002), suggesting that the initial colonisation of substrates by polychaetes is important in facilitating the subsequent processes.

Acknowledgements

I should like to thank Alicia Osorno for supplying some of the illustrations on Figures 2 and 3, and Kate Attwood for help in preparing the figures. The paper also benefitted from constructive comments from two reviewers and I would also like to thank the editors for inviting me to contribute this chapter.

References

Augener H (1914) Polychaeta II, Sedentaria. In: Michaelsen W, Hartmeyer R (eds) Die Fauna Südwest-Australiens. Ergebnisse der Hamburger südwest-australischen Forschungsreise 1905. Fischer, Jena, 5:65-304

Beesley PL, Ross GJB, Glasby CJ (2000) Polychaetes & Allies: The southern synthesis. Fauna of Australia 4A. Polychaeta, Myzostomida, Pogonophora, Echiura, Sipuncula. CSIRO Publ, Melbourne, 465 pp

Bellwood DR, Choat JH (1990) A functional analysis of grazing in parrotfishes (family Scaridae): the ecological implications. Environ Biol Fish 28:189-214

Blainville HM de (1825) Oursins. Dictionnaire Sci Nat 37:59-102

Blake JA (1969) Systematics and ecology of shell boring polychaetes from New England. Amer Zoologist 9:813-820

Blake JA, Evans JW (1973) *Polydora* and related genera as borers in mollusks shells and other calcareous substrates. Veliger 15:235-250

Bosc LAG (1802) Histoire naturelle des vers, contenant leur description et leurs moeurs, avec figures dessinées d'après nature. Deterville, Paris, 324 pp

Bush KJ (1904) Tubiculous annelids of the tribes Sabellides and Serpulides from the Pacific Ocean. Harriman Alaska Exped 12:169-355

Chazottes V, Le Campion-Alsumard T, Peyrot-Clausade M (1995) Bioerosion rates on coral reefs: interaction between macroborers, microborers and grazers (Moorea, French Polynesia). Palaeogeogr Palaeoclimatol Palaeoecol 113:189-198

Chughtai I, Knight-Jones EW (1988) Burrowing into limestone by sabellid polychaetes. Zool Scripta 17:231-238

Conand C, Heeb M, Peyrot-Clausade M, Fontaine MF (1998) Bioerosion by the sea urchin *Echinometra* on La Reunion reefs (Indian Ocean) and comparison with Tiahura reefs (French Polynesia). In: Mooi R, Telford M (eds) Echinoderms. Balkema, Rotterdam, pp 609-615

Cuvier G (1817) Le règne animal distribué d'après son organisation, pour servir de base à l'histoire naturelle des animaux et d'introduction à l'anatomie comparée. Deterville, Paris, vol 2, 532 pp

Dana JD (1846) Zoophytes. U S Explor Exped 1838-1842, 7:1-120 and 709-720

Day JH (1967) A monograph on the Polychaeta of Southern Africa. Part 1. Publ Brit Mus Nat Hist London 656, 458 pp

Dorsett DA (1961) The behaviour of *Polydora ciliata* (Johnst.). Tube building and burrowing. J Mar Biol Assoc UK 56:649-674

Evans JW (1969) Borers in the shell of the sea scallop, *Placopecten magellanicus*. Amer Zoologist 9:775-782

Fitzhugh K, Rouse GW (1999) A remarkable new genus and species of fan worm (Polychaeta: Sabellidae: Sabellinae) associated with marine gastropods. Invertebrate Biol 118:357-390

Fulton CJ, Bellwood DR (2005) Wave induced water motion and the functional implications for coral reef fish assemblages. Limnol Oceanogr 50:255-264

Giard A (1893) Sur un type nouveau et aberrant de la famille des Sabellides (*Caobangia billeti*). C R Soc Biol Paris, Ser 9, 5:473-476

Gibson PH (1977) Reproduction in the cirratulid polychaetes *Dodecaceria concharum* and *D. pulchra*. J Zool London 182:89-102

Gibson PH, Clark RB (1976) Reproduction of *Dodecaceria caulleryi* (Polychaeta, Cirratulidae). J Mar Biol Assoc UK 41:577-590

Glynn P (1997) Bioerosion and coral-reef growth: A dynamic approach. In: Birkeland C (ed) Life and death of coral reefs. Chapman and Hall, New York, pp 68-95

Grube AE (1870) Bemerkungen über Anneliden des Pariser Museums. Arch Natgesch Berlin 36:281-352

Haigler S (1969) Boring mechanisms of *Polydora websteri* inhabiting *Crassostrea virginica*. Amer Zoologist 9:821-828

Hartman O (1943) Description of *Polydora websteri* n.sp. In: Loosanoff VL, Engle J (ed) *Polydora* in oysters suspended in the water. Biol Bull 85:70-72

Holmes KE, Edinger EN, Hariyadi, Limmon GV, Risk MJ (2000) Bioerosion of live massive corals and branching coral rubble on Indonesian coral reefs. Mar Poll Bull 40:606-617

Hutchings PA (1974) A preliminary report on the density and distribution of invertebrates living in coral reefs. Proc 2nd Int Coral Reef Symp 2:285-296

Hutchings PA (1983) Cryptofaunal communities of coral reefs. In: Barnes DJ (ed) Perspectives in coral reefs. AIMS, Townsville, pp 200-208

Hutchings PA (1986) Biological destruction of coral reefs - A review. Coral Reefs 4:239-252

Hutchings PA, Murray A (1982) The establishment of polychaete populations to coral substrates at Lizard Island, Great Barrier Reef - an experimental approach. Austral J Mar Freshwater Res 33:1029-37

Hutchings PA, Peyrot-Clausade M (2002) The distribution and abundance of boring species of polychaetes and sipunculans in coral substrates in French Polynesia. J Exper Mar Biol Ecol 269:101-121

Hutchings PA, Kiene WE, Cunningham RB, Donnelly C (1992) Experimental investigation of bioerosion at Lizard Island, Great Barrier Reef. Part 1. Patterns in the distribution and extent of non-colonial, boring communities. Coral Reefs 11:23-31

Hutchings PA, Peyrot-Clausade M, Osorno A (2005) Influence of land runoff on rates and agents of bioerosion of coral substrates. Mar Poll Bull 51:438-447

Kiene WE, Hutchings PA (1994) Experimental investigations on patterns in the rates of bioerosion at Lizard Island, Great Barrier Reef. Coral Reefs 13:91-98

Knight-Jones P (1983) Contributions to the taxonomy of Sabellidae (Polychaeta). Zool J Linn Soc 79:245-295

Johansson KE (1926) Bemerkungen über die Kinbergschen Arten der Familien Hermelidae und Sabellidae. Ark Zool Stockholm 18A(7):1-28

Jones ML (1969) Boring of shell by *Caobangia* Giard, 1893 in freshwater snails of southeast Asia. Amer Zoologist 9:829-835

Jones ML (1974) On the Caobangiidae, a new family of the Polychaeta, with a redescription of *Caobangia billeti* Giard. Smithson Contr Zool 175:1-55

Jumars PA (1974) A generic revision of the Dorvilleidae (Polychaeta) with six new species from the deep North Pacific. J Linn Soc London Zool 54:101-135

Lamarck J-B de (1816) Histoire naturelle des animaux sans vertèbres, présentant les caractères généraux et particuliers de ces animaux, leur distribution, leurs classes, leurs familles, leurs genres, et la citation des principales espèces qui s'y rapportent; précédée d'une introduction offrant la détermination des caractères essentiels de l'animal, sa distinction du végétal et des autres corps naturels, enfin, l'exposition des principes fondamentaux de la zoologie. Verdière, Paris, 3:1-586

Lazar B, Loya Y (1991) Bioerosion of coral reefs: A chemical approach. Limnol and Oceanogr 36:377-383

Le Campion-Alsumard T, Romano J-C, Peyrot-Clausade M, Le Campion J, Paul L (1993) Influence of some coral reef communities on the calcium carbonate budget of Tiahura reef (Moorea, French Polynesia). Mar Biol 115:685-693

Le Campion-Alsumard T, Golubic S, Priess K (1995) Fungi in corals: symbiosis or disease? Interaction between polyps and fungi causes pearl-like skeleton biomineralization. Mar Ecol Progr Ser 117:137-147

Link HF (1807) Beschreibung der Naturalien-Sammlung der Universität zu Rostock. Adler, Rostock, pp 1-38

Linnaeus C (1758) Systema naturae per regna tria naturae, secundum classes, ordines, genera, species, cum characteribus, differentiis, synonymis, locis. Tomus I. Editio decima, reformata. Laurentii Salvii, Holmiae, 824 pp

Londoño-Cruz E, Cantera JR, Toro-Farmer G, Orozco C (2003) Internal bioerosion by macroborers in *Pocillopora* spp. In the tropical eastern Pacific. Mar Ecol Progr Ser 265:289-295

Mao Che L, Le Campion-Alsumard T, Boury-Esnault N, Payri C, Golubic S, Bezac C (1996) Biodegradation of shells of the black pearly oyster *Pinctada margaritifera* var. *cumingii*, by microborers and sponges of French Polynesia. Mar Biol 126: 509-519

Musso BM (1994) Internal bioerosion of in situ living and dead corals on the Great Barrier reef. James Cook Univ North Queensland, 212 pp

Nishi E, Nishihira M (1999) Use of annual density banding to estimate longevity of infauna of massive corals. Fish Sci 65:48-56

Oken L (1815) Okens Lehrbuch der Naturgeschichte. Dritter Theil. Zoologie. Mit vierzig Kupfertafeln. Erste Abtheilung. Fleischlose Thiere. Reclam, Leipzig, 850 pp

Örsted AS (1843) Annulatorum danicorum conspectus fasc. J Maricolae Copenhagen, 52 pp

Osorno A, Peyrot-Clausade M, Hutchings PA (2005) Patterns and rates of erosion in dead *Porites* across the Great Barrier Reef, Australia after 2 and 4 years of exposure. Coral Reefs 24:292-303

Pari N, Peyrot-Clausade M, Le Campion-Alsumard T, Hutchings PA, Chazottes V, Golubic S, Le Campion T, Fontaine MF (1998) Bioerosion of experimental substrates on high islands and atoll lagoons (French Polynesia) after 2 years of exposure. Mar Ecol Progr Ser 166:119-130

Pari N, Peyrot-Clausade M, Hutchings PA (2002) Bioerosion of experimental substrates on high islands and atoll lagoons (French Polynesia) during 5 years of exposure. J Exp Mar Biol Ecol 276:109-127

Perkins TH (1984) Revision of *Demonax* Kinberg, *Hypsicomus* Grube, and *Notaulax* Tauber, with a review of *Megalomma* Johansson from Florida (Polychaeta: Sabellidae). Proc Biol Soc Wash 97: 285-368

Perry CT (1998) Macroborers within coral reef framework at Discovery Bay, north Jamaica: species distribution and abundance, and effects on coral preservation. Coral Reefs 17:277-287

Peyrot-Clausade M, Hutchings PA, Richard G (1992) Successional patterns of macroborers in massive *Porites* at different stages of degradation on the barrier reef, Tiahura, Moorea, French Polynesia. Coral Reefs 11:161-166

Peyrot-Clausade M, Le Campion-Alsumard T, Hutchings PA, Le Campion J, Payri C, Fontaine MF (1995) Initial bioerosion of experimental substrates in high islands and atolls lagoons (French Polynesia). Oceanol Acta 18:531-541

Sacco F (1897) I molluschi dei terreni terziarii del Piemonte e della Liguria. Pt. 23, Ostreidae, Anomiidae, and Dimyidae. Turin, 66 pp

Salvat B, Hutchings PA, Aubanel A, Tatarata M, Dauphin C (2001) The status of the coral reefs and marine resources of French Polynesia, New Caledonia. Global Coral Reef Monitoring Network, Fond Nat Polynesia, Papeete, Tahiti

Sato-Okoshi W, Okoshi K (1993) Microstructure of scallop and oyster shells infested with boring *Polydora*. Nippon Suisan Gakkaishi 59:1243-1247

Savigny JS (1818) Section on Annelida in Lamarck JB de (1818). Hist Nat Animaux sans Vertebras 5, 618 pp

Schönberg CHL, Wilkinson CR (2001) Induced colonization of corals by a clionid bioeroding sponge. Coral Reefs 20:69-76

Sebens KP, Johnson AS (1991) Effects of water movement on prey capture and distribution of reef corals. Hydrobiologia 226:91-102

Simkiss K, Tyler C (1957) A histochemical study of the organic matrix of hen egg shells. Quart J Microsc Sci 98:19-28

Simon CA, Ludford A, Wynne S (2006) Spionid polychaetes infesting cultured abalone, *Haliotis midae*, in South Africa. African J Mar Sci 28:167-171

Tauber P (1879) Annulata Danica. En kritisk revision af de i Danmark fundne Annulata, Chaetognatha, Gephyrea, Balanoglossi, Discophorae, Oligochaeta, Gymnocopa og Polychaeta. Reitzel, Copenhagen, 144 pp

Tribollet A, Decherf G, Hutchings PA, Peyrot-Clausade M (2002). Spatial large scale variability in bioerosion of experimental coral substrates on the GBR (Australia): Importance of microborers. Coral Reefs 21:424-432

Tribollet A, Langdon C, Golubic S, Atkinson M (2005) Endolithic microflora are major primary producers in dead carbonate substrates of Hawaiian coral reefs. Coral Reefs 24:422-434

Vogel K (1993) Bioeroders in fossil reefs. Facies 28:109-114

Wilkinson C (2004) Status of coral reefs of the world: 2004. Austral Inst Mar Sci, Townsville, 301 pp

Wilson R (2000) Family Spionidae. In: Beesley PL, Ross GJB, Glasby CJ (eds) Polychaetes & Allies: The Southern Synthesis. Fauna of Australia 4A. Polychaeta, Myzostomida, Pogonophora, Echiura, Sipuncula. CSIRO Publ, Melbourne, pp 196-200

Zottoli RA, Carriker MR (1974) Burrow morphology, tube formation, and microarchitecture of shell dissolution by the spionid polychaete *Polydora websteri*. Mar Biol 27:307-316

Parapholas quadrizonata (Spengler, 1792), dominating dead-coral boring bivalve from the Maldives, Indian Ocean

Karl Kleemann[1]

[1] Centre for Earth Sciences, University of Vienna, 1090 Vienna, Austria, (karl.kleemann@univie.ac.at)

Abstract. Endolithic bivalves were studied to establish abundance ranking and bioerosive impact on Maldivian coral reefs. In samples of dead coral, *Parapholas quadrizonata* (Spengler, 1792) was by far the most numerous bivalve borer. The assemblages of variously sized and aged specimens, their spatial situation and the resulting behaviour are described and discussed. The morphology of shells and boreholes is illustrated. Intra-species competition for space may have restricted the individual growth but was not necessarily lethal. The borings show unique internal features that allow distinction from those of *Gastrochaena* species, whereas the borehole apertures are undistinguishable between the two genera.

Keywords. Bioerosion, boring bivalve, *Parapholas quadrizonata*, Martesiinae, Pholadidae, coral reef, Maldives

Introduction

The present paper is based on observations made in the Indian Ocean during the Coral Reef Expedition Maldives 2007, conducted from March 15 to 29, under the leadership of Karen and Wolfgang Loch (http://www.expeditionmaldives2007.org). During the cruise four atolls were examined in the central part of the republic (Fig. 1). The main goal of the expedition was to compare the state of live coral cover of the reefs of today and half a century ago and its recovery by resettlement after the severe coral bleaching event of 1998 (Scheer 1958, 1972, 1974; Pillai and Scheer 1976; Loch et al. 2002, 2004; Schuhmacher et al. 2005; Loch and Loch 2006; Wallace and Zahir 2007). The present study focused on collecting samples of endolithic bivalves and their substrate to establish a preliminary ranking of their abundance and to assess the presumed bioerosive impact on the reef frame-builders. Bioerosion is a major threat for the maintenance of stressed reefs, particularly if live coral cover drops below 50%. Therefore, a preliminary study of the present situation in the Maldivian reefs was undertaken (http://bufus.sbg.ac.at/info/Info%2037/Info37-home.htm).

M. Wisshak, L. Tapanila (eds.), *Current Developments in Bioerosion*. Erlangen Earth Conference Series, doi: 10.1007/978-3-540-77598-0_14, © Springer-Verlag Berlin Heidelberg 2008

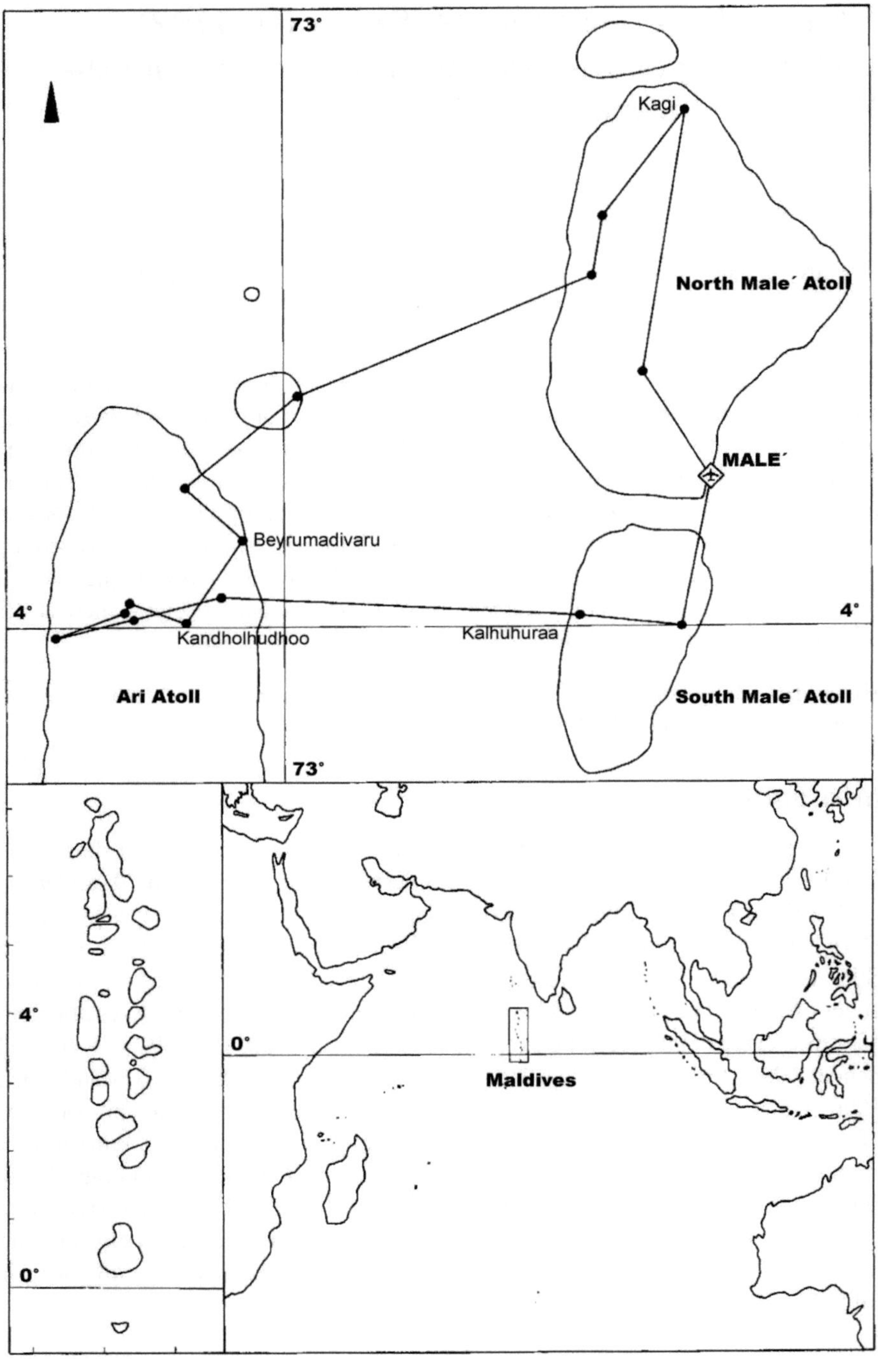

Fig. 1 Location of the Maldives in the Indian Ocean and the route of the expedition in the central area of the archipelago, starting from Male northwards on March 15th, 2007

Methods

Dead coral surfaces with apertures of endolithic bivalves were selected from the reef top down to 12 m depth using SCUBA. From sites of special interest, images were taken with a digital camera (Camedia 350 Zoom) in an under-water-housing before and after opening the dead coral with hammer and chisel. The coral was split open in several steps in order to preserve as many of the bivalves as possible undamaged and to observe more details of the endolithic community structure. All dislodged specimens were put into small mini-grip plastic bags, which were secured in a wide-necked plastic bottle and attached to the dive jacket for safe transport to the vessel, where further treatment followed. Small specimens were preserved in pure ethanol for DNA analysis. Others were put in 70% ethanol or 5-10% formaldehyde in seawater. In the laboratory, bivalves and coral samples were investigated in more detail using a dissecting microscope, dental surgery tools, callipers and photographic equipment (Coolpix 4500). Corals were cleaned with household bleach and H_2O_2.

Results

General observations and species found

In the field, the relatively large, 2-4 cm wide and dumbbell-shaped borehole openings of the mytilid bivalve *Leiosolenus obesus* (Philippi, 1847) were very conspicuous, especially when in small groups in dead massive coral (Fig. 2A). This species was the largest lithophagine bivalve in the collected material, with specimens reaching almost 10 cm in length (Kleemann 1984; Owada 2007). Next in length was *Lithophaga nigra* (Orbigny, 1842), known from the Maldives and other Indo-Pacific reefs under the junior synonym *Lithophaga teres* (Philippi, 1846); (see Kleemann 1984). Only a few specimens occurred in the samples of dead coral and the largest measured 65.2 / 16.0 / 13.0 mm in length / height and width (diameter), respectively. Further species found in this material consisted of quite common specimens of *Leiosolenus malaccanus* (Reeve, 1857), a single individual of the crenellinine *Gregariella coralliophaga* (Gmelin, 1791), two petricolids *Petricola lapicida* (Gmelin, 1791), and a few specimens of the gastrochaenid *Gastrochaena cuneiformis* (Spengler, 1783), *G. dentifera* (Dufo, 1840) and unidentified species (Carter et al. in press). Unexpectedly, the martesiinid *Parapholas quadrizonata* (Spengler, 1792) was most numerous and occurred in very dense assemblages. The external coral surfaces are riddled with the paired apertures which lead to the flask-shaped chambers occupied by the shells themselves. Each paired aperture may be figure-of-eight shaped and lined with a calcareous sheath (equivalent to *Teredolithus* Bartsch, 1930; see Kelly and Bromley 1984: 796; Fig. 2B), a configuration generally well-known from species of *Gastrochaena* Spengler, 1783 (Carter 1978; Nielsen 1986; Kleemann 1998).

Population density of boring bivalves in general seemed to be distinctly higher (1) on the reef flat near the reef edge and the shallow reef slope to about 3 m depth than in deeper samples, and (2) in massive dead coral versus those in ramose corals and coral rubble. No quantitative samples were taken.

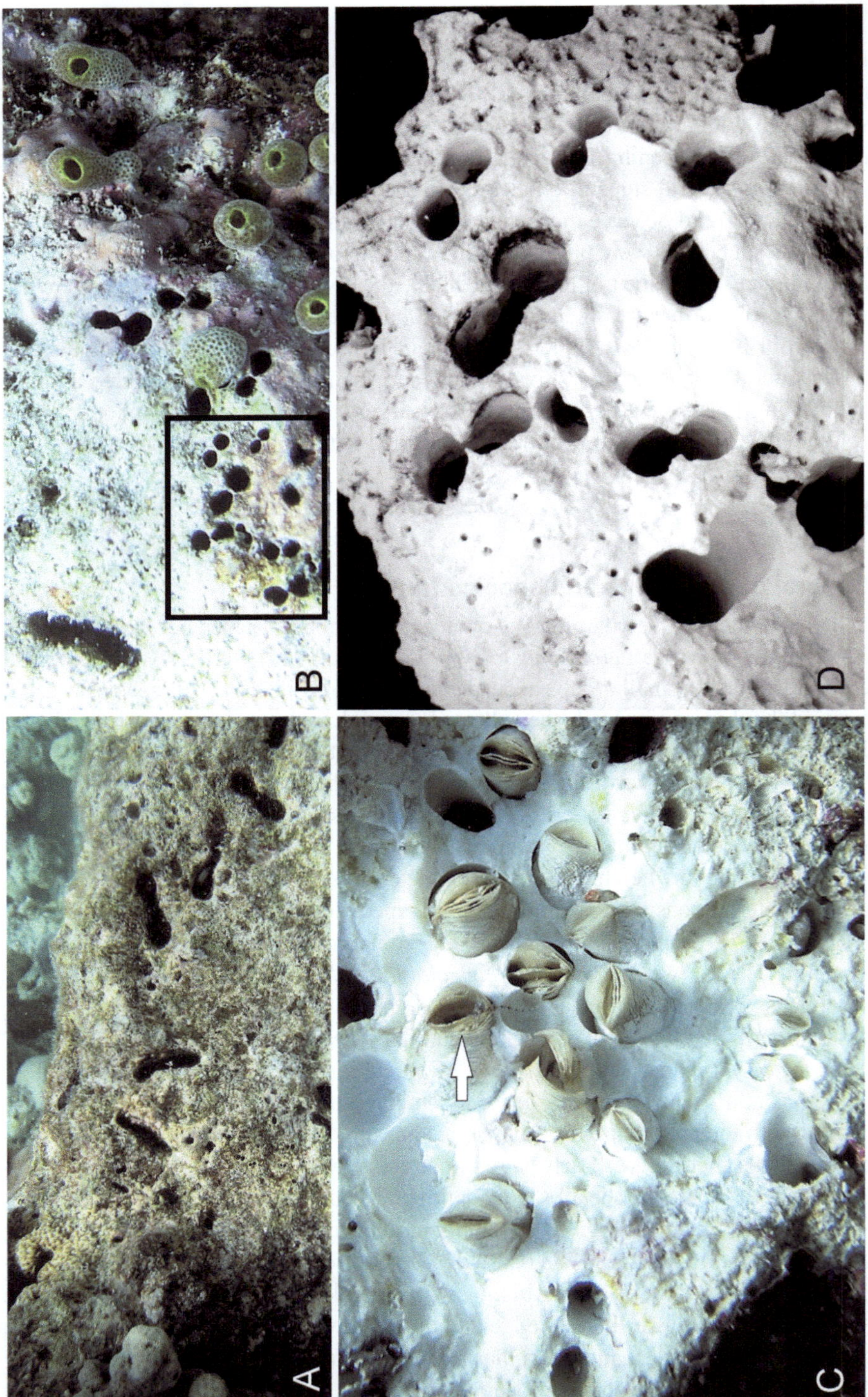
A
B
C
D

The shell morphology

The living shell of *Parapholas quadrizonata* and its borings are shown in Figures 2C and 3D-G. The general morphology is explained in Figure 4 and measurements are given in Table 1.

According to Turner (1998: 375), *Parapholas* Conrad, 1848, as a member of Martesiinae, has an apophysis, mesoplax, metaplax and hypoplax, and the presence of the callum is the most distinctive feature of the subfamily. Nielsen (1986) pointed out the commarginal sculpture of fine wavy ridges anteriorly, with the posterior part being covered by overlapping, fan-like protrusions of the straw-coloured periostracum. Additional features include the *'wide, drop-shaped pedal gape, which in the adult phase becomes closed by a callum from each valve meeting in the middle and united by a narrow zone of thin, brown periostracum.'*. He did not mention the middle lateral part of the shell, lying between two separate lines, which run different lengths from the umbo posteriorly to the ventral shell edge, yielding an oblique and arcuate, slender triangular field, named *'disk'* (Kelly 1988: 348; Fig. 3D). Anteriorly, the surface of this field is smoother than the anterior part of the shell, and posteriorly it becomes somewhat more coarsely sculptured at the beginning of the posterior part of the shell, resulting in an elevated stripe (Fig. 3D). Dorso-posteriorly of the middle part, the calcareous shell surface is still sculptured by commarginal but more pronounced and spaced ridges (Fig. 3D), and is covered by protruding periostracal lamellae (Figs. 2C, 3G). They are closely arranged like short scales in a series of increasing width and length towards the posterior. The periostracal lamellae cover the actual posterior shell surface and prolong its rim up to several millimetres forming a collar, e.g., a calcareous shell measuring 28.5 mm in length is extended by 4.5 mm giving a total of 33 mm by measuring the periostracum. The collar extends into the narrow space between the wall of the shell-chamber and the obliquely protruding constriction at the beginning of the siphonal tube (see below). In household bleach, the periostracum may be dissolved completely and therefore the lamellae are missing in Figures 3D-F and the line-drawing Figure 4.

Fig. 2 **A** Dumbbell-shaped borehole apertures of aggregated *Leiosolenus obesus* in massive dead coral, ~3 m depth, Kagi, North Male Atoll (N 04°40.5' E 73°30.8'). Frame size ~40 x 20 cm. **B** Mainly figure-of-eight-shaped apertures of predominating *Parapholas quadrizonata* and, in the upper left corner, one of *L. obesus*, in massive dead coral; the synascidian *Didemnum* Savigny, 1816 specimens at right indicate a steeply inclined surface. Water depth ~2 m, Beyrumadivaru, Ari Atoll (N 04°07.3' E 72°56.5'). Whole frame size ~15 x 7.5 cm, secondary frame ~45 x 30 mm (enlarged in D). **C** Assemblage of *Parapholas quadrizonata* in borings, with siphons completely retracted, in anterior view in freshly broken dead coral depicted in B. The serial periostracal lamellae form a collar at the shell posterior; the last-formed are not yet the longest. Note also the varying orientations of specimens. From the largest empty shell one arm of an ophiuroid (arrow) is protruding. Frame size ~15 x 11 cm. **D** Coral fragment from framed area in B. Note the different sizes of figure-of-eight apertures of *Parapholas quadrizonata*, rarely separated into two openings of the same calcareous tube (two at the right). This indicates the repeated, probably more than annually occurring infestation of the same spot by successive generations of veligers. Frame size ~45 x 30 mm

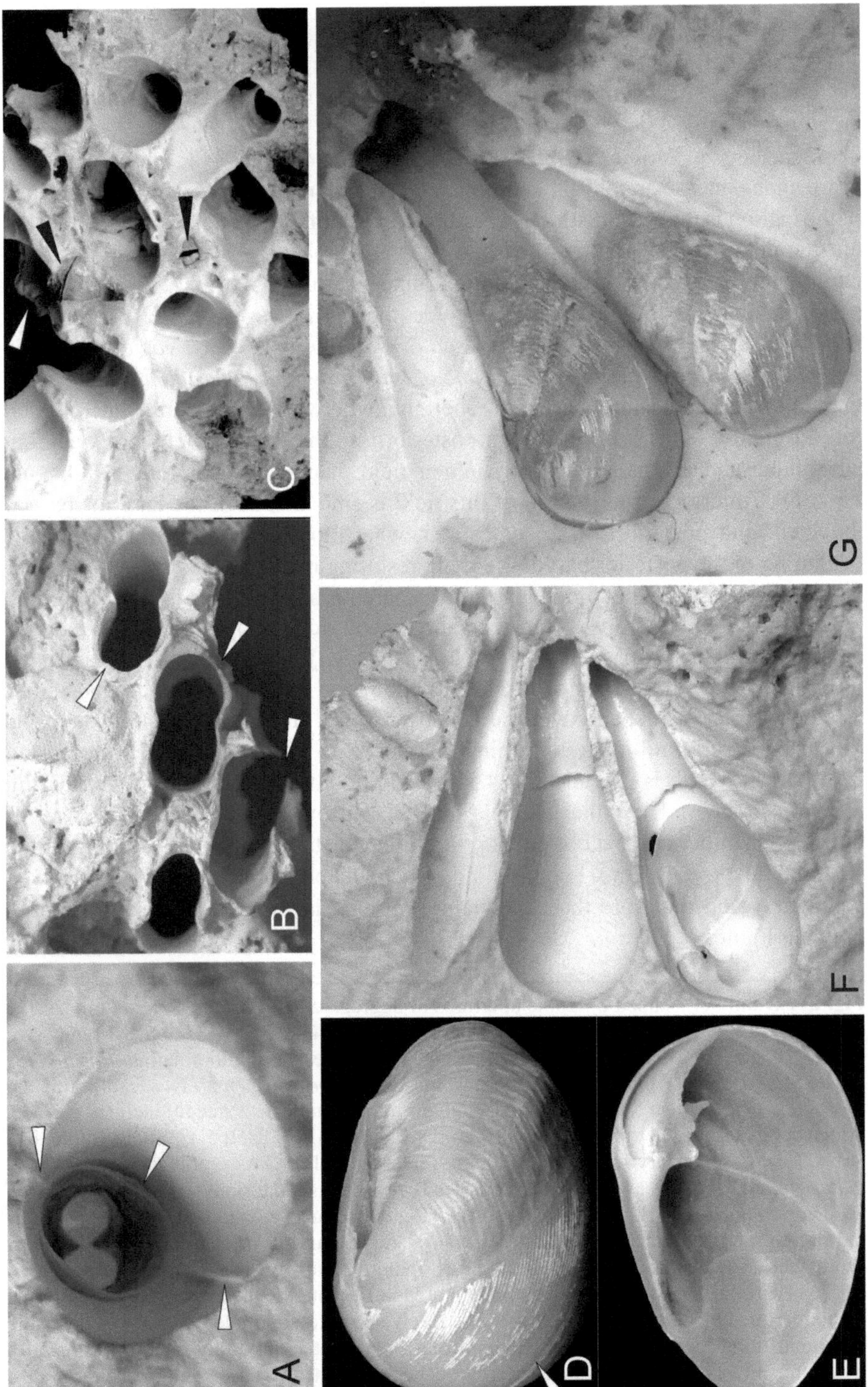

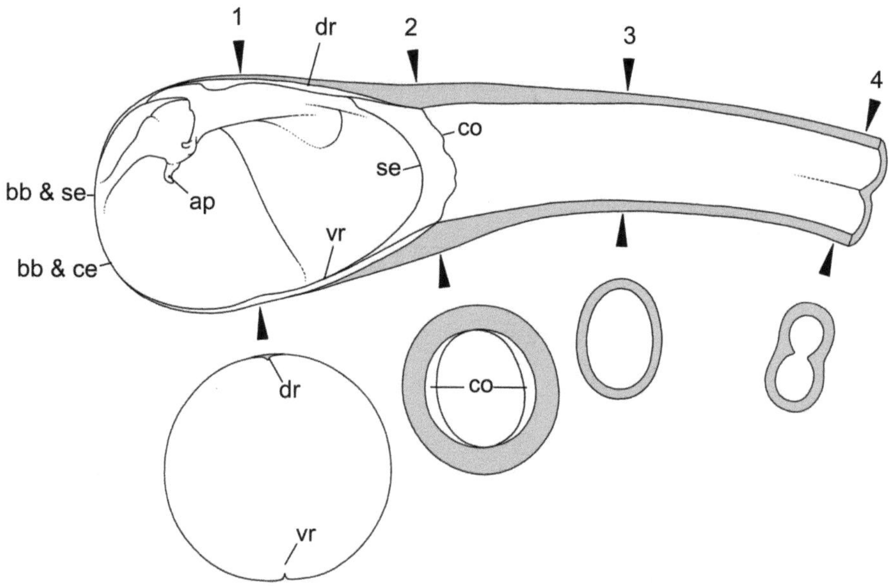

Fig. 4 Line drawing of the right valve of *P. quadrizonata* in its boring in lateral, internal view. Numerals 1-4 indicate the location of cross-sections, the pairs of black pointers show their respective position along the boring. ap = apophysis, bb = borehole base, ce = callum edge, co = constriction, dr = dorsal ridge, se = shell edge, vr = ventral ridge. Note the variation in thickness of the calcareous lining (shaded) between cross-sections 1-4. Note the three unique features of the boring, which are a dorsal and a ventral ridge and the velum-like constriction (Fig. 3A, C, F). The valve lacks periostracal lamellae on its posterior shell edge due to dissolution in commercial bleach (compare with Figures 2C and 3G)

Fig. 3 A Transversely opened *P. quadrizonata* boring revealing the posterior part of the shell-chamber with the unique features, the dorsal and ventral ridge, both leading to the velum-like constriction (pointers). Note that the tube opening in the background is turned right about 45° from the sagittal plane of the chamber. Frame size ~3 x 2.5 cm. **B** External view of apertures of calcareous-lined siphonal tubes of several *P. quadrizonata* individuals penetrating dead coral. The exterior surface is slightly eroded. Three apertures are from borings illustrated in **G** (pointers). Length of central aperture 12.5 mm, frame size ~35 x 25 mm. **C** Lower surface of coral chip figured in figure 2D, with maximal thickness of 20 mm. Note two bivalves (black pointers), in situ in borings between densely packed adjacent borings; parts of the callum halves are exposed. These had no free space ahead, but survived long enough to construct a callum. The encrustations in one tube indicate the constructor died (white pointer). Frame size ~45 x 25 mm. **D** Left valve of *Parapholas quadrizonata*, 34.5 mm length, in lateral external view; showing the callum (pointer) – the typical feature of Martesiinae – but lacking the periostracum with lamellae through bleaching. See E for interior and G for life position. **E** Same specimen as in D in lateral internal view. For similar view of right valve F, lower left corner. **F** Similar view to G with most of the shell material and soft parts removed and specimen bleached; one right valve of *P. quadrizonata* remains in the lower boring showing the interior aspect. For left valve see D-E; see text and compare with line-drawing (Fig. 4). Frame size ~75 x 90 mm. **G** Freshly split longitudinally opened borings containing live *P. quadrizonata*. Note the tight fit of the shells in the chambers and the inhalent and exhalent siphon showing lengthwise fusion except at the distal ends. Frame size ~75 x 70 mm

Table 1 Shell dimensions of 42 *Parapholas quadrizonata*, collected from ~25 x 25 cm of the dead coral shown in Figures 2B-D and 3F-G. Measurements are given in millimetres for length (L; with lamellae, to the closest half mm), height (H) and width (W). Missing measurements, caused by fractured specimens, are indicated by '?'. Specimens marked with '*' indicate measurements of three specimens with an atypical calcareous girdle at their maximum diameter

Specimens with callum (n = 13, mean length 30.0 mm)			Specimens without callum (n = 29, mean length 22.4 mm)		
L	H	W	L	H	W
44.0	24.0	~24.0	34.5	16.5	18.4
41.5	23.2	23.2	33.0	15.8	17.8
40.0	22.6	22.1	33.0	14.9	16.2
37.0	~20.5	?	32.0	16.3	?
37.0	20.3	?	32.0	15.0	?
34.0	20.0	~20.0	31.0	15.6	15,8*
29.0	14.0	?	30.1	13.0	?
26.0	15.2	14.4	30.0	14.3	16.7
26.0	14.4	14.1	29.0	15.4	17.1
22.6	11.1	11.5	28.0	~13.5	?
19.0	10.7	10.3	~27.0	14.1	15.3
19.0	10.5	10.8*	22.5	11.5	12.2
15.0	8.2	7.7	22.0	10.6	10.9
			21.0	?	11.3
			20.3	?	?
			20.0	8.7	9.1
			19.5	10.5	10.6*
			18.5	9.0	10.3
			18.5	8.8	10.2
			18.5	8.8	9.1
			17.0	9.0	?
			~17.0	?	?
			16.0	8.5	9.6
			15.0	7.6	8.2
			15.0	7.0	7.3
			13.0	6.8	7.4
			12.0	6.1	7.1
			12.0	5.8	6.9
			11.0	5.3	5.9

In respect to the soft body, the large, drop-shaped base of the foot is worth noting, mirrored by the gape between the shells (as long as no callum halves are developed, separated by periostracum of individually differing width). Also, the longitudinally fused siphons, separated only at the distal end, form a paired opening with figure-of-eight shaped aperture formed by the calcareous tube they secrete (Figs. 2B, D, 3A-B).

The borehole morphology

These are very typical clavate bivalve borings, comparing well with those described by Kelly and Bromley (1984). From the outer coral surface into the substratum, the clavate borings are composed of a siphonal tube and a shell chamber, separated from the tube by a velum-like constriction, slightly protruding into the cavity on both longer sides of the slightly oval-shaped cross-section (Figs. 3A, C, F, 4). This structure is part of the solid lining of the borehole wall, and is not perpendicular but inclined towards the shell-chamber (Fig. 3A, C, F).

The right valve in Figure 3F fits so tightly into the chamber, despite the missing periostracum, that it is prevented from falling out by two ridges in the longitudinal, dorso-ventral plane. The dorsal ridge appears to be part of the calcareous lining of the boring by the bivalve, while the ventral one seems to be a remnant of the dissolved substrate, in this case the dead coral *Porites* Link, 1807. As shown in the drawing (Fig. 4), the shell-chamber, being distinctly wider than the siphonal tube and more or less circular in cross-section, has a hemispherical anterior and a somewhat conical posterior half. This closely mirrors the shape of the slightly smaller constructor due to the developed callum (Figs. 3D, G, 4). The siphonal tube reaches from the velum-like constriction (Fig. 3A), which separates it from the shell-chamber, to the outer substrate surface. There it is usually figure-of-eight-shaped, rarely separated into two rounded openings, of which the exhalent one is circular (Figs. 2B, D, 3A-B). Except near the outer surface, tube cross-sections are oval, widest near the chamber and narrowing gently towards the entrance to the boring, before flaring at the exterior surface (Fig. 4). Length / width of the apertures in Figure 2D is 4.5 / 2 mm, 5.2 / 2.4 mm, 6.7 / 3.2 mm, 7.5 / 3.5 mm, 8.0 / 4.5 mm, 9.0 / 4.6 mm, and 10.1 / 5.0 mm. Figure 2D shows a paired siphonal opening partly overgrown and constricted by post-mortal growth of coralline algae and therefore strictly should not be included, however estimated measurements are: 12.5 / 5.5 mm.

The assemblages

The assemblages were composed of groups of specimens belonging to different size-classes, ranging from 11 to 44 mm in length. These values were also reflected in the sizes of the boreholes and their openings. The latter reach up to 14 mm in maximum length and 7.5 mm in maximum width of the usually undivided siphonal apertures (Figs. 2B, D, 3A-B). The distance between adjacent openings was often very narrow to minimal (Fig. 3B). Consequently, the space between the respective borings was even narrower (Fig. 3C), as their diameter increases with depth. Thus borings sometimes lie in direct contact with neighbours, but the calcareous linings do not generally intersect (Figs. 2C, 3B-C, F-G). In other borers, such as *Lithophaga* Röding, 1798 and *Leiosolenus* Carpenter, 1856, borings without and with calcareous linings do intersect (Kühnelt 1930; Kleemann 1974, 1980; Owada 2007). The largest assemblages of *P. quadrizonata* were found in massive coral, although smaller individuals were present in various dead corals including rubble.

Discussion

Regarding its biology, *Parapholas quadrizonata* is a poorly known bivalve, with few records in the literature, although its wide geographical distribution is Indo-Pacific (Oliver 1992: 203, noting it from Aden, Red Sea). Nielsen (1986) described the species from Phuket, Thailand, in association with the coral *Porites*. Turner (1998: 375) mentioned *Parapholas* as a member of Martesiinae occurring in Australia. I have observed only 21 specimens of *P. quadrizonata* (ranging from 7 to 30 mm in length) at Lizard Island, Great Barrier Reef (Australia) in three weeks, but none in the northern Red Sea during several months of study. According to the present findings, peak population densities probably occur at the Maldives.

The largest *P. quadrizonata* in the present study, measuring 44 mm in length, is only 6 mm short of the reported maximum size of 50 mm (Oliver 1992); the smallest specimen is 11 mm. On the other hand, Nielsen (1986) noted a maximum diameter of 34 mm (probably because shell length, without periostracal lamellae, is difficult to measure) in comparison with 24 mm in the Maldives samples; presumably, his 34 mm wide specimens from Phuket, Thailand, reached about 60 mm in total length including the periostracum.

The four zones of the shell, which give rise to the specific name *quadrizonata*, include the smooth callum (or the wide gape, if not developed), the fine, commarginally sculptured anterior part, the middle part between the two lateral oblique lines, and the dorso-posterior section, covered by the flexible lamellae (Fig. 3D, E, G).

As in gastrochaenids and various other boring bivalves, the calcareous lining, which changes markedly in thickness along the siphonal tubes (Figs. 3F-G, 4), is supposedly secreted from the bivalve's siphons and mantle tissue. The velum-like constriction (Fig. 3A, C, F) differs distinctly from structures found in some *Eufistulana* Eames, 1951, *Gastrochaena* and *Spengleria* Tryon, 1861 species (Gohar and Soliman 1963; Soliman 1973; Carter 1978; Morton 1983; Nielsen 1986).

The different size-classes of shells in the aggregations of *P. quadrizonata* may be primarily the result of successive generations, probably annual, of veligers infesting the same site. A second possibility must be considered due to intra-species space competition. When specimens become crowded by neighbours of probably the same age (Fig. 3C), they stop growing and build a callum as so-called stenomorphs (Bartsch 1923). Stenomorphs are adults of dwarf size but not necessarily immature. They appear to survive for years, feeding and reproducing, but unable to grow. Overlaps between adjacent borings in live *P. quadrizonata* were not observed. At Phuket, Nielsen (1986) noted a 15 mm shell as the smallest with a callum, the same size as one individual in the present material. A further reason to develop a callum before being fully grown could be if an individual has lengthy pauses between phases of growth or reproduction. There were, however, no indications for this based on shell structure. This may be worth investigating in more detail. There may also be clues hidden in the layered structure of the lining. Specimens less than ~10 mm in length may have been overlooked in the few samples available, or may be missing owing to collecting date (perhaps before or too soon after a new settling season).

The siphonal tube usually extends up to or a little above the substrate surface, but is not usually chimney-like and not separated into two openings, as stated by Nielsen (1986) in his findings at Phuket. No prominent elevations and only a few isolated ('closed' figure-of-eight) apertures were noted (Fig. 2B, D).

The borehole morphology of *P. quadrizonata* fits best to a mixture of *Gastrochaenolites dijugus* Kelly and Bromley, 1984 and *G. turbinatus* Kelly and Bromley, 1984 (Kelly and Bromley 1984: fig. 3). It shows some unique and thus characteristic features for a determination and differentiation from the otherwise very similar borings of certain *Gastrochaena* species, e.g., *G. dentifera* (Dufo, 1840). Alas, the fossilisation potential of these fine structures is poor and therefore the probability to establish a new ichnospecies is minimal. The size of intact tube apertures are suitable as a rough age indicator (because stenomorphs and *Gastrochaena* specimens are mostly rare), once a relation between size and age is established. So far only estimates can be proposed. Considering an estimated growth rate of 10 mm a^{-1} (no smaller specimens were found, but may have been overlooked), an estimated age of at least 4 years for the largest specimens (and apertures) should be on the conservative side because growth in length usually decreases with age.

The three unique borehole features of a dorsal and ventral ridge in the shell chamber, and the constriction at the beginning of the siphonal tube, are probably better recognisable in the cross-sections (1 and 2, respectively) than in the schematic longitudinal section of the boring with the right valve in the chamber preserved (Fig. 4, compare with Fig. 3A, C, F). These features are not always well developed, and three of the collected specimens have a calcareous girdle at their largest diameter. I have only a hypothesis for the former situation and no explanation for the latter circumstance. Presumably, growth is not a continuous process and the likely seasonal increase of a boring necessarily shifts the position of velum-like constriction. When the shell-chamber increases, the above features may become (partly) dissolved. The constriction of the boring probably functions as a holdfast in conjunction with the lamellae of the shell during hydraulic and / or boring activities, perhaps even as a lock in case of a major disturbance. At times of stress the bivalves can retract the extendable siphons completely (Figs. 2C, 3G). The cross-sections 3 and 4 (Fig. 4) through the siphonal tube are similar to or indistinguishable from those of certain *Gastrochaena* species (Gohar and Soliman 1963; Nielsen 1986; Kleemann 1998).

Considering (1) the common, dense aggregations of boring bivalves in dead coral rock at shallow depth in several Maldivian reefs, which probably promoted spawning success and yielded abundant larval recruits trying to find appropriate settling places, (2) the decrease of substrate volume up to ~50%, from its surface to ~2 cm depth (Figs. 2C, 3B-C), and the borehole dimensions of *P. quadrizonata* reaching 80 mm in length and 25 mm in diameter near the deep end and an estimated volume of up to ~20 ml, and (3) an estimated life-span of 3-8, probably 10 years, the impact of these bivalves on reefs with less than 50% live coral cover may be severe and accelerate their decline. Particularly accelerated by boring sponges, parrotfish and other bio-erosive organisms, competing for available space as well as food, destroy more substrate than frame-builders are able to construct at the same time. Therefore, the situation is already tense at some reefs with poor coral recovery since 1998, e.g., at Kalhuhuraa, South Male Atoll, while some others are in astonishingly good shape, e.g., at Kandholhudoo, Ari Atoll (Fig. 1; Wallace and Zahir 2007).

The biology – including life-span, local population density and depth distribution, growth, boring and succession rates – of this dominant boring bivalve needs to be investigated in detail for a better understanding of its impact on the reef system. This should provide answers to key questions such as: is spawning and reproduction a brief or a prolonged seasonal annual event or more or less permanent? Are there recognisable shifts in population density and local distribution within a few years?

Conclusions

- *Parapholas quadrizonata* is the most abundant dead-coral boring bivalve in the Maldives.
- The largest specimen measured 44 mm in length and 24 mm in width, the smallest 15 mm (with callum).
- Successive recruitment on massive dead coral lead to dense assemblages of variously sized and oriented individuals.
- No intra-specific competition was observed.
- Boreholes were up to 80 mm in length and 25 mm in diameter and borehole volume may reach 20 ml.
- Aperture size in borings may originally indicate size / age of occupant, but may be modified by bioerosion of outer surface by other organisms or by accretion.
- Borehole morphology and the presence of dorsal and ventral ridges and the velum can be used to distinguish the vacant boreholes of *P. quadrizonata* from those of *Gastrochaena* species.
- Boreholes resemble a mixture of *Gastrochaenolites dijugus* and *G. turbinatus*.

Acknowledgements

Thanks are due to the Ministry of Fisheries, Republic of Maldives, several German sponsors (especially the Fundus Group), to Matthias Seidel, to whom this article is dedicated, Wolfgang and Karen Loch for organising and leading the expedition, and the rest of the team. Michael Stachowitsch and Simon Kelly substantially improved the English text. Kelly and an anonymous reviewer assisted with critical comments and helpful suggestions. Figures 1 and 4 were prepared by N. Frotzler and R. Gold.

References

Bartsch P (1923) Stenomorph, a new term in taxonomy. Science 53:330

Bartsch P (1930) *Teredolithus*, a new collective group name. Science 71:460-461

Carpenter PP (1855-57) Catalogue of the collection of Mazatlan shells in the British Museum, collected by Frederick Reigen. London, pp 1-120 [1855] 121-444 [1856] 445-552 [1857]

Carter JG (1978) Ecology and evolution of the Gastrochaenacea (Mollusca, Bivalvia) with notes on the evolution of the endolithic habitat. Bull Peabody Mus Nat Hist, Yale Univ 41, 92 pp

Carter JG, McDowell T, Namboodiri N (in press) The identity of *Gastrochaena cuneiformis* Spengler, 1783, and the evolution of *Gastrochaena, Rocellaria*, and *Lamychaena* (Mollusca, Bivalvia, Gastrochaenoidea). J Paleont

Conrad TA (1848) Observations on the Eocene formation, and descriptions of one hundred and five new fossils of that period, from the vicinity of Vicksburg, Mississippi; with an appendix. J Acad Nat Sci Philadelphia, Ser 2, 1:111–134

Dufo MH (1840) Observations sur les Mollusques marins, terrestres et fluviatiles des îles Séchelles et des Amirantes. II. Ann Sci Nat Paris (Zoologie) 14:45-80 and 166-221

Eames FE (1951) A Contribution to the study of the Eocene in Western Pakistan and Western India: B. Description of the Lamellibranchia from standard sections in the Rakni Nala and Zinda Pir areas of the Western Punjab and in the Kohat District. Phil Trans Roy Soc London B 235:311-482

Gmelin JF (1791) Caroli a Linné, systema naturae per regna tria naturae. Editio decima tertia. Beer, Lipsiae, Vol 1, Pars 6, pp 3021-3910

Gohar HF, Soliman GN (1963) On the rock-boring lamellibranch *Rocellaria rüppeli* (Deshayes). Publ Mar Biol Stn Al-Ghardaqa 12:145-157

Kelly SRA (1988) Cretaceous wood-boring bivalves from western Antarctica with a review of the Mesozoic Pholadidae. Palaeontology 31:341-372

Kelly SRA, Bromley RG (1984) Ichnological nomenclature of clavate borings. Palaeontology 27:793-807

Kleemann KH (1974) Raumkonkurrenz bei Ätzmuscheln. Mar Biol 26:361-364

Kleemann KH (1980) Boring bivalves and their host corals from the Great Barrier Reef. J Mollusc Stud 46:13-54

Kleemann KH (1984) *Lithophaga* (Bivalvia) from dead coral from the Great Barrier Reef, Australia. J Mollusc Stud 50:192-230

Kleemann KH (1998) Superfamily Gastrochaenoidea. In: Beesley PL, Ross GJB, Wells A (eds) Mollusca: The southern synthesis. Fauna of Australia 5A, CSIRO Publ, Melbourne, pp 367-370

Kühnelt W (1930) Bohrmuschelstudien I. Paläobiologica 3:53-91

Link HF (1806-1808) Beschreibung der Naturalien-Sammlung der Universität zu Rostock. Adler, Rostock, pp 1-160 [1806], pl 1-30, pp 1-38 [1807], pl 1-38 [1808]

Loch K, Loch W (2006) Änderung der Steinkorallen-Vermehrungsstrategie durch Klimawandel? Biol Unserer Zeit 36(3):148-149

Loch K, Loch W, Schuhmacher H, See WR (2002) Coral recruitment and regeneration on a Maldivian reef 21 months after the coral bleaching event of 1998. PSZN Mar Ecol 23:219-236

Loch K, Loch W, Schuhmacher H, See WR (2004) Coral recruitment and regeneration on a Maldivian reef four years after the coral bleaching event of 1998. Part 2: 2001-2002. PSZN Mar Ecol 25:145-154

Morton B (1983) The biology and functional morphology of *Eufistulana munia* (Bivalvia: Gastrochaenacea). J Zool 200:381-404

Nielsen C (1986) Fauna associated with the coral *Porites* from Phuket, Thailand. Part 1: Bivalves with description of a new species of *Gastrochaena*. Phuket Mar Biol Cent, Res Bull 42:1-24

Oliver PG (1992) Bivalved seashells of the Red Sea. Hemmen, Wiesbaden, 330 pp

Orbigny AD d' (1842) Mollusques. In: Sagra R de la (ed) Histoire physique, politique et naturelle de l'île de Cuba. Bertrand, Paris, Atlas

Owada M (2007) Functional morphology and phylogeny of the rock-boring bivalves *Leiosolenus* and *Lithophaga* (Bivalvia: Mytilidae): a third functional clade. Mar Biol 150:853-860

Philippi RA (1842-1850) Abbildungen und Beschreibungen neuer oder wenig gekannter Conchylien. Fischer, Cassel, 1:1-20 [1842] 21-76 [1843] 77-186 [1844] 187-204 [1845], 2:1-64 [1845] 65-152 [1846] 153-232 [1847], 3:1-50 [1847] 51-82 [1848] 1-88 [1849] 89-138 [1850]

Pillai CSG, Scheer G (1976) Report on the stony corals from the Maldive Archipelago. Zoologica 43(126):1-83

Reeve LA (1857-58) Monograph of the genus *Lithodomus*. In: Reeve LA (ed) Conchologia iconica or illustrations of the shells of molluscous animals. Reeve Brothers, London, vol 10, 5 pls

Röding PF (1798) Museum Boltenianum sive catalogus cimeliorum e tribus regnis naturae quae olim collegerat Joa. Fried. Bolten, M.D.p.d. per XL. annos proto physicus Hamburgensis. Pars secunda continens conchylia sive testacea univalvia, bivalvia et multivalvia. JC Trappius, Hamburg, 199 pp

Savigny JC (1816) Recherches anatomiques sur les ascidies composées et sur les ascidies simples - Système de la classe des Ascidies. Mémoires sur les Animaux sans Vertèbres, G Dufour, Paris, 2:1-239

Scheer G (1958) In den Korallenriffen der Malediven. Jena Rundsch 3:156-158

Scheer G (1972) Investigation of coral reefs in the Maldive Islands with notes on lagoon patch reefs and the method of coral sociology. Proc 1st Int Coral Reef Symp, Mar Biol Assoc India, Mandapam Camp/India, pp 87-120

Scheer G (1974) Investigations of coral reefs at Rasdu atoll in the Maldives with the quadrat method according to phytosociology. Proc 2nd Int Coral Reef Symp, Brisbane, 2:655-670

Schuhmacher H, Loch K, Loch W (2005) The aftermath of coral bleaching on a Maldivian reef - a quantitative study. Facies 51:85-97

Soliman GN (1973) On the structure and behaviour of the rock-boring bivalve *Rocellaria retzii* (Deshayes) from the Red Sea. Proc Malacol Soc London 40:313-318

Spengler L (1783) Beskrivelse over an nye Slaegt af toskallede Muskeler, som kan kaldes *Gastrochaena*, I tre foranderlige Arter, hvoraf hver boer i et forskielligt Ormehuus. N Saml Kongel Danske Vidensk Selsk Skr 2:174-183

Spengler L (1792) Betragtninger og Anmaerkninger ved den Linneiske Slaegt *Pholas* blant de mangeskallede Muskeler, med dens hidindtil bekiendte gamle og nye Arter, samt den dermed i Forbindelse staaende Slaegt *Teredo* Linn. Skr Naturhist Selsk Copenhagen 2:72-106 [An English translation is found in: Hylleberg J, Knudsen J (2001) Translations into English of Lorenz Spengler's papers on bivalves (1783-1798). Part 4. 1792: The genera *Pholas* and *Teredo*. Phuket Mar Biol Cent, Spec Publ 25:557-570]

Tryon GW (1861) On the Mollusca of Harper's Ferry, Virgina. Proc Acad Nat Sci Philadelphia 13:396-399

Turner RD (1998) Superfamily Pholadoidea. In: Beesley PL, Ross GJB, Wells A (eds) Mollusca: The southern synthesis. Fauna of Australia 5A, CSIRO Publ, Melbourne, pp 371-378

Wallace C, Zahir H (2007) The "Xarifa" expedition and the atolls of the Maldives, 50 years on. Coral Reefs 26:3-5

Echinometrid sea urchins, their trophic styles and corresponding bioerosion

Ulla Asgaard[1], Richard G. Bromley[1]

[1] Department of Geography and Geology, University of Copenhagen, 1350 Copenhagen K, Denmark, (rullard@geol.ku.dk)

Abstract. The Echinometridae is a diverse, largely tropical family of echinoids. Several species are active borers in shallow water, at and below the low-tide line, especially in coral reef and beachrock substrates, commonly in high-energy situations. In the Caribbean and Atlantic Ocean, *Echinometra lucunter* erodes two types of boring. Most commonly, the widely reported elongate grooves are produced. In these traps, drifting fragments of algae are caught on the incoming tide. Drifting algae are also caught actively by 'chopsticks-like' manoeuvres of the spines. Some simple gardening is undertaken on the boring walls and floor, where short algal turf and endolithic algae are harvested. Juveniles start by making simple cup-shaped borings. In some cases, however, *E. lucunter* retains this form of bioerosion to adulthood. Cup-shaped borings must indicate emphasis on alga-catching in high-energy environments and less on the gardening, grazing trophic style.

In the Indo – West Pacific realm, *Echinometra mathaei* undertakes similar trophic activities to those of *E. lucunter*, especially in the high-energy environment outside barrier reefs. However, in protected shoreward lagoons, another trophic activity is practiced that has largely been overlooked in echinoids. In this oceanic region, a recently identified food-source is available in back-barrier lagoons, i.e., flocs of organic particle aggregates that, like the marine snow of the open sea, are swept into the lagoon on the rising tide from the reef crest, which is subaerially exposed at low tide. Crowded *E. mathaei* on beachrock surfaces, each in a cup-shaped boring (*Circolites* isp.), show alga-catching and within-boring algal turf gardening. But in addition, marine snow is collected on the spines at each rising tide.

Echinostrephus molaris, a small echinoid that practices alga-catching on the outer side of the barrier reef, shows extreme specialisation for particle-aggregate collection in back-barrier sites. *E. molaris* and *E. aciculatus* are the only deeply boring echinoids, producing thumb-sized borings (*Trypanites* isp.). Very attenuated, mucus-coated aboral spines are extended from the mouth of the boring as the nutrient flux of the tidal surge climaxes. Quantities of marine snow are collected from the spines by searching tube-feet.

Keywords. Echinometrid echinoids, bioerosion, mucus trap, particle aggregates, marine snow, *Trypanites*, *Circolites*

M. Wisshak, L. Tapanila (eds.), *Current Developments in Bioerosion*. Erlangen Earth Conference Series, doi: 10.1007/978-3-540-77598-0_15, © Springer-Verlag Berlin Heidelberg 2008

Introduction

The second half of the 19th century and the beginning of the 20th saw a wealth of controversial literature on the behaviour of the bioeroding, echinid echinoid *Paracentrotus lividus* (Lamarck, 1816) (review in Otter 1932). On rocky shores, this species bores depressions that vary from cup-shaped to deep pockets having a narrow entrance opening. Märkel and Meier (1967) finally, and very convincingly, showed that this echinoid uses the teeth of its Aristotle's lantern to bore and not its spines, and that the animal was not imprisoned within its boring, but was able to leave it at will by lowering its spines in order to glide out through the narrow entrance. The boring activity of this species is strongest on the northwestern coasts of Europe, which have a prevailing west wind and high tidal amplitude, whereas in the Mediterranean Sea, which hardly has a tide, the species is widespread but generally does not bore (Otter 1932 and own observations in the eastern Mediterranean). However, Martinell (1981) and JM de Gibert (pers. comm. 2007) observed that in the western Mediterranean, *P. lividus* is commonly found in its cup-shaped borings in extremely shallow rockgrounds of very different lithology (igneous, metamorphic and sedimentary).

Although the camarodont family Echinometridae contains some of the most destructive borers on tropical coasts, these do not feature in western culinary cultures (in contrast to the Echinidae), and this may have led to the neglect of the ecology of this family, as can be seen clearly in the monograph by Mortensen (1943). Though Mortensen knew the earlier literature and even quoted the different opinions at length, he stubbornly maintained that it was the spines that assist the echinoid in fitting natural cavities in rock and corals for better protection. Apart from in grazing, active dental bioerosion was not considered. Many authors have been of the same opinion (e.g., McLean 1967; Neumann 1968; Ormond and Campbell 1971; Dart 1972). Previous research on the type species, *Echinometra lucunter* (Linnaeus, 1758), in the Caribbean and Atlantic, confirms that this species is a true borer of cup-shaped pits and elongate grooves, that it browses the algal turf and coralline algae on the walls of the boring and furthermore, that it catches drifting algal fragments with its spines.

However, when we turn our attention to the Indo – West Pacific echinometrids, there are new aspects of ecology to add to the previously known grazing and occasional suspension feeding activities, namely gardening and the trapping of flocs of organic aggregates, similar to 'marine snow', from the flooding of the coral reef crest by the incoming tide. The influence of these trophic styles on the bioerosional work of these echinoids is the subject of this paper.

Echinometrid importance in bioerosion

Since Neumann (1966, 1968) defined bioerosion, numerous attempts have been made to quantify the bioerosional impact of different animal groups. However, when it comes to the bioerosion by echinoids, the importance of the borings for the breakdown of rock is rarely quantified. The sediment content of the intestine is weighed and the focus is on grazing, often to such a degree that borings are not mentioned at all. Much of the research lumps together grazing by diadematid echinoids with that of the echinometrids.

During the last 30 years, natural catastrophes, climate changes, pollution, quarrying and overfishing have led to a rapid decline in coral reefs followed by an explosion of articles on the delicate balance between carbonate bioaccretion and bioerosion. Here the grazing echinoids, and not least the echinometrids, have been considered as culprits. Only a relevant fraction of this literature is referred to here.

Echinometra lucunter (Linnaeus, 1758)

Observations by McLean (1967), assuming that the grazing went on for 24 hours a day, led him to the conclusion that an individual 3 cm in diameter on Barbados could enlarge its boring in beachrock with 14 cm^3/year. On Bermuda, Hunt (1969) estimated that a single specimen in a population of 25 individuals/m^2 would remove 280 g of aeolianite from its boring per year. This can be calculated to 7000 g/m^2/year. Ogden (1977) calculated from gut contents that 100 individuals/m^2 during a year could remove 3900 g of $CaCO_3$ from the walls of their borings. Hoskin et al. (1986) made controlled experiments on limestone blocks on the Little Bahama Bank and found that *Echinometra lucunter* removed 0.667 g/cm^2/year. By excluding sea urchins from an area of aeolianite they found that the associated rock-boring fauna removed 0.183 g/cm^2/year. The sediment produced by the echinoids consisted of 46% sand and 54% mud by weight and was considered to be a major constituent of the carbonate 'mud apron' deposited at depths from 300 m to >600 m. Bak (1994) reviewed earlier work on echinoid bioerosion in the Caribbean and West Atlantic. During a period of 21 months, Allouc et al. (1996) measured the erosion of individual borings of *E. lucunter* in volcanic rocks of the Senegal coast, Africa. They found that in the lowermost part of the intertidal zone, where the population was densest, the rate of bioerosion could reach 2050 g/m^2/year. This corresponds to a deepening of 1 mm/year of each boring, although borings in the associated ferruginous crusts deepened by only 0.4 mm/year.

Echinometra viridis (Agassiz, 1863)

Griffin et al. (2003) estimated the bioerosion on coral reefs in Puerto Rico by *Echinometra viridis*. An average carbonate bioerosion rate was calculated to 114-4140 g/m^2/year (highest for shallow water <3 m). Brown-Saracino et al. (2007) found that echinoid bioerosion in Belize recently had become dominated by *E. viridis*. Bioerosion estimated for several sites from 200 g/m^2/year at protected reefs to 2000 g/m^2/year at disturbed reefs. There is not a word about borings.

Echinometra mathaei (Blainville, 1825)

Russo (1980) was convinced that *Echinometra mathaei* reused pits bored by others and that the removed $CaCO_3$ found in the gut mainly originated from feeding. On the contrary he thought that the $CaCO_3$ in the gut of *Echinostrephus aciculatus* represented boring activity. He presented a joint bioerosion result for the two species of 80-325 g $CaCO_3$/m^2/year. McClanahan (1988) found that a guild of diadematids and *E. mathaei* could be endangered when overfishing of carnivorous fish resulted in over-crowding of *E. mathaei* and a rapid decline in population due to bioerosion reducing the habitat. Bak (1990) had a pessimistic view of the future of Pacific atolls, calculating that construction of reef frame by *Porites* spp. was approximately 2100 g/m^2/year, while destruction by two diadematid echinoids and *E. mathaei* was 4550 g/m^2/year. On a fringing reef in French Polynesia, Mills et al. (2000) found that the echinoids fed by night on the algal turf and digested in the daytime. The average removal of $CaCO_3$/day/individual was 0.32 g. McClanahan and Kurtis (1991) had reported 0.42 g/day from Kenya. Carreiro-Silva and McClanahan (2001) documented that overfishing of predatory fish from unprotected reefs would lead to a dominance of *E. mathaei* as a producer of carbonate sediment and thus to a lower population density of the diadematids. Intraspecific competition for algal turf and competition from herbivorous fish would finally lead to a collapse of the *E. mathaei* population. Newly protected reefs showed a slowing down of the process and long protected reefs showed a balance between predators and herbivores and a more stable rate of bioaccretion / bioerosion. Chazottes et al. (2004) studied the carbonate particles produced by *E. mathaei* and showed that the grain size depended on the structure of the coral skeletons and that the fine-grained sediment was suspended and removed from the area by waves and currents. However, the bioeroding echinoids have some advocates. Dart (1972) pointed out that removal of algal turf by grazing echinoids and fish might help coral larvae to settle and start new colonies. Highsmith (1980) showed how important boring sponges and *E. mathaei* are in making massive coral heads and microatolls weak and thus more readily broken up in storms to rejuvenate and flourish in new niches.

Previous work on Caribbean and tropical Atlantic *Echinometra* spp.

Echinometra lucunter has been used as a character-species for the border between the sublittoral and eulittoral zones (the infralittoral margin) on rocky coasts (Loenhoud and Sande 1977). Brattström (1980) placed *E. lucunter* in the uppermost part of the sublittoral (zone of algal mat) though at many of his localities it is also found in the lowermost mixed algal zone of the littoral. Kaye (1959: 89 and fig. 34) described cup-shaped borings by *E. lucunter* in the intertidal zone in marine-cemented aeolianites but added that they might also be able to produce elongate grooves.

For 5 months, McLean (1967) studied a population of *E. lucunter*, members of which were boring in beachrock in Six Men's Bay, Barbados (Fig. 1). He considered, erroneously, that the spines were used to prise out grains in the boring process. After a severe storm he found that the 2-3 mm long algal turf usually present in the groove-shaped borings was totally scraped off and a blue-green calcareous paste was found in the guts of dissected specimens.

During a three-week study of the behaviour of *E. lucunter* on Bermuda, Hunt (1969) found very few borings unoccupied. Small individuals inhabited cup-shaped borings. It seemed clear that the mature individuals, as they created their long, groove-shaped borings, bored slowly forward, grazing on algae. However, the specimens positioned at the end of their borings did not move during the three weeks of observation. What did they feed on until their 'lawn' had recovered? She noted that the Aristotle's lantern was large and could be extended 'over an inch' away from the test and might operate in an area more than 5 cm in diameter.

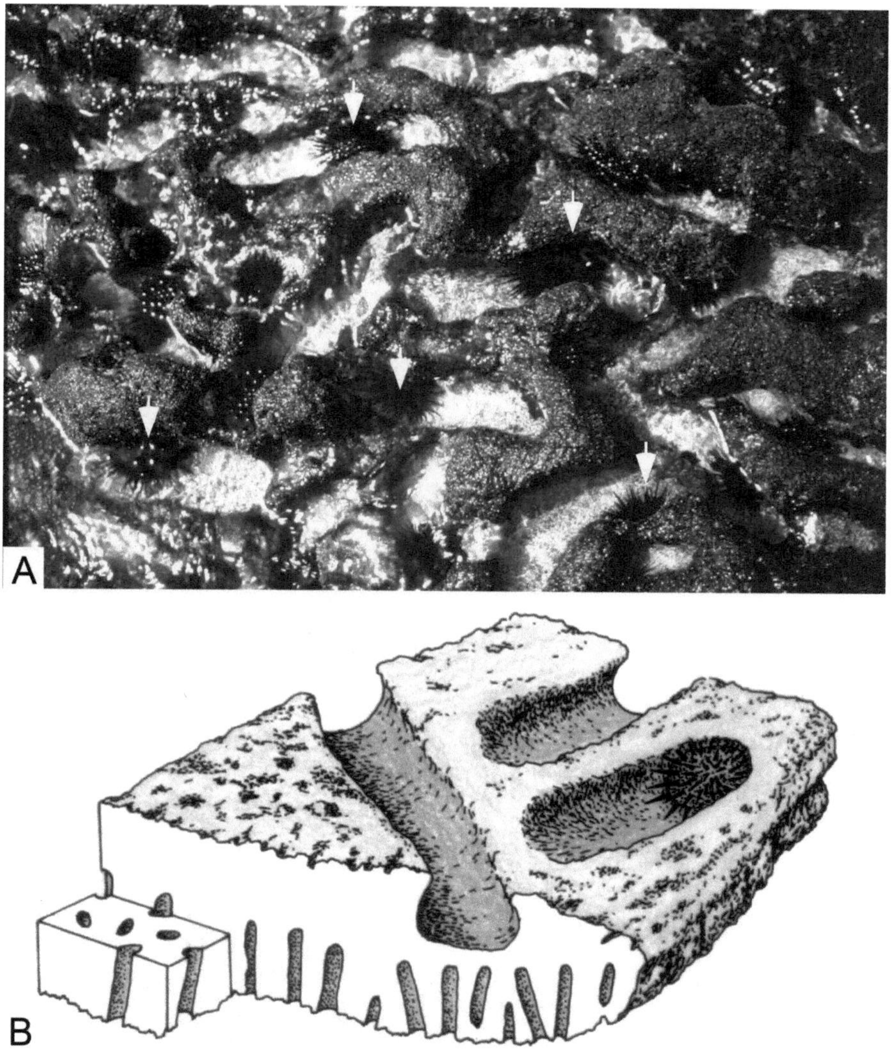

Fig. 1 Surface of beachrock in the infratidal zone, at sea-level at low tide, at Six Men's Bay, Barbados. **A** Detail of the surface showing pale-coloured boring grooves of *Echinometra lucunter*. The sea urchins are about 3 cm long, one in each boring (five indicated by arrows). **B** Sketch of the beachrock emphasising the form of the borings. On the well illuminated, unprotected upper surface of the beachrock are the echinoids; on the shaded, overhanging undersurface, the plankton-feeding thoracican barnacle *Lithotrya* sp. produces abundant *Trypanites* isp. Modified after Ekdale et al. (1984: fig. 16-3)

The large, groove-boring individuals had the oval, elongate test shape characteristic of the species (Fig. 2A). This has been considered to be connected with a restricted, directional locomotion, resulting in the groove-shaped borings (Hunt 1969; Bromley 1978). However, Hunt (1969) observed that the juvenile individuals, producing cup-shaped borings, lacked the elongate form of the adults.

Ogden (1977) reported on *E. lucunter* from the U.S. Virgin Islands. Individuals moved a few centimetres a day as they grazed the walls of the groove-shaped boring. He noted that the algal turf outside the boring was very rarely touched by the echinoids. An estimated 55% of the diet consisted of drifting algae caught by the spines.

Fig. 2 Two ovoid species of *Echinometra*. **A** The Atlantic *E. lucunter*, from Bermuda, showing its powerful spines and ovoid shape. Aboral view of the test, left, and adoral view with spines showing the large teeth. **B** The Pacific *E. oblonga*, from the New Hebrides, seen in aboral views. T. Mortensen collection, Zoological Museum, University of Copenhagen, Denmark

Focke (1977) studied *E. lucunter* in the Dutch Antilles on the windward side of Curaçao on the surf platform, where porous bioaccretions mostly constructed of coralline algae border a terraced system of pools. Here and in the subtidal lower notch, up to 200 *E. lucunter*/m² penetrated into Pleistocene limestone. Individual borings too were bordered with algal bioaccretion up to 15 cm thick. There was no mention of the shape of the borings. Observations and experiments made by Steneck and Adey (1976) indicated that the hydrodynamic and light conditions in and around cavities and echinoid borings led to an increase in the bioaccretion.

Hoskin et al. (1986) studied *E. lucunter* populations by day and night in the Bahamas using SCUBA; the sea urchins did not leave their borings.

Ogden et al. (1989) confirmed earlier observations that *E. lucunter* stays within its boring. When a square metre was cleared of echinoids only 6% of the borings were occupied 10 days later.

Hunt (1969) observed no aggression in *E. lucunter*, but Grünbaum et al. (1978) reported agonistic behaviour between individuals. By planting a conspecific intruder into the boring of another individual, these authors observed a strong reaction in the form of pushing and biting off of spines, especially but not uniquely, by the owner of the boring. The intruder invariably lost the fight where it was smaller than the owner. The authors also suggested that the agonistic behaviour ensured that the owner of a boring was alone, and thereby had a better opportunity to catch a larger supply of drifting algae, than if sharing the boring with others. It was pointed out that the behaviour was supported by the unusual muscular development in the lantern remarked upon by Mortensen (1943).

Similar studies by Shulman (1990), involving the introduction of several Caribbean genera to each other, led to aggression. When *E. lucunter* and *E. viridis* individuals encountered each other, they interacted in 46-79% of trials, nearly always by pushing, but rarely followed by biting. The owner of the boring won when bigger or equal in size. He found the same aggression pattern in the two congeners in slightly deeper water where the specimens were not producing borings. He concluded that the special construction of the lantern of *E. lucunter* could not be an adaptation for aggression alone, but did not offer any alternative suggestions.

We have not observed *E. viridis* during our study, but Jones and Sefton (1979: 33) figured a group of *E. viridis* in their borings without further comments. McGehee (1992) found *E. viridis* on backreefs in Puerto Rico among pebbles and sand. It became more abundant than *E. lucunter* at >3 m. It did not bore and its presence was negatively correlated with water movement. She found that the stouter spines of *E. lucunter* (compared to those of *E. viridis*) likewise suggest an adaptation to shallow and rough waters.

Hendler and Pawson (2000) noticed that *E. viridis* in the quiet waters of the Rhomboidal Cays of Belize did not seek shelter, but foraged in the open on the sediment.

Griffin et al. (2003) described bioerosion by *E. viridis* on coral reefs of Puerto Rico; they mentioned borings without describing them. Brown-Saracino et al. (2007) described the erosive powers of *E. viridis* when discussing the $CaCO_3$

budget on reefs depleted in carnivorous fish by overfishing, but did not mention the borings. Where *E. lucunter* and *E. viridis* overlap in distribution they do not interbreed (McCartney et al. 2000).

In a thorough study of *E. lucunter* boring into volcanic rocks on the West-African (Senegal) coast, Allouc et al. (1996) found that here the borings consist of perfect cups; no elongated grooves were found. Some Quaternary borings were figured from raised beaches (Allouc et al. 1996: fig. 10). We suggest that the hemispherical borings, as opposed to grooves, made by *E. lucunter* in volcanic rocks on the west coast of Senegal result from a larger tidal range than that seen in the Caribbean and West Atlantic. Allouc et al. (1996) found that *E. lucunter* bored rapidly into dolerite because endolithic fungi attack and break down augite and other dark minerals, thus making it easy for the echinoids to prize out the remaining brittle crystals. Maybe the fungi also could be an important part of their diet?

In summary, *E. lucunter* on western Atlantic coasts predominantly bores groove-shaped borings in carbonate rocks at the infralittoral margin. In this high-energy environment, the echinoid is protected within its boring and can catch floating fragments of organic debris, especially algae, using its spines. Grazing the turf of algae growing on the boring walls and floor further augments the diet. Borings of cup shape are mentioned, but these may generally be the work of juveniles.

On the eastern Atlantic coast of Senegal, however, Allouc et al. (1996) documented *E. lucunter* boring in volcanic rocks, assisted by fungal breakdown of dark minerals (and perhaps feeding on these). The borings here were exclusively cup-shaped, not grooves.

Own observations

Caribbean and Bermudan Echinometridae

During several short periods during the 1970s, we have observed *Echinometra lucunter* on Bermuda (Bromley 1978) and in Six Men's Bay on the leeward side of Barbados. We have always found the species in its groove-shaped borings. We can confirm the earlier observations that the borings are systematically browsed and deepened. Catching of drifting macroalgae by means of pairs or groups of spines was observed, the 'chopsticks method'. We here suggest that the specialisation of the Aristotle's lantern supports the unusual mobility of this apparatus as described by Hunt (1969). The mobility and muscular strength are necessary for this echinoid when rotating around its long axis in order to graze on the turf of its boring. We did not observe any specimens leaving their borings to graze on the lush algal growth outside.

It was therefore surprising to find in T. Mortensen's collection (Zoological Museum, University of Copenhagen) some large, unpublished slabs of limestone containing boring individuals of *E. lucunter*, collected by himself in 1906 from Long Point on the south coast of Saint Croix in the Virgin Islands of the Caribbean Sea. *E. lucunter* borings ('shallow holes') at this locality are referred to by Mortensen (1943:

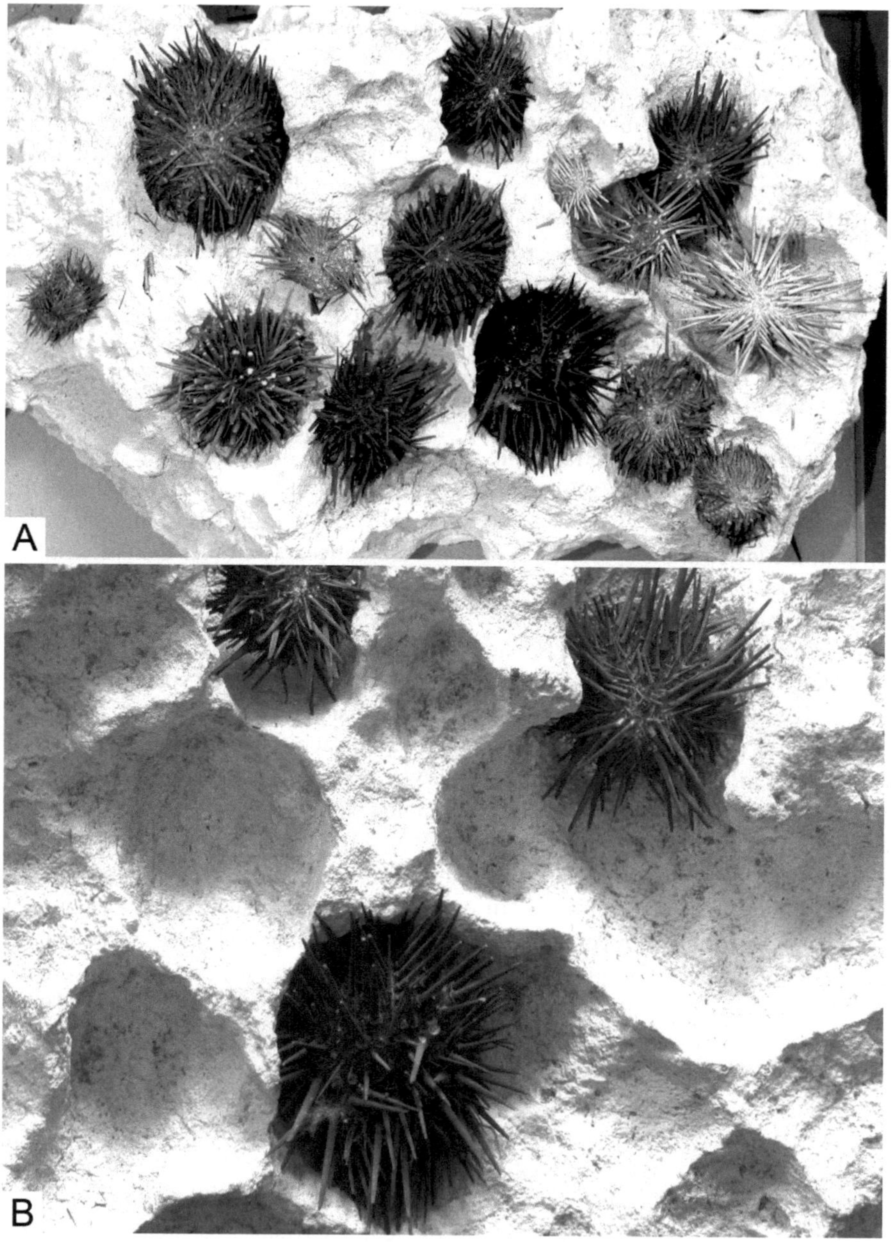

Fig. 3 Slab of cemented carbonate lagoon sediment, Virgin Islands, West Indies. **A** Fourteen individuals of *Echinometra lucunter* in close proximity in oval, cup-shaped borings. Note that the longest diameter of most of the echinoids parallels the side-line of the figure. **B** Close view of right centre of A, most echinoids removed to show oval shape of the borings, *Circolites* isp. T. Mortensen collection, Zoological Museum, University of Copenhagen, no. ZMUC ech-584

364). As Mortensen did not use diving gear, collection of these large specimens must have occurred in paddling depth at low tide. The largest slab (no. ZMUC ECH-584) contains 14 individuals within their borings (Fig. 3A). Removal from this specimen of removable, in situ echinoids, revealed the typical irregular borings often seen in the western Atlantic and the Caribbean in beachrock and limestone. But these borings, like their inhabitants, have a distinctly oval outline, not circular cups, not grooves (Fig. 3B). The shape of the ovoid echinoids indicates that these are conspecific with those that bore grooves, and that the different modes of behaviour must be due to environmental differences. It is likely that this behaviour, where the possibility of turf-algal harvesting is much reduced, can only be sustained where copious quantities of detrital algae are present.

Indo – West Pacific (IWP) Echinometridae

Echinometra oblonga (Blainville, 1825)

The IWP equivalent of *Echinometra lucunter* may be said to be *Echinometra oblonga* (Fig. 2B). The two species are physically very similar, including the ovoid form of the test, but the spines of *E. oblonga* are noticeably stouter than those of *E. lucunter*.

Thus it might be expected that, like *E. lucunter*, *E. oblonga* should be well suited as a bioeroder of grooves. Why else should regular echinoids, which have no preferred direction of movement, develop a bilateral morphology, than as a specialisation for movement forward and backward within a groove-shaped habitat?

Grabowsky (1994) attempted to prove that oblong echinometrids had no preferred direction of movement. However, the aquarium experiment was too ill-designed to produce any reliable information. The echinoids were considered to both bore and to cling to their substrate using the spines.

However, *E. oblonga* has a rather limited research literature, and there is no mention of it bioeroding grooves. In only a single case, cup-shaped and groove borings are mentioned in connection with *E. mathaei* and *E. oblonga* (see Ogden et al. 1989), but it is not quite clear which of these echinoids makes the grooves. A closer investigation of this matter would be most welcome. Mortensen (1943: 395) wrote: *'The characteristic upright position of the spines in typical specimens of oblonga indicates that this form is particularly a rock borer, or at least lives in holes in the rocks or corals. I regret to have no remembrance of having observed this, although I have collected a good number of specimens at various places.'*

Russo (1977) noticed that *E. oblonga* had coarser spines than *E. mathaei*. This was an advantage in the heavy surf, while *E. mathaei* was found in calmer water.

The black type D that was described as found in intertidal surge areas in Okinawa and Hawaii (Palumbi and Metz 1991) has later been recognised as *E. oblonga* (see Appana et al. 2004).

Sri Lankan echinometrids

Our research area was a coral reef at the village of Hikkaduwa on the west coast of Sri Lanka (90 km south of Colombo). The reef was visited three times in December to January between 1982 and 1986. Here we met an underwater world significantly different from the coral reefs of the Caribbean and Atlantic.

At that time the narrow barrier reef was protected from fishing at the village in order to boost the growing tourist industry, at a time of unrest in the Northern Province around Jaffna, where reefs are very well developed. South of Hikkaduwa, towards Galle, the reef thins out and disappears. Along the coast to the north, the reef had been destroyed in many places by fishing with dynamite and blasted for burning lime in small village kilns. At Hikkaduwa itself, two bare offshore rocks of pegmatite protect the shore from ocean rollers. Here the reef joins the rocks from south and north, thus protecting a promontory of beachrock on a long sandy beach and a wide triangular lagoon containing patch reefs.

The lagoon is narrow to the north of the rocks and the water is only a few metres deep on the inside of the barrier. There were no patch reefs here but instead, a sandy bottom and healthy coral thickets and coral rubble. At several places along the landward coast of the lagoon, there were sand-free exposures of beachrock at low-tide mark.

The tidal range is small, about 50 cm at spring tide. We were unable to skin dive on the ocean side of the reef on account of a strong current running from south towards north. According to the fishermen, leaving in their catamarans at sunset to return in the morning, the current is a narrow one, quickly crossed in a spurt of vigorous paddling.

During the total of eight weeks we spent in the area the wind only blew from the ocean during the night. During the day the sea was calm and the tides gently came and went. Neighbouring reefs near Galle on the southwest corner of Sri Lanka were not associated with further exposures of beachrock.

Nutrient fluxes, particle aggregates and marine snow

At paddling depth the lagoon did not show much active life at low tide, when the crest of the reef was laid bare, but a few deposit feeders were visible (Fig. 4A). However, with the gentle flood tide came extensive floats of scum, foam with grey, organic-rich detritus, somewhat related to the marine snow of open water. For about twenty minutes, there was feverish benthic activity. From every crevice and under every boulder, the ophiuroid *Ophiocoma scolopendrina* (Lamarck, 1816) crawled up to sweep the coral mucus and particle aggregates from the surface, waving its free arms and only holding on to the substrate with one or two arms (Fig. 4B), as described by Chartock (1983) and Oak and Scheibling (2006). Eventually, as the surface rose out of reach, the ophiuroids dropped to the bottom and for some time deposit-fed on the settled and decomposing marine snow. After about half an hour of suspension and deposit feeding the ophiuroids retired to their crevices as the larger carnivores and deposit feeders arrived with the tide, and there they stayed hidden until the next tidal flood surged in. The enormous role played by the copious amounts of mucus produced by the corals on the exposed reefs of the IWP has recently been reviewed by Wild et al. (2004, 2005) and Huettel et al. (2006). Especially the last-mentioned paper describes the situation we witnessed where bubbly floats of mucus packed with detritus were caught at the surface and later, after gaseous release, the decomposing floats sank and were seized in the water column or picked up from the substrate.

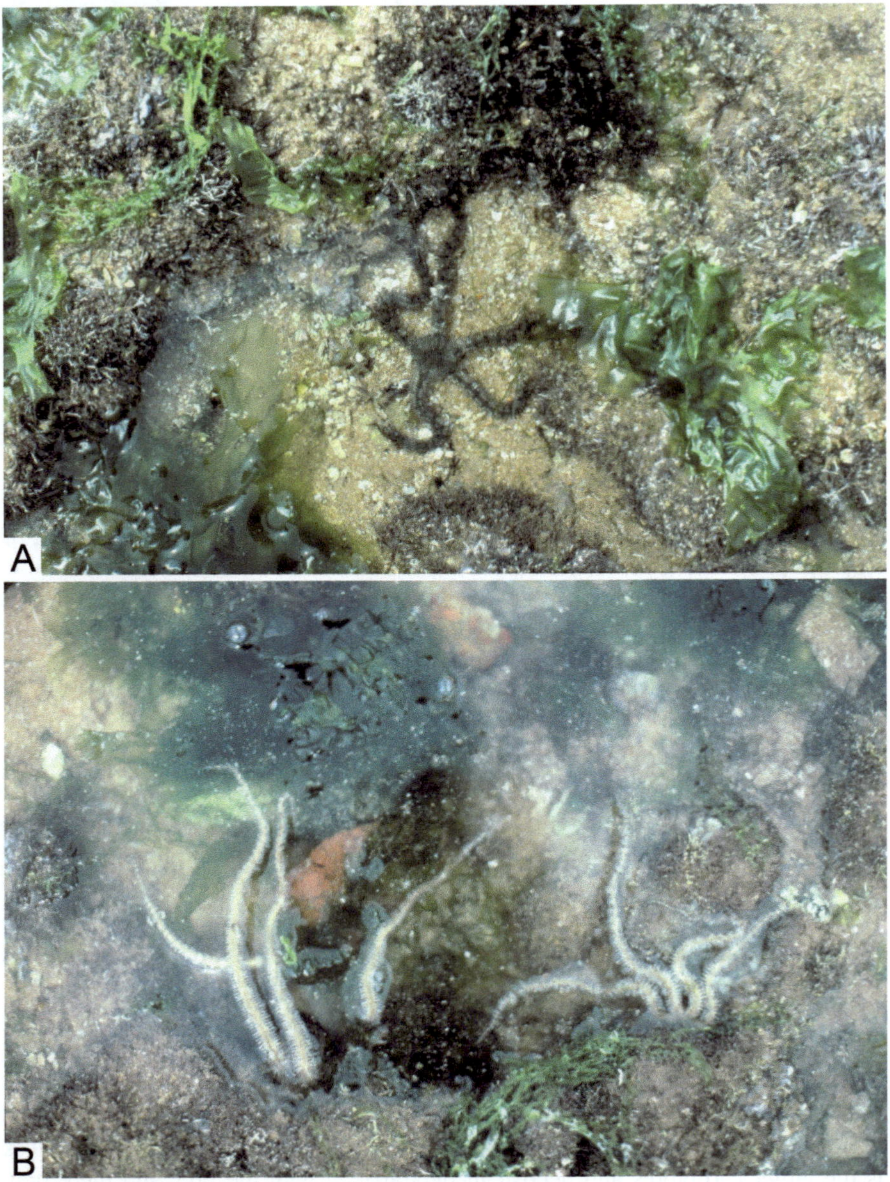

Fig. 4 The effect of the nutrient flux on the ophiuroid *Ophiocoma scolopendrina* near the shore of the backreef lagoon, Hikkaduwa, Sri Lanka. **A** Low tide: the carnivorous fish leave the shore area at low tide and the ophiuroids are deposit feeding on the seafloor. **B** The nutrient flux has brought organic scum across the lagoon. The ophiuroids climb to the water surface and vigorously sweep the scum. The individual to the right hangs onto rock substrate with a single arm, thereby freeing 4 arms for scum-collection

Echinoid utilisation of the nutrient fluxes

Among the members of the reef community, the boring echinoids also utilise the marine snow suspended in the water column. On the Hikkaduwa reef, two echinometrid species were active in this respect.

Echinometra mathaei (Blainville, 1825)

This common IWP echinoid has recently been subjected to molecular research and this has indicated that in fact four sibling species are present (e.g Tsuchiya and Nishihira 1984; Palumbi and Metz 1991; McCartney et al. 2000; Kinjo et al. 2004; Appana et al. 2004). However, no formal species have been erected and apparently only biochemical features separate them, and vague differences in colour of test and spines. Until these results are translated into morphological characters, the name we apply to our material must be regarded as 'sensu lato'.

E. mathaei is the dominant boring echinometrid in the IWP. At Hikkaduwa it was only seen on the beachrock where, at low tide, the crowded population was isolated in shallow pools, each individual in its well polished hemispherical boring (Fig. 5). Some individuals, lacking protective borings, and living on the fringe of the 'echinoid city', on the vertical seaward side of the beachrock (Fig. 6B) may be left subaerially exposed at low tide and are commonly preyed upon by the Indian town crows. The echinoids were uniformly coloured and nearly black, having an overall diameter of about 7 cm. Their tests did not show the elongation seen in the type species, *E. lucunter*.

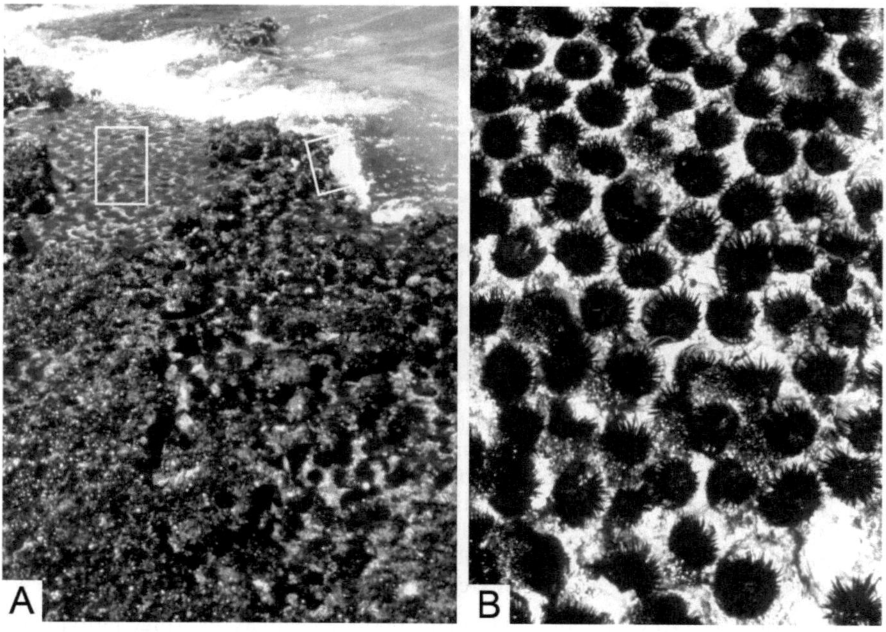

Fig. 5 Beachrock at Hikkaduwa, Sri Lanka. **A** General view at low tide, the white surf indicating the seaward edge of the beachrock platform. Outlines of the areas detailed in B (left) and Fig. 6B (right) indicated. **B** Close-up view of the population of *Echinometra mathaei*, individuals are about 7 cm in diameter

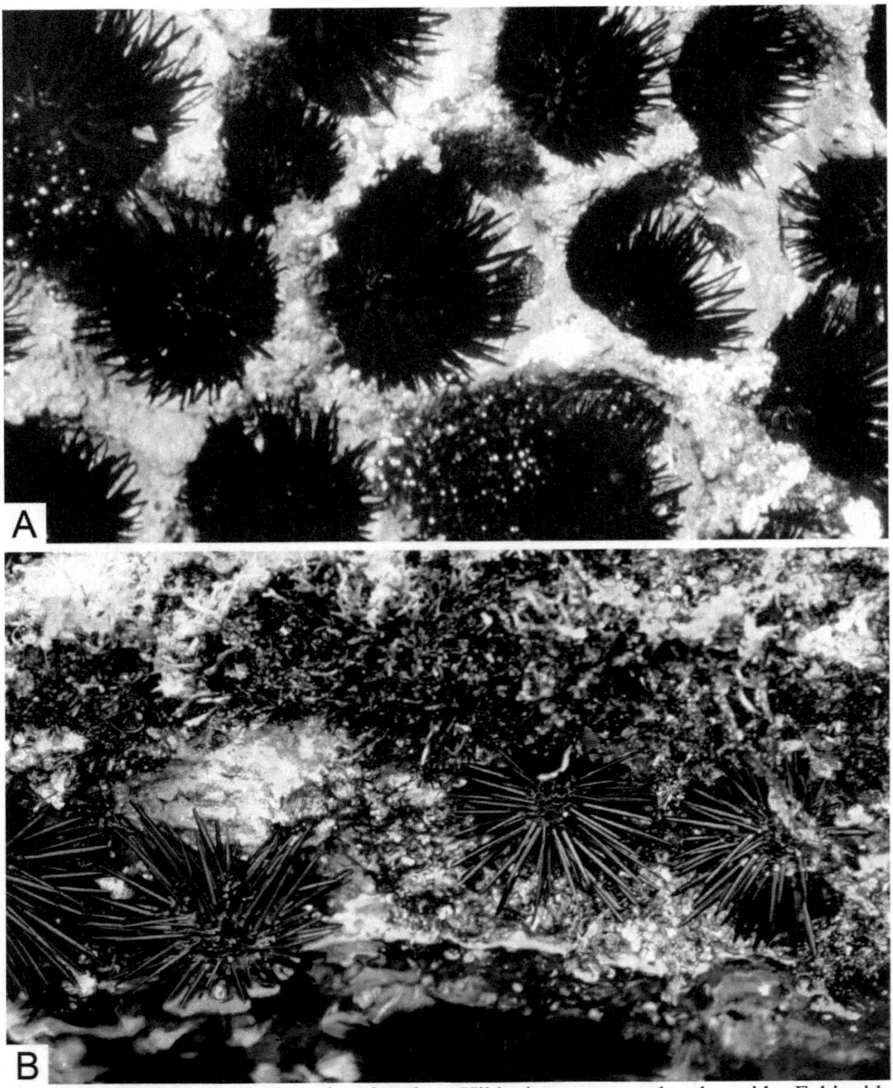

Fig. 6 *Echinometra mathaei* on beachrock at Hikkaduwa, exposed at low tide. Echinoids about 7 cm across. **A** Close-up view of left marked area of Figure 5A showing echinoids in their hemispherical borings. The white points of reflected light over some individuals are caused by raised spines breaking the water surface. All spines are raised in position to capture drifting algal fragments in the in-coming tidal waters. **B** Homeless individuals, stranded above the low-tide waterline (see right-hand area marked in Fig. 5A). Subaerial exposure is dangerous, as air may enter the water vascular system with lethal results. Drifting algae are out of reach, and the upper spines are lowered into a horizontal position

At low tide, the crowded individuals in the pools, as well as the fringe individuals on the vertical surfaces, were inactive, most having their spines lowered. Many took up a resting position on the sides of the boring. This position exposed the floor of the cup-shaped borings to maximum insolation, like small greenhouses in the tropical sunshine, encouraging the growth of epi- and endolithic algae ready to be harvested at night. This must be regarded as a simple process of gardening, comparable to that of *E. lucunter*.

With the first surge of the incoming tide, the echinoids moved to the bottom of their borings and raised their spines to the maximum, even to break the surface of the water (Fig. 6A). Soon the surface of the spines looked dusty with the adhering detritus from the marine snow. The growth of algae around and between the borings in the pools seems to indicate that the echinoid borers do not forage outside their borings. The garden within the boring, however, is regularly grazed, almost to the degree of polishing.

Others' observations on E. mathaei

It is difficult to compare our observations with those of other authors, as their work described populations from rough waters at unprotected coasts or on the oceanward side of reefs (e.g., Bertram 1936; Otter 1937; Khamala 1971; Ormond and Campbell 1971; Dart 1972; Campbell et al. 1973). The last-mentioned paper seems to be the first to report that, in choppy and rough seas, *E. mathaei* catches drifting algae by the chopsticks method as seen in *E. lucunter*. This lifestyle has also been recorded by Russo (1977, 1980) and later authors. All the above papers agree that *E. mathaei* lives in 'crevices', which it might widen somewhat by using its spines (sic), rather than in cup-shaped borings.

Observations by Khamala (1971) indicated that some populations of *E. mathaei* on inner reefs (not exposed at low tides) did not home for the same 'crevice' when moved. The species furthermore could thrive on coral rubble and under massive coral colonies and coral slabs (Dart 1972; Highsmith 1980).

Ogden (pers. comm. in Russo 1980: 104) had noticed that *E. mathaei* defended their borings against conspecific intruders. McClanahan and Kurtis (1991) also noted agonistic behaviour.

Ogden et al. (1989) on Hawaii studied some *E. mathaei* in their borings and found that the gut contents were algae growing in and at the opening of the boring. The bottom of the borings was polished like the Sri Lankan borings observed by us.

On Hawaii, Hart and Chia (1990) placed in aquaria, blocks containing borings inhabited by *E. mathaei*. The specimens that were offered drift-algae did not leave their borings to feed, but were 'sedentary' as Russo (1980) called it, whereas starved specimens left their borings regularly to graze on the algal turf outside. Similarly starved specimens were seen to leave their borings by night to graze on the glass of the aquaria (Russo 1980).

Tsuchiya and Nishihira (1984) found two types of populations of *E. mathaei* on the Japanese island Okinawa: the one with individuals in cup-shaped borings that they did not leave, the other a gregarious population grazing the surfaces peacefully. The first type was found on the reef crest, the other was found in calm back-reef conditions in crevices or under corals, even on bare rock or on rock covered with sand and coral rubble. It was concluded that the difference in behaviour in the two populations was connected with the hydrodynamic conditions. Prince (1995) found two similar populations on intertidal rock platforms at Rottnest Island, Western Australia. One population was found on a 50 m wide platform subject to a strong southwest swell. Here a population of up to a hundred specimens/m^2 sat in their individual cups, which they did not leave. The borings occupied only about 40% of the platform. The other population was of a similar size and inhabited a >100 m wide platform that was protected from the regular swell most of the time but received storms from the northwest. Here *E. mathaei* was found in groups in large shallow 'barren' depressions that covered about 90% of the platform. After a molecular test was made on the two populations it was concluded that hydrodynamic conditions, not genetics, governed the behaviour.

Echinostrephus molaris (Blainville, 1825)

The genus contains two very small echinometrid species, *Echinostrephus molaris*, studied here, and the type species *E. aciculatus* Agassiz, 1863. According to Mortensen (1943) and Russo (1980), *E. aciculatus* is restricted to the West Pacific.

All authors have agreed that *E. molaris* lives in finger-shaped borings of its own making (Fig. 7). Mortensen (1943: 287, 315) gave a vivid description of its functional morphology and suspension feeding on the basis of fieldwork on the reefs of the island Amboina (Indonesia). He studied the contents of its intestine and found detritus, plant remains, and fragments of calcareous skeletons. Campbell et al. (1973) confirmed Mortensen's observations from the choppy environment on the outer side of the reef in the Sudanese Red Sea where *E. molaris* catches drifting algae with its spines (chopsticks method) and, by use of its long aboral podia, drops the material into its boring for later ingestion.

Echinostrephus Agassiz, 1863 is the only known echinoid genus that is adapted to 100% suspension feeding, to which its remarkable anatomy bears witness. The test is barrel-shaped and the echinoid can only rotate in its boring about its oral – aboral axis; it could not survive outside its boring. The long aboral spines, the very short spines on the sides of the test to provide secure anchorage within and movement along the boring, its large peristome and Aristotle's lantern for boring with, and curved adoral spines to hold large particles towards the teeth are its specialised equipment. The deep, dark boring is not meant for gardening.

We found these small echinoids in their characteristic finger-shaped borings just inside the crest of the reef. At ebb they were found less than 40 cm below the water surface in dead corals. Most of the time they were positioned at the bottom of their borings and the tips of their tuft of aboral spines could just be seen far down within the tightly fitting hole. Since most of the time the echinoids remain inactive, they are easily overlooked (Lawrence 1970). However, just as the particle aggregates started to drift in with the tide, the echinoids quickly moved to the opening of the boring and spread their aboral spines out to form a hemispherical 'net' (Fig. 8) where marine snow was trapped as drops of detritus-laden mucus gliding down the spines (Fig. 9). These drops of detritus were picked up by podia that constantly snaked their way up and down along the spines. When the nutrient supply came to an end, the echinoids retreated to the bottom of their borings.

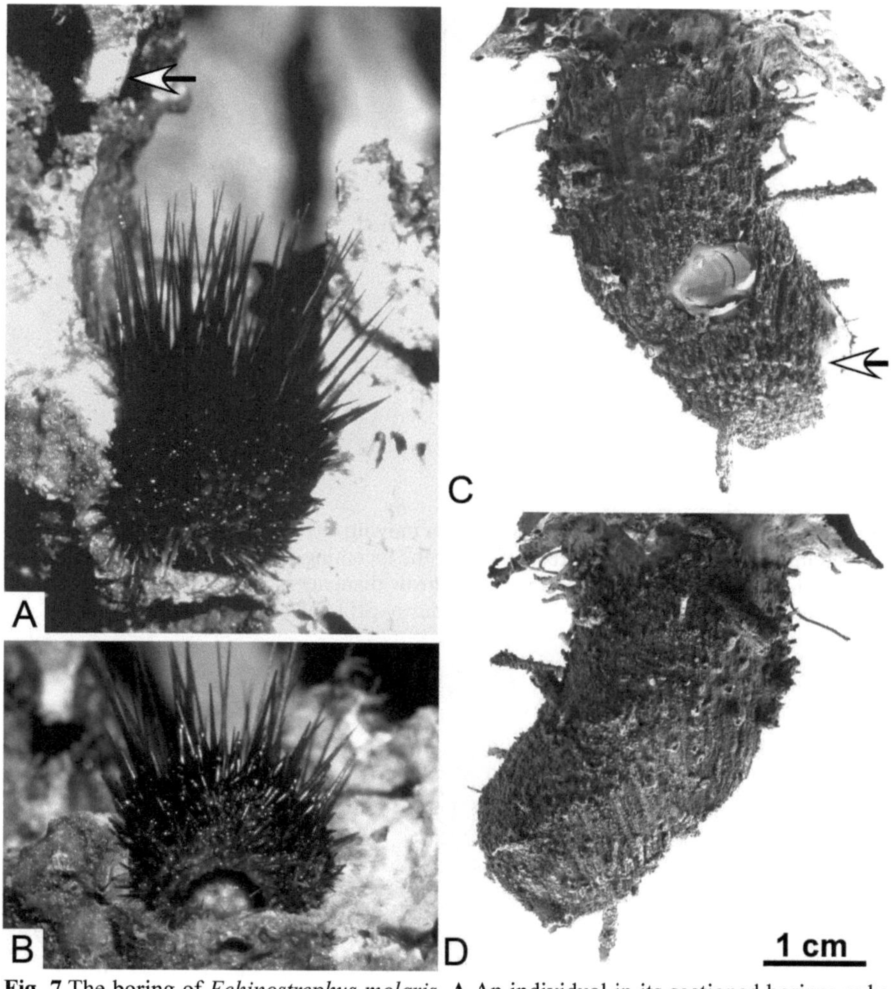

Fig. 7 The boring of *Echinostrephus molaris*. **A** An individual in its sectioned boring; only the left wall to the aperture is preserved (arrow). Note the short spines for gripping the wall of the boring, and the long, aboral spines bundled together while inside the boring. **B** Same individual seen from below, showing the extremely large Aristotle's lantern, designed for bioerosion. **C-D** Latex cast of an *E. molaris* boring. The white patches on the wall (one arrowed) are encrusting bivalves. Elsewhere, the coral structure provides a xenoglyphic ornament to the wall. Several slender *Trypanites* extend into the wall, probably the work of polychaete annelids. Scale bar: 1 cm. MGUH 28732

Protected within their narrow borings, the small echinoids carry a remarkable armament of pedicellaria for protection against settling larvae of would-be cohabitants of the boring (Fig. 10). The long, aboral crown spines possess deep longitudinal grooves (Fig. 11B) that possibly may house mucus-secreting glandular tissue, although this needs to be tested. Out of reach of the pedicellaria, some symbionts have nevertheless gained a foothold in the long aboral spines, producing minute borings (Fig. 11A). The nature of this tracemaker is unknown.

Fig. 8 Two individuals of *Echinostrephus molaris* moving up their borings to the aperture, and coming into feeding position, in preparation for the incoming tidal surge. The left individual is nearly ready to begin suspension feeding. Burrow diameters 1.5 cm. Lagoonal side of the barrier reef, low-tide water depth 0.5 m. Hikkaduwa, Sri Lanka

Fig. 9 Close-up view of an *Echinostrephus molaris* suspension feeding in the first surge of water as the tide turns. The water is rich in suspended organic particles, seen here as white spots, many out of focus. Some of these are caught on the long, mucus-coated aboral spines. Rapidly moving, long tubefeet (not visible in the figure) constantly search the surfaces of the spines, collecting the particles and transporting them into the boring. Width of figure about 3 cm

Fig. 10 SEM scans of a small individual of *Echinostrephus molaris* from the Mortensen collection, from the island Amboina, Indonesia. Geological Museum, University of Copenhagen, MGUH 28733. **A** General view, showing the long, adapical spines and the short spines for gripping onto and climbing up the boring. A remarkable density of pedicellaria is seen among the short spines. **B** Close view showing a diversity of pedicellaria. The spines have a variety of ornaments. Some are smooth, others have a spiny ornament

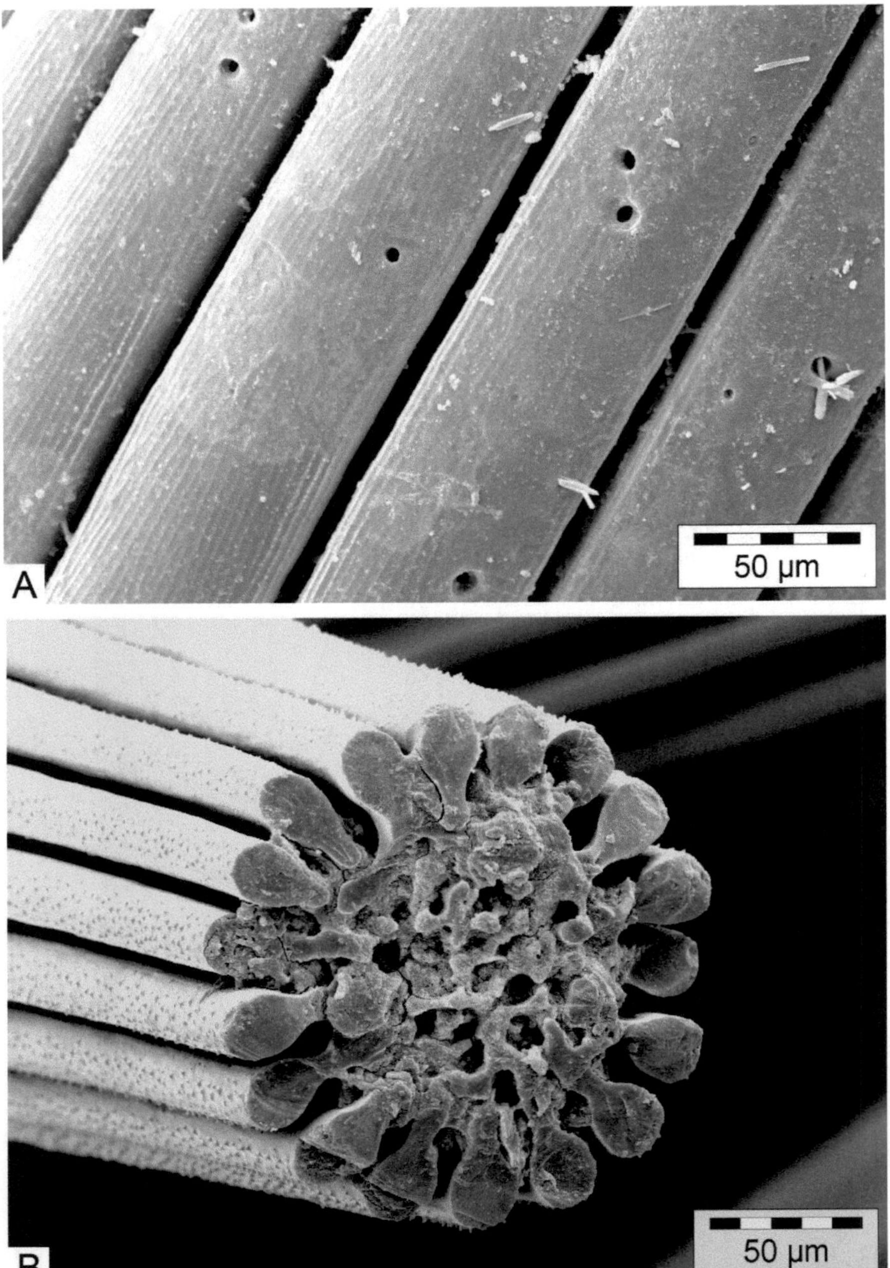

Fig. 11 SEM scans of the same individual as Figure 10. **A** A long, aboral spine having a nearly smooth surface. This has been bored in vivo by unknown organisms, representing symbionts. **B** A spinose surface ornament on a long spine. The significance of these different surface textures is unknown. All spines have in common a set of deep grooves running the length of the spine. Possibly the glands for secreting the copious mucus may reside within these grooves. MGUH 28733

The borings that extend from the main boring wall (Fig. 7C-D) are probably post mortem, having succeeded the death of the echinoid; the echinoid boring was empty when collected. Owing to the extreme shortness of the spines on the ambitus (greatest horizontal circumference of the test) and adoral parts, the pedicellaria would be able to reach all parts of the boring wall as the echinoid moved up and down its boring, and attempts at colonisation by other organisms would have been terminated. Thus, although the echinoid could not have survived outside its boring, its morphology is well adapted to rapid movement within its boring, conferring a high degree of safety, the efficient exploitation of the environment for food, and freedom from competition from other organisms.

Others' observations on *Echinostrephus* spp.

Bak (1990) noted that *Echinostrephus molaris* at Tiahuara Atoll was found only in eroded, flat, and dead corals. Russo (1980) reported *Echinostrephus aciculatus* from Enewetak Atoll where it occurred consistently with *Echinometra mathaei*. Both were boring and feeding on drift algae.

Conclusions

Several species of the regular echinoid family Echinometridae are active bioeroders in tropical and subtropical, shallow marine environments.

In the Atlantic Ocean and Caribbean region, *Echinometra lucunter* produces borings having two distinct morphologies. Elongated grooves are widely reported, but cup-shaped borings also occur, corresponding to ichnogenus *Circolites* Mikuláš, 1992. A specimen in the T. Mortensen collection, Zoological Museum, Copenhagen, Denmark, from the West Indies, shows *E. lucunter* in oval borings, corresponding to the ovoid shape of the borer. It is confirmed that *E. lucunter* normally does not leave its boring, and derives nutrition from catching drifting organic material, chiefly algal fragments brought into the boring by the rising tidal surge. These are either passively trapped in the boring groove, or actively plucked from the current by the echinoid using spines in a chopsticks-like fashion. In addition, the walls and floor of the boring are cropped and eroded by powerful teeth for short turf algae and endolithic algae in a simple gardening process.

Another ovoid species of *Echinometra*, *E. oblonga*, occurs in the Indo – West Pacific realm. Although comparison with *E. lucunter* indicates that *E. oblonga* very likely produces groove-shaped borings, this has not been conclusively demonstrated.

In recent years, it has been shown that coral reefs of the Indo – West Pacific realm possess a nutrient flux system that is connected with the fact that the reef crest is exposed subaerially at low tide, in contrast to the Atlantic coral reef realm. As the reef crest is drowned with the initial phase of the rising tide, a nutrient flux is swept into the lagoonal and shore waters.

We studied two species of echinometrid bioeroders in the 1980s in the barrier reef and lagoon system at Hikkaduwa, Sri Lanka: *Echinometra mathaei* and *Echinostrephus molaris*. Behind the barrier reef, the shore locally exposes beachrock or Pleistocene reef rock. This is colonised by crowded communities of *E. mathaei* individuals, each occupying a cup-shaped boring. Trophic styles are as in *E. lucunter*, i.e., trapping and catching of drifting algae, and gardening of algal turf within the boring. In addition, however, particle aggregates from the nutrient flux are caught on the spines of *E. mathaei* and collected by tubefeet.

On the lagoonal side of the barrier reef at Hikkaduwa, the small echinoid *E. molaris* bores thumb-sized borings, ichnogenus *Trypanites* Mägdefrau, 1932, into the coral rock. The animal fits the diameter of its boring and never leaves it. It catches drifting algal fragments with its spines. However, its chief trophic style in the lagoonal side of the barrier crest is the collection of suspended particle aggregates in the tidal flux. As the tide turns, the echinoid climbs to the entrance of its boring and spreads its long, slender aboral spines as a hemispherical array. The spines are drenched in mucus and marine snowflakes adhere to them. Long tubefeet actively collect this matter and pass it down into the safety of the boring for later ingestion.

A remarkable armament of pedicellaria is present on the body of the echinoid. The extreme shortness of the ambital and adoral spines would allow the pedicellaria to have access to the walls of the boring as the echinoid passed up and down between termination and aperture, cleaning the surface of organisms.

Acknowledgements

We are most grateful to Christine Schönberg (Oldenburg) for help with the particle aggregate literature, Hans Jørgen Hansen (Copenhagen) for applying his SEM skills and Margit Jensen and Claus Nielsen (Copenhagen) for help with the Theodor Mortensen Collection. Referees Gerhard Cadée and Jordi M. de Gibert provided insightful suggestions for improvement of the paper. In memory of Wimala, the gifted batik artist, and catamaran fishermen Ari and Sonny, lost in the 2004 tsunami; we couldn't have done it without them!

References

Agassiz A (1863) List of the echinoderms sent to different institutions in exchange for other specimens. Bull Mus Comp Zool 1:17-28 [not seen]

Allouc J, Le Campion-Alsumard T, Leung Tack D (1996) La bioérosion des substrats magmatiques en milieu littoral: l'exemple de la presqu'île du Cap Vert (Sénégal Occidental). Geobios 29:485-502

Appana SD, Vuki VC, Cumming RL (2004) Variation in abundance and spatial distribution of ecomorphs of the sea urchins, *Echinometra* sp. nov. A and *E.* sp. nov. C on a Fijian reef. Hydrobiologia 518:105-110

Bak RPM (1990) Patterns of echinoid bioerosion in two Pacific coral reef lagoons. Mar Ecol Prog Ser 66:267-272

Bak RPM (1994) Sea urchin bioerosion on coral reefs: place in the carbonate budget and relevant variables. Coral Reefs 13:99-103

Bertram GL (1936) Some aspects of the breakdown of coral at Ghardaqa, Red Sea. Proc Zool Soc London 106:1011-1026

Blainville HMD de (1825) Dictionnaire des Sciences Naturelles [ed not indicated]. Levrault, Strasbourg, 37:88 and 94-95 [not seen]

Brattström H (1980) Rocky-shore zonation in the Santa Marta area, Colombia. Sarsia 65:163-226

Bromley RG (1978) Bioerosion of Bermuda reefs. Palaeogeogr Palaeoclimatol Palaeoecol 18:169-197

Brown-Saracino J, Peckol P, Curran HA, Robbart M (2007) Spatial variation in sea urchins, fish predators, and bioerosion rates on coral reefs of Belize. Coral Reefs 26:71-78

Campbell AC, Dart JKG, Head SM, Ormond RFG (1973) The feeding activity of *Echinostrephus molaris* (de Blainville) in the central Red Sea. Mar Behav Physiol 2:155-169

Carreiro-Silva M, McClanahan TR (2001) Echinoid bioerosion and herbivory on Kenyan coral reefs: the role of protection from fishing. J Exper Mar Biol Ecol 262:133-153

Chartock MA (1983) Habitat and feeding observations on species of *Ophiocoma* (Ophiocomidae) at Enewetak. Micronesia 19:131-149

Chazottes V, Chevillotte V, Dufresne A (2004) Caractérisation de la production particulaire par l'oursin *Echinometra mathaei* sur les résifs Indo-Pacifiques: influence des peuplements coralliens et implications sur la dynamique sédimentaire. Geobios 37:13-22

Dart JKG (1972) Echinoids, algal lawn and coral recolonization. Nature 239:50-51

Ekdale AA, Bromley RG, Pemberton SG (1984) Ichnology. Trace fossils in sedimentology and stratigraphy. SEPM Short Course 15, 317 pp

Focke JW (1977) The effect of a potentially reef-building vermetid-coralline algal community on an eroding limestone coast, Curaçao, Netherlands Antilles; a preliminary report. Proc 3rd Int Coral Reef Symp, Miami, Florida 1:239-255

Grabowsky GL (1994) Symmetry, locomotion, and the evolution of an anterior end: a lesson from sea urchins. Evolution 48:1130-1146

Griffin SP, Garcia RP, Weil E (2003) Bioerosion in coral reef communities in southwest Puerto Rico by the sea urchin *Echinometra viridis*. Mar Biol 143:79-84

Grünbaum H, Bergman G, Abbott DP, Ogden JC (1978) Intraspecific agonistic behavior in the rock-boring sea urchin *Echinometra lucunter* (L.) (Echinodermata: Echinoidea). Bull Mar Sci 28:181-188

Hart LJ, Chia F-S (1990) Effect of food supply and body size on the foraging behavior of the burrowing sea urchin *Echinometra mathaei* (de Blainville). J Exper Mar Biol Ecol 135:99-108

Hendler G, Pawson DL (2000) Echinoderms of the Rhomboidal Cays, Belize: Biodiversity, distribution, and ecology. Atoll Res Bull 466-480:275-299

Highsmith RC (1980) Passive colonization and asexual colony multiplication in the massive coral *Porites lutea* Milne Edwards & Haime. J Exper Mar Biol Ecol 47:55-67

Hoskin CM, Reed JK, Mook DH (1986) Production and off-bank transport of carbonate sediment, Black Rock, southwest Little Bahama Bank. Mar Geol 73:125-144

Huettel M, Wild C, Gonelli S (2006) Mucus trap in coral reefs: formation and temporal evolution of particle aggregates caused by coral mucus. Mar Ecol Prog Ser 307:69-84

Hunt M (1969) A preliminary investigation of the habits and habitat of the rock-boring urchin *Echinometra lucunter* near Devonshire Bay, Bermuda. In: Ginsburg RN, Garret P (eds) Reports of research 1968. Seminar on organism - sediment interrelationships. Bermuda Biol Stn Res Spec Publ 2:35-40

Jones A, Sefton N (1979) Marine life of the Caribbean. Macmillan Educ, London, 90 pp

Kaye CA (1959) Shoreline features and Quaternary shoreline changes Puerto Rico. U S Geol Surv Prof Pap 317-B, 140 pp

Khamala CPM (1971) Ecology of *Echinometra mathaei* (Echinoidea, Echinodermata) at Diani Beach, Kenya. Mar Biol 11:167-172

Kinjo S, Shirayama Y, Wada H (2004) Phylogenetic relationships and morphological diversity in the family Echinometridae (Echinoida, Echinodermata). In: Heinzeller T, Nebelsick J H (eds) Echinoderms: München. Taylor & Francis, London, pp 527-530

Lamarck J-B de (1816) Histoire naturelle des animaux sans vertèbres, présentant les caractères généraux et particuliers de ces animaux, leur distribution, leurs classes, leurs familles, leurs genres, et la citation des principales espèces qui s'y rapportent; précédée d'une introduction offrant la détermination des caractères essentiels de l'animal, sa distinction du végétal et des autres corps naturels, enfin, l'exposition des principes fondamentaux de la zoologie. Verdière, Paris, 3:1-586 [not seen]

Lawrence JM (1970) *Echinostrephus molaris* (Blainville) (Echinodermata, Echinoidea) in the Gulf of Elat. Israel J Zool 19:175-176

Linnaeus C (1758) Systema naturae per regna tria naturae, secundum classes, ordines, genera, species, cum characteribus, differentiis, synonymis, locis. Tomus I. Editio decima, reformata. Laurentii Salvii, Holmiae, 824 pp

Loenhoud PJ van, Sande JCPM van de (1977) Rocky shore zonation in Aruba and Curaçao (Netherlands Antilles), with the introduction of a new general scheme of zonation. I, II. Mar Ecol C 80:437-474

Mägdefrau K (1932) Über einige Bohrgänge aus dem Unteren Muschelkalk von Jena. Paläont Z 14:150-160

Märkel K, Maier R (1967) Beobachtungen an lochbewohnenden Seeigeln. Natur Museum 97:233-243

Martinell J (1981) Actividad erosiva de *Paracentrotus lividus* (Lmk.) (Echinodermata, Echinoidea) en el litoral gerundense. Oecol Aquat 5:219-225

McCartney MA, Keller G, Lessios HA (2000) Dispersal barriers in tropical oceans and speciation in Atlantic and eastern Pacific sea urchins of the genus *Echinometra*. Molec Ecol 9:1391-1400

McClanahan TR (1988) Coexistence in a sea urchin guild and its implications to coral reef diversity and degradation. Oecologia 77:210-218

McClanahan TR, Kurtis JD (1991) Population regulation of the rock-boring sea urchin *Echinometra mathaei* (de Blainville). J Exper Mar Biol Ecol 147:121-146

McGehee A (1992) Distribution and abundance of two species of *Echinometra* (Echinoidea on coral reefs near Puert Rico). Caribbean J Sci 28:173-183

McLean RF (1967) Erosion of burrows in beachrock by the tropical sea urchin, *Echinometra lucunter*. Canad J Zool 45:586-588

Mills SC, Peyrot-Clausade M, Fontaine MF (2000) Ingestion and transformation of algal turf by *Echinometra mathaei* on Tiahura fringing reef (French Polynesia). J Exper Mar Biol Ecol 254:71-84

Mikuláš R (1992) Early Cretaceous borings from Stramberk (Czechoslovakia). Casopis pro Mineralogii a Geologii, Roč 37:297-312

Mortensen T (1943) A monograph of the Echinoidea 3(3). Camarodonta II. Echinidae, Strongylocentrotidae, Paraseleniidae, Echinometridae. Reitzel, Copenhagen, 446 pp

Neumann AC (1966) Observations on coastal erosion in Bermuda and measurements of the boring rate of the sponge *Cliona lampa*. Limnol Oceanogr 11:92-108

Neumann AC (1968) Biological erosion of limestone coasts. In: Fairbridge RW (ed) Encyclopedia of geomorphology. Book Corp, New York, 75-81 pp

Oak T, Scheibling RE (2006) Tidal activity pattern and feeding behaviour of the ophiuroid *Ophiocoma scolopendrina* on a Kenyan reef flat. Coral Reefs 25:213-222

Ogden JC (1977) Carbonate-sediment production by parrot fish and sea urchins on Caribbean reefs. SEPM Stud Geol 4:281-288

Ogden NB, Ogden JC, Abbott IA (1989) Distribution, abundance and food of sea urchins on a leeward Hawaiian reef. Bull Mar Sci 45:539-549

Ormond RFG, Campbell AC (1971) Observations on *Acanthaster planci* and other coral reef echinoderms in the Sudanese Reed Sea. Symp Zool Soc London 28:433-454

Otter GW (1932) Rock-burrowing echinoids. Biol Rev 7:89-107

Otter GW (1937) Rock-destroying organisms in relation to coral reefs. Sci Rep Great Barrier Reef Exp 1928-29, Brit Mus Nat Hist, London, 1(12):323-352

Palumbi SR, Metz EC (1991) Strong reproductive isolation between closely related tropical sea urchins (genus *Echinometra)*. Molecular Biol Evol 8:227-239

Prince J (1995) Limited effects of the sea urchin *Echinometra mathaei* (de Blainville) on the recruitment of benthic algae and macroinvertebrates into intertidal rock platforms at Rottnest Island, Western Australia. J Exper Mar Biol Ecol 186:237-258

Russo AR (1977) Water flow and the distribution and abundance of echinoids (Genus *Echinometra*) on an Hawaiian reef. Austral J Mar Freshwater Res 28:693-702

Russo AR (1980) Bioerosion by two rock boring echinoids (*Echinometra mathaei* and *Echinostrephus aciculatus*) on Enewetak Atoll, Marshall Islands. J Mar Res 38:99-110

Shulman MJ (1990) Aggression among sea urchins on Caribbean coral reefs. J Exper Mar Biol Ecol 140:197-207

Steneck RS, Adey WH (1976) The role of environment in control of morphology in *Lithophyllum congestum* a Caribbean algal ridge builder. Bot Marina 19:197-215

Tsuchiya M, Nishihira M (1984) Distribution of two types of the sea-urchin *Echinometra mathaei* (Blainville) on Okinawan reef flat. Galaxea 3:131-143

Wild C, Rasheed M, Werner U, Franke U, Johnstone R, Huettel M (2004) Degradation and mineralization of coral mucus in reef environments. Mar Ecol Prog Ser 267:159-171

Wild C, Woyt H, Huettel M (2005) Influence of coral mucus on nutrient fluxes in carbonate sands. Mar Ecol Prog Ser 287:87-98

III

Symbiotic interactions

Boring a mobile domicile: an alternative to the conchicolous life habit

Christian Neumann[1], Max Wisshak[2], Richard G. Bromley[3]

[1] Museum für Naturkunde, Humboldt University, 10115 Berlin, Germany, (christian.neumann@museum.hu-berlin.de)
[2] Institute of Palaeontology, Erlangen University, 91054 Erlangen, Germany
[3] Department of Geography and Geology, University of Copenhagen, 1350 Copenhagen K, Denmark

Abstract. Conspicuous borings, belonging to the new ichnotaxon *Trypanites mobilis* isp. n., recorded in the bulbous spines of the echinoid *Tylocidaris* and in the spherical calcareous sponge *Porosphaera globularis*, are common in Late Cretaceous (Cenomanian) to Early Paleocene (Danian) strata of Central and NW Europe. It is suggested that the borings were produced post mortem by sipunculan worms to create a lightweight and mobile domicile, offering effective shelter and protection. This strategy, which is hitherto not known from any other animal group, represents an alternative to the conchicolous life habit. Both habits evolved almost simultaneously in sipunculans at the end of the Early Cretaceous, suggesting escalation as a response to increased predation. Moreover, it bears a number of advantages such as superior substrate availability, the avoidance of replacing the substrate during ontogeny, as well as comparatively lightweight domiciles. On the other hand, the high degree of dependence on only very few suitable specific host taxa is a major drawback of this strategy. It ultimately led to the disappearance of this mode of life in favour of the still widespread conchicolous habit at the end of the Danian when the hosts vanished together with the chalk-sea ecosystem.

Keywords. Sipuncula, *Trypanites mobilis* isp. n., Echinoidea, Porifera, bioerosion, conchicolous habit, palaeoecology

Introduction

The predator-avoidance strategy of using discarded shells for shelter after the original builder has died is known as the conchicolous habit (Vermeij 1987). It occurs in various animal groups and dates back to the Early Cambrian (Unal and Zinsmeister 2006). However, in most groups the conchicolous habit evolved not before the Mesozoic. Hermit crabs (Paguridae) for instance, the most well-known among all conchicoles, appeared in the Early Jurassic and first used empty ammonite shells for shelter (Jagt et al. 2006). Another group of which the conchicolous habit is known are the sipunculans (peanut worms), a small phylum

M. Wisshak, L. Tapanila (eds.), *Current Developments in Bioerosion*. Erlangen Earth Conference Series, doi: 10.1007/978-3-540-77598-0_16, © Springer-Verlag Berlin Heidelberg 2008

of non-segmented coelomate worms that have existed since the Cambrian (Huang et al. 2004). Today, the sipunculans number slightly more than 150 species, which are widespread throughout the oceans from intertidal to abyssal depths. Sipunculans are either suspension feeders, surface browsers or deposit feeders and dwell in cryptic habitats (burrows in sand or mud), in rock crevices, or they bore in calcareous rocks where they form a significant component of the bioeroder communities (Rice 1970; Cutler 1994; Hutchings and Peyrot-Clausade 2002). Conchicolous habit is known from three extant genera (*Aspidosiphon* Diesing, 1851, *Phascolion* Théel, 1875 and *Golfingia* Lankester, 1885; see Vermeij 1987), which utilise either discarded gastropod or scaphopod shells. Conchicolous habit in sipunculans often occurs together with a symbiotic relationship with scleractinian corals of the genus *Heterocyathus* Milne-Edwards and Haime, 1848 or *Heteropsammia* Milne-Edwards and Haime, 1848, which overgrow the inhabited gastropod shell (Schindewolf 1959; Yonge 1974). The overgrowth by the coral may completely engulf the gastropod shell, but the sipunculan maintains an open canal by processes of bioclaustration, and may even increase in size as its substrate accretes (Scheidegger 1972; Arnaud and Thomassin 1976; Kohn and Arua 1999). This kind of symbiosis dates back to the Albian and suggests that the conchicolous life habit in sipunculans appeared at least by the end of the Early Cretaceous (Stolarski et al. 2001).

Here we report on bored psychocidarid echinoid spines and calcareous sponges, which co-occur in Late Cretaceous to Early Paleocene soft-bottom sediments of NW Europe. In 1900, Grönwall described spines of the psychocidarid echinoid *Tylocidaris vexillifera* (Schlüter, 1892) from the Danian of Denmark, which showed large, conspicuous borings. Grönwall avoided giving an interpretation of how the borings were formed but clearly stated their post mortem origin. His paper, however, subsequently became a classic example of fossil evidence of parasitism (e.g., Mortensen 1928; Tasnádi-Kubacska 1962; Boucot 1990), suggesting a syn vivo symbiotic relationship between the borer and the echinoid. Our study supports Grönwalls statement that the borings were produced after the echinoid died when the spines already had fallen off the echinoid's test. Comparison of the borings with those found in the co-occurring sponge *Porosphaera globularis* (Phillips, 1829) suggests that they were made by the same trace maker. Bored *P. globularis* skeletons were used by humans as pearls since the Palaeolithic (Bednarik 2005; but see Rigaud 2007) and thus represent the world's oldest known beads. Borings in *P. globularis* skeletons have been studied earlier by Nestler (1961) and Müller (1970).

Porosphaera globularis (Calcarea; Minchinellidae) is a roughly spherical sponge ranging from 1 to >35 mm. It is widely distributed in (but not restricted to) the boreal Late Cretaceous and first occurs in the Turonian (*Inoceramus labiatus* Zone; Wood 2002). It is most common in the chalk facies, e.g., it is name-giving to the '*Porosphaera* beds' of the Early Maastrichtian (*Belemnella lanceolata* Zone) of Norfolk. Less commonly, it occurs in other facies types such as marls and bioclastic limestone. Borings in *Porosphaera* have also been observed from chalks of the northern Tethys margin (Maastrichtian of northern Bulgaria, personal observations). In the Danian, *P. globularis* is replaced by the much smaller *P. adhaerens* Brünnich-Nielsen, 1926.

The psychocidarid genus *Tylocidaris* Pomel, 1883 is an echinoid with bulbous to club-shaped spines. It first appeared in the Albian and is widely distributed throughout the Late Cretaceous and Palaeogene. Only one Recent psychocidarid species (*Psychocidaris oshimai* Ikeda, 1935) is known from deeper waters of the tropical Indo-Pacific.

Our study suggests that the boreholes, assigned to the trace fossil *Trypanites* Mägdefrau, 1932, were produced by sipunculan worms for use as mobile shelters. We suggest that the bored spines and sponges served as domiciles and therefore represent an alternative to the conchicolous life habit which originated simultaneously in sipunculans during the mid-Cretaceous.

Material and methods

For a comprehensive study of the borings an extensive survey of museum collections was carried out. It turned out that borings were restricted to calcareous sponges of the genus *Porosphaera* Steinmann, 1878 and to the bulbous spines of the psychocidarid echinoid *Tylocidaris*. Bored *Porosphaera* and *Tylocidaris* spines were widely distributed in the Cretaceous and Paleocene of Europe and our study focuses on occurrences in the North Sea Basin (N-German and Danish subbasins). The material yielding positive findings is listed in Table 1. Additional echinoid taxa possessing comparable spine morphologies but lacking similar borings are listed in Table 2.

The richest material used for our study stems from *Porosphaera globularis* from the Early Maastrichtian of Rügen, N Germany and from spines of *Tylocidaris baltica* (Late Maastrichtian) and *Tylocidaris vexillifera* (Late Danian) of Denmark. These were consequently chosen for statistical treatment and population analyses.

Metric measurements (length and maximum diameter of the spines, the maximum diameter of the sponges, maximum diameter and length of the borings) were taken with the aid of precision sliding callipers to an accuracy of 0.1 mm. The angle of the mean axis of the borings in relation to that of the median axis of the spines was determined with a setsquare. Basic statistical analyses were carried out with Microsoft EXCEL.

Digital imaging of the echinoid spines and sponges was undertaken after coating with ammonium chloride. For the study of the traces, the spines were embedded under vacuum in epoxy resin (Araldite BY158 + Aradur 21), subsequently cut alongside the boring, and the skeletal matrix of the spines removed with diluted hydrochloric acid (~5%). The resulting casts of the spines were then carefully ground down to the median line with the aid of a hand-held precision drilling machine in order to allow the best view of the borings. The casts were photographed after coating with ammonium chloride and by SEM after sputter coating with gold.

Institutional abbreviations:
MNHB = Museum für Naturkunde, Humboldt University Berlin, Germany
MGUH = Geological Museum, Copenhagen University, Denmark

Table 1 Number, taxa, localities and proposed palaeoenvironment of the investigated bored *Porosphaera* skeletons and *Tylocidaris* spines from the Cretaceous and Paleocene of Europe (* = only counted, not measured)

Species	Locality / stratigraphy	Environment (reference)	Bored specimen	Measured specimen
Tylocidaris sp.	Kassenberg / Germany Early Cenomanian	Coastal, carbonatic (Scheer and Sottrop 1995)	2	2
Tylocidaris sorigneti	Essen / Germany Cenomanian	Neritic, siliciclastic sands (Frieg et al. 1990)	2	2
Tylocidaris aff. *clavigera*	Ahaus / Germany Turonian	Carbonate shoal (Ernst et al. 1998)	6	6
Tylocidaris gosae	Lengede / Germany Santonian	Neritic, siliciclastic mud (Beck 1920)	3	3
Tylocidaris cf. *clavigera*	Kent / England Turonian – Santonian	Nannoplankton ooze (Mortimore 1997)	1	1
Tylocidaris sp.	Tempelhof / Germany Late Cretaceous	(no reference)	1	1
Tylocidaris baltica	Stevns Klint / Denmark Late Maastrichtian	Nannoplankton ooze (Surlyk et al. 2006)	43	43
Tylocidaris baltica	Dania, Mariager Fjord / DK Late Maastrichtian	Nannoplankton ooze (Thomsen 1995)	1	1
Tylocidaris baltica	Kongsdal, Hobro / Denmark Late Maastrichtian	Nannoplankton ooze (Thomsen 1995)	1	1
Tylocidaris sp.	Lundergaard / Denmark Late Danian	Nannoplankton ooze (Thomsen 1995)	1	1
Tylocidaris vexillifera	Kiel / Germany Late Danian	Glacial erratics (no reference)	2	2
Tylocidaris vexillifera	Faxe / Denmark Late Danian	Bryzoan bioherms (Bernecker and Weidlich 1990)	9	9
Tylocidaris vexillifera	Herfølge / Denmark Late Danian	Nannoplankton ooze (Thomsen 1995)	356*	650*
			total: 72	**total: 72**
Porosphaera globularis	Misburg / Germany Late Campanian	Marly calcisphere ooze (Ernst 1963)	0	4
Porosphaera globularis	Lüneburg / Germany Late Campanian	Nannoplankton ooze (Heinz 1926)	4	7
Porosphaera globularis	Rügen / Germany Early Maastrichtian	Nannoplankton ooze (Reich and Frenzel 2002)	67	722
Porosphaera globularis	Hemmoor / Germany Late Maastrichtian	Nannoplankton ooze (Schmid et al. 2004)	1	1
Porosphaera globularis	Maastricht / Netherlands Late Maastrichtian	Carbonate shoals (Felder 1996)	10	54
			total: 82	**total: 788**

Table 2 Echinoid species having glandiform spines but no *Trypanites* borings (n = number of measured spines)

Species	Stratigraphy	Locality	n
Caenocidaris cucumifera (Agassiz, 1840)	Bajocian	Buxieres / France	115
Pseudocidaris mammosa (Agassiz, 1840)	Kimmeridgian	Ile de Ré / France	103
Cidaris cherennensis Savin, 1905	Hauterivian	St. Pierre de Cherenne / France	78
Tylocidaris strombecki (Desor, 1858)	Cenomanian	Tarmove / Bulgaria	34
Balanocidaris glandaria (Fraas, 1878)	Cenomanian	Tripoli / Lebanon	11
Tylocidaris squamifera (Schlüter, 1897)	Campanian	Ignaberga / Sweden	197
Tylocidaris abildgaadi Ravn, 1928	Lower Danian	Store Klint / Denmark	133
Tylocidaris bruennichi Brotzen, 1959	Middle Danian	Faxe / Denmark	57
Tylocidaris oedumi Brünnich-Nielsen, 1938	Middle Danian	Stevns Klint / Denmark	38

Results

Systematic ichnology

Ichnogenus *Trypanites* Mägdefrau, 1932

Diagnosis: Single-entrance, cylindrical, unbranched borings in lithic substrates, having circular cross-sections throughout length. The axes of the borings may be straight, curved or irregular (modified after emended diagnosis in Bromley and D'Alessandro 1987).

Trypanites mobilis isp. n.

Figs. 1-5

1900 Borrade ekinidtaggar.- Grönwall, 33-36, figs. 1-9
1928 Spines of *Tylocidaris vexillifera* Schlüter, attacked by a boring organism.- Mortensen, 40-42, fig. 26, [figure reproduced from Grönwall 1900]
1961 Angebohrte *Porosphaera globularis*.- Nestler, 39-42, table 10/2
1962 Angeschwollene Stacheln des Seeigels *Stereocidaris indica*.- Tasnádi-Kubacska, 94-100, fig. 128, [figure reproduced from Grönwall 1900]
1964 Seeigelstachel von *Tylocidaris*, mit Öffnung zu dem Lebensraum eines Einwohners.- Gripp, 70, table 11/4, [non table 11/5]
1970 *Porosphaera globularis* mit Perforationen.- Müller, 708-720, fig. 1, table 1-2
1990 Deformed and swollen *Stereocidaris indica* spines.- Boucot, 75-76, fig. 66, [figure reproduced from Grönwall 1900]
1993 *Porosphaera globularis* mit Durchbohrungen.- Müller, 184-185, fig. 176
2005 *Porosphaera globularis*.- Bednarik, 540-542, figs. 3-4

Diagnosis: Short and wide *Trypanites* having a cylindrical or slightly curved form, emplaced within but never completely penetrating a subspherical bioclast (calcareous sponge or psychocidarid echinoid radiole).

Differential diagnosis: T. mobilis isp. n. is distinguished from all other *Trypanites* ichnospecies by its diagnostic and exclusive occurrence in mobile sub-spherical bioclasts. The type ichnospecies *T. weisei* Mägdefrau, 1932 is mostly smaller in size and more slender. *T. fimbriatus* (Stephenson, 1952) is similar to the present traces except for its hardground occurrence and a slightly club-shaped morphology. In contrast to all other *Trypanites* ichnospecies *T. solitarius* (Hagenow, 1840) is oriented (sub-)horizontally in respect to the substrate surface and may show sinuosity. Like *T. weisei*, *T. fosteryeomani* Cole and Palmer, 1999 is more slender.

Etymology: mobilis, Latin = movable

Type material, locality and horizon: The holotype is a trace in a cast *Tylocidaris baltica* spine from the Late Maastrichtian (*casimirovensis*-Zone) of Stevns Klint, Denmark (Figs. 1A, 3; MGUH 28417). Paratypes comprise all other specimens, both in *Tylocidaris* spines as well as *Porosphaera* sponges, depicted on Figures 1 through 5 except for 2C (MGUH 28418 to MGUH 28423; MNHB MB.E 5717, 5719, 5720, 5721.1, 5721.6, 6202; MNHB MB.PO 2532.1 to 2532.8, 2533).

Description: The borings in question are simple vertical to obliquely oriented tubes, perfectly circular in cross-section, of near-constant diameter throughout their length with a rounded blunt termination. The unlined tube walls are rough and reflect the microstructure of the porous host substrate.

The bored *Tylocidaris* spines (Figs. 1-3) measure 8.0 to 39.7 mm in length (mean = 22.3 ± 7.6 mm) and 3.9 to 20.0 mm in maximum width (mean = 11.4 ± 4.0 mm). The *Trypanites* borings measure 1.6 to 7.9 mm in diameter (mean = 4.0 ± 1.4 mm) and a length of 0.8 to 26.5 mm (mean = 12.3 ± 6.7 mm). Most traces enter the spines at the apex (71% within 30° of the length axis of the spines; Figs. 1A-B, 3A, 4A-C), some at the side (26% between 30° and 150°; Fig. 2A, D) and few from or near the shaft (3% within 150° and 180° from the apex; Figs. 2B, E, 4D). The traces are straight where they are entering the spines at the very apex or shaft. The further the boring is located off the length axis, the more they may be bent towards it. In most specimens, the boring is almost penetrating right through the substrate, leaving only a very thin wall of substrate (Figs. 1A-D, 3A, 4E), which is commonly eroded or broken open (Figs. 1A, 2A). Only 2 out of 66 and respectively 4 out of the total of 422 investigated bored spines exhibited two borings (Fig. 2A). Few initial traces have been observed, which are much wider than deep (Fig. 2D).

In *Porosphaera* (Fig. 5), the borings measure 1.0 to 6.9 mm in diameter (mean = 3.0 ± 1.2 mm) at a length of 0.5 to 28.9 mm (mean = 7.4 ± 4.4 mm) and are always straight. As in the bored *Tylocidaris* spines, the tunnels almost penetrate the substrate, unless opened at their termination by erosion (Fig. 5G). No sponges were encountered with multiple borings. Few initial traces were observed (Fig. 5A-C).

Provenance and stratigraphic range: Early Cenomanian to Danian sediments of Europe.

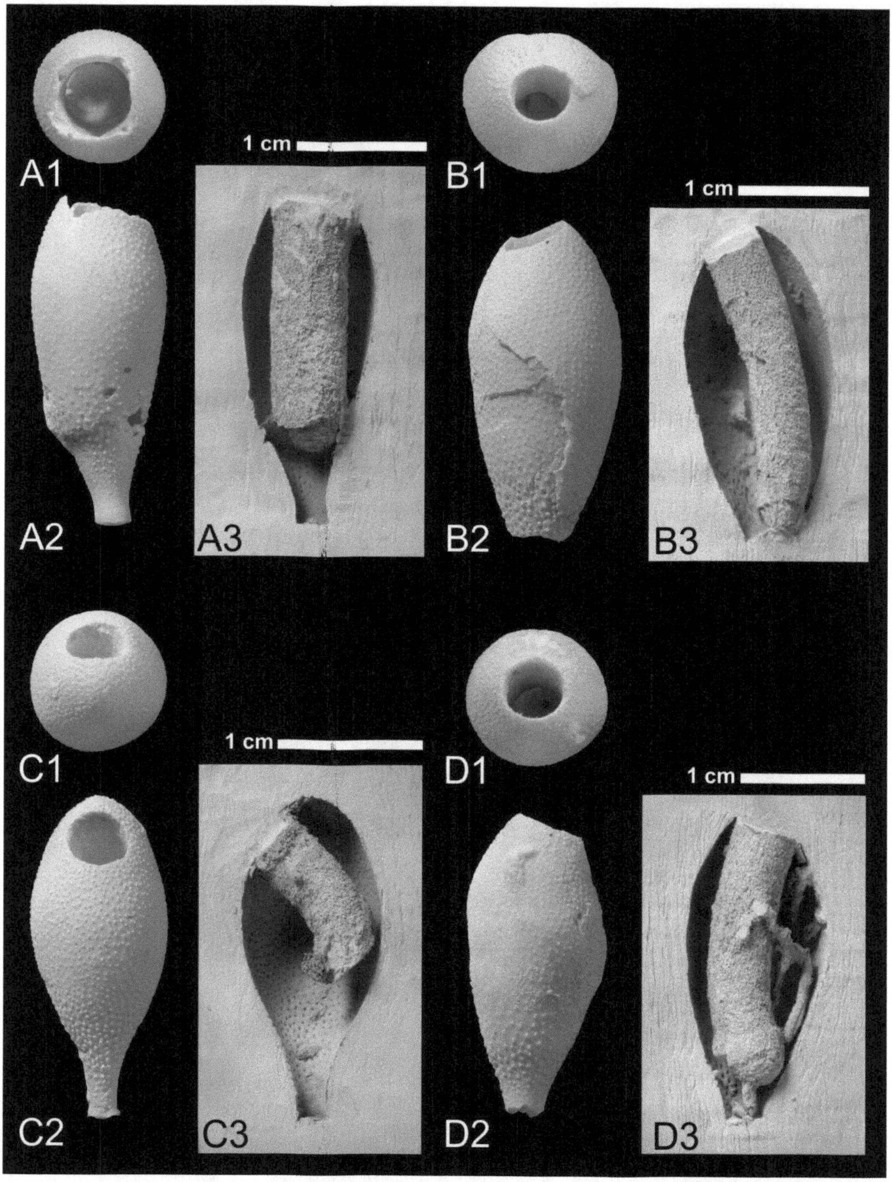

Fig. 1 *Tylocidaris baltica* echinoid spines from the Late Maastrichtian of Stevns Klint, Denmark, with *Trypanites mobilis* isp. n. borings, each in apical view, lateral view, and lateral view of epoxy resin cast. **A** Holotype of *Trypanites mobilis* isp. n. entering the spine at the tip (MGUH 28417). **B** Slightly bent specimen entering the substrate slightly off the tip (MGUH 28418). **C** Moderately bent specimen entering the spine further off the tip (MGUH 28419). **D** Slightly bent trace in association with a *Talpina* boring system (MGUH 28420)

Fig. 2 *Tylocidaris baltica* echinoid spines from the Late Maastrichtian of Stevns Klint, Denmark, with *Trypanites mobilis* isp. n. borings. **A** One of the only two specimens featuring two borings (MGUH 28421). **B** Boring entering the spine from near the shaft (MGUH 28422). **C** Bored spine with pronounced predation traces (bite traces) (Alice Rasmussen collection, no inventory number). **D** Spine with very initial boring (MNHB-MB.E6202). **E** Spine with boring at the shaft removing the latter (MGUH 28423)

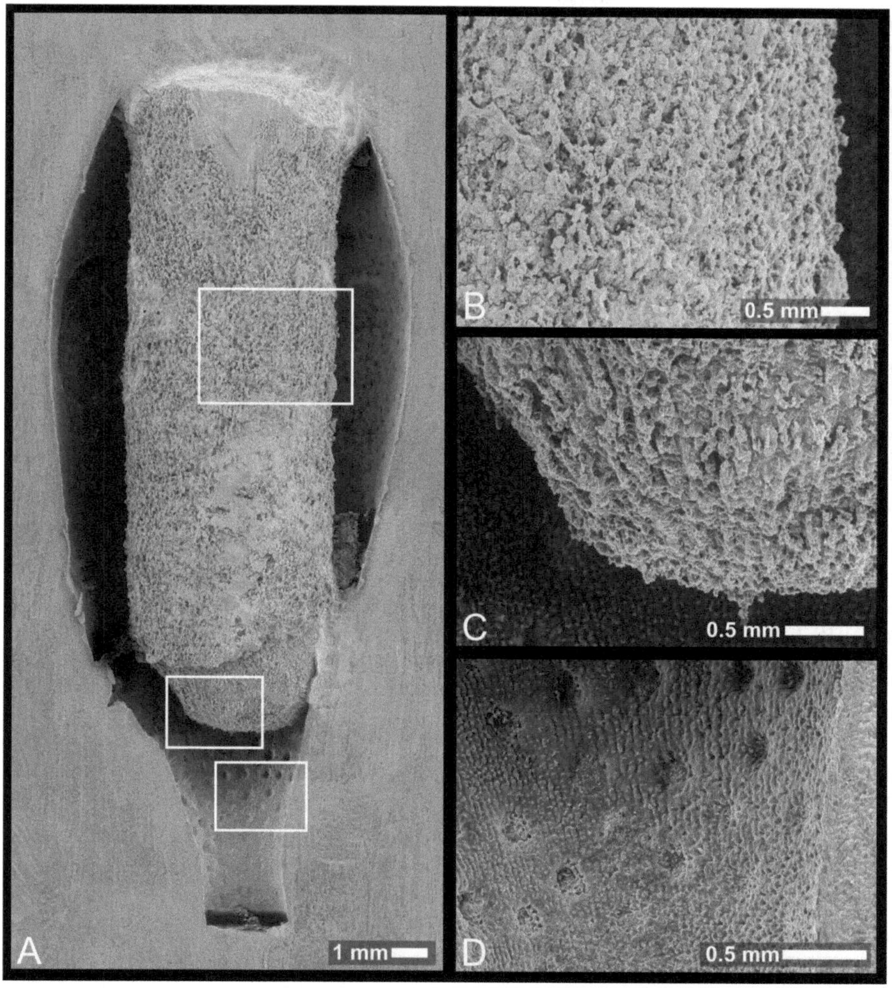

Fig. 3 A SEM image of an epoxy resin cast of a *Tylocidaris baltica* echinoid spine from the Late Maastrichtian of Stevns Klint, Denmark with the holotype of *Trypanites mobilis* isp. n. (same specimen as Fig. 1A; MGUH 28417). **B** Close-up of wall structure. **C** Close-up of wall structure with xenoglyphic ridges. **D** Pristine outer surface of the echinoid spine lacking microborings

Population statistics

In order to evaluate population characteristics of bored and unaffected *Porosphaera globularis*, a random sample of 722 specimens from one locality (Rügen, Germany), among which, 67 (= 9.3%) exhibited the boring, were measured and statistically analysed (Fig. 6). The diameter of all sponges ranges from 1.4 to 31.8 mm with a mean of 7.5 ± 3.8 mm. Among these, the bored specimens range from 4.3 to 24.9 mm in diameter with a mean of 10.2 ± 3.7 mm. The size-class frequency distribution (Fig. 6A) shows a positively skewed Gaussian distribution. The relative

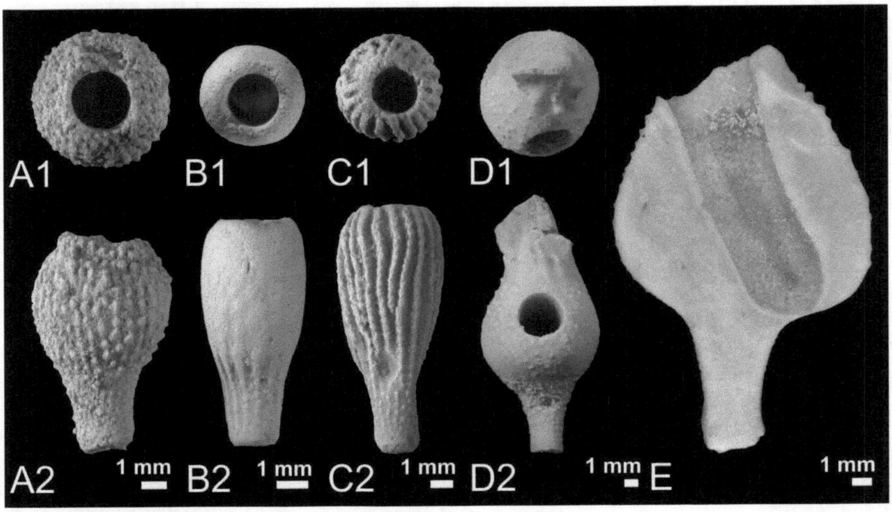

Fig. 4 **A** *Tylocidaris sorigneti* from the Cenomanian of Essen, Germany, with *Trypanites mobilis* isp. n. boring from the tip (MNHB-MB.E 5719). **B** *Tylocidaris* sp. from the Early Cenomanian of Kassenberg, Germany, with boring from the tip (MNHB-MB.E 5717). **C** *Tylocidaris gosae* from the Santonian of Lengende, Germany, with boring from the tip (MNHB-MB.E 5720). **D** Lateral boring in *Tylocidaris vexillifera* from the Late Danian of Faxe, Denmark (MNHB-MB.E 5721.6). **E** Sectioned *Tylocidaris vexillifera* from the Late Danian of Faxe, Denmark, with slightly bent tunnel almost penetrating the spine (MNHB-MB.E 5721.1)

percentage of bored specimens per size class is increasing with size, whereas the highest total number of bored sponges is found in the 6-8, 8-10 and 10-12 mm size classes (Fig. 6A). The diameter-class frequency distribution of the borings also exhibits a positively skewed Gaussian distribution and the relative percentage of apparently penetrating borings increases with larger diameters (Fig. 6B). The diameter of the bored sponges versus that of the borings (Fig. 6C) shows a moderate positive correlation ($r^2 = 0.64$) with a mean ratio of 3.5 ± 1.3 (min. 2.3, max. 6.3). The boring length versus diameter also shows a moderate positive correlation ($r^2 = 0.61$) with a ratio of 2.5 ± 0.9 (min. 0.3, max. 4.3).

In the case of *Tylocidaris*, only bored specimens were statistically evaluated, as there were 37 bored *Tylocidaris baltica* spines from the Late Maastrichtian of Stevns Klint, Denmark.

Only part of the spines were complete with a mean length of 27.4 ± 4.9 mm (min. 21.6, max. 39.7) but the spine body was complete in most cases with a mean length of 23.6 ± 4.6 mm (min.15.0, max. 34.2). The maximum diameter of the bored spines ranges from 9.9 to 20.0 mm with a mean of 13.9 ± 2.9 mm. Only one out of the 37 spines contained more than one boring and in 15 of the 37 spines the boring showed apparent complete penetration. The mean boring diameter was 4.9 ± 1.1 mm (min. 2.1, max. 7.9) and the boring depth measured a mean 16.2 ± 5.5 mm (min. 0.2, max. 6.3). With the smaller number of samples, the correlation between spine diameter and boring diameter (Fig. 6E) was weaker if compared to the sponge

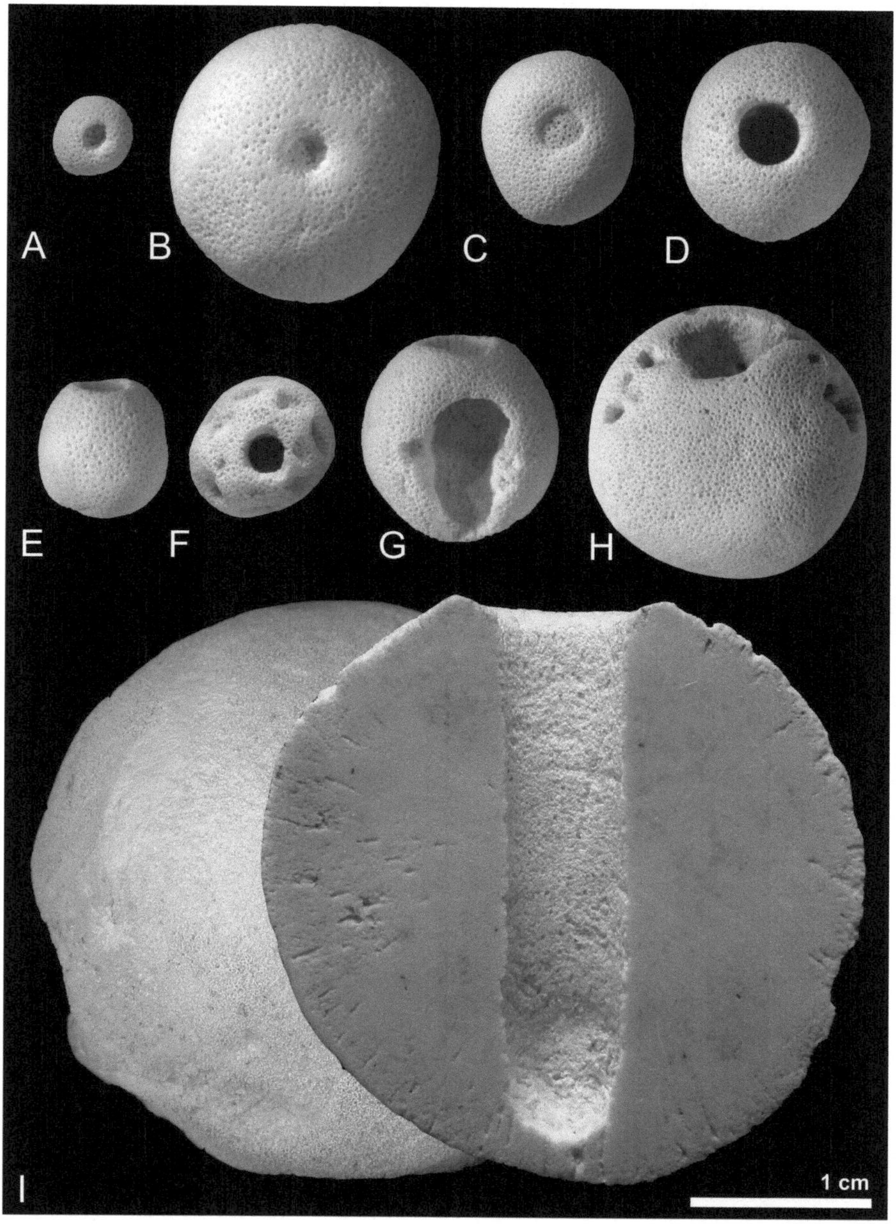

Fig. 5 *Porosphaera globularis* with *Trypanites mobilis* isp. n. from the Early Maastrichtian of Rügen, Germany (A-H: MNHB-MB.PO 2533) and Hemmoor, Germany (I: MNHB-MB.PO 2532). **A-C** Initial borings. **D-E** Borings almost penetrating the sponge. **F** Bored sponge with pronounced predation (bite) traces. **G** Eroded sponge with apparently penetrating boring. **H** Large sponge with boring affected by moderate predation traces. **I** Very large sponge with large boring seen in the cross section

host ($r^2 = 0.43$) with a mean ratio of 2.9 ± 0.6 (min. 1.9, max. 4.8). The same applies to the ratio of boring length versus diameter (Fig. 6F; $r^2 = 0.26$) with a mean ratio of 3.4 ± 1.1 (min. 0.2, max. 6.3).

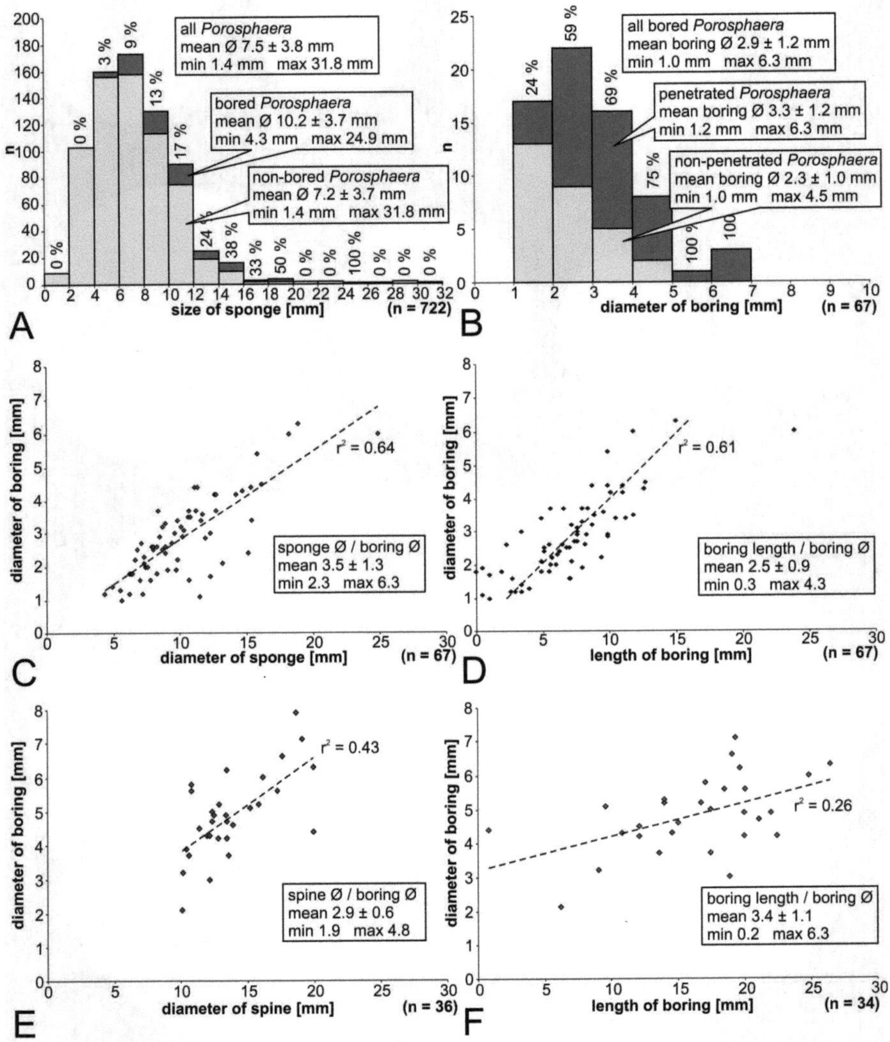

Fig. 6 Statistical population analysis of a random *Porosphaera globularis* sample containing 722 specimens from the Early Maastrichtian of Rügen, Germany (A-D) and 37 bored *Tylocidaris baltica* spines from the Late Maastrichtian of Stevns Klint, Denmark (E-F). **A** Size-class frequency histogram showing a positively skewed Gaussian distribution and the relative percentage of bored specimens per size class. **B** Diameter-class frequency distribution of the borings also showing a positively skewed Gaussian distribution and the relative percentage of apparently penetrating borings. **C** Moderate correlation of the diameter of the bored sponges versus that of the borings. **D** Moderate correlation of boring length versus diameter. **E** Weak correlation of the diameter of the spines versus that of the borings. **F** Weak correlation of boring length versus diameter

Temporal and spatial occurrence patterns

The oldest records are from the *Tylocidaris* sp. from the shallow-marine Early Cenomanian deposits of Kassenberg near Mühlheim (Westphalia) and from *Tylocidaris sorigneti* (Early Cenomanian of Essen). The fossil record is relatively scarce throughout the Late Cretaceous until the Maastrichtian. In the Maastrichtian, where the majority of investigated specimens have been obtained, both *Porosphaera* and spines of *Tylocidaris baltica* commonly contained the borings. A gap of the evidence for bioerosion is apparent in the Early and Middle Danian. In the Late Danian, the boring frequency is again high with more than half of the spines (54.7%, n = 650) of *Tylocidaris vexillifera* containing a boring.

Discussion

Sipunculans as the producer of *Trypanites mobilis* isp. n.

Müller (1970) suggested that the borings in *Porosphaera* were produced by sipunculans that utilised the sponges as domiciles or camouflage. Referring to modern conchicolous sipunculans living in symbiosis with the corals *Heteropsammia* and *Heterocyathus*, he argued that thick-shelled and heavyweight bivalves, brachiopods and echinoids were unsuitable domiciles for the sipunculans. Despite the lack of body fossil evidence (typical for most *Trypanites* occurrences), we herein follow the interpretation of the borings in question being the probable work of sipunculan worms and extend this interpretation also to the bored echinoid spines. This is in accordance with the general interpretation of *Trypanites* where sipunculans are commonly referred to as a potential trace maker (e.g., Bromley 1972, 1994). For the present material, this interpretation is particularly likely considering the close resemblance of the proposed conchicolous life style to recent ecological strategies of these worms (Fig. 7).

Little is known about the boring process of the sipunculans. Both mechanical and chemical mechanisms have been advocated (Rice and McIntyre 1972). As summarised by Williams and Margolis (1974), the presence of abrasive structures as hooks, spines and papillae on the introvert and body wall are clues for mechanical boring while the presence of glands in the epidermis and the restriction to predominantly calcareous substrates argue for a chemical mechanism.

Host-specific bored mobile shelters: benefits and constraints

There is no doubt that the borings have been produced after the death of the host organisms. Syn vivo infestation would be expected to stimulate growth responses such as regeneration, callus formation or embedment of the intruder. However, skeletal tissue response has never been observed, either in the stereom of the cidarid spines or in the skeleton of *Porosphaera*. Additional evidence comes from the orientation of the borings: in some cases, borings in *Tylocidaris baltica* spines have been initiated from the proximal limit of the spine, thereby removing the shaft (Fig. 4E1-E2). This would have not been possible while the spines were attached to the living host echinoid for obvious reasons. Thus, we support Grönwall's original

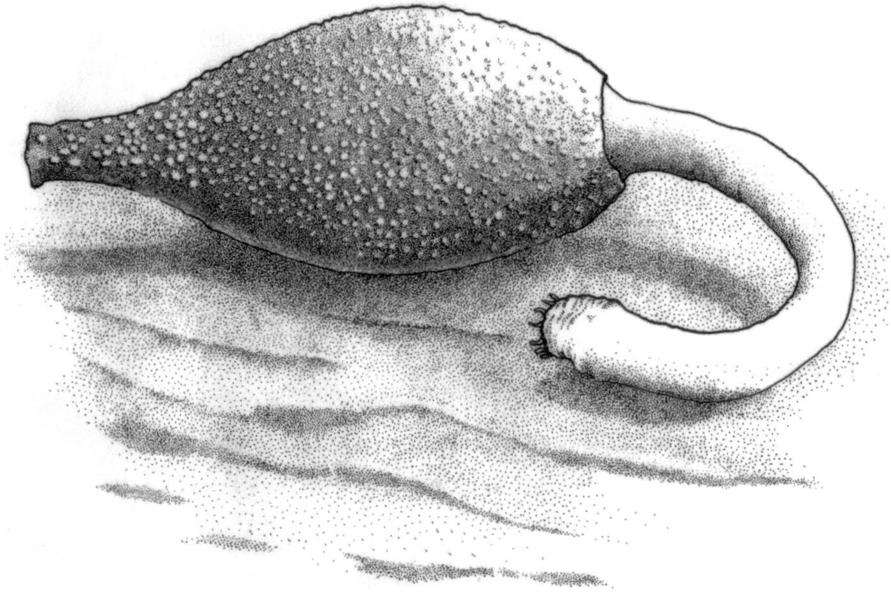

Fig. 7 Reconstruction of a living sipunculan worm using a bored *Tylocidaris baltica* spine as a domicile on a soft-bottom habitat

interpretation and disprove subsequent assumptions of various authors (Mortensen 1928; Tasnádi-Kubacska 1962; Boucot 1990) of a syn vivo parasitic relationship.

We suggest that the same tracemaker utilised both *Tylocidaris* spines and *Porosphaera globularis* skeletons as mobile shelters. The restriction of borings to these taxa shows that the borer was highly dependent on the supply of suitable specific host substrates. *Porosphaera globularis* and spines of *Tylocidaris* perfectly meet the requirements for a bored mobile domicile with many advantages if compared to the alternative conchicolous mode of life: within their geographical and stratigraphical range, both were highly abundant organisms, especially in the chalk facies. Thus, supply of suitable bioclasts was high, promoting the evolution of this boring strategy. In this respect, it is interesting to note that cidarid spines (in contrast to the spines of other echinoids) do not grow with the test but cidarids have to replace their spines from time to time by larger spines. They do so by shedding the spines with the help of calcite resorption along 'Prouho's membrane', a truncation plane across in the spine's shaft (Prouho 1887; Märkel and Röser 1983). Thus, cidarid echinoids continuously shed spines as substrates available for borers. The calcareous nature of the sponge skeleton and echinoid spines allows a chemical boring mechanism where etching is involved. Moreover, the borer benefited from skeletal structure of the chosen bioclasts since the porous sponge skeleton and the echinoid stereom meshwork make the substrate extremely lightweight but stable at the same time. An echinoid spine or sponge is lighter than an equally-sized gastropod shell of the same stability, representing a further advantage of this boring strategy if compared to the conchicolous habit since carrying around a heavyweight domicile is more energy-consuming. In contrast to utilising discarded gastropod shells, boring in cidarid spines bears another important advantage: Hylleberg (1974) reports that 90% out of 700 gastropod shells in a haul were inhabited by the sipunculan *Phascolion*, suggesting a high degree of intraspecific competition. In conchicolous sipunculans, the supply of empty shells is thus the principal factor in determining distribution. Vermeij (1987) points out that continuous supply of empty gastropod shells is a limiting factor for the distribution and abundance of conchicoles, that are thus highly dependent on gastropod predators or mortality events such as storms to guarantee sufficient supply of available domiciles. Since the aragonitic shells of gastropods are generally not preserved in the Late Cretaceous and Danian chalk, the abundance of potential gastropod host shells remains an unknown figure. Another factor is found when considering the ontogeny of the conchicolous sipunculans. Their syringe should exactly fit with the open shell space in contracted condition, requiring host substrate change with progressing growth. Thus, intraspecific competition and competition with other conchicoles for suitable shells may have been very high. Boring sipunculans in contrast construct their own perfectly fitting shelter and they can enlarge it as they grow. Hence, interspecific competition with other conchicoles would not have occurred.

The borer showed a preference for globular substrates. Irregularly or cushion-shaped species of *Porosphaera* (e.g., *P. patelliformis* Hinde, 1904 or *P. arreta* Hinde, 1904) or the long-shafted spines of *Tylocidaris clavigera* and other cidarids have been avoided. The echinoid spines have mostly been bored along the long axis. This strategy enables the longest boring and bored volume with respect to the spine shape. This behavioural pattern also infers that the sipunculan must have had a remarkable sense for recognising the shape of the host substrate and boring orientation. Also, it was able to detect the stage at which the boring almost completely penetrated the substrate in order to stop its boring activity. The preferred longitudinal orientation of the boring relative to substrate shape conveys another advantage in providing a better avoidance of smothering by sediment when the spine was moved by hydrodynamic forces. In an elongate but thick spine, the sipunculan worm was always kept above the sediment / water interface when the spine was rolling across the sea floor, whereas this was not favoured by slim spines or a lateral position of the boring. No such advantage was provided by the spherical sponges.

Evolutionary implications

Whereas many organisms in hard-substrate communities settle on the surface (epibionts), others bore into the substrate for protection (domichnia). A major drawback of this mode of life is immobility and thus, most of these organisms belong to the filter feeder guild. Deposit feeding, in contrast, requires mobility. This prerequisite can only be matched by specialised forms. Deposit feeders that lack a protective skeleton are generally more vulnerable to predation and thus mostly burrow beneath the sediment surface. Whereas most boring sipunculans have been restricted to hardgrounds, the capability of utilising discarded shells or to bore into mobile bioclasts gives access to soft sediment as a nutrient source and allows mobility which is essential for surface-feeding.

The ability of boring into calcareous bioclasts enabled certain sipunculans to create their own microhabitat, which offered protection from predators and allowed a mobile life habit on the surface of the sea floor. It is quite possible that these sipunculans also possessed an operculum-like anterior shield that plugged the opening of the hole and protected against predation as can be seen in many of the Recent boring species (Cutler 1994). However, did the bored domiciles consequently represent predator-free microhabitats? Recent sipunculans are well-documented prey items for turrid and muricid gastropods and teleost fishes. In fact, for many fish species, they form an important component of their diet and some even depend in significant ways on sipunculans as prey (Cutler 1994). In fact, the sheltered domiciles were occasionally also targeted by predators. Bored and subsequently crushed spines revealing tooth traces of fish predators have been observed in *Tylocidaris baltica* (n = 5, 12.8%) and *T. vexillifera* (n = 2, 0.5%; Fig. 2C), proving that durophagous fishes successfully preyed on sheltered sipunculans. Corresponding predation traces have also been observed in *Porosphaera globularis* (n = 2, 2.4%; Fig. 5F, H). However, the relatively low percentage of specimens showing signs of predation suggests an effective although not ironclad protection from predators. The appearance of spine- and subsequently sponge-boring behaviour in sipunculans in the Early Cenomanian, shortly after the conchicolous life habit appeared in the Late Albian (Stolarski et al. 2001) suggests escape strategies from increasing predatory pressure and supports Vermeij's (1977) escalation hypothesis of the Mesozoic Marine Revolution.

In the Late Cretaceous of the North Sea Basin, the record of bored echinoid spines and *Porosphaera* skeletons is relatively scarce until the Maastrichtian. Before this, the spines of relatively small *Tylocidaris* species (e.g., *T. sorigneti*, *T. gosae*) were the only available substrates. Boring frequency considerably increased during the Early Maastrichtian when *Porosphaera* became more abundant and in the Late Maastrichtian, when spines of *Tylocidaris baltica* were available. *T. baltica* possessed exceptionally large spines, attracting boring sipunculans. At the end of the Cretaceous, *T. baltica* became extinct and was replaced by other species with much smaller spines (*T. oedumi*, *T. abildgaadi* and *T. bruennichi,* see Wind 1954; Brotzen 1959; Gravesen 1993). Due to their smaller size and long shafts (in the case of *T. oedumi*), the spines probably did not match the requirements of the sipunculans. At the end of the Cretaceous period, *Porosphaera globularis* also became extinct and was replaced by much smaller Danian species such as *P. adhaerens* Brünnich-Nielsen, 1926 in the Danish Basin. The latter reached a maximum size of less than 10 mm and thus also was below the size preferred by the borer. This suggests an evolutionary trend of increasing size of the borer towards the Maastrichtian which parallels the size evolution of *Tylocidaris*. In the Late Danian, with *Tylocidaris vexillifera*, again a large-spined *Tylocidaris* species was available and boring rates again increased considerably to more than 50%.

Our observations suggest a shift from neritic towards hemipelagic environments. From the Cenomanian to the Santonian, all records are from shallow-marine deposits. From the Campanian towards the Late Danian, all positive records are from hemipelagic marl and chalk facies. For instance, in the shallow-marine

Campanian deposits of Ignaberga (Scania, Sweden), *Tylocidaris squamifera* is an abundant species, but borings have not been observed.

The high degree of dependence on only two host taxa supplying suitable substrates appears to represent the weakest point in the mobile-boring mode of life. When the chalk sedimentation stopped at the end of the Danian, the benthic ecosystems of the North Sea Basin witnessed a dramatic change. Both host taxa, *Porosphaera* and *Tylocidaris* disappeared and so did the mobile-boring life habit of the sipunculans (the only surviving extant member of the psychocidarid echinoids, *Psychocidaris oshimai* from the Indo-Pacific Ocean, has never been reported to be bored by sipunculids). Thus, eventually the conchicolous life habit, which is less dependent on single host taxa, proved to be the more successful, despite the disadvantages previously discussed.

Conclusions

- Borings belonging to the ichnogenus *Trypanites,* and named herein *T. mobilis* isp. n., are common in the Late Cretaceous and Danian of NW Europe, where they are restricted to the spines of the echinoid genus *Tylocidaris* and the sponge *Porosphaera globularis*. Thereby they show closely reminiscent morphology and were thus presumably produced by the same tracemaker.
- We assume that the borings were produced post mortem by sipunculan worms and there are strong arguments that the bored bioclasts served as mobile shelters, analogous to empty gastropod shells in the conchicolous life habit. The mobile-boring life habit is not known from any other animal group. There are, however, several reports of sipunculans following a conchicolous life habit in ancient and modern seas.
- Both, tylocidarid spines and *Porosphaera* skeletons offered lightweight and continuously available substrates. Boring shelters in bioclasts reduced inter- and intraspecific competition.
- Both strategies allowed the sipunculans to exploit the surfaces of soft bottoms, possibly making use of a deposit feeding mode of nourishment, and providing effective protection from predators. However, there is positive evidence of a certain degree of predation.
- In sipunculans, the conchicolous life habit and the behaviour of boring into bioclasts evolved almost simultaneously in the mid-Cretaceous (Albian and Cenomanian, respectively) suggesting escalation as a response to increased predation. Thus, our observation supports the hypothesis of the Mesozoic Marine Revolution.
- Evolutionary trends can be observed towards increasing size of borings and bored substrates and a habitat shift from shallow to deeper environments.
- With the vanishing of the chalk-sea ecosystem at the end of the Danian and the extinction (or disappearance) of substrate-supplying host taxa, the mobile-boring life habit disappeared in sipunculans, whereas the conchicolous life habit, which is not dependent on specific hosts, still exists and has proven to be the more successful.

Acknowledgements

We would like to thank Christina Franzén-Bengtson (Stockholm, Sweden), David Harper and Sten Lennart Jakobsen (both Copenhagen, Denmark) for support and the loan of material in their care. Sten Jakobsen is especially acknowledged for the permit of preparing epoxy resin casts of echinoid spines. For generous donation of material from their collections, we give our thanks to Alice Rasmussen (Faxe, Denmark) and Volker Thiel (Düsseldorf, Germany). Jan Müller-Edzards (Berlin, Germany) prepared the reconstruction drawing. Robert Bednarik (Caulfield South, Australia) provided clues on the use of bored *Porosphaera* as Palaeolithic pearls. We kindly acknowledge Marianne Falk (Berlin, Germany) who generously translated Swedish literature. Leif Tapanila (Pocatello, USA) helped improving the quality of the manuscript and Geerat J. Vermeij (Davis, USA) is gratefully acknowledged for his particularly valuable comments. CN wishes to thank the European Union funding programme SYNTHESYS (DK-TAF-2388, SE-TAF-1753) for financial support.

References

Agassiz L (1840) Catalogus systematicus Ectyporum Echinodermatum fossilium Musei Neocomiensis, secundum ordinem zoologicum dispositus; adjectis synonymis recentioribus, nec non stratis et locis in quibus reperiuntur. Sequuntur characteres diagnostici generum novorum vel minus cognitorum. Petitpierre, Neuchâtel, 20 pp

Arnaud PM, Thomassin BA (1976) First records and adaptive significance of boring into a free-living scleractinian coral (*Heteropsammia michelini*) by a date mussel (*Lithophaga lessepsiana*). Veliger 18:367-374

Beck G (1920) Tektonische und paläogeographische Untersuchungen im Gebiet zwischen Hildesheim und Braunschweig. Abh Preuss Geol Landesanst, N F 85:1-126

Bednarik RG (2005) Middle Pleistocene beads and symbolism. Anthropos 100:537-552

Bernecker M, Weidlich O (1990) The Danian (Paleocene) coral limestone of Fakse, Denmark: A model for ancient aphotic, azooxanthellate coral mounds. Facies 22:103-138

Boucot AJ (1990) Evolutionary paleobiology of behavior and coevolution. Elsevier, Amsterdam, 725 pp

Bromley RG (1972) On some ichnotaxa in hard substrates, with a redefinition of *Trypanites* Mägdefrau. Paläont Z 46:93-98

Bromley RG (1994) The palaeoecology of bioerosion. In: Donovan SK (ed) The palaeobiology of trace fossils. Wiley and Sons, Chichester, pp 134-154

Bromley RG, D'Alessandro A (1987) Bioerosion of the Plio-Pleistocene transgression of southern Italy. Riv Ital Paleont Stratigr 93:379-442

Brotzen F (1959) On *Tylocidaris* species (Echinoidea) and the stratigraphy of the Danian of Sweden (with a bibliography of the Danian and Paleocene). Sveriges Geol Unders C 571 (Årsb 54):1-81

Brünnich-Nielsen K (1926) Kalken paa Saltholm. Danm Geol Unders 4:1-23

Brünnich-Nielsen K (1938) Faunen i ældere Danium ved Korporalskroen. Medd Dansk Geol Foren 9:117-126

Cole AR, Palmer TJ (1999) Middle Jurassic worm borings, and a new giant ichnospecies of *Trypanites* from the Bajocian/Dinantian unconformity, southern England. Proc Geol Assoc 110:203-209

Cutler E (1994) The Sipuncula – Their systematics, biology and evolution. Cornell Univ Press, Ithaca, 453 pp
Desor E (1858) Synopsis des Echinides fossils. Reinwald, Paris, 490 pp
Diesing KM (1851) Systema Helminthum. Braumüller, Wien, 2, 588 pp
Ernst G (1963) Zur Feinstratigraphie und Biostratonomie des Obersanton und Campan von Misburg und Höver bei Hannover. Mitt Geol Staatsinst Hamburg 32:128-147
Ernst G, Seibertz E, Wood CJ (1998) Cenomanian-Turonian of Wüllen near Ahaus. In: Mutterlose J, Bornemann A, Rauer S, Spaeth C, Wood CJ (eds) Key localities of the northwest European Cretaceous. Bochumer Geol Geotechn Arb 48:157-164
Felder PJ (1996) Late Cretaceous (Santonian-Maastrichtian) sedimentation rates in the Maastricht (NL), Liège/Campine (B) and Aachen (D) area. Ann Soc Géol Belg 117:311-319
Fraas O (1878) Geologisches aus dem Libanon. Jh Ver Vaterl Natkd Württemb 34:257-391
Frieg C, Hiss M, Käver M (1990) Alb und Cenoman im zentralen und südlichen Münsterland (NW-Deutschland) – Stratigraphie, Fazies, Paläogeographie. N Jb Geol Paläont Abh 181:325-363
Gravesen P (1993) Early Danian species of the echinoid genus *Tylocidaris* (Cidaridae, Psychocidarinae) from eastern Denmark. Contr Tert Quatern Geol 30:41-73
Gripp K (1964) Erdgeschichte von Schleswig-Holstein. Wachholz, Neumünster, 411 pp
Grönwall KA (1900) Borrade Echinidtaggar från Danmarks Krita. Medd Dansk Geol Foren 6:33-36
Hagenow F (1840) Monografie der Rügen'schen Kreide-Versteinerungen. II. Abteilung: Radiarien und Annulaten. N Jb Mineral Geogn Geol Petrefaktenkd 1840:631-672
Heinz R (1926) Beitrag zur Kenntnis der Stratigraphie und Tektonik der oberen Kreide Lüneburgs. Mitt Mineral Geol Staatsinst Hamburg 8:1-109
Hinde GJ (1904) On the structure and affinities of the genus *Porosphaera* Steinmann. J Roy Microscop Soc 1:1-25
Huang D-Y, Chen J-Y, Vannier J, Saiz Salinas JI (2004) Early Cambrian sipunculan worms from southwest China. Proc Roy Soc London B 271:1671-1676
Hutchings PA, Peyrot-Clausade M (2002) The distribution and abundance of boring species of polychaetes and sipunculans in coral substrates in French Polynesia. J Exper Mar Biol Ecol 269:101-121
Hylleberg J (1974) On the ecology of the sipunculan *Phascolion strombi* (Montagu). In: Rice ME, Todorovic M (eds) Proceedings of the International Symposium on the biology of the Sipuncula and Echiura. Naucno Delo Press, Belgrade, pp 241-250
Ikeda H (1935) Preliminary report on a new cidaroid sea-urchin from the western Pacific. Proc Imp Acad Japan 11:386-388
Jagt JWM, van Bakel BWM, Fraaije RHB, Neumann C (2006) In situ fossil hermit crabs (Paguroidea) from northwest Europe and Russia. Preliminary data on new records. Rev Mex Cienc Geol 23:364-369
Kohn AJ, Arua I (1999) An Early Pleistocene assemblage from Fiji: gastropod faunal composition, paleoecology and biogeography. Palaeogeogr Palaeoclimatol Palaeoecol 146:99-145
Lankester ER (1885) Protozoa. In: Encyclopaedia Britannica. Vol 19, 9th ed, London, pp 830-866
Mägdefrau K (1932) Über einige Bohrgänge aus dem Unteren Muschelkalk von Jena. Paläont Z 14:150-160
Märkel K, Röser U (1983) Calcite resorbtion in the spine of the echinoid *Eucidaris tribuloides*. Zoomorphology 103:43-58
Milne-Edwards H, Haime J (1848) Recherches sur les polypiers. 2. Mémoire: Monographie des Turbinolides. Ann Sci Nat Zool, Ser 3, 9:211-344

Mortensen T (1928) A monograph of the Echinoidea. I. Cidaroidea. 5. Reitzel, Copenhagen, 51 pp
Mortimore RN (1997) The chalk of Sussex and Kent. Geol Assoc Field Guide 57, 139 pp
Müller AH (1970) Über *Porosphaera* (Porifera, Calcarea) und ihr Endolithion. Mber Akad Wiss Berlin 12:708-720
Müller AH (1993) Lehrbuch der Palaeozoologie, II(1), Protozoa - Mollusca 1. Fischer, Jena, 685 pp
Nestler H (1961) Spongien aus der weißen Schreibkreide (Untermaastricht) der Insel Rügen (Ostsee). Paläont Abh 1:1-70
Phillips J (1829) Illustrations of the geology of Yorkshire; or a description of the strata and organic remains of the Yorkshire Coast. Part 1. Author's Edition, York, 192 pp
Pomel A (1883) Classification méthodique et genera des échinides vivants et fossiles. Jourdan, Alger, 132 pp
Prouho H (1887) Recherches sur le *Dorocidaris papillata* et quelques autres échinides de la Méditerranée. Arch Zool Exper Gén 15:213-280
Ravn JPJ (1928) De regulære echinider i Danmarks Kridtaflejringer. Kongel Danske Vidensk Selsk Skr, Natv Math Afd, Ser 9, 1:1-62
Reich M, Frenzel P (2002) Die Fauna und Flora der Rügener Schreibkreide (Maastrichtium, Ostsee). Arch Geschiebekd 3:73-284
Rice ME (1970) Introduction. In: Rice ME, Todorovic M (eds) Proceedings of the International Symposium on the biology of the Sipuncula and Echiura. Naucno Delo Press, Belgrade, pp XI-XIX
Rice ME, McIntyre IG (1972) A preliminary study of sipunculan burrows in rock thin sections. Caribbean J Sci 12:41-44
Rigaud S (2007) Révision critique des *Porosphaera globularis* interprétées comme éléments de parure acheuléens. unpublished masters thesis, Université Bordeaux, 63 pp
Savin L (1905) Révision des échinides fossiles du département de l'Isère. Bull Soc Stat Sci Nat Isère 4:5-220
Scheer U, Stottrop U (1995) Die Kreide am Kassenberg. In: Weidert W (ed) Klassische Fundstellen der Paläontologie 3. Goldschneck, Korb, pp 140-141
Scheidegger G (1972) *Heterocyathus*, an ahermatypic madreporarian in the *Halophila* environment of the Gulf of Elat, Red Sea. Heinz Steinitz Mar Biol Lab Sci Newsl 2:8-9
Schindewolf OH (1959) Würmer und Korallen als Synöken. Zur Kenntnis der Systeme *Aspidosyphon*/*Heteropsammia* und *Hicetes*/*Pleurodictium*. Abh Akad Wiss Lit, Math-Natwiss Kl, 6:259-327
Schlüter C (1892) Die regulären Echiniden der norddeutschen Kreide. II. Cidaridae, Salenidae. Abh K Preuss Geol Landesanst, N F 5:1-243
Schlüter C (1897) Über einige exocyclische Echiniden der baltischen Kreide und deren Bett. Z Dt Geol Ges 49:18-50
Schmid F, Schulz M-G, Wood C (2004) The Maastrichtian sections of Hemmoor and Kronsmoor – Retrospect, stocktaking and bibliography. Geol Jb, A 157:11-22
Steinmann G (1878) Über fossile Hydrozoen aus der Familie der Coryniden. Palaeontographica 25:102-124
Stephenson LW (1952) Larger invertebrate fossils of the Woodbine Formation (Cenomanian) of Texas. US Geol Surv Prof Pap 242, 226 pp
Stolarski J, Zibrowius H, Löser H (2001) Antiquity of the scleractinian-sipunculan symbiosis. Acta Palaeont Pol 46:309-330
Surlyk F, Damholt T, Bjerager M (2006) Stevns Klint, Denmark: Uppermost Maastrichtian chalk, Cretaceous-Tertiary boundary, and lower Danian bryozoan mound complex. Bull Geol Soc Denmark 54:1-48

Tasnádi-Kubacska A (1962) Paläo-Pathologie. Fischer, Jena, 269 pp
Théel H (1875) Le *Phascolion (Phascolosoma) strombi* (Mont.). Bihang Kungliga Svenska Vetensk Akad Handl 3:1-7
Thomsen E (1995) Kalk og kridt i den danske undergrund. In: Nielsen B (ed) Danmarks geologie fra Kridt til i dag. Geol Inst Aarhus Univ, Aarhus, pp 32-67
Unal E, Zinsmeister WJ (2006) Earliest record of sheltering strategy for predator avoidance from the Early Cambrian of the Marble Mountains, California. Philadelphia Annu Meet, 22-25 October 2006, Geol Soc Amer, Abstracts with Programs, Paper 230-3, 38(7):550
Vermeij G (1977) The Mesozoic marine revolution: Evidence from snails, predators and grazers. Paleobiology 3:245-258
Vermeij G (1987) Evolution and escalation. An ecological history of life. Princeton Univ Press, Princeton, 527 pp
Williams JA, Margolis SV (1974) Sipunculid burrows in coral reefs: Evidence for chemical and mechanical excavations. Pacific Sci 28:357-359
Wind J (1954) *Tylocidaris* piggene som ledeforsteninger i vort øvre Senon og Danien. Medd Dansk Geol Foren 12:481-486
Wood R (2002) Sponges. In: Smith AB, Batten DJ (eds) Fossils of the chalk. Palaeont Assoc, London, pp 27-41
Yonge CM (1974) A note on mutualism between sipunculans and scleractinian corals. In: Rice ME, Todorovic M (eds) Proceedings of the International Symposium on the biology of the Sipuncula and Echiura. Naucno Delo Press, Belgrade, pp 305-311

Biogeographical distribution of *Hyrrokkin* (Rosalinidae, Foraminifera) and its host-specific morphological and textural trace variability

Lydia Beuck[1], Matthias López Correa[1], André Freiwald[1]

[1] Institute of Palaeontology, Erlangen University, 91054 Erlangen, Germany, (lydia.beuck@pal.uni-erlangen.de)

Abstract. The parasitic foraminifer *Hyrrokkin sarcophaga* predominantly infests the cold-water coral *Lophelia pertusa* and the co-occurring bivalve *Acesta excavata*, showing a commensal or parasitic behaviour. It occurs also on some other corals (e.g., *Caryophyllia sarsiae*), bivalves (e.g., *Delectopecten vitreus*) and sponges (*Geodia* sp.), typically in aphotic environments. The aim of the study is to describe its traces from various host substrates, to characterise its parasitic behaviour and to map the geographical distribution of the genus *Hyrrokkin*. Epoxy-resin casts of *H. sarcophaga* traces in *A. excavata*, *C. sarsiae*, *D. vitreus* and *L. pertusa,* and of *H. carnivora* traces in *A. excavata*, were SEM analysed. The boring pattern is in all cases characterised by a shallow groove of up to 7 mm in diameter (max. 2 mm deep), from which 'whip'-shaped extensions protrude vertically into the substrate. In *A. excavata* the foraminifer can penetrate the entire valve to the mantle cavity, producing a thick shaft of fused 'whips'. This parasitic attack is answered by a strong callus formation of the mollusc. One individual foraminifer can repeatedly bypass the organic-rich callus, resulting in a thick aragonite pinnacle. The trace surface texture is xenoglyph and changes with the penetrated host-microstructures. This is especially obvious on deeply penetrating trace portions (e.g., 'whip'-shaped filaments) and is a strong indication for a chemical penetration mode (etching). The trace of *Hyrrokkin* is described as *Kardopomorphos polydioryx* igen. n., isp. n. On the substrates without the shaft, related to parasitic behaviour, *Hyrrokkin* might feed directly on adjacent external host tissue. *H. sarcophaga* is known along the North Atlantic continental margin from polar to subtropical latitudes and *H. carnivora* occurs on the continental margin of Mauritania, Congo and Guinea. In the Mediterranean we could document the parasitism of *H. sarcophaga* from Last Glacial *A. excavata.* Traces or detached foraminifer tests occur in Early Pleistocene cold-water coral deposits on Sicily and Rhodes. Recent *H. sarcophaga* has not been observed above 11°C and is scarce near 5°C water temperature. *Hyrrokkin* sp. was reported from the Canadian Pacific on fossil sponges and was observed on *Acesta patagonica* in the Beagle Channel (Chile).

Keywords. Parasitism, cold-water corals, bioerosion, foraminifer, *Lophelia*, *Acesta*, *Hyrrokkin*

M. Wisshak, L. Tapanila (eds.), *Current Developments in Bioerosion*. Erlangen Earth Conference Series, doi: 10.1007/978-3-540-77598-0_17,

Introduction

Twenty-three species of foraminifera are known to make cavities into hard substrates, mainly from turbulent, warm, shallow-water coral reef environments (see Cedhagen 1994; Vénec-Peyré 1996; Santos and Mayoral 2006) and their temporal occurrence ranges from Jurassic to Recent (Voigt and Bromley 1974; Baumfalk et al. 1982; Cherchi and Schroeder 1991, 1994; Plewes et al. 1993). However, just a few cold-water species are known to date, which are *Cibicides refulgens* de Montfort, 1808 (see Alexander and DeLaca 1987), *Rosalina globularis* d'Orbigny, 1826 (see DeLaca and Lipps 1972), *Globodendrina monile* Plewes, Palmer and Haynes, 1993 (see Cherchi and Schroeder 1991; Plewes et al. 1993; Beuck and Freiwald 2005; Wisshak et al. 2005), *Hyrrokkin carnivora* (Todd, 1965) and *H. sarcophaga* Cedhagen, 1994 (e.g., see compilation in Cedhagen 1994; Freiwald and Schönfeld 1996). Several hypotheses (feeding, protection, test-building) have been proposed to explain this mode of life and the involved boring process. Some species make microborings, in which they anchor themselves (Poag 1971), some produce attachment scars (DeLaca and Lipps 1972; Alexander and DeLaca 1987), and others make holes, which they occupy throughout life (e.g., Todd 1965; Vénec-Peyré 1991). Primary reasons for bioeroding behaviour of foraminifera are considered feeding (Bromley 1970; Warme 1975) and / or protection from turbulence, as most species have been reported from shallow-water environments (e.g., Cherchi and Schroeder 1991). Since bioeroding foraminifera are sessile and believed to capture food by means of their pseudopodia they depend on the availability of resources in close vicinity to them whether they are grazers, suspension feeders, parasites, predators or they take up dissolved material (Vénec-Péyre 1996). Furthermore, it is supposed by Cherchi and Schroeder (1991) that marine substrata may support bacterial or protistan biotas, which can serve as a food source for foraminifers and a mobile substrate that is elevated above the sediment water interface may as well facilitate their nourishment, as supposed by Smyth (1988) for *Cymbaloporella tabellaeformis* (Brady, 1884).

Ichnotaxonomic studies on foraminiferan borings comprise five ichnogenera, which are *Camarichnus* Santos and Mayoral, 2006, produced by adnate lituolids and cibicidids, *Canalichnus* Santos and Mayoral, 2006 as traces of saccamminids, *Semidendrina* Bromley, Wisshak, Glaub and Botquelen, 2007, generated by *Globodendrina monile*, *Diplatulichnus* Nielsen and Nielsen, 2001 and *Stellatichnus* Nielsen and Nielsen, 2001.

Hyrrokkin sarcophaga and *Hyrrokkin carnivora*

Hyrrokkin sarcophaga (Cedhagen, 1994) is known to preferentially colonise the bivalves *Acesta excavata* (Fabricius, 1779) and additionally *Delectopecten vitreus* (Gmelin, 1791), but also sponges of the family Geodiidae (*Isops phlegraei* Sollas, 1880, *Geodia* sp., *Geodia macandrewi* Bowerbank, 1858, *Geodia barretti* Bowerbank, 1858) and Ancorinidae (*Stelletta normani* Sollas, 1880), the scleractinian corals *Lophelia pertusa* (Linnaeus, 1758) (preferentially), *Madrepora oculata* Linnaeus, 1758 and *Caryophyllia sarsiae* (Zibrowius, 1974), the octocoral *Primnoa resedaeformis* (Gunnerus, 1763), other foraminiferans, like *Paromalina coronata* (Parker and Jones, 1865), and rocks (see additionally Cedhagen 1994; Freiwald and Schönfeld 1996). Tests of this large foraminifer can be up to 7 mm in diameter (Freiwald and Schönfeld 1996). *H. sarcophaga* is sinistrally coiled, flat trochospiral, low-spired, bilamellar, planoconvex or concavoconvex and has numerous areal apertures covering up to half of the umbilical side area (Cedhagen 1994). It is attached to the host with its involute umbilical side, which is planar or concave. The apical side is evolute and has a low spire with a convex surface. The rounded perimeter of the test tends to more lobate chambers in adult specimens (Cedhagen 1994). In contrast, *H. carnivora*, earlier known as *Rosalina carnivora*, is reported from 380-951 m water depth in the Gulf of Guinea parasitising on the bivalve *Acesta angolensis* Adam and Knudsen, 1955 (see Todd 1965; Marche-Marchad 1979; Cedhagen 1994), from shallow-water environments attached to seaweed as well as solid objects, unattached in bottom sediments (Todd 1965). *A. angolensis* is the southern 'neighbour' of *A. excavata* and at the same time a closely related species (Boss 1965). Both limid bivalves can hardly be distinguished by their biometric parameters (e.g., width and height) and the same is true for the two parasitic foraminifera, which are also closely related (see discussion in Cedhagen 1994).

The genus *Hyrrokkin* is known to penetrate the body wall of its host to feed on its soft tissues (Todd 1965; Marche-Marchad 1979; Cedhagen 1994). Todd (1965) reports from *H. carnivora* that on the external surface of *Acesta* the attachment scars created by the foraminifer are roughly circular depressions; on the interior the penetration scars are gall-like raised areas of thickened shell material, in the centre of which a scab of brownish material fills a small depression and besides the major opening that extends through the bivalve shell, there are several deep cavities plus numerous small pits and craters. Furthermore, she remarks that on the interior of this valve only one to two holes appear to have penetrated completely through the clam shell and the diameter is 0.13 mm on exterior but only 0.08 mm on the interior. She concludes that the bivalve had apparently tried to repair the injury by secreting numerous layers of mother-of-pearl on the inside of the corroded area but these layers were also corroded. The major opening is positioned on centre of attachment beneath the umbilicus of the test (Todd 1965). Cedhagen (1994) remarks that the diameter of *H. sarcophaga* pits on *Lophelia pertusa* are slightly larger than those of the foraminiferan tests. The foraminifer does not move from its pit but often builds an additional, irregular chamber in the direction facing the prey. Cedhagen (1994) reported that *Hyrrokkin* is not seen to respond to epizoans, which colonise the apical part of the test, since pseudopodia are active under and around the test and do not penetrate minute pores at its top. The rare occurrence on rocks is indicating that the foraminifer can survive on food sources other than that from host specimens, at least for some time (Cedhagen 1994).

Aim of study

The analysis of trace casts of *Hyrrokkin* borings in Recent and fossil *Lophelia pertusa*, *Acesta excavata* and *Delectopecten vitreus*, Recent *Caryophyllia sarsiae* and *Geodia* sp. makes it possible to give a general morphological characterisation of the trace with the aim to establish a new ichnospecies and to summarise the currently known temporal and geographical distribution. We focus on the morphological variability of *Hyrrokkin* traces, as well as on their surface texture variations with regard to its host skeletal microstructure and characterise the defence mechanisms of living hosts.

Materials and methods

The analysed *Hyrrokkin* substrates (Fig. 1; SEM samples listed in Appendix) originate from the Norwegian Shelf (*Acesta excavata, Delectopecten vitreus* and *Lophelia pertusa*), the Porcupine Seabight (*Lophelia pertusa, Caryophyllia sarsiae*),

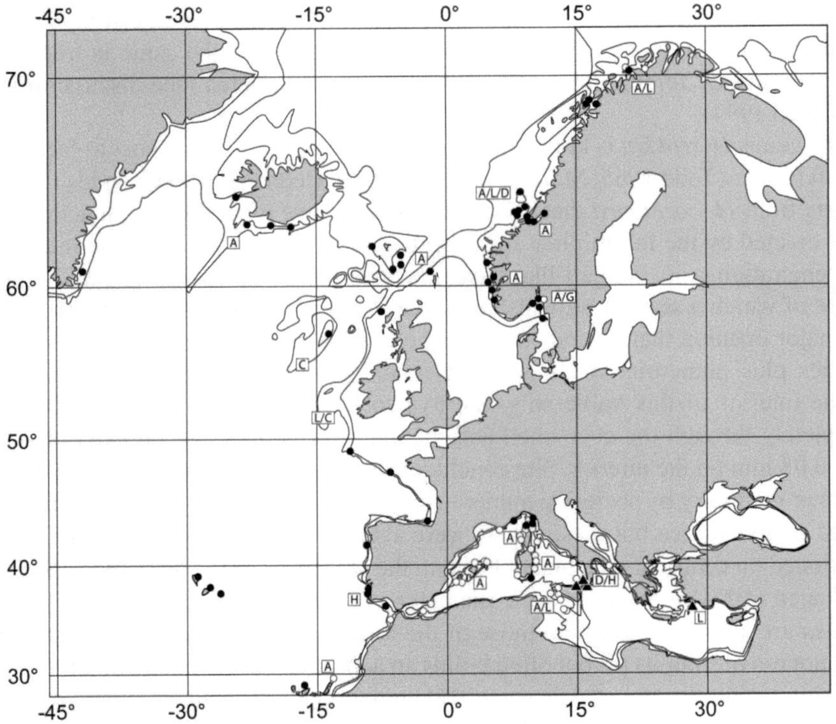

Fig. 1 The map shows the distribution of *Acesta excavata* (modified from López Correa et al. 2005), which is the principal host for *Hyrrokkin sarcophaga*. The parasite is known from all Recent localities. Black circles denotes Recent occurrences, white circles indicate Late Pleistocene to Holocene Mediterranean sites and black triangles stand for onshore outcrops of Early to Mid-Pleistocene age. Letters indicate the substrate for *Hyrrokkin* for mentioned or studied in this paper; A = *Acesta excavata*, L= *Lophelia pertusa*, D = *Delectopecten vitreus*, C = *Caryophyllia sarsiae*, G = *Geodia* sp., H = *Hyrrokkin sarcophaga* tests isolated in sediment

the Mauritanian margin (*Acesta excavata*), the Mediterranean Sea (Last Glacial and Late Pleistocene *Acesta excavata*, Last Glacial *Lophelia pertusa*) and Early to Mid-Pleistocene onshore outcrops on Sicily (*Delectopecten vitreus*) and Rhodes (*Lophelia pertusa*). Apart from *Caryophyllia sarsiae* samples from the Porcupine Seabight, which were fixed in 70% ethanol, all other samples were air-dried at room temperature after sampling.

H. sarcophaga has not yet been documented from the present-day Mediterranean (see also Cedhagen 1994; Freiwald and Schönfeld 1996) but boring traces preserved in Pleistocene *A. excavata* shells and Early Pleistocene cold-water coral deposits on Rhodes (Bromley 2005; Titschack and Bromley 2005) instead suggest that the foraminifer must have persisted in the Mediterranean as well (Figs. 1-6). Unfortunately, none of the fossil samples had the parasite still attached. But foraminiferan tests of *H. sarcophaga* were found in Salice (Sicily, Italy), 15 km from Messina, at the locality Contrada Coilare. The deposit is known as 'sabbie gialle' (yellow sands) and of Lower Pleistocene age. The sediment was coarse-stratified and very rich in cirripeds and *Delectopecten vitreus* (Vertino 2003).

Endobiont casts

The degree of internal shell degradation by boring organisms was analysed with a vacuum cast-embedding technique (see Beuck and Freiwald 2005). The resulting epoxy-resin casts exhibited the endobiontic borings as positive structure and were subsequently investigated with scanning electron microscopy (SEM). Fossil traces in Mediterranean *Acesta excavata* still contained many organic matrix sheets of the bivalve's repair layers, and received an additional treatment with hydrogen peroxide (5% H_2O_2) in order to remove these organics.

Thin sections

Thin sections were prepared along and across the main growth axes of *Acesta excavata*, to study internal architecture and layering of the shell with a high-resolution light-optical microscope (Zeiss Axiophot) under normal and polarised light. The fluorescence mode of the microscope was used to visualise organic matter in thin sections, emitting a strong greenish or bluish light. Single valves were embedded in a two-compound epoxy-resin (100 parts Araldit and 28 parts Araldur) to ensure sawing stability. Air inclusions were avoided with a 10 minutes exposure to a high vacuum and curing took place at ambient room-temperature. A low-speed carbonate saw was employed to separate slabs, which were then polished (up to grit powder 1000) and fixed under vacuum to a carrier-glass with a two-component glue (Biresin). Slabs were abraded down to a final thickness of <25 µm with an automatic abrasive device. Some polished shell sections were etched with dilute hydrochloric acid (0.1 N HCl) for 30 to 45 s, then rinsed in deionised water and dried. The ideal etching time for microstructure investigations was 45 seconds.

Best insights on the shell macro- and microstructure were achieved with sections along the main growth axes. Supplementary orthogonal and oblique sections granted a 3D-confirmation for the observed layers and structures. Microstructures of *Acesta excavata* are described in detail in López Correa et al. (2005).

Scanning electron microscope (SEM)

Endobiont casts and etched slabs for microstructural analyses were sputter-coated with gold for the investigation with the CamScan scanning electron microscope (SEM). Samples were mounted in an Edwards S150B sputter-coater, with an ideal distance of 4 cm (min. 1 cm) between target and gold plate. Maximum current was set at 20 mA, with a constant voltage of 10 kV. Sputtering was applied for 1.5 min, using argon gas (Ar_2) as transmitting medium. Vacuum was kept constant at 7 mbar, ensuring a constant gas flux, to receive a homogenous gold film.

Sample photography

All samples were photographically documented with a digital camera (Nikon CoolPix). Close-ups were carried out with a Leica M420 macroscope, offering a magnification range of up to 70 x. The Kodak-Ektachrom 64T was used for macroscopic and microscopic imaging. The Olympus DP10 camera, alternatively mounted to the macroscope, provided digital images. Morphometric analyses were carried out on digital images using the software Adobe Photoshop.

Results

Valves of *Acesta excavata* and dead skeleton portions of *Lophelia pertusa* provide a substrate for various epibionts in the bathyal environment. Recent shells and skeletons exposed to the water column often show dense overgrowth (Figs. 2A, 3A, 4A-B, 5B). Among the most common organisms are serpulids, bryozoans, foraminifers and sponges, but also cold-water corals and saddle-oysters (for details see López Correa 2004). Traces of boring foraminifers, especially of *H. sarcophaga* are highly abundant on Recent Norwegian *Acesta excavata* and on Pleistocene material from the Mediterranean. Adult foraminifera produce a narrow pit that penetrates the entire valve and enters the inner soft-tissue part, causing a host reaction, which is manifested in a callus development (Cedhagen 1994; Freiwald and Schönfeld 1996). Abandoned and fossil attachment grooves are often secondarily colonised by endobionts, like boring fungi, worms and sponges.

Tests grown on sutures or other irregularly-shaped areas of *Acesta* valves, may show a host-surface dependent morphology. Tests on the outer surface of *Acesta* have a quite planar to plano-concave umbilical side. On geodiid sponges *Hyrrokkin* tests are more bloated / flattened-spheroid in shape and the spiral height approaches almost the diameter of the umbilical side. In concert with the tests, the *Hyrrokkin* traces vary amongst different host species.

Lophelia pertusa biodegradation processes generally initiate on tissue-free parts of the skeleton with an infestation by fungi and successively by larger sized borings, which can be, depending on environment for example, boring foraminifera, sponges, bryozoans and annelids and resulting in a substrate porosity of 70% or even more (Beuck and Freiwald 2005; pers. observations). *Hyrrokkin sarcophaga* on *Lophelia pertusa* was often found to be attached in the live zone of polyps in close vicinity to the tentacles (Fig. 2C). Other positions along the dead coral framework were frequent as well (Fig. 2A-B, D-F). In contrast, *Hyrrokkin sarcophaga* traces analysed from live *Caryophyllia sarsiae* were positioned directly below the polyp's edge zone (Fig. 3A-D). In all cases, when collected, live foraminifers on host substrates were marginally surrounded by excreted organic residues of brown colour (Fig. 3A). Within live coral portions just few species have already been reported, amongst epibionts (e.g., hydrozoans;

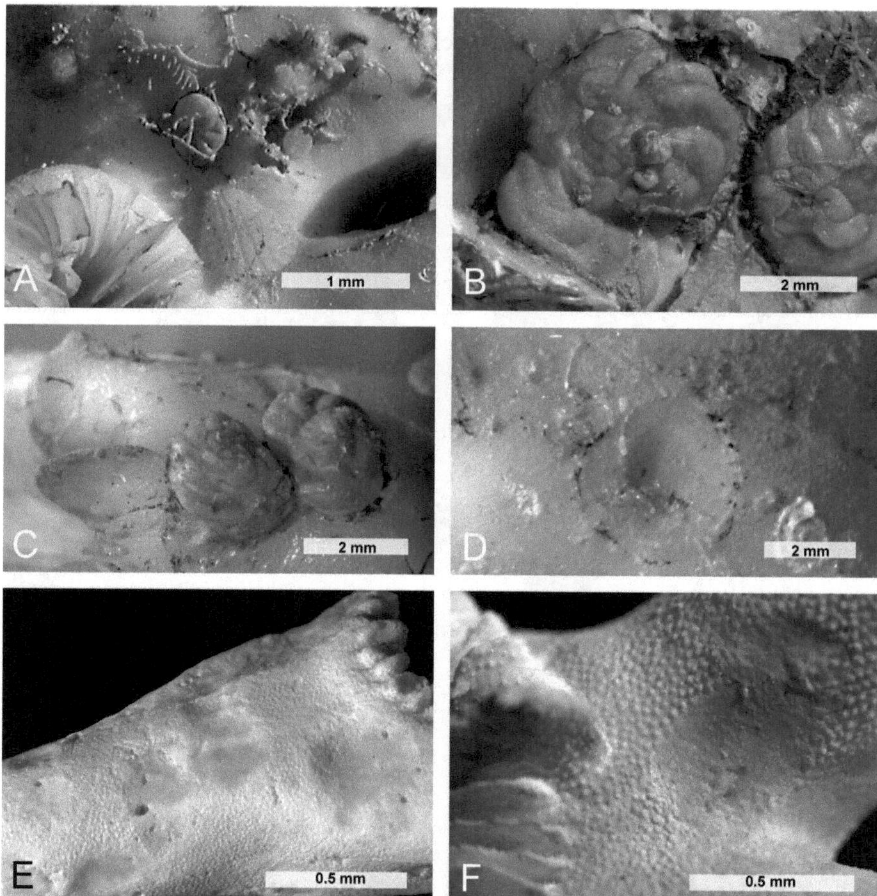

Fig. 2 Microscopic images of Recent (A-D) and fossil (E-F) *Lophelia pertusa* samples. **A** *Hyrrokkin sarcophaga* colonising dead framework in association with the brachiopod *Terebratulina retusa*, epizoic bryozoans and serpulids (Pos228-200-3). **B** Two *Hyrrokkin sarcophaga* tests colonised by bryozoans and other foraminifers; note the partial embedding of the left individual (top) by the coral (JH5-99-4); image with courtesy of K. Kaszemeik. **C** *Hyrrokkin* individuals preferentially colonise upper calyx portions (Pos228-200-3). **D** *Hyrrokkin* groove on recently dead skeleton (LI94-163). **E** Fossil, Last Glacial *L. pertusa* with numerous potential *Hyrrokkin* traces (CS96-64). **F** Same coral as in E with potential *Hyrrokkin* groove in vicinity to the polyp; note the well preserved coral surface

see Taviani et al. 2005) and some boring sponges (see Beuck et al. 2007) causing aberrant coral morphologies, however, the presence of *H. sarcophaga* does not cause any morphological influence on the skeleton of the host coral; though the branching pattern of *Lophelia pertusa* and the resulting topography on the coral framework lead to a strongly variable form of the foraminiferan test with a tendency to concavely-shaped umbilical sides (Figs. 6A-E, 7C-D, G-H; see also Cedhagen 1994). Traces of *H. sarcophaga* and *H. carnivora* cannot be distinguished morphologically. The trace characteristics are described in the following and led to establish a new ichnospecies based on borings on fossil *Acesta excavata*.

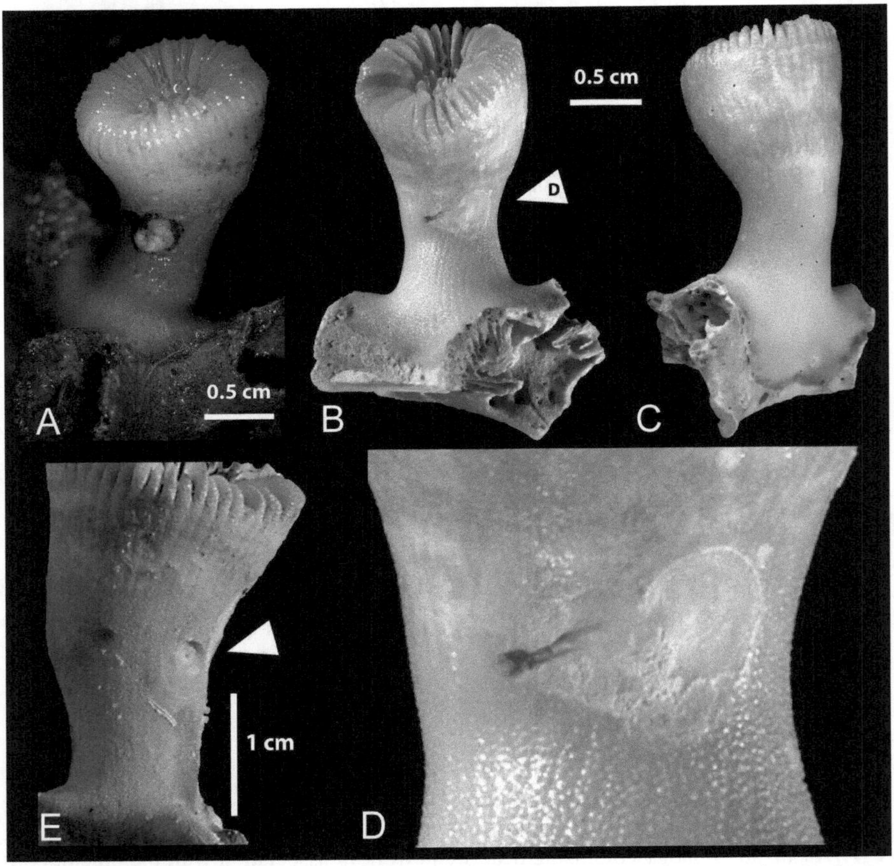

Fig. 3 Overview images of *Caryophyllia sarsiae* samples from the Porcupine Seabight; samples M61-3-603-1 (A-D) and M61-3-603-2. (E). **A** Live *Caryophyllia sarsiae* on dead *Lophelia pertusa* framework with live *Hyrrokkin sarcophaga* at the polyp's basal edge; note the brownish organic residue skirting the foraminiferan test. **B** Sample cleaned from organic remains and the *Hyrrokkin* test, now exposing *Kardopomorphos polydioryx* igen. n., isp. n. **C** The 'backside' of the corallite is free from foraminiferal traces. **D** Close-up of the shallow pit showing that the protoplasm seems to form a shallow depression (left) apically to the latest chamber. **E** Sample M61-3-603-2. Abandoned attachment scar of *Hyrrokkin sarcophaga* on a dead *Caryophyllia sarsiae,* exposing a central canal (white arrow)

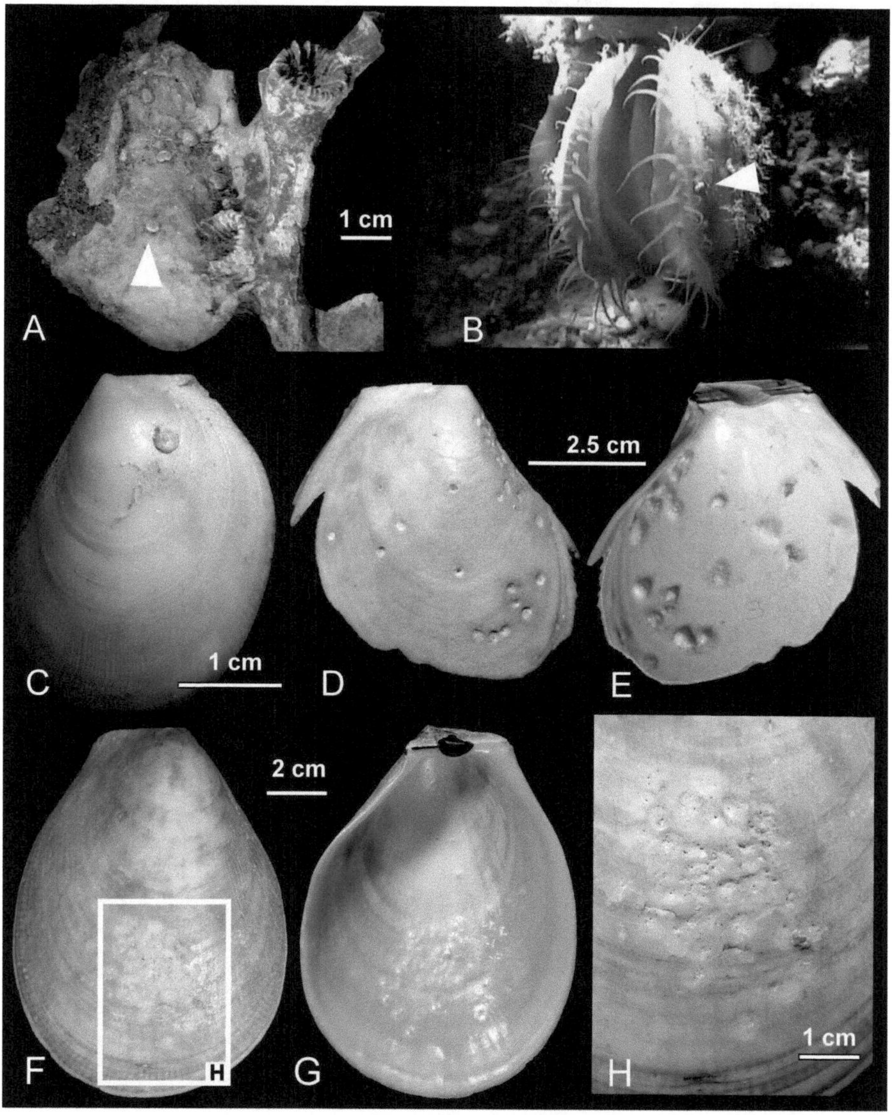

Fig. 4 Overview images of *Hyrrokkin* substrates. **A** Numerous individuals of *Hyrrokkin sarcophaga* (see white arrow) infesting a geodiid sponge, which has overgrown a dead *Lophelia pertusa* branch (LI94-163). **B** Submersible image of a filtering *Acesta excavata* with a large *H. sarcophaga* specimen attached close to the valve margin (Sula Ridge, Norway). **C** Juvenile *A. excavata* with a large *H. sarcophaga* specimen (Pos228-215). **D** Outer surface of a Last Glacial Mediterranean *A. excavata* shell (COR2-39); note the high amount of grooves. **E** Inner side of D showing the protuberances as reaction of the bivalve against infestation (aragonitic callus formation). **F** Recent *A. excavata* (Mauritania) with a multitude of attachment scars produced by *Hyrrokkin carnivora.* **G** Inner side of F, equivalent callus formation as observed for Norwegian *A. excavata* infested by *H. sarcophaga*. **H** Close-up of F showing high attachment scar densities and part of a *H. carnivora* test

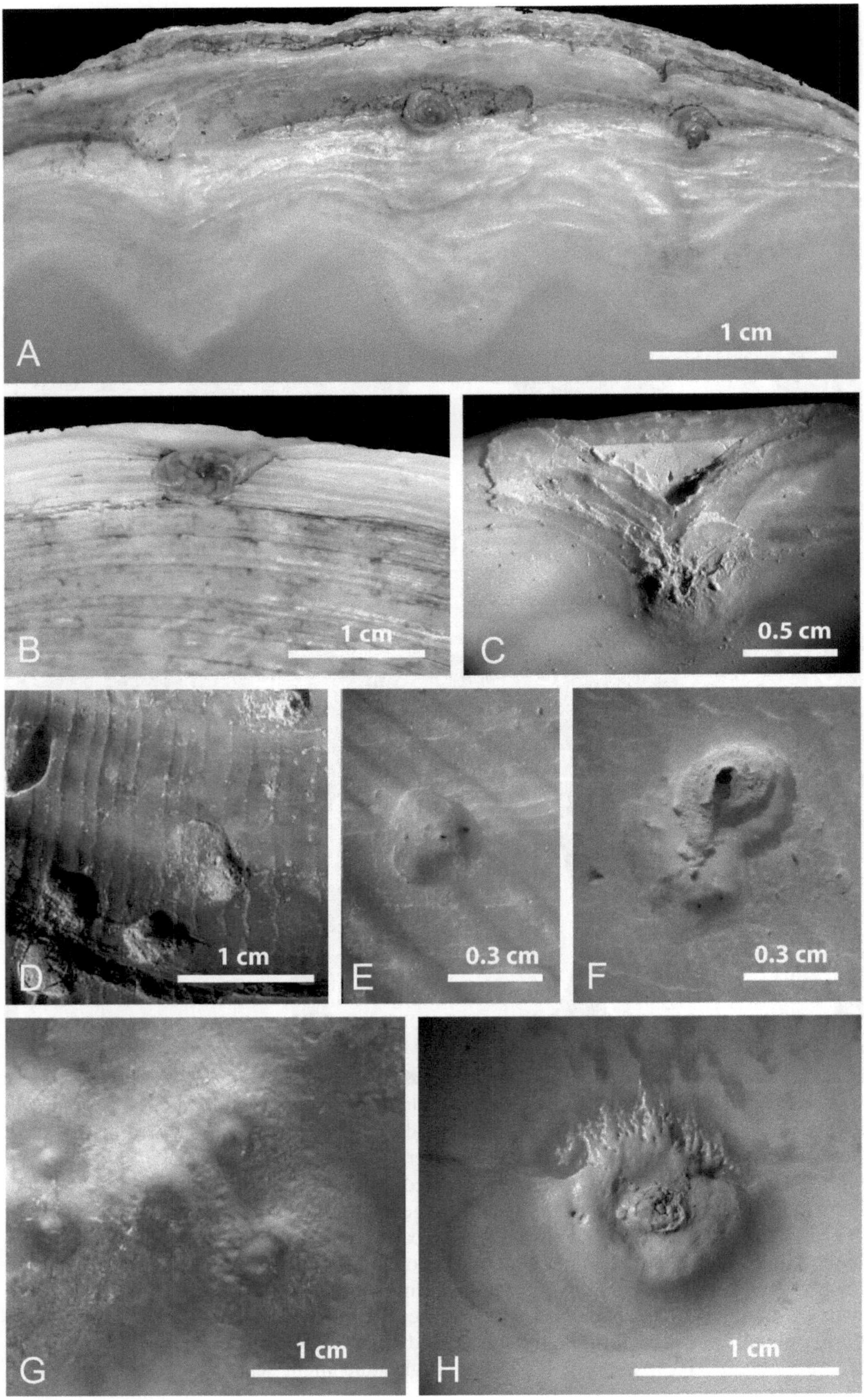
A
1 cm
B
1 cm
C
0.5 cm
D
1 cm
E
0.3 cm
F
0.3 cm
G
1 cm
H
1 cm

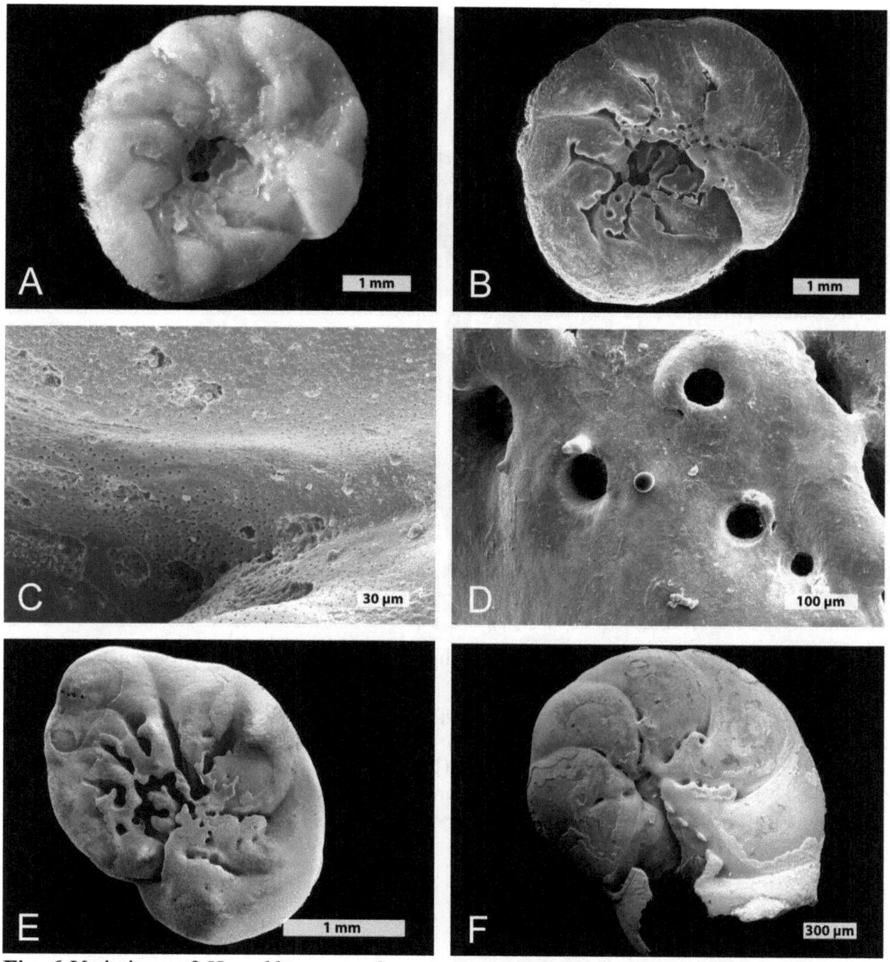

Fig. 6 Variations of *Hyrrokkin sarcophaga* test morphology (A: microscopic image, B-H: SEM). **A** Umbilical side of *H. sarcophaga* on *Lophelia pertusa* (Pos228-200-3). **B** SEM image of A accents the distribution of apertures. **C** Close-up of B showing numerous pores. **D** Close-up of the larger test apertures, where the protoplasm exits to form the 'whip'-shaped canals in the substrate. **E** Test from *L. pertusa*, irregular substrate morphologies influence the shape of the foraminifer (Pos228-200-3). **F** Test from geodiid host (POS228-201).

Fig. 5 *Hyrrokkin* spp. and its traces on *Acesta* spp. and the bivalve's callus formation. **A** Three attachments on the inner side of *A. patagonica* (Beagle Channel) with two in situ tests; note the retracted mantle precipitates and the partial embedding of the test to the right. This is the only sample with *Hyrrokkin* being attached to the internal side of *Acesta*. **B** *H. carnivora* close to the margin of *A. excavata* (Mauritania). **C** Irregular callus resulting from a retracted mantle, due to *H. sarcophaga* close to the margin of *A. excavata* (ALK232-BG-1066). **D** *H. sarcophaga* scars on a Late Pleistocene *A. excavata* (COR2-8). **E-F** Varying scar morphology depending on the penetrated mineralogical layers in *A. excavata* (E: calcite, F: calcite and crossed-lamellar aragonite underneath, COR2-39. **G** Inside of D with callus pinnacles. **H** Recent callus in *A. excavata* at the pallial line, note the sealed central canal (ALK232-1066-BG)

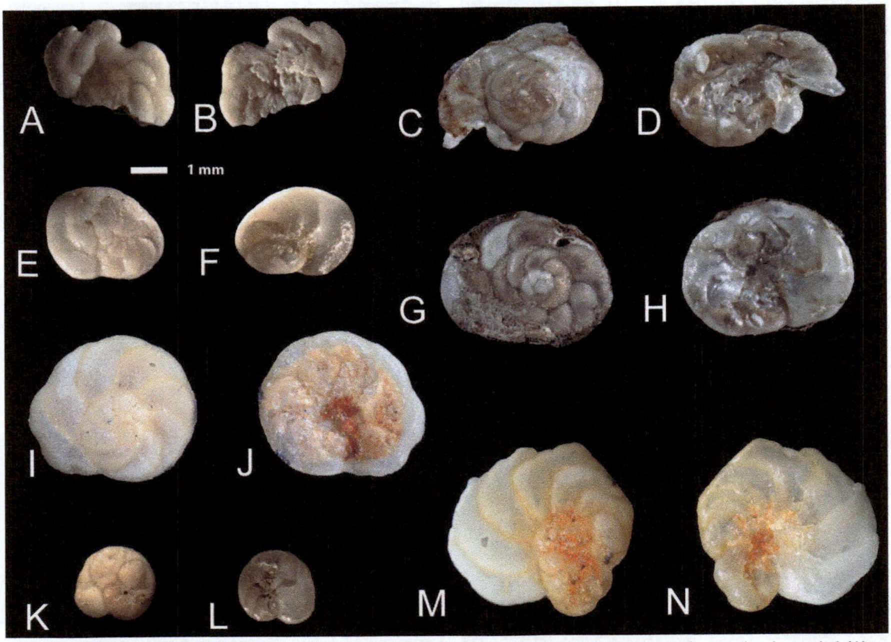

Fig. 7 Recent *Hyrrokkin* spp. from various hosts show a high morphological variability, depending on their substrate topography (A-B: *Hyrrokkin carnivora*, Mauritania; C-N: *Hyrrokkin sarcophaga*, mid-Norway; upper side and umbilical side pictured adjacent; the same scale bar applies to all specimens). **A-B** *H. carnivora* (specimen from Fig. 4I), host *Acesta excavata*. **C-D** Host *Lophelia pertusa* (Pos228-228). **E-F** Host *Caryophyllia sarsiae* (M61-3-603-1). **G-H** like C-D, note the brown organic residuals skirting the test. **I-J** Host *A. excavata* (VH95-168). **K-L** Host geodiid sponge (POS228-201). **M-N** like I-J

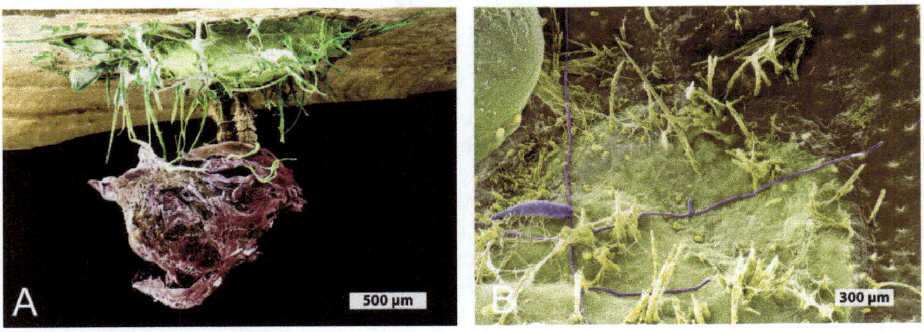

Fig. 8 Substrate variability of *Kardopomorphos polydioryx* igen. n., isp. n. (false-coloured SEM). The parts in green, consisting of a shallow elevation and 'whip'-shaped filaments, are obligatory for mature trace stages in contrast to the central canals (brown) and apical callus cast, which may exist depending on host substrate. **A** Inside the host *Acesta excavata* (ALK232-1066-BG) the trace consists of a shallow depression, 'whip'-shaped filaments (green), two distinct central canals (brown) and the callus cast (purple). **B** Traces by two *Hyrrokkin* individuals in green (upper left corner and centre) inside a Recent *Lophelia pertusa* (cast surface: brownish; Pos228-200-3). The presence of a zooid of the stoloniferous bryozoan *Spathipora* (violet) on the surface of *Kardopomorphos polydioryx* igen. n., isp. n. (green) indicates that the pit has been abandoned by the foraminifer

Systematic ichnology

Kardopomorphos igen. n.

Diagnosis: Shallow disk-shaped pit of rounded shape, often spiral-shaped with a shallower lateral extension.

Etymology: kárdopomorphós (κάρδοπομορφός; greek) = trough-shaped

Type ichnospecies: Kardopomorphos polydioryx isp. n.

***Kardopomorphos polydioryx* isp. n.**
Figs. 2-5, 8-14

2005 *Hyrrokkin sarcophaga* traces.- Bromley, 901, fig. 4A-B
2005 'multitude of 'whip'-like canals'.- Beuck and Freiwald, 926
2005 *Podichnus centrifugalis*.- Wisshak et al., 994, fig. 10A
2006 'Foraminiferan-form 3'.- Wisshak, 94, fig. 29E

Diagnosis: Shallow, often spiral-shaped depression, from which 'whip'-shaped, occasionally branched canals can be extended deeply into the substrate. Facultative, one or few central canals of larger diameter can occur. The surface texture of the entire trace is xenoglyph.

Etymology: polydióryx (πολυδιωρυξ; greek) = many canals

Type material, locality and horizon: Holotype - epoxy resin cast (Fig. 11G-H) of a trace from a Pleistocene radiocarbon-dated (39,960 ± 820 ^{14}C-years BP) *Acesta excavata* (see López Correa et al. 2005). Type locality is Urania Bank, Strait of Sicily (Canale di Sicilia), Italy, collected with R/V Urania at station COR2-08 at N36°51.43' / E013°08.12' / 181-631 m / dredge sample (López Correa 2004; López Correa et al. 2005). Further studied fossil trace material comes from R/V Urania station COR2-39 from a suite of Last Glacial *A. excavata*, one of which had been radiocarbon-dated to 25,026 ± 90 ^{14}C-years BP (AMS-^{14}C, GIF-50141; López Correa and Taviani - unpublished data). COR2-39 - N36°23.48' / E013°15.17' / 523 m / grab sample (López Correa 2004; López Correa et al. 2005).

The holotype is deposited at the Museum für Naturkunde of the Humboldt-University in Berlin, Germany (see Appendix for inventory numbers) alongside all pictured casts of Recent *Hyrrokkin sarcophaga* and *Hyrrokkin carnivora*, comprising characteristic morphologies of *Kardopomorphos polydioryx* igen. n., isp. n. (see Fig. 8). Further, uncasted Recent traces of *Hyrrokkin sarcophaga* in its most typical hosts *Acesta excavata*, *Lophelia pertusa* and *Caryophyllia sarsiae* are stored in the same collection.

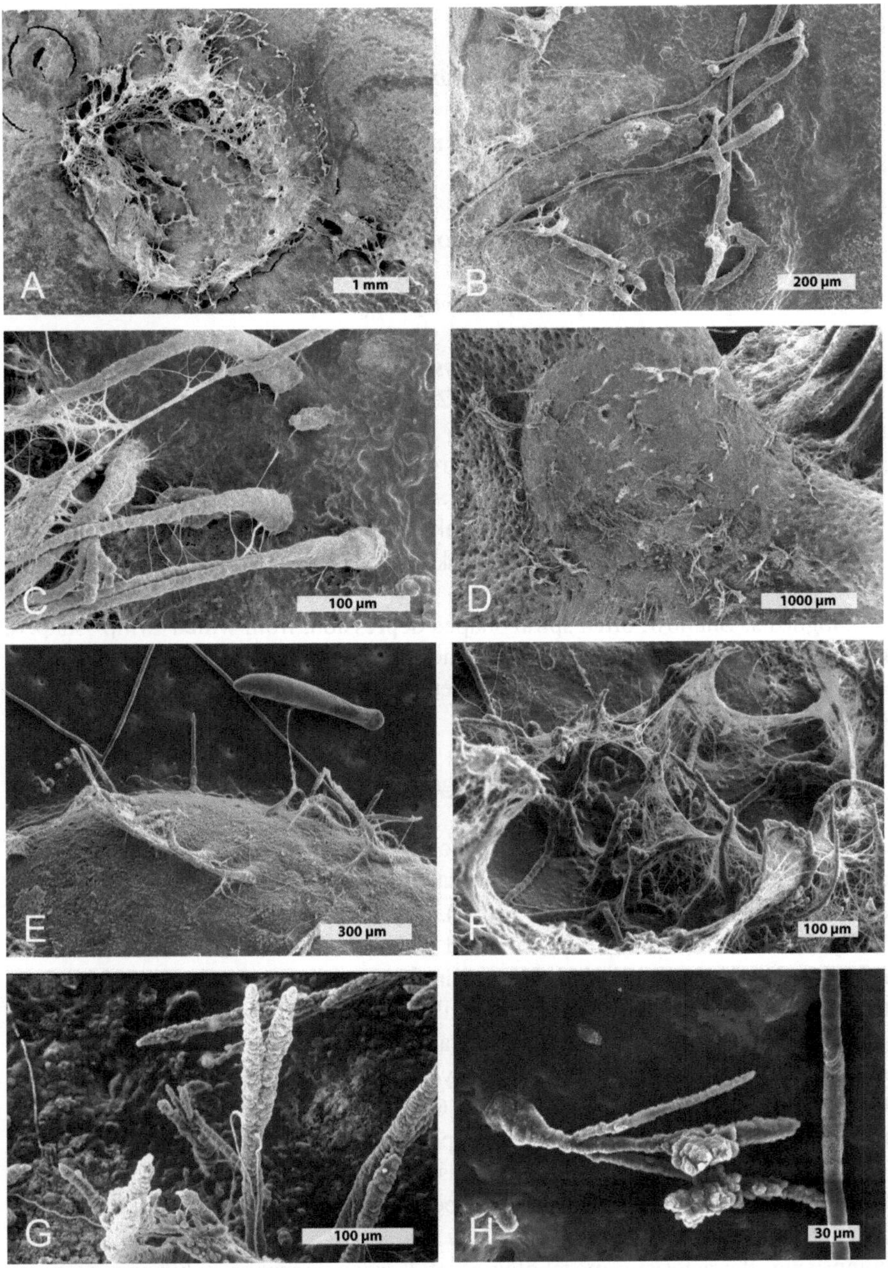

Fig. 9 SEM images of *Hyrrokkin sarcophaga* traces inside *Lophelia pertusa.* **A** Overview (Pos228-200-3); compare with Fig. 7A-C. **B** Close-up of A; note the up to 1.5 mm long 'whip'-shaped filaments. **C** Detail of A showing the base of the filaments measuring about 40 µm in diameter; the surface texture is influenced by the host (xenoglyph). **D** Trace cast in vicinity to septa (right) with irregularly distributed filaments (LI94-163). **E** *Kardopomorphos polydioryx* igen. n., isp. n. (lower part) and *Spathipora* isp.; note the sharp boundary (Pos228-

Description: In early ontogenetic stages the foraminifer produces a shallow, circular depression of 0.5-1.5 mm in diameter, from which 'whip'-shaped filaments with an approximately circular cross-section start to protrude mainly perpendicularly into the substrate. Very rarely the opposite is documented that the depression is established after the filaments. Occasionally the filaments are branched and show fused bases (see Fig. 9F-H). The length of the tapering filaments increases constantly with infestation time of the foraminifer. Thus, the growth of the foraminifer parallels a vertical penetration into the host substrate and the trace starts to vary in morphology between host species (see Fig. 8). Adult grooves are often negatives of the individual chambers of the foraminifer test and are constantly growing throughout the lifetime of the foraminifer, etching into the substrate.

Older stages in *Lophelia pertusa* resemble in morphology ontogenetic stages, however, the dimension of the depression is larger sized, e.g., maximal diameters measured have been 5.5 x 3.9 mm and a penetration depth of up to 500 μm has been observed (Fig. 9A, D-E). In mature stages, numerous 'whip'-shaped filaments are present and can reach penetration depths of 1.5 mm (Fig. 9A-C). In general, their individual diameters at the base range between 29 and 40 μm in contrast to the narrower tips, which are between 5-18 μm. The groove may have a shallow extension outside of the latest foraminifer chamber. Filaments seem to be more concentrated along the sutures of foraminifer tests and rarely outside the clearly defined depression. Some isolated ones may appear which are within 1 mm from the main depression. Individual filament casts show an incremented appearance of regular segments of about 4.5 μm and rarely up to 15 μm in length (Fig. 9C, 13D).

H. sarcophaga grooves analysed from dead *Caryophyllia sarsiae* (Fig. 3E) show much deeper extensions of up to 2.4 mm into the extra-thecal aragonite and penetrating at their deepest point into the septal aragonite (Fig. 10B, D) as reported from a trace with diameters of 4.8 x 3.7 mm. In contrast, maximal penetration depths of a trace occurring in vicinity to the live coral tissue are about 300 μm (see Fig. 10A) but have just slightly smaller trace diameters (4.3 x 3.5 mm). In general, apically the latest foraminiferan chamber in the area of the next chamber the protoplasm can have formed a shallow depression of about 700-1,000 μm wide and 100 μm deep (exclusively within the theca), and also some small protruding filaments can already occur. 'Whip'-shaped filaments have been documented with a basal diameter ranging between 29 to 91 μm and then quickly tapering towards the tip to a diameter of 9-22 μm. Their lengths are generally about 240 μm (Fig. 10C-E). Frequently the filaments appear in clusters and their bases can be fused. Filaments become more frequent at the highest position of the 'bowl'-shaped elevation and can be clustered forming a larger sized canal in one case of about 596 x 341 μm in diameter (Fig. 10D-G), reaching the inner calyx area. The infestation causes no kind of callus or similar abnormal growth or deformation on the coral. Rarely, it is observed that the polyp embeds the test of the foraminifer partly at its edge (see Fig. 3B).

200-3). **F** *H. sarcophaga* trace cast camouflaged by a network of fungal traces (Pos228-200-3). **G** *Hyrrokkin* trace cast filaments variously branched; note the verrucous surface texture (Pos228-200-3). **H** Young stage of a highly branched filament (Pos228-200-3)

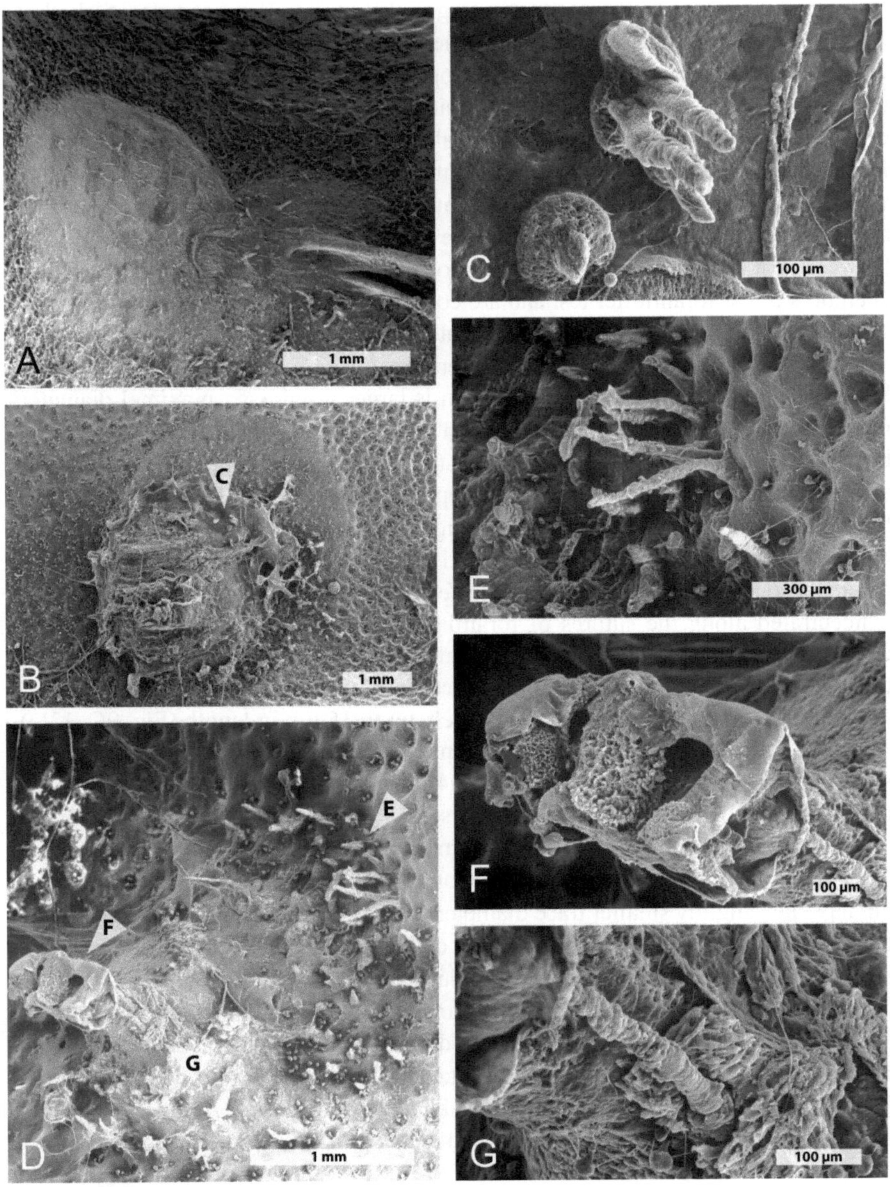

Fig. 10 SEM images of *Hyrrokkin sarcophaga* trace casts in *Caryophyllia sarsiae* from the Porcupine Seabight. **A** *Kardopomorphos polydioryx* igen. n., isp. n. within sample M61-3-603-1 (compare Fig. 3). **B** Overview of larger sized trace forming a central canal on the deepest part of the depression; note the shallow depression most probably resulting from the latest chamber of the test. **C** Detail of B showing the 'whip'-shaped filaments. **D** Trace cast resulting from a deep groove forming a central canal. **E-G** Details of D. **E** 'Whip'-shaped filaments at the trace periphery. **F** Close-up of the deeply penetrating central canals. **G** The abandoned trace surface serves as substrate for secondary settlers, here the fungus *Dodgella priscus* (trace: *Saccomorpha clava*). B-G (M61-3-603-2)

The adult morphology of *Kardopomorphos polydioryx* igen. n., isp. n. in live *Acesta excavata* resembles those in *Caryophyllia sarsiae*, however, it is characterised by a host reaction where the foraminifer successfully penetrated the valve causing an aragonitic callus to close the hole and is thus easily recognisable as pinnacle on the shell interior (Figs. 4C-H, 5). The depressions reach maximal penetration depths of about 500 µm, individual filaments can reach 50 µm in diameter and documented central canals can be up to 420 µm in diameter (Figs. 4H, 5D-F, 11, 12). The repair zone (pinnacle) has a typical diameter of 0.5 to 1 cm and consists of thin aragonite laminae, which are plastered onto the crossed-lamellar aragonite and are enriched in organic matter. However, these organic-rich laminae do not grow in optical continuity with the adjacent crossed-lamellae. Up to 20 laminae have been counted within individual blisters. Single laminae do not exceed 20 µm and away from the pit their thickness decreases rapidly (Fig. 13A-B).

Up to 56 pits have been observed on a single valve on fossil *Acesta excavata* (Canale di Sicilia) and up to 69 pits on Recent ones (Sula Reef). Recent valves occasionally have parasites still attached in life-position. In general, trace casts in *Acesta* vary in shape depending on age of the trace, on the implementation of a complete penetration as well as on the pinnacle size. In adult traces the morphology of the foraminiferan groove may also be a negative of the several foraminiferan chambers and often secondary settlers are seen to colonise abandoned traces, giving the trace an even more complex shape (Fig. 14).

In *Lophelia pertusa* and *Caryophyllia sarsiae* the surface of the depression is mostly smooth in the thecal aragonite and sometimes mimics the 'feather'-shaped aragonite crystal bundles. In case the depression is reaching the septal aragonite, the surface shows ridges in direction of the coral septa. The perimeter of the trace is very sharp. The surface texture of 'whip'-shaped filaments within *Lophelia pertusa* and *Caryophyllia sarsiae* is verrucous to granular (Fig. 9G-H, 10E, G).

Acesta excavata has a bimodal mineralogy with an external calcite layer and an internal aragonite layer. SEM images revealed three major microstructure units: The calcitic layer shows a subdivision into two sub-layers, an outer fibrous-prismatic zone and a micro-granular zone below; crossed-lamellar structure is characteristic for the internal aragonite. Crossed-lamellar aragonite builds the interior shell-layer and also makes up the entire hinge (Fig. 13A). Detailed microstructural observations were made in anticipation of isotopic studies to provide a background on shell structure and its mineralogical composition (López Correa et al. 2005). The external valve side shows clear and regularly spaced growth rings. Increment widths vary from 0.4-0.9 cm. The shell reaches its maximum height (10-13 cm) early during its lifetime. About 20-25 wide increments are seen in adult shells, while all subsequent external increments (up to about 50) are crowded and not clearly distinct. The aragonitic hinge by contrast exhibits a minute banding (0.5 mm) of very regular ribs, which correspond to the number of external bands. Due to the xenoglyphic nature of *Hyrrokkin* borings, the trace casts from *Acesta excavata* clearly exhibit their original spatial extension within the individual shell layers by their characteristic surface texture.

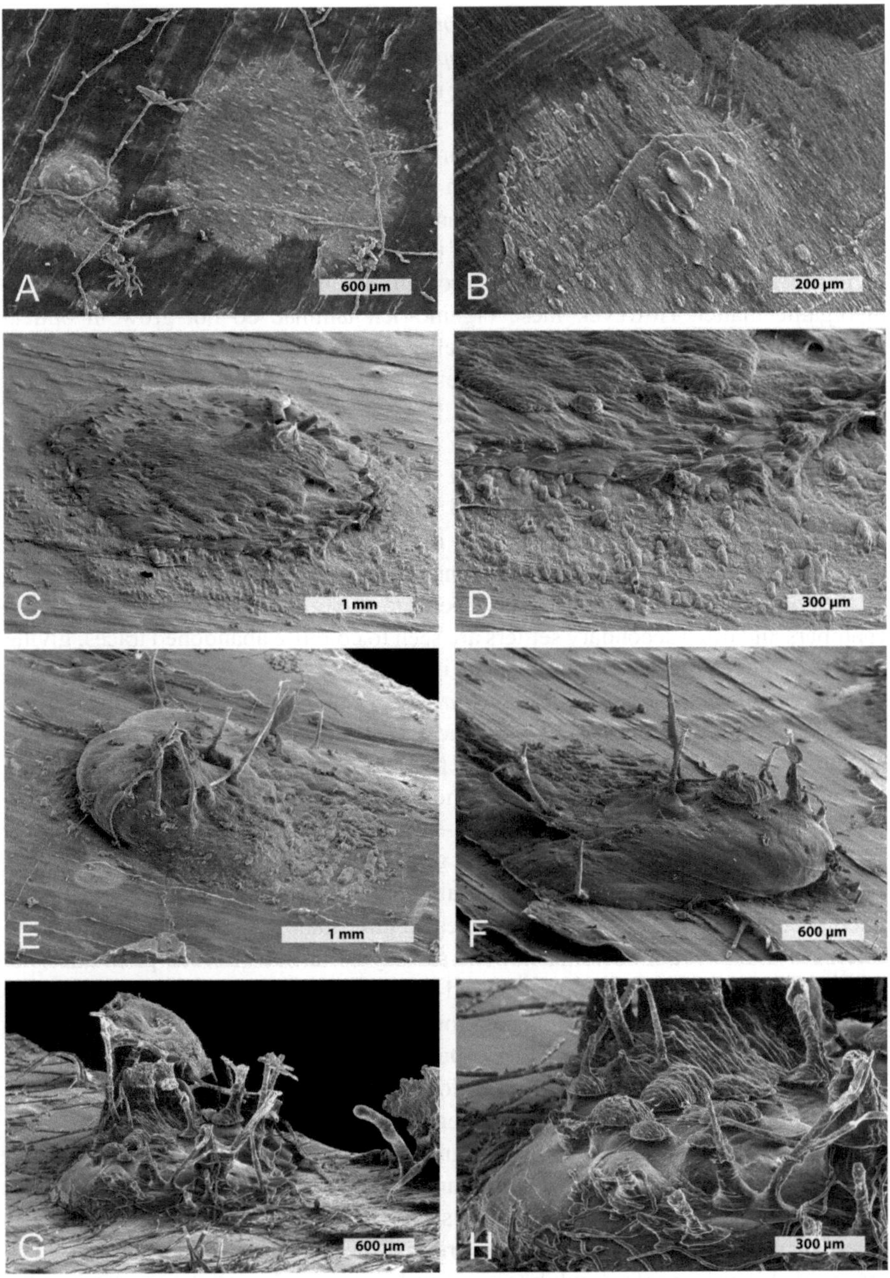

Fig. 11 SEM images of *Hyrrokkin sarcophaga* trace casts (*Kardopomorphos polydioryx* igen. n., isp. n.) in *Acesta excavata* high-lighting different surface textures and trace morphologies (A-H Pleistocene, Strait of Sicily, COR2-08). **A** Shallow etch groove of a young infestation stage, next to *Orthogonum* isp. traces. Deeper portions are elongated within the axes of the external fibrous prismatic calcite. **B** Detail of a young infestation stage within the fibrous

Differential diagnosis: Kardopomorphos igen. n. differs from the circular borings of *Oichnus* Bromley, 1981 in shape. At first glimpse incomplete *Oichnus* (see Nielsen et al. 2003), with a hemispherical shape, may be mistaken for *Kardopomorphos* igen. n., but the latter shows in addition often a spiral array, which corresponds to the individual foraminifer chambers. While incomplete *Oichnus* comprises a single smooth-walled hemispherical depression with a roughly circular diameter (e.g., Bromley 2005: fig. 3), the sub-circular depression of *Kardopomorphos* igen. n. is usually shallower and more dish-shaped, with numerous canals exiting the depression in *Kardopomorphos polydioryx* igen. n., isp. n. Moreover, facultatively the main depression may show a shallow lateral extension, which lacks in *Oichnus*.

The early stages of *Anellusichnus* Santos, Mayoral and Muñiz, 2005, produced by barnacles, consist of a circular shallow groove with a flat bottom often highlighted as discoloured area. The initial groove shows no further morphological sub-distinction, while *Kardopomorphos* igen. n. can show a spiral array of sub-depressions, corresponding to individual chambers of the trace-making foraminifer. Fully developed *Anellusichnus* ispp. (Santos et al. 2005) show one or more concentric external furrows, sometimes undulating to subpolygonal, around the initial shallow groove, which make them clearly distinct from *Kardopomorphos* igen. n. Further, the trace described herein differs from the shallow etchings of *Centrichnus concentricus* Bromley and Martinell, 1991, attributed to verrucid barnacles, by its hemispherical and often spiral shape, lacking the arcuate concentrically to ring-shaped arranged sub-grooves.

Sessile foraminifera, like *Cibicides* spp., may produce similar grooves as *Hyrrokkin* spp. with spiral segmentation and may thus become a second ichnospecies of *Kardopomorphos* igen. n., which is lacking in the characteristic 'whip'-shaped filaments and the facultative central canal of *Kardopomorphos polydioryx* igen. n., isp. n. At present no fossil material is available to us, to establish this potential new subspecies; in Recent cold-water coral habitats *Cibicides lobatulus* (Walker and Jacob, 1798) produces this trace, described as 'Foraminiferan-form 1' in Wisshak (2006). Similarly the 'Foraminiferan-form 2' (see Wisshak 2006), produced by *Gypsina vesicularis* (Parker and Jones, 1860), has a hemispherical shape with a perfectly circular outline of up to 1 mm in diameter and with a characteristic central depression (Wisshak 2006). Further, the few and occasionally radiating grooves are lacking in *Kardopomorphos* igen. n.

prismatic calcite, orthogonal lines are the external bivalve growth increments. **C** Like A and B but with the groove also penetrating into the underlying microgranular calcite. Very few 'whips' and one central canal (cast broken off). **D** Close up of C showing the transition between the two calcitic microstructures. **E** Adult groove with a higher number of 'whips', smooth surface within the microgranular calcite. **F** Same as E note the 'overhang' of the groove. **G** Fully developed trace of *Kardopomorphos polydioryx* igen. n., isp. n., with numerous canals and 'whips' penetrating through the crossed lamellar aragonite, covered by organic matrix of the callus (holotype). **H** Close-up of the groove at the transition from microgranular calcite (smooth texture) to crossed-lamellar aragonite (ribbed texture)

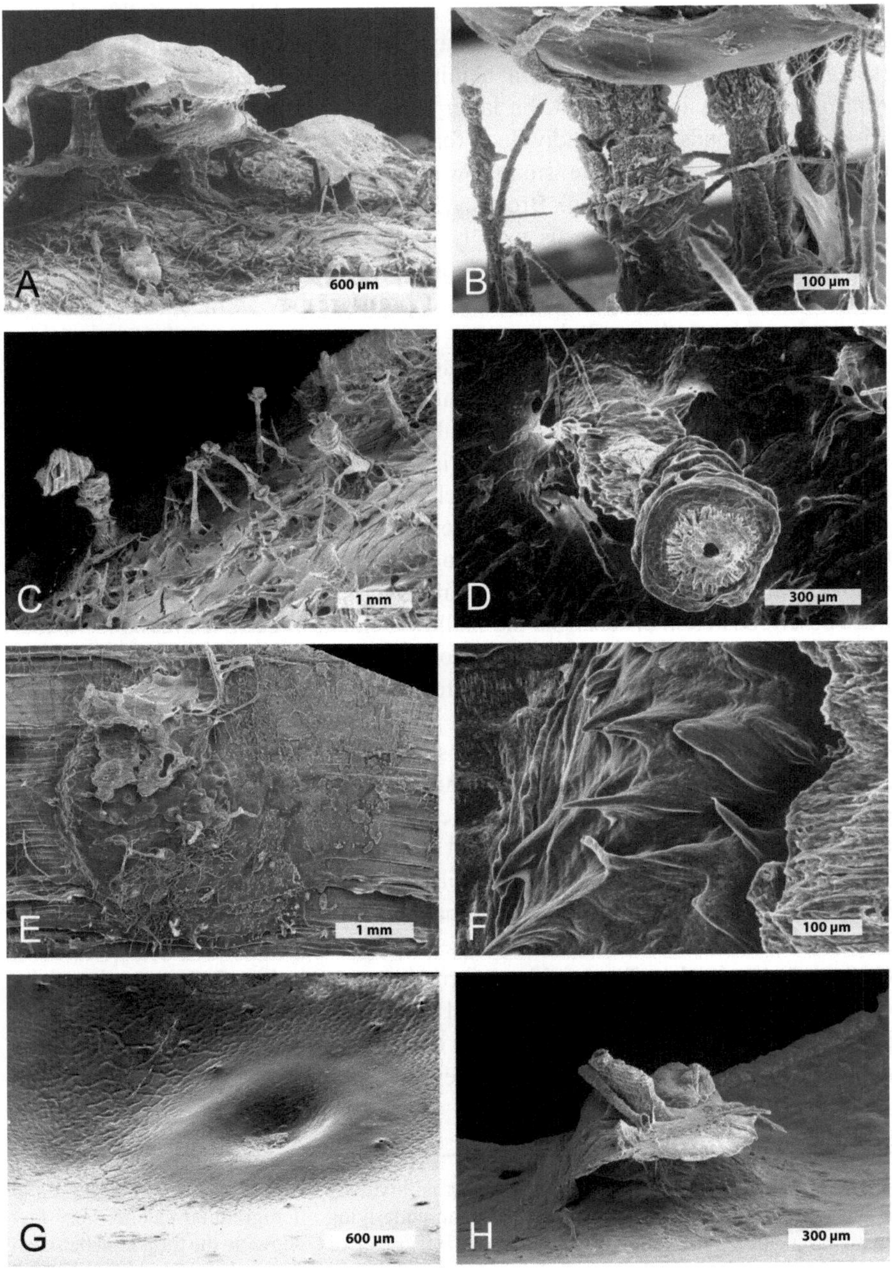

Fig. 12 SEM images of *Hyrrokkin sarcophaga* trace casts in *Acesta excavata* (A: Recent, Sula Ridge, Pos228-200-3; B: Recent, W-Oslofjord, ALK232-BG-1066; C-D: Recent, Mauritania, MAU 2; E-H: Pleistocene, Strait of Sicily, COR2-08). **A** Large canals of adult traces, corresponding to the parasitic feeding behaviour. The large sheets are organic matrix of the bivalves foliated repair layer (callus). **B** Two canals (detail of Fig. 8A) which reached the mantle cavity, sealed off by an organic matrix sheet (above) of the callus. Their surface

Remarks: The parasitic foraminifer *H. sarcophaga* is preferentially attached to the bivalve *Acesta excavata* and to the azooxanthellate coral *Lophelia pertusa*. On the Norwegian Shelf and in the Scandinavian fjords, both hosts are highly abundant and the parasite is virtually present at all localities (see also Burden-Jones and Tambs-Lyche 1960; Jensen and Frederiksen 1992). The northernmost locality is at Stjernsund at 70°N, which comprises also the lowest temperature of 5 to 6°C and the abundance of the parasite is markedly reduced. *Acesta excavata* has also been reported from Newfoundland and from the Iceland Shelf, but yet the presence of the foraminifer could not be verified. Cedhagen (1994) noted that *A. excavata* from the Kosterfjord, although some hundred specimens were collected, had not shown an infestation by *H. sarcophaga*. Samples of *A. excavata* collected during the R/V Alkor cruise 232 to the Kosterfjord and the Skagerrak in November 2003 showed instead traces of the foraminifer at most sampled sites, including Kosterfjord. Both, subfossil and recently dead shells were infested.

Lophelia pertusa is widespread along the Celtic margin and especially within the Porcupine Seabight the parasitic foraminifer is highly abundant. *Acesta excavata* instead is only occasionally reported south of the Faroe-Shetland channel (López Correa et al. 2005), but the parasite has not been mentioned to be associated with the bivalve in that area. From the Porcupine Seabight and the Rockall Bank *H. sarcophaga* could further be documented on *Caryophyllia sarsiae*. The *Hyrrokkin* species from the Portuguese margin and from the Mediterranean can be assigned to *H. sarcophaga* (see also Schönfeld 1997).

The fossil traces from the Pleistocene Mediterranean clearly point to the parasitic boring activity of *Hyrrokkin sarcophaga* in particular within its host *Acesta excavata*. *A. excavata* is present in the Recent Mediterranean, but extremely scarce (López Correa et al. 2005). Their shells do not carry the parasite traces and the foraminifer itself has not been reported from the present-day basin. Likewise radiocarbon-dated *A. excavata* shells from the Early to mid-Holocene Mediterranean, which experienced similar climatic conditions as today (López Correa and Taviani, unpublished ^{14}C-data), do not contain traces of the foraminifer. Last glacial and older Pleistocene valves (López Correa and Taviani, unpublished geochronological data) on the contrary show a dense infestation by the foraminifera, indicating a high parasite abundance. From the last glacial period, only one *Lophelia pertusa* has been

texture perfectly mimics the cross-lamellar microstructure of the internal aragonite layer of *Acesta excavata*. This xenoglyphic texture is also seen imprinted on the smaller 'whips'. **C** Smooth groove surface within the microgranular calcite layer, with numerous 'whip'-shaped filaments distributed over the entire surface, vertically penetrating the crossed-lamellar aragonite underneath. Canals which reached the mantle cavity show a larger diameter, their thicker tips relate to brocken-off callus casts. **D** Detail of a 'whip' from C. **E** Sharp outline of the groove, with the typical shallow extension below the latest test chamber (right). Only few 'whip' are out side the main groove. Their varying shapes depend on the penetrated microstructures of the bivalve shell. **F** Smooth surface texture of the groove within the microgranular calcite and ribbed surface within the crossed-lamellar aragonite (right). **G** Cast of the smooth aragonitic callus surface, showing the sealed central canal. **H** Complex cast structure within the crossed-lamellar aragonite and the callus, the foliated structure corresponds to insoluble organic matrix sheets incorporated in the layered callus

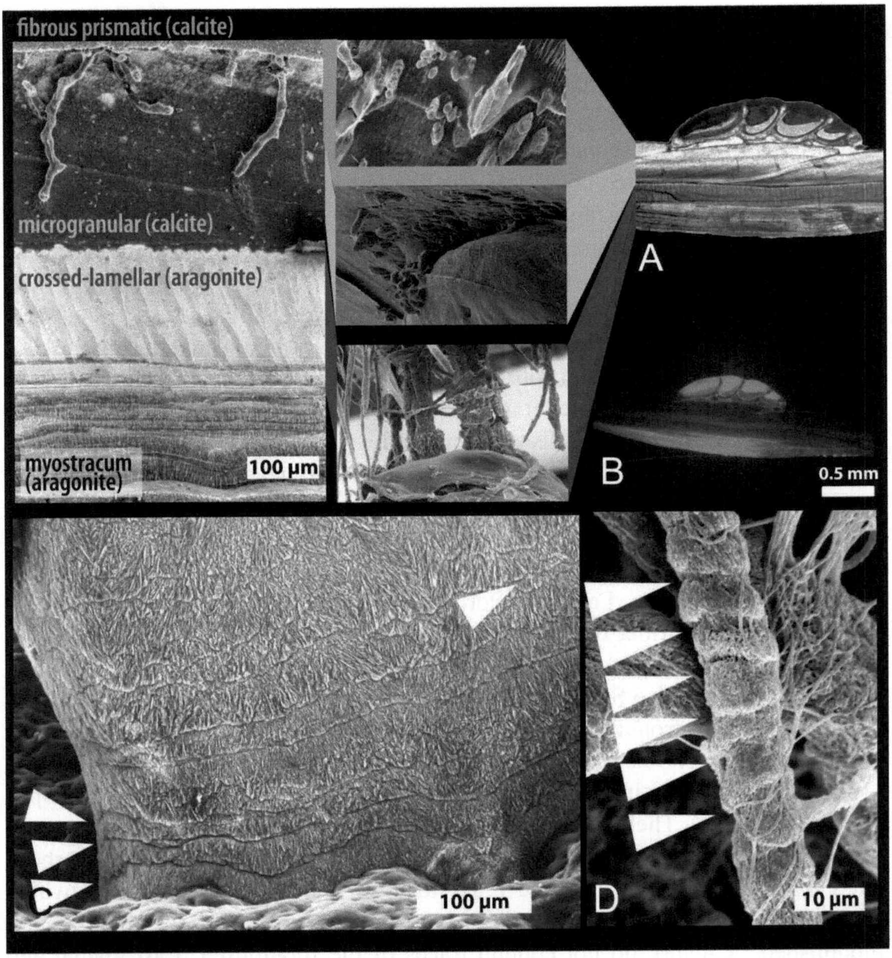

Fig. 13 A SEM-image of the microstructure layers in *Acesta excavata* – fibrous prismatic external calcite, microgranular calcite and crossed lamellar aragonite. The trace surface texture of *Kardopomorphos polydioryx* igen. n., isp. n. shows a clear dependence from these microstructures. This indicates an etching boring strategy, which leads to a xenoglyphic surface texture. Thin-section under transmitted light showing the three principal shell layers of *A. excavata*, the fourth innermost layer is the callus formed during and after the parasitic attack of *Hyrrokkin sarcophaga*. This aragonitic shell repair layer is composed of foliated organic-rich strata, which had been quickly precipitated. **B** Under fluorescent light the organic portions within the Recent foraminifer and of the callus exhibit a strong luminescence. **C** Microstructure (SEM-image) of the thecal aragonite in *Lophelia pertusa*, with bands composed of aragonite bundles. **D** The incremental growth of the 'whip'-shaped filaments of *Kardopomorphos polydioryx* igen. n., isp. n. might be related to this microstructure of *L. pertusa* (Pos228-215)

Fig. 14 SEM images of secondary settlers of *Hyrrokkin* traces in *Acesta excavata* (A-F: Pleistocene, Strait of Sicily, COR2-08; G-H: Mauritania, MAU 2). **A** Colonisation by a boring worm (centre) and by a boring sponge. **B** Initial infestation of a 'whip'-shaped

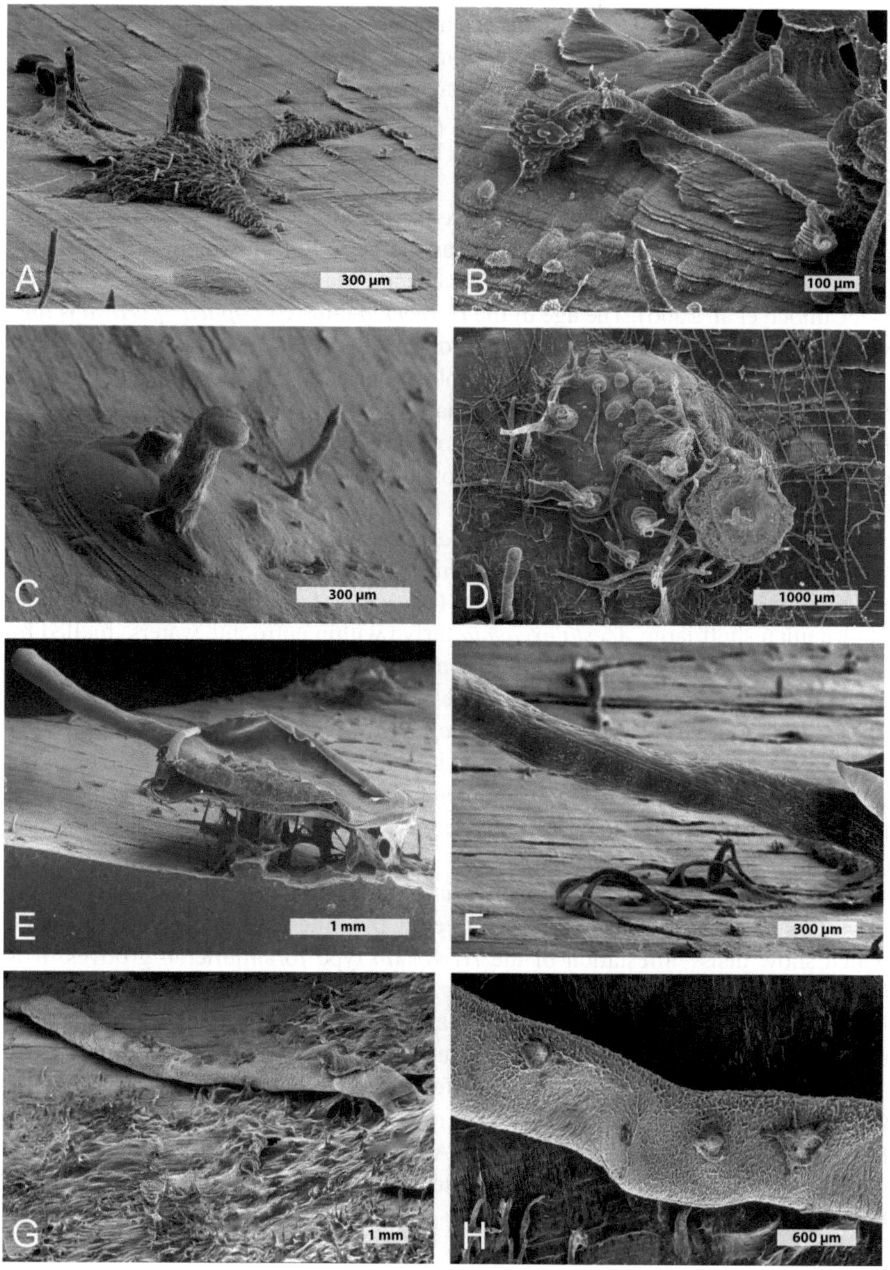

filament by a boring sponge. **C** Juvenile *Hyrrokkin* trace stage infested by a boring annelid. **D** Occurrence of the fungal trace *Orthogonum tubulare* at the foraminiferan trace periphery; note the annelid trace *Caulostrepsis* isp. on a small-sized *Hyrrokkin* trace (lower left corner). **E** Surface parallel orientation of a *Trypanites weisei* boring. **F** Close-up of E; note the imprint of the shell texture on the trace surface. **G** Surface paralleling *Caulostrepsis* isp. **H** Detail of G; note the apical portions of *Hyrrokkin* 'whips' on the *Caulostrepsis* isp. cast surface observed with traces of *H. sarcophaga*, while last glacial shells of *A. excavata* from the same sites show a massive infestation. It is noteworthy to mention that fossil *L. pertusa* with the foraminiferan trace have been found in the Lower Pleistocene deposits on Rhodes (Bromley 2005) and *Hyrrokkin* tests have further been found in similar deposits on Sicily (collected by authors). From the submerged pre-Last Glacial cold-water coral habitats *A. excavata* from Cialdi Seamount (López Correa et al. 2005) showed to be infested by *H. sarcophaga*. Ongoing U / Th-dating efforts reveal the Younger Dryas as the period of maximum growth of the bathyal Mediterranean corals, while Last Glacial Maximum ages are absent so far (McCulloch et al. 2006). The age of the *Hyrrokkin* infested *Lophelia pertusa* is thus very likely from the Younger Dryas period as well.

Intense sampling of the lush Recent cold-water coral colonies at Santa Maria di Leuca (Apulia, S-Italy), with conventional grab and dredge sampling, as well as ROV-examinations during the R/V Meteor M70/1 cruise yielded a wealth of live coral samples. However, none of them proved to be infested by *H. sarcophaga*, nor have their traces been found. This picture is in line with the Holocene and present-day absence of *H. sarcophaga* on *Acesta excavata*.

On the western African margin the same parasitic behaviour is documented for the foraminifer *H. carnivora* on its host *Acesta angolensis* and on the present-day Mauritanian shelf large *Acesta excavata* thrive within eutrophic waters, fuelled by the strong up-welling system. This Recent population is infested by the parasitic foraminifer *H. carnivora*, analogues to the situation on the Norwegian Shelf, where *Acesta excavata* is infested by *H. sarcophaga*. However, *Lophelia pertusa* and other deep-sea corals have not been found to be infested, but this lack is likely to be based on the scarce availability of coral samples from this area. Furthermore, *Acesta* sp. from the Mid-Atlantic Ridge also showed traces of the foraminiferan infestation (R.G. Bromley pers. comm. 2005).

On the deep continental margin of Guinea and Congo (Sao Tome), fossil valves of *Acesta* cf. *excavata* have been recovered by American expeditions in 1963, which show dense infestation of the *Hyrrokkin* type foraminiferan grooves. It is likely that these correspond to fossil *H. carnivora*. But absolute radiometric data are lacking and we thus base the description and establishment of the new ichnogenus and ichnospecies on the fossil Mediterranean material. Fossil *Acesta* cf. *excavata* material from Jamaica does not show infestation by *Hyrrokkin*-type foraminifera. Guilbault et al. (2006: fig. 10.15-10.17) reports Recent occurrences of the genus from the West Canadian Shelf and Heron-Allen and Earland (1922) from Antarctica. Similar traces have furthermore been described by Neumann and Wisshak (2006) from Late Cretaceous sea urchins.

Discussion

The parasitic habit

The xenoglyphic surface texture of *Hyrrokkin* indicates a carbonate removal from the host skeleton by chemical dissolution (corrosion). The distinct increments on 'whip'-shaped filaments in *Lophelia pertusa* could further be interpreted as an incremental growth, however, these structures can also derive from the host specific skeletal microstructure (compare Fig. 13C-D). A parasitic behaviour has already been described in detail by Cedhagen (1994) for *H. sarcophaga* on *Acesta excavata*, showing that pseudopodia were found to project through the etched canal. The mantle tissue of the bivalve close to the pseudopodia, particularly in the epithelium contained a larger number of nuclei than non-infested parts of the mantle indicating high rates of metabolism and cell division in the infested parts. Further, Cedhagen reports that some parts of the mantle epithelium were destroyed by the foraminifer to allow its pseudopodia to penetrate the sub-epithelial parts of the mantle as well. However, these kinds of completely penetrating, larger-sized canals could not be documented for *Hyrrokkin* traces within *Lophelia pertusa*. Consequently there is no evidence of active etching to reach the internal polyp, to attain an active parasitic life-style as in *Acesta* spp. Nevertheless, the organic content within the host skeleton can serve the foraminifer as food source. The attack of foraminifera on the coral polyp has no lethal effect and the polyp does not precipitate an aberrant skeleton as could be expected. The depression as well as the 'whip'-shaped filaments thus can be primarily interpreted as an anchor function to enhance the attachment strength. However, on a live *Caryophyllia sarsiae* sample a larger sized individual of *H. sarcophaga* is positioned in close vicinity to the host tissue and its protoplasm is concentrated towards the lower border of the polyp (Fig. 3A). Its trace cast consists of a depression and only few, shallow penetrating 'whip'-shaped filaments, giving no evidence for a parasitic behaviour. The external shallow depression of the foraminiferan trace produced by the latest chamber shows *Saccomorpha clava* and *Orthogonum tubulare* trace remains, indicating that these skeleton areas have previously not been covered by the host tissue. But since the foraminiferan test has grown quite large, it can be considered that the juvenile test must have been positioned at least within the polyp's edge zone if not in the live portions and since the test does not show any partial embedding by the polyp, which is a quite common defence mechanism against epizoans amongst corals (see Harmelin 1990), it can be taken into account that *H. sarcophaga* possesses the ability to feed on the bordering host tissue. The same can be concluded for individuals that are positioned in live portions of *Lophelia pertusa* skeletons, presuming that the adjacent tissues serve as their main food source. However, *Hyrrokkin* traces on dead *Caryophyllia sarsiae* hosts show that the foraminifer has formed in later stages on its deepest point of the depression an increased number of vertical 'whip'-shaped filaments forming a larger size canal and protruding deeply into the substrate, reaching the inner calyx area. This trace variation documented within one host species might indicate a change to a vertical extension of the trace when loosing the contact to neighbouring host tissue, attempting to reach the inner calyx area, which contains the polyp tissue, since the host polyp is tending to grow upwards. All *Caryophyllia* samples studied show bent corallite morphologies, the polyps most probably face the main direction of food supply and thus the spatial occurrence of *H. sarcophaga* might indicate a preference or a facilitated attachment on current exposed sites of hosts. Faster bioerosion successions of these sides are already reported by Beuck and Freiwald (2005). Cedhagen (1994) reports further trace variations such as the amount of pits per individual, since more pits have been counted on sponges than on bivalves and relates it to the differences in response and nutritional content of the hosts like local necrosis of the sponge tissue.

We assume that the degree of influence on the test morphology of *H. sarcophaga* by the host substrates analysed (see also Cedhagen 1994) is lowest for the geodiid sponges, which might be due to their relatively soft skeleton, in contrast to individuals that colonise hard substrates, like coral skeletons. Thus, it can be considered that the shape of tests on geodiid sponges might resemble the true isomorphic test shape (Fig. 6F) and planar to concave umbilical sides derived from the individual skeletal host topographies (see Fig. 7A-J, M-N).

The life-time of *Acesta* exceeds that of the foraminifer by far, and adult shells of *Acesta excavata* often show a large number of *H. sarcophaga* traces, but few still contain the foraminifer in place. In contrast, Cedhagen (1994) reports *Delectopecten vitreus* specimens with fatal infestation, showing that a small host is unfavourable for *H. sarcophaga*, because such a host's nutritional contents cannot support the parasite long enough. Abandoned grooves are sites of subsequent secondary bioerosion and can thus be overprinted by traces such as *Trypanites weisei* Mägdefrau, 1932, *Caulostrepsis* isp., *Spathipora* isp., *Entobia* ispp., *Orthogonum tubulare* Radtke, 1991, and *Saccomorpha clava* Radtke, 1991 (see Fig. 14).

Geographical and temporal distribution of *Hyrrokkin*

The present-day bathyal temperature of 13.9°C throughout the Mediterranean basin is most likely too warm for the foraminifer. The temporal distribution pattern of *Hyrrokkin* may be related to the different temperature regimes throughout past climatic glacial-interglacial climate changes. Considering the Holocene and present-day absence of *H. sarcophaga* on *Acesta excavata* in the warm Mediterranean a temperature dependence could also explain why the foraminifer is not recorded from the Recent Moroccan margin and very scarce on the Atlantic Iberian margin, while the glacial Mediterranean shows numerous occurrences. The Gulf of Cadiz and the adjacent Atlantic regions are affected by the warm and salty Mediterranean Outflow Water (MOW), which has a typical temperature of about 12°C. To the north and the south, the foraminifer is highly abundant where temperatures are below 10°C, like in the Porcupine Seabight, or, for example on the Mauritanian continental slope. The biogeographic distribution of *H. sarcophaga* and *H. carnivora* forms two clearly separated zones today, with *H. sarcophaga* occurring from northern Norway to Portugal and *H. carnivora* from Mauritania to Congo. Traces of the two species cannot be distinguished. The glacial presence of a *Hyrrokkin* species within the Mediterranean basin is indicated by the abundance of its traces on the typical hosts *Acesta excavata* and *Lophelia pertusa*. Although none of these fossil host samples had the parasitic foraminifer still attached, we can assign those traces to *H. sarcophaga*, since it has been found within the same sedimentary units on the seabed or in outcrops. Their planar to plano-concave umbilical sides indicate an attachment to planar to plano-convex-shaped host substrates, e.g., bivalve shells.

The available set of radiometric data (McCulloch et al. 2006) shows a strong dependence of faunal shifts from palaeoceanographic changes. Together with the absence of *H. sarcophaga* from interglacial deposits, it allows us to use this foraminifer in the Mediterranean as an indicator of glacial conditions. It has to be tested with geochemical methods, if this holds true also for Early and Mid-Pleistocene climatic cycles, when the bathyal cold-water coral association showed much higher degrees of similarity with the present-day Atlantic coral association.

Acesta excavata is extremely rare south of the Faroe-Shetland Channel and *H. sarcophaga* is only present on its coral host *Lophelia pertusa* and rarely on *Caryophyllia* spp. in the Porcupine Seabight. It is noteworthy that the limid bivalve *Lima marioni* Fischer, 1882 occupies the same ecological niche as *Acesta excavata*, by nestling within the growing coral framework. The bivalve is byssally attached to the tissue-free coral skeleton just below their active growth zone. *Lima marioni* is present from the Gulf of Cadiz to the Rockall Bank, but is lacking in the Mediterranean. *Lima marioni* is very frequent on the Galicia Bank (off north-western Portugal), settling on *Madrepora oculata*. Despite these similarities, *Lima marioni* has not been observed to be a host for the parasitic foraminifer *H. sarcophaga*. This is even more evident in the Porcupine Seabight, where *Lophelia pertusa* is heavily infested and *Lima marioni* from the same site is free from the parasitic foraminifer. Potentially the spiny morphology of the shell inhibits the growth of the foraminifer.

Lima marioni is present on the Mauritanian shelf within fossil cold-water coral deposits, together with *Acesta excavata*. Equal abundances of *Acesta excavata* and *Lima marioni* occur within the same stratigraphic horizons in cores from the Moroccan Shelf, which have been assigned to an age of >200 ka (N. Frank pers. comm.). These valves show no infestation by *Hyrrokkin*, which may point to an interglacial deposition with palaeotemperatures beyond the suitable range for the parasite, similar to the lack of *Hyrrokkin* in the present-day Mediterranean.

Conclusions

- *Kardopomorphos polydioryx* igen. n., isp. n. contains two obligate trace morphology features, which are a depression and multiple 'whip'-shaped filaments, and possesses two facultative features, namely one or more deeply penetrating larger sized canals as known to occur in *Caryophyllia* and *Acesta* and a callus cast, which is uniquely attributed to *Acesta*, resulting from a defence mechanism against the parasitic habit of the host.
- *Kardopomorphos polydioryx* igen. n., isp. n. is made by *Hyrrokkin*.
- The high morphological variability of *K. polydioryx* isp. n., isp. n. depends on the host, host substrate and the feeding behaviour of *Hyrrokkin*: the trace morphology is substrate-size dependent (stenomorph) and its texture varies with penetrated substrate mineralogy and microstructure (xenoglyph).
- The depression as well as the 'whip'-shaped filaments are interpreted to be primarily anchor structures in contrast to the deeply penetrating centrally arranged canals that are seen as true parasitic structures.

- The present-day bathyal temperature of 13.9°C throughout the Mediterranean basin is most likely too warm for the foraminifer. The temporal distribution pattern of *Hyrrokkin* may be related to the different temperature regimes throughout past glacial-interglacial climate changes.

Acknowledgements

M. Taviani (CNR-ISMAR, Bologna) kindly provided fossil *A. excavata* shells from several sites in the Mediterranean Sea for this study and moreover permitted the storage of the type material in the trace fossil collection of the Natural History Museum of the Humboldt-University in Berlin, Germany. We are very much indebted to H. Zibrowius (Marseille, France) and M. Wisshak (IPAL, University of Erlangen, Germany) for inspiring discussions. We are grateful to I. Di Geronimo and A. Vertino (University of Catania, Italy) for showing the outcrop in Salice and to provide us with detailed information about the outcrop. L. Schäffer (University of Tübingen, Germany) provided the trace name suggestions and the name etymology. Fig. 2B is included with courtesy of K. Kaszemeik (Bremen, Germany). Our reviewers T. Cedhagen (Århus, Denmark) and J. Schönfeld (IFM-GEOMAR, Kiel, Germany), provided constructive comments, which helped to improve the manuscript. This is a contribution to WP2 within the HERMES-Project (Contract No: GOCE-CT-2005-511234-1) of the European Commission.

References

Adam W, Knudsen J (1955) Note sur quelques espèces de mollusques marins nouveaux ou peu connus de l'Afrique occidental. Bull Inst Roy Sci Nat Belg 31(61):1-25

Alexander SP, DeLaca TE (1987) Feeding adaptations of the foraminiferan *Cibicides refulgens* living epizoically and parasitically on the Antarctic scallop *Adamussium colbecki*. Biol Bull 173:136-159

Baumfalk YA, Fortuin AR, Mok RP (1982) *Talpinella cunicularia* n. gen., n. sp., a possible foraminiferal parasite of Late Cretaceous *Orbitoides*. J Foram Res 12:185-196

Beuck L, Freiwald A (2005) Bioerosion patterns in a deep-water *Lophelia pertusa* (Scleractinia) thicket (Propeller Mound, northern Porcupine Seabight). In: Freiwald A, Roberts JM (eds) Cold-water corals and ecosystems. Springer, Berlin Heidelberg, pp 915-936

Beuck L, Vertino A, Stepina E, Karolczak M, Pfannkuche O (2007) Skeletal response of *Lophelia pertusa* (Scleractinia) to bioeroding sponge infestation visualised with micro-computed tomography. Facies 53:157-176

Boss KJ (1965) Note on *Lima (Acesta) angolensis*. Nautilus 79:54-58

Bowerbank JS (1858) On the anatomy and physiology of the Spongiadae. Part I. On the spicula. Phil Trans Roy Soc London 148:279-332

Brady HB (1884) Report on the foraminifera dredged by H.M.S. Challenger during the years 1873-1876. Rep Sci Results Voyage H.M.S. Challenger, 1873-1876, Zool 9:1-814

Bromley RG (1970) Borings as trace fossils and *Entobia cretacea* Portlock, as an example. In: Crimes TP, Harper JC (eds) Trace fossils. Geol J Spec Issue 3:49-90

Bromley RG (1981) Concepts in ichnotaxonomy illustrated by small round holes in shells. Acta Geol Hisp 16:55-64

Bromley RG (2005) Preliminary study of bioerosion in the deep-water coral *Lophelia*, Pleistocene, Rhodes, Greece. In: Freiwald A, Roberts JM (eds) Cold-water corals and ecosystems. Springer, Berlin Heidelberg, pp 895-914

Bromley RG, Martinell J (1991) *Centrichnus*, new ichnogenus for centrically patterned attachment scars on skeletal substrates. Bull Geol Soc Denmark 38:243-252

Bromley RG, Wisshak M, Glaub I, Botquelen A (2007) Ichnotaxonomic review of dendriniform borings attributed to foraminifera: *Semidendrina* igen. nov. In: Miller III W (ed) Trace fossils: concepts, problems, prospects. Elsevier, Amsterdam, pp 518-530

Burdon-Jones C, Tambs-Lyche H (1960) Observations on the fauna of the North Brattholmen stone-coral reef near Bergen. Årb Univ Bergen 29:1-24

Cedhagen T (1994) Taxonomy and biology of *Hyrrokkin sarcophaga* gen. et sp. n.: a parasitic foraminiferan (Rosalinidae). Sarsia 79:65-82

Cherchi A, Schroeder R (1991) Perforations branchues dues à des foraminifers cryptobiotiques dans des coquilles actuelles et fossils. C R Acad Sci Paris, Sér II 312:111-115

Cherchi A, Schroeder R (1994) Schizogony of an early Barremian cryptobiotic miliolid. Boll Soc Paleont Ital, Spec Vol 2:61-65

DeLaca TE, Lipps JH (1972) The mechanism and adaptative significance of attachment and substrate pitting in the foraminiferan *Rosalina globularis* D'Orbigny. J Foram Res 2:68-72

Fabricius JC (1779) Reise nach Norwegen mit Bemerkungen aus der Naturhistorie und Oekonomie. Bohn, Hamburg, 388 pp

Fischer P (1882) Diagnoses d'espèces nouvelles de mollusques recueillis dans la cours des expéditions scientifiques de l'aviso le Travailleur (1880 et 1881). J Conchyl 30:49-53

Freiwald A, Schönfeld J (1996) Substrate pitting and boring pattern of *Hyrrokkin sarcophaga* Cedhagen, 1994 (Foraminifera) in a modern deep-water coral reef mound. Mar Micropaleont 28:199-207

Gmelin JF (1791) Caroli a Linné, systema naturae per regna tria naturae. Editio decima tertia. Beer, Lipsiae, Vol 1, Pars 6, pp 3021-3910

Guilbault J-P, Krautter M, Conway KW, Barrie JV (2006) Modern Foraminifera attached to hexactinellid sponge meshwork on the West Canadian shelf: comparison with Jurassic counterparts from Europe. Palaeont Electr 9(1), 48 pp [http://www.palaeo-electronica.org/2006_1/sponge/issue1_06.htm]

Gunnerus JE (1763) Om en Søevext, allevegne ligesom besat med Frøehuse, *Gorgonia resedæformis*. Trondhiem Selsk Skr 2:321-329

Harmelin JG (1990) Deep-water crisiids (Bryozoa: Cyclostomata) from the northeast Atlantic Ocean. J Nat Hist 24:1597-1616

Heron-Allen E, Earland A (1922) Protozoa, Part II. Foraminifera. Brit Antarct ('Terra Nova') Exped 1910, Zool 6(2):25-268

Jensen A, Frederiksen R (1992) The fauna associated with the bank-forming deepwater coral *Lophelia pertusa* (Scleractinaria) on the Faroe shelf. Sarsia 77:53-69

Linnaeus C (1758) Systema naturae per regna tria naturae, secundum classes, ordines, genera, species, cum characteribus, differentiis, synonymis, locis. Tomus I. Editio decima, reformata. Laurentii Salvii, Holmiae, 824 pp

López Correa M (2004) Evaluation of *Acesta excavata* from fossil Mediterranean and Recent North Atlantic sites as a new tool for palaeoenvironmental studies. MSc Thesis, Univ Tübingen, 69 pp

López Correa M, Freiwald A, Hall-Spencer J, Taviani M (2005) Distribution and habitats of *Acesta excavata* (Bivalvia: Limidae), with new data on its shell ultrastructure. In: Freiwald A, Roberts JM (eds) Cold-water corals and ecosystems. Springer, Berlin Heidelberg, pp 173-205

Mägdefrau K (1932) Über einige Bohrgänge aus dem unteren Muschelkalk von Jena. Paläont Z 14:150-160

Marche-Marchad I (1979) Sur le foraminifère *Rosalina carnivora* Todd, 1964 (Discorbidae) parasite d'un Lamellibranche. Bull IFAN, Ser A/1, 41:103-111

McCulloch MT, Taviani M, Montagna P, Mortimer G, Remia A (2006) Proliferation and demise of Mediterranean deep-sea corals. Geochim Cosmochim Acta, 70(18), Suppl 1, p A407

Montfort D de (1808) Conchyliologie systématique, et classification méthodique des coquilles; offrant leurs figures, leur arrangement générique, leurs descriptions caractéristiques, leurs noms; ainsi que leur synonymie en plusieurs langues. Ouvrage destiné à faciliter l'étude des coquilles, ainsi que leur disposition dans les cabinets d'histoire naturelle. Coquilles univalves, cloisonnées. Schœll, Paris, Vol 1, 409 pp

Neumann C, Wisshak M (2006) A foraminiferal parasite on the sea urchin *Echinocorys*: ichnological evidence from the Late Cretaceous (Lower Maastrichtian, Northern Germany). Ichnos 13:185-190

Nielsen KSS, Nielsen JK (2001) Bioerosion in Pliocene to late Holocene tests of benthic and planktonic foraminiferans, with a revision of the ichnogenera *Oichnus* and *Tremichnus*. Ichnos 8:99-116

Nielsen KSS, Nielsen JK, Bromley RG (2003) Palaeoecological and ichnological significance of microborings in Quaternary foraminifera. Palaeont Electr 6(2), 13 pp [http://www.palaeo-electronica.org/2003_1/ichno/issue1_03.htm]

Orbigny AD de (1826) Tableau méthodique de la classe des Céphalopodes. *Ann Sci Nat Paris, Ser I,* 7:245-314

Parker WK, Jones TR (1860) On the nomenclature of foraminifera. IV. The species enumerated by Lamarck. Ann Mag Nat Hist 3:29-40

Parker WK, Jones TR (1865) On some foraminifera from the North Atlantic and Arctic Oceans, including Davis Strait and Baffin's Bay. Phil Trans Roy Soc London 155:325-441

Plewes CR, Palmer TJ, Haynes JR (1993) A boring foraminiferan from the Upper Jurassic of England and Northern France. J Micropalaeont 12:83-89

Poag CW (1971) Notes on the morphology and habit of *Vasiglobulina alabamensis* (Foraminiferida). J Paleont 45: 961-962

Radtke G (1991) Die mikroendolithischen Spurenfossilien im Alt-Tertiär West-Europas und ihre palökologische Bedeutung. Courier Forschinst Senckenberg 138:1-185

Santos A, Mayoral E (2006) Bioerosive structures of sclerozoan foraminifera from the Lower Pliocene of Southern Spain: a contribution to the palaeoecology of marine hard substrate communities. Palaeontology 49:719-732

Santos A, Mayoral E, Muñiz F (2005) Bioerosion scars of acorn barnacles from the southwestern Iberian peninsula, Upper Neogene. Riv Ital Paleont Stratigr 111:181-189

Schönfeld J (1997) The impact of the Mediterranean Outflow Water (MOW) on benthic foraminiferal assemblages and surface sediments at the southern Portuguese continental margin. Mar Micropaleont 29:211-236

Smyth MJ (1988) The foraminifer *Cymbaloporella tabellaeformis* (Brady) bores into gastropod shells. J Foram Res 18:277-285

Sollas WJ (1880) The sponge-fauna of Norway; a report on the Rev. A.M. Norman's collection of sponges from the Norwegian Coast. Ann Mag Nat Hist, Ser 5, 5:130-144, 241-259, 396-409

Taviani M, Remia A, Corselli C, Freiwald A, Malinverno E, Mastrototaro F, Savini A, Tursi A (2005) First geo-marine survey of living cold-water *Lophelia* reefs in the Ionian Sea (Mediterranean basin). Facies 50:409-417

Titschack J, Bromley RG (2005) Plio-Pleistocene cliff-bound, wedge-shaped, warm-temperate carbonate deposits from Rhodes (Greece): sedimentology and facies. Sediment Geol 180:29-56

Todd R (1965) A new *Rosalina* (Foraminifera) parasitic on a bivalve. Deep-Sea Res 12:831-837

Vénec-Peyré MT (1991) Distribution of living benthic foraminifera on the back-reef and outer slopes of a high island (Moorea, French Polynesia). Coral Reefs 9:193-203

Vénec-Peyré MT (1996) Bioeroding foraminifera: a review. Mar Micropaleont 28:19-30

Vertino A (2003) Sclerattiniari Plio-Pleistocenici ed attuali del Mediterraneo (Sistematica, Biostratinomia e Paleoecologia). PhD Thesis, Univ Messina, 306 pp

Voigt E, Bromley RG (1974) Foraminifera as commensals around clionid sponge papillae: Cretaceous and recent. Senckenbergiana Marit 6:33-45

Walker G, Jacob E (1798) In: Kanmacher F (ed) Adams' essays on the microscope. Dillon and Keating, London, p 642 [Note: this is a reference to an untitled essay]

Warme JE (1975) Borings as trace fossils, and the process of marine bioerosion. In: Frey RW (ed) The study of trace fossils. Springer, Berlin Heidelberg, pp 181-227

Wisshak M (2006) High-latitude bioerosion: the Kosterfjord experiment. Lect Notes Earth Sci 109:1-202

Wisshak M, Freiwald A, Lundälv T, Gektidis M (2005) The physical niche of bathyal *Lophelia pertusa* in a non-bathyal setting: environmental controls and palaeoecological implications. In: Freiwald A, Roberts JM (eds) Cold-water corals and ecosystems. Springer, Berlin Heidelberg, pp 979-1001

Zibrowius H (1974) *Caryophyllia sarsiae* n. sp. and other recent deep-water *Caryophyllia* (Scleractinia) previously referred to little known fossil species (*C. arcuata*, *C. cylindracea*). J Mar Biol Assoc UK 54:769-784

Appendix - SEM sample details

Type	Substrate	Age	Remarks	Figures	Museum #
Sula Ridge (Mid-Norway), Pos228-200-3, N64°05' / E008°04'/ 273 m					
H. sarcophaga	*L. pertusa*	Recent	1 test *H. sarcophaga*	2A, 6E	MB.W 3000
epoxy-cast	*L. pertusa*	Recent	alive	not pictured	MB.W 3001
epoxy-cast	*L. pertusa*	Recent	alive	not pictured	MB.W 3002
epoxy-cast	*L. pertusa*	Recent	alive	8B; 9E; 9G-H	MB.W 3003
epoxy-cast	*L. pertusa*	Recent	alive	not pictured	MB.W 3004
epoxy-cast	*L. pertusa*	Recent	alive	9F	MB.W 3005
epoxy-cast	*L. pertusa*	Recent	alive	not pictured	MB.W 3006
epoxy-cast	*L. pertusa*	Recent	alive	not pictured	MB.W 3007
H. sarcophaga	*L. pertusa*	Recent	1 test *H. sarcophaga*	not pictured	MB.W 3008
Sula Ridge (Mid-Norway), Pos228-215, N64°06' / E008°07' / 276 m					
H. sarcophaga	*L. pertusa*	Recent	1 test *H. sarcophaga*	6A-D	MB.W 3009
epoxy-cast	*L. pertusa*	Recent	alive; test MB.W 3009	9A-C; 13D	MB.W 3010
epoxy-cast	*A. excavata*	Recent	outside	not pictured	MB.W 3011
epoxy-cast	*A. excavata*	Recent	inside	not pictured	MB.W 3012
Sula Ridge (Mid-Norway), LI94-163, N64°05' / E008°02' / 283 m					
epoxy-cast	*L. pertusa*	Recent	alive	not pictured	MB.W 3013
epoxy-cast	*L. pertusa*	Recent	alive	9D	MB.W 3014
Sula Ridge (Mid-Norway), POS228-201, N64°05' / E008°01' / 260 m					
H. sarcophaga	*Geodia* sp.	Recent	6 tests *H. sarcophaga*	4A, 6F, 7K-L	MB.W 3015

Type	Substrate	Age	Remarks	Figures	Museum #
W-Oslofjord (S-Norway), ALK232-1066 BG, N59°03'/E010°44'					
epoxy-cast	*A. excavata*	Recent	margin/inside	5C; cast not pictured	MB.W 3016
epoxy-cast	*A. excavata*	Recent	pallial/outside	5H; cast not pictured	MB.W 3017
epoxy-cast	*A. excavata*	Recent	centre	8A, 12B	MB.W 3018
Franken Mound, Rockall Bank, M61-3-603-1, N56°30' / W017°18' / 629 m					
H. sarcophaga	*C. sarsiae*	Recent	1 test *H. sarcophaga*	3A-D; 7E-F	MB.W 3019
epoxy-cast	*C. sarsiae*	Recent	alive; test MB.W 3019	3A-D; 10A	MB.W 3020
epoxy-cast	*C. sarsiae*	Recent	Dead	3E; 10B-C	MB.W 3021
epoxy-cast	*C. sarsiae*	Recent	Dead	3E; 10D-G	MB.W 3022
Mauritania, N19° / W017° / ~600 m					
epoxy-cast	*A. excavata*	Recent	margin/alive; test MB.W 3024	5B; cast not pictured	MB.W 3023
H. carnivora	*A. excavata*	Recent	1 test *H. carnivora*	5B	MB.W 3024
H. carnivora	*A. excavata*	Recent	1 test *H. carnivora*	4F-H; 7A-B	MB.W 3025
epoxy-cast	*A. excavata*	Recent	outside; test MB.W 3025	4F-H; 12C-D; 14G-H	MB.W 3026
epoxy-cast	*A. excavata*	Recent	inside; test MB.W 3025	4F-H; not pictured	MB.W 3027
Strait of Sicily (Italy), COR2-08, N36°51' / E013°08' / ~631 m					
epoxy-cast	*A. excavata*	Fossil	outside	5D; 5G; 14A; 14E-F	MB.W 3028
epoxy-cast	*A. excavata*	Fossil	outside	11G-H; 12E-F; 14D	MB.W 3029
epoxy-cast	*A. excavata*	Fossil	outside	not pictured	MB.W 3030
epoxy-cast	*A. excavata*	Fossil	outside	not pictured	MB.W 3031
epoxy-cast	*A. excavata*	Fossil	outside	11A-D	MB.W 3032
epoxy-cast	*A. excavata*	Fossil	outside	not pictured	MB.W 3033
epoxy-cast	*A. excavata*	Fossil	outside	11E-F	MB.W 3034
epoxy-cast	*A. excavata*	Fossil	inside	12G-H	MB.W 3035
Strait of Sicily (Italy), COR2-39, N36°23' / E013°15' / 523 m					
epoxy-cast	*A. excavata*	Fossil	outside	not pictured	MB.W 3036
epoxy-cast	*A. excavata*	Fossil	outside	not pictured	MB.W 3037
epoxy-cast	*A. excavata*	Fossil	outside	not pictured	MB.W 3038
epoxy-cast	*A. excavata*	Fossil	inside	not pictured	MB.W 3039
Strait of Sicily (Italy), CS96-64, N36°26' / 12°36' / 380 m					
epoxy-cast	*L. pertusa*	Fossil	untypical trace	2E-F; not pictured	MB.W 3040
Salice, Sicily (Italy), N38°15' / E15°29' / onshore outcrop					
H. sarcophaga	*D. vitreus*	Fossil	2 tests *H. sarcophaga*	not pictured	MB.W 3041

Endolithic sponge versus terebratulid brachiopod, Pleistocene, Italy: accidental symbiosis, bioclaustration and deformity

Richard G. Bromley[1], Lydia Beuck[2], Emma Taddei Ruggiero[3]

[1] Department of Geography and Geology, University of Copenhagen, 1350 Copenhagen K, Denmark, (rullard@geol.ku.dk)
[2] Institute of Palaeontology, Erlangen University, 91054 Erlangen, Germany
[3] Dipartimento di Scienze della Terra, Università di Napoli Frederico II, 80138 Napoli, Italy

Abstract. The dorsal valve of a Pleistocene terebratulid brachiopod, *Terebratula scillae* Seguenza, 1871, has developed a malignant cyst due to colonisation in vivo by an endolithic sponge. This trace fossil is a compound boring and bioclaustration structure, representing a boring that has grown in unison with the growth of the cyst. The brachiopod has grown to adult size and growthlines indicate that it was colonised by the sponge when about half grown. Malformation of the shell may not have caused the death of the brachiopod and the sponge does not appear to have outlived its host; both symbionts seem to have died more or less simultaneously. This minus-minus relationship of two symbionts is considered to be a case of 'accidental symbiosis'.

Keywords. Bioerosion, cyst-formation, bioclaustration, terebratulid brachiopod, endolithic sponge, 'accidental symbiosis', micro-CT

Introduction

Endolithic sponges generally colonise brachiopod shell to a lesser degree than they do mollusc shell substrates. This may be related to the generally thick periostracum that covers the calcitic brachiopod shell. Thus, sponge-bioeroded brachiopods generally show signs of having become colonised post mortem, when the inner surface of the shell becomes available. Nevertheless, in vivo bioerosion of brachiopod shell does occur (e.g., MacKinnon and Biernat 1970). Brachiopod shells commonly carry attachment scars, *Podichnus* Bromley and Surlyk, 1973, of other brachiopods, (e.g., Bromley and Heinberg 2006: fig. 12). In many such cases, the substrate shell shows no reaction to the emplacement of the scar, possibly suggesting that the substrate was dead, or living but thick enough to accommodate the scar, and the structure is a pure bioerosion trace fossil. In certain cases, however, the substrate has reacted to the scar and wound tissue is present (Taddei Ruggiero 1989, 1991, 1999; Taddei Ruggiero and Bitner in press). The trace fossil in this case is classifiable as a compound boring and bioclaustration (e.g., Tapanila and Ekdale 2007).

M. Wisshak, L. Tapanila (eds.), *Current Developments in Bioerosion*. Erlangen Earth Conference Series, doi: 10.1007/978-3-540-77598-0_18,

Such biogenic structures produced within living substrates are also known as 'embedment structures' (Bromley 1970; Ekdale et al. 1984), where a skeletonised organism ('host symbiont') is colonised by and overgrows another living organism ('settler symbiont'), which maintains a chamber within the overgrowing organism's skeleton (Tapanila 2005). If the growth rate of the embedded organism (settler symbiont) exceeds that of the overgrowing host organism it may bioerode the skeleton in order to enlarge its chamber (Ekdale et al. 1984).

Palmer and Wilson (1988) introduced the term 'bioclaustration' for this process by which one skeleton-secreting organism overgrows another, calling the chamber in the skeleton a 'pseudoboring' (Taylor 1990). Commonly, however, the chamber is a genuine boring, not a pseudoboring, or it may contain elements of both categories. Consequently, the recent claim by Bertling et al. (2006) that bioclaustration structures are not trace fossils needs revision. The present paper describes a remarkable example of a bioclaustration where the settler symbiont is represented by a true boring.

Methodology

The results presented herein are based on macro- and microscopic observations as well as x-ray and micro-CT analyses. In order to reconstruct the spatial distribution of the trace galleries within the brachiopod valve a high-resolution micro-CT scanner (Institute of Medical Physics, Erlangen University, Germany) was used (see also Kalender 2005). The samples were scanned at a tube voltage of 130 kV, a power of 10 W and a nominal voxel size of 40 μm, creating 720 projections. The data were visualised in syngo® CT.3D with 2D and 3D post processing tools. The generation of images, 2D serial section movies (2d1.mpg; 2d2.mpg; 2d3.mpg; see doi:10.1594/PANGAEA.667142 for all 2D and 3D animations), 3D rotating sample reconstruction movies (3d1.mpg; 3dtrans1.mpg; 3dtrans2.mpg) and movies for anaglyph glasses (rb1.mpg; rbtrans1.mpg; rbtrans2.mpg) was carried out according to the methodology described in Beuck et al. (2007). Ichnometry was done with Adobe Photoshop® CS software.

A malignant growth containing a sponge boring

A specimen of *Terebratula scillae* Seguenza, 1871, from the Early Pleistocene of Gallina (Calabria, Italy) shows a case (Fig. 1) of bioerosion by a sponge that has caused malignant growth of shell material on the left side of the brachial valve (see also Taddei Ruggiero 1989: pl. 6, fig. 6). The growth measures 18 x 20 mm in plan view and about 8 mm in thickness. Its weight would have caused serious imbalance of the shell and the growth has even extended beyond the commissural plane of the shell, which would have interfered with the articulation of the valves, almost certainly preventing their closure (the ventral valve is missing).

X-radiography revealed that the growth contains the boring of an endolithic bioeroding sponge, so the specimen was studied more closely using micro-CT scanning. This shows that the boring is strongly camerate (see movies 2d1.mpg;

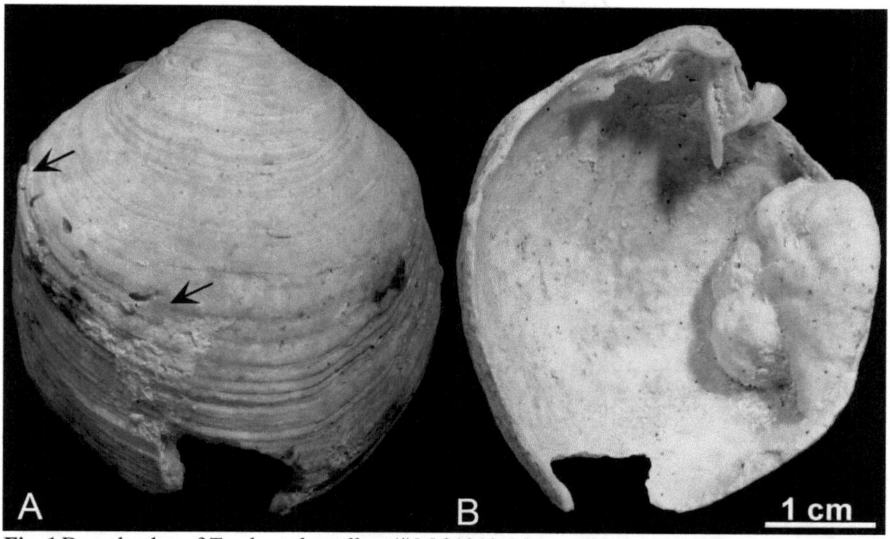

Fig. 1 Dorsal valve of *Terebratula scillae*. (# M.21910; Museo di Paleontologia, Università degli Studi di Napoli Frederico II). **A** External view. There is a slight change in the growth direction at the growthline, which is most marked between the arrows, corresponding to the edge of the cyst. **B** Internal view showing the cyst at right. The notch in the commissure is modern damage

2d2.mpg; 2d3.mpg) and measures 15-20 mm in diameter (2d1.mpg; 2d2.mpg; 2d3.mpg; 3d1.mpg; 3dtrans1.mpg; 3dtrans2.mpg). The individual chambers have very irregular morphology from oval to lobate shaped and can be up to 14 mm in length (Fig. 2D-F). They are separated by slight constrictions of about 0.9 mm in diameter. The trace is connected with the exterior via slightly conical shaped aperture canals of 1-2 mm in length, which narrow from 0.6-1.2 mm at the chamber to 0.3-0.4 mm at the aperture. Documented search threads are up to 6 mm long and between 70-100 µm in diameter. They ramify at angles between 45-75°. This shows that the boring is strongly camerate, but that the individual chambers have very irregular morphology and size range (Fig. 2D-F). Within its self-imposed prison, the sponge shows extreme stenomorphy, rendering ichnotaxonomic evaluation difficult. Nevertheless, the larger chambers resemble those of *Entobia paradoxa* (Fischer, 1868), with which it may be compared. Crowding may not be the sole reason for irregular development of the boring. The continuous growth of the substrate, as the host added new shell material, will have allowed the sponge to modify its boring in a way not available in a dead substrate.

No attempt was made to search for preserved spicules. The chances of opaline silica preservation would be negligible in the poorly cemented, carbonate-dominated sediment, and the chambers themselves remain largely empty of sediment (cf. Bromley and Schönberg this volume).

The external surface of the shell (Fig. 3A) outside the malformation bears about 25 small apertures 0.2-0.9 mm in diameter, and a single one 1.4 mm wide. These are clearly apertures that housed the inhalent and exhalent papillae for water circulation within the sponge (Fig. 2E). On the inward surface of the growth

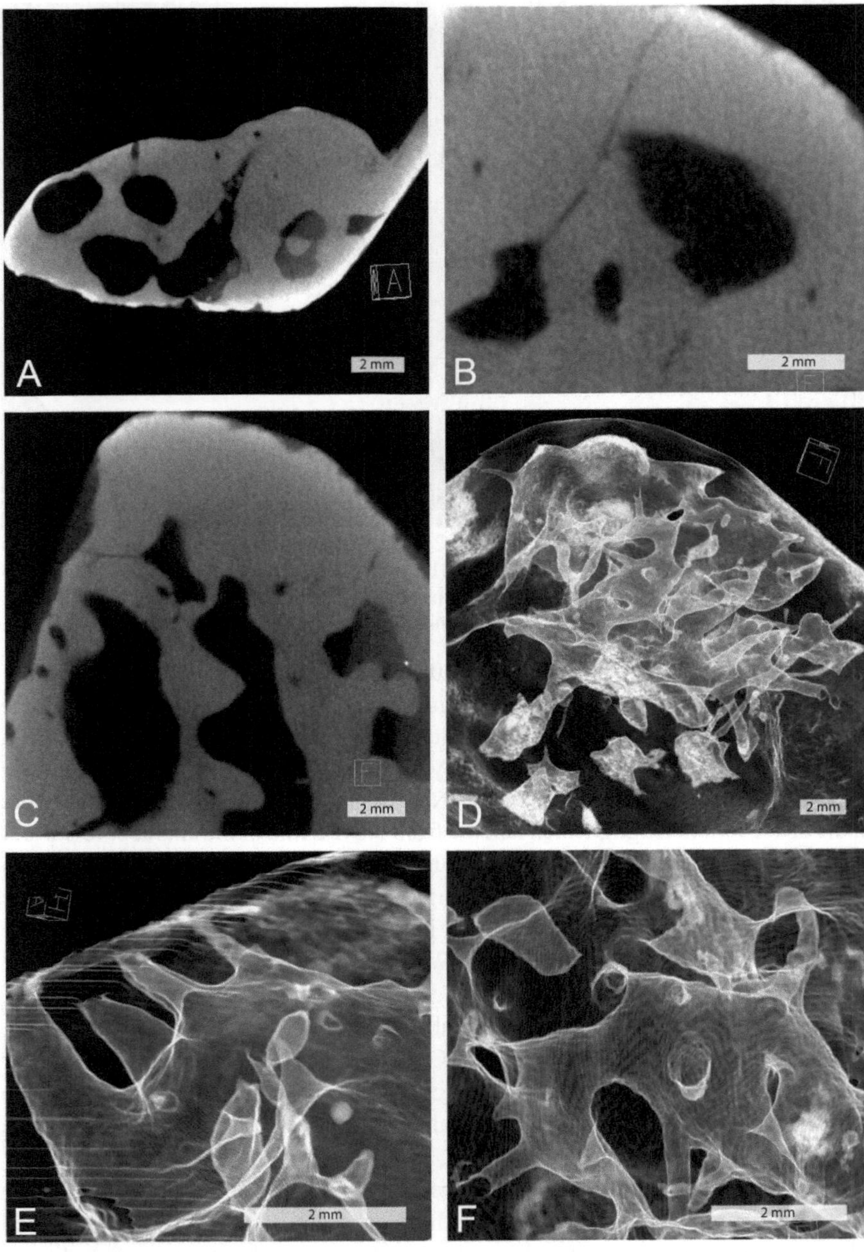

Fig. 2 Micro-CT analysis of the cyst (A-C: virtual 2D sections). **A** The shell is sectioned on the right. Five largely empty chambers of the sponge boring are seen but the central one contains a little geopetal sediment and that to the right is nearly filled. **B** An unusually long exploratory thread running from chamber to the surface. **C** Irregular chambers and a branched exploratory thread. The chamber to the right contains geopetal sediment. **D** Overview of the entobian. **E** Close view showing four slightly conical apertural canals extending from a large chamber to the surface. **F** Irregular chambers connected by cylindrical intercameral canals

Fig. 3 The sponge-created cyst. **A** Left side of the valve viewed externally. The cyst is seen to extend beyond the commissure (white arrowhead). Papillar apertures penetrate the shell surface, two indicated by black arrows. The black arrowheads at top left indicate the line along which the growth direction of the valve slightly changes. **B** The cyst on the left side of the brachiopod valve. Cardinal process to the left. The surface of the cyst is punctured by many fine pores representing the ends of exploratory threads

(Fig. 3B), much smaller perforations occur, scattered rather evenly, mostly less than 0.1 mm wide (Fig. 2A-C). These must represent terminations of exploratory threads (sensu Bromley and D'Alessandro 1984), for monitoring the location of the ever-growing substrate surface. The mantle of the brachiopod intimately overlaid this surface and water would not have been available to the sponge on this side.

Discussion

The sponge has not attempted to invade the thin, remaining, unmalformed part of the shell of the brachiopod, but restricted itself to within the calcitic cyst that the brachiopod has been forced to provide for it. Living within the shell of a live brachiopod, the sponge would have had the advantage of having its substrate kept free of sediment by the brachiopod. In normal brachiopod orientation, it would have occupied the downward-facing (dorsal) valve, further preventing sediment fouling.

The brachiopod shows good size and regular growth for most of its life, and may have reproduced many times. It was apparently only colonised by the sponge when about half grown. This is indicated by a slight change in direction of growth of the shell, most marked where the cyst is developed (Figs. 1A, 3A).

The brachiopod may have died first. The extra energy required in excessive construction of shell material would have been a physiological burden. The growing cyst would have interfered increasingly with the function of the lophophore and ultimately seems to have prevented complete closure of the valves (Fig. 3A). The sponge may therefore have caused the death of its host. Boring sponges commonly kill host oysters when their bioerosion weakens the hinge area to breaking point (e.g., Korringa 1954; Yonge 1960). In such cases, following the death of the host, the sponge may continue to extend its boring and can now place papillar apertures on the inner surface of the shell. However, no such postmortal boring is seen in the present case. Papillar apertures have not been emplaced in the inner surface of the cyst and the boring has not been extended beyond the margins of the malformation into the thin shell.

On the other hand, it is possible that the sponge did not cause the death of its host. Shell deformity due to overcrowding is quite commonplace in brachiopods, and specimens thus affected commonly manage to survive to adult maturity (D. I. MacKinnon pers. comm.).

Neither is there indication that the sponge died first, because it would be expected that the surviving brachiopod would then have covered over the terminal pores of the exploratory threads on the inner surface of the cyst, and this has not happened. If the brachiopod host died first, the sediment-shedding advantages would cease, and burial in sediment might have caused the death of the insurgent sponge. Alternatively, a sedimentary event might have killed both animals simultaneously.

In an attempt to determine the type of symbiosis represented here, it seems that neither symbiont has clearly benefited from the partnership. Possible benefits might include sediment-shedding by the brachiopod, which would benefit the sponge; and maybe the presence of the sponge might somehow deter predators from attacking the brachiopod. But such small benefits would be far outweighed by the negative effect of the very restricted substrate for the sponge and the shell deformity of the brachiopod. Thus, this is a minus-minus association that might be termed antagonism, involving an unusual form of competition for growth space (cf. Clarke 1954; Ager 1963). Perhaps Boucot (1990), following Kuris (1974), would compare the sponge with a parasitoid, a parasite-type organism the activities of which normally result in the death of the host. But this terminology is more suited for categories of behaviour that are normalised. For true parasites, commensals, mutualists, etc., symbiosis is a normal way of life. The right species of wasp seeks out the right species of spider to lay its egg on. In the present case we may instead speak of an ‘accidental symbiosis’, where the proximity of the two animals brought no benefits.

Conclusions

A larval sponge managed to secure itself in the calcitic shell of a brachiopod. The growing sponge irritated the brachiopod and was sealed off by the secretion of wound tissue. Gradually a calcite cyst was secreted, into which the sponge simultaneously bioeroded. Gradually the life processes of the brachiopod were interfered with, as the mantle cavity was asymmetrically filled with the growing cyst until the valves no longer could be closed. There is evidence that both symbionts died more or less simultaneously. The very stenomorphic boring created by the embedded sponge is a bioclaustration and a trace fossil. The minus-minus relationship of the two organisms may be considered an ‘accidental symbiosis’.

Acknowledgements

Referees Ole Tendal (Copenhagen) and David I. MacKinnon (Christchurch) are thanked for insightful comments for improving the manuscript. We express our gratitude to Willi Kalender's working group, especially to Marek Karolszak (Institute of Medical Physics, Erlangen University) for their generation of the micro-CT scan. Rainer Sieger (Alfred Wegener Institute, Bremerhaven) is thanked for the data input on PANGAEA. The study was funded by HERMES Project GOCE-CT-2005-511234 (Hotspot Ecosystem Research on the Margins of European Seas).

References

Ager DV (1963) Principles of paleoecology. McGraw-Hill, New York, 371 pp

Bertling M, Braddy SJ, Bromley RG, Demathieu GR, Genise J, Mikuláš R, Nielsen JK, Nielsen KSS, Rindsberg AK, Schlirf M, Uchman A (2006) Names for trace fossils: a uniform approach. Lethaia 39:265-286

Beuck L, Vertino A, Stepina E, Karolczak M, Pfannkuche O (2007) Skeletal response of *Lophelia pertusa* (Scleractinia) to bioeroding sponge infestation visualised with micro-computed tomography. Facies 53:157-176

Boucot AJ (1990) Evolutionary paleobiology of behavior and coevolution. Elsevier, Amsterdam, 725 pp

Bromley RG (1970) Borings as trace fossils and *Entobia cretacea* Portlock, as an example. In: Crimes TP, Harper JC (eds) Trace Fossils. Geol J Spec Issue 3:49-90

Bromley RG, D'Alessandro A (1984) The ichnogenus *Entobia* from the Miocene, Pliocene and Pleistocene of southern Italy. Riv Ital Paleont Stratigr 90:227-296

Bromley RG, Heinberg C (2006) Attachment strategies of organisms on hard substrates: a palaeontological view. Palaeogeogr Palaeoclimatol Palaeoecol 232:429-453

Bromley RG, Schönberg CHL (this volume) Borings, bodies and ghosts: spicules of the endolithic sponge *Aka akis* sp. nov. within the boring *Entobia cretacea*, Cretaceous, England. In: Wisshak M, Tapanila L (eds) Current developments in bioerosion. Springer, Berlin, pp 235-248

Bromley RG, Surlyk F (1973) Borings produced by brachiopod pedicles fossil and recent. Lethaia 6:349-365

Clarke GL (1954) Elements of ecology. Wiley, New York, 534 pp

Ekdale AA, Bromley RG, Pemberton SG (1984) Ichnology – trace fossils in sedimentology and stratigraphy. SEPM Short Course 15, 317 pp

Fischer MP (1868) Recherches sur les éponges perforantes fossiles. Nouv Arch Mus Hist Nat Paris 4:117-173

Kalender WA (2005) Computed Tomography. Fundamentals, System Technology, Image Quality, Applications. Publicis, Erlangen, 306 pp

Korringa P (1954) The shell of *Ostrea edulis* as a habitat. Arch Néerland Zool 10:32-152

Kuris AM (1974) Trophic interactions. Similarity of parasitic castrators to parasitoids. Quart Rev Biol 49:130-148

MacKinnon DI, Biernat G (1970) The probable affinities of the trace fossil *Diorygma atrypophilia*. Lethaia 3:163-172

Palmer TJ, Wilson MA (1988) Parasitism of Ordovician bryozoans and the origin of pseudoborings. Palaeontology 31:939-949

Seguenza G (1871) Studi paleontologici sui brachiopodi terziari dell'Italia meridionale. Boll Malacol Ital 6:1-79

Taddei Ruggiero E (1989) Evidenze di predazione e parassitismo su gusci di brachiopodi. Atti III simposio ecologia e paleoecologia delle comunità bentoniche, Univ Catania, pp 615-631

Taddei Ruggiero E (1991) A study of damage evidence in brachiopod shells. In: MacKinnon DI, Lee DE, Campbell JD (eds) Brachiopods through time. Balkema, Rotterdam, pp 203-210

Taddei Ruggiero E (1999) Bioerosive processes affecting a population of brachiopods (Upper Pliocene, Apulia). Bull Geol Soc Denmark 45:169-172

Taddei Ruggiero E, Bitner MA (in press) Bioerosion on brachiopod shells – Cenozoic perspective. Trans Roy Soc Edinburgh, Alwyn Williams Memorial Volume: Brachiopod Research into the Third Millennium

Tapanila L (2005) Palaeoecology and diversity of endosymbionts in Palaeozoic marine invertebrates: trace fossil evidence. Lethaia 38:89-99

Tapanila L, Ekdale AA (2007) Early history of symbiosis in living substrates: trace fossil evidence from the marine record. In: Miller W (ed) Trace fossils: Concepts, problems, prospects. Elsevier, Amsterdam, pp 345-355

Taylor PD (1990) Preservation of soft-bodied and other organisms by bioimmuration. Palaeontology 33:1-17

Yonge CM (1960) Oysters. Collins, London, 209 pp.

IV

Spectrum of substrates

Micro-bioerosion in volcanic glass: extending the ichnofossil record to Archaean basaltic crust

Nicola McLoughlin[1], Harald Furnes[1], Neil R. Banerjee[2], Hubert Staudigel[3], Karlis Muehlenbachs[4], Maarten de Wit[5], Martin J. Van Kranendonk[6]

[1] Centre for Excellence in Geobiology and the Department of Earth Science, University of Bergen, 5007 Bergen, Norway, (nicola.mcloughlin@geo.uib.no)
[2] Department of Earth Sciences, University of Western Ontario, London, Ontario N6A 5B7, Canada
[3] Scripps Institution of Oceanography, University of California, La Jolla, CA 92093-0225, USA
[4] Department of Earth and Atmospheric Sciences, University of Alberta, Edmonton, Alberta T6G 2E3, Canada
[5] AEON and Department of Geological Sciences, University of Cape Town, Rondebosch 7701, South Africa
[6] Geological Survey of Western Australia, Western Australia 6004, Australia

Abstract. Microbial bioerosion of volcanic glass produces conspicuous ichnofossils in oceanic crusts that are a valuable tracer of sub-surface microorganisms. Two morphologically distinct granular and tubular ichnofossils are produced. The 'Granular form' consists of individual or coalescing, spherical bodies with average diameters of ~0.4 µm. The 'Tubular form' are straight, sometimes branched, to curving and spiralled tubes with average diameters of 1-2 µm and lengths of up to ~200 µm. A biogenic origin for these structures is confirmed by: the concentration of DNA that binds to biological stains in recent examples; enrichments in C, N and P along their margins in both recent and ancient examples; and systematic C isotope shifts measured upon disseminated carbonate in the surrounding glass. The constructing microorganisms are thought to include heterotrophs and chemolithoautotrophs that may utilise Fe and Mn from basaltic glass as electron donors and derive carbon sources and electron acceptors from circulating fluids. These microbial ichnofossils are found at depths of up to 550 metres in the oceanic crust in the glassy rims of pillow basalts and interpillow breccias. A diverse spectrum of ichnofabrics is created by overlapping phases of granular and tubular bioerosion; banded abiotic dissolution; and the precipitation of phyllosilicates, zeolites and iron-oxy-hydroxides. The resulting ichnofabrics have been documented from in situ oceanic crust spanning the youngest to the oldest oceanic basins (0 to 170 Ma). Their geological record extends to include meta-volcanic glass in oceanic crustal fragments from Phanerozoic to Proterozoic ophiolites. Examples infilled by the mineral titanite ($CaTiSiO_4$) have also been found in Palaeo- to Mesoarchaean

M. Wisshak, L. Tapanila (eds.), *Current Developments in Bioerosion*. Erlangen Earth Conference Series, doi: 10.1007/978-3-540-77598-0_19, © Springer-Verlag Berlin Heidelberg 2008

pillow basalts from the Barberton Greenstone Belt of South Africa and the East Pilbara Terrane of Western Australia. Direct ^{206}Pb / ^{238}U radiometric dating of the Australia examples has confirmed their Archaean age and thus they represent the oldest candidate ichnofossils on Earth.

Keywords. Bioerosion, volcanic glass, ichnofossils, oceanic crust, deep biosphere, origins of life

Historical perspective on bioerosion in volcanic glass

Ichnofossils formed by the bioerosion of volcanic glass are a tracer of sub-surface microbial activity which constitutes one of the largest and least explored portions of the modern and especially ancient biosphere. In contrast to the study of bioerosion in sedimentary and biological substrates the bioerosion of volcanic glass is a relatively new field of research. One of the earliest reports of biological etching of glass is the description of surface pitting on church window-panes near growing lichens (Mellor 1922; see also Krumbein et al. 1991 for a review). The bioerosion of natural glasses was reported somewhat later, with the finding of surface grooves on glass shards in Miocene tephras that were likened to fungal borings in carbonate grains (Ross and Fisher 1986). In addition, algal microborings in beach rocks have been found to show comparable morphologies in both carbonate grains and adjacent man-made glass shards (Jones and Goodbody 1982). The terminology describing these various ecological niches found within rocks was revised by (Golubic et al. 1981) and microorganisms that actively penetrate rock substrates are termed euendoliths, whereas those that inhabit pre-existing fractures are termed chasmoendoliths and those that dwell within pre-existing cavities are termed cryptoendoliths.

Early studies of euendoliths in natural glass did not offer a convincing mechanism of how these microorganisms bioerode. More recent studies of sub-glacial volcanic breccias from Iceland found bacteria within pits on the surface of the volcanic glass and these led Thorseth et al. (1992) to propose that the microbes locally modify the pH and thereby accelerate glass dissolution. This phenomenon was subsequently experimentally investigated by Thorseth et al. (1995), and Staudigel et al. (1995, 1998) who found that volcanic and synthetic glasses inoculated with microbes develop etch pits and surface alteration rinds under laboratory conditions. Numerous studies have followed to document the widespread occurrence of comparable microbial bioerosion textures in volcanic glass from Ocean Drilling Project (ODP) and Deep Sea Drilling Project (DSDP) drill cores from in situ oceanic crust (e.g., Fisk et al. 1998; Furnes et al. 2001d). Several groups have also worked to develop criteria for establishing the biogenicity and antiquity of these textures (e.g., Staudigel et al. 2006; Banerjee et al. 2007; McLoughlin et al. 2007a) and to extend their geological record to include fragments of ancient oceanic crust preserved in ophiolites and greenstone belts (e.g., Furnes et al. 2004, 2007b and references therein).

In this article the term 'bioalteration' is used to describe the suite of textures produced by microbial bioerosion and subsequent mineralisation in volcanic glass. It is proposed that the resulting structures are de facto trace fossils produced by the actions of euendolithic microorganisms residing within the glass. The main lines of evidence that have been argued to support their biogenicity are summarised and the likely microbial metabolisms of the constructing organisms are discussed. In the Appendix we describe and illustrate the morphologies of these ichnofossils and make brief comparisons to known ichnofossils from sedimentary substrates. A model for how these microorganisms bioerode volcanic glass is also proposed. Lastly, the geological record of these bioalteration textures is reviewed, focusing in particular on the oldest putative examples and what these may tell us about the origins of life on Earth. Morphologically comparable textures that have recently been described from within silicate phenocrysts such as olivine and pyroxene are not discussed in detail here and are described elsewhere, e.g., Fisk et al. 2006.

Evidence for the bioerosion of volcanic glass

The aqueous alteration of basaltic glass produces a pale yellow to dark brown material referred to as palagonite. This appears as banded material on either side of fractures with a relatively smooth interface between the fresh and altered glass. Palagonite can be divided into two types: early stage amorphous gel-palagonite that matures to form fibro-palagonite which consists of clays, zeolites and iron-oxy-hydroxides (Peacock 1926). Palagonitisation is a complex and continuous aging process involving incongruent and congruent dissolution accompanied by precipitation, hydration and pronounced chemical exchange that occurs at low to high-temperatures (e.g., Thorseth et al. 1991; Stroncik and Schmincke 2001; Walton and Schiffman 2003; Walton et al. 2005). Palagonitisation has traditionally been regarded as a purely physio-chemical phenomenon, however, several observations made upon recent volcanic glass point towards microbial mediation of these processes. Briefly these include: the discovery of bacteria in or near etch pits on the surface of the volcanic glass (e.g., Thorseth et al. 2001); the observation of bacteria shaped moulds in porous palagonite and zeolite filled fractures and alteration rims (e.g., Thorseth et al. 2003); and the presence of organic material at the interface of fresh and altered glass that binds to targeted DNA probes (e.g., Giovannoni et al. 1996; Torsvik et al. 1998; Banerjee and Muehlenbachs 2003). These and additional lines of evidence that support biological participation in the alteration of volcanic glass are explained further below. We concentrate here upon the ramified alteration fronts found at the interface of fresh and altered glass where granules and tubes project into the glass and are argued to represent the traces of microbial dissolution (e.g., Furnes et al. 2007b and references therein). These are distinct from the near-planar alteration fronts produced by the chemical modification of volcanic glass. We will focus here upon the ichnofossils found in meta-volcanic glass rather than the mineral encrusted bacterial shaped structures found in palagonite filled fractures, because the latter appear to have a lower preservation potential in the geological record.

One of the first techniques to show that biological material is concentrated at the ramified interface between fresh and altered glass was the application of DAPI (4, 6 diamino-phenyl-indole) dye which binds to nuclei acids, along with fluorescent oligonucleotide probes that target bacterial and archeal RNA (e.g., Giovannoni et al. 1996; Torsvik et al. 1998: fig. 2; Banerjee and Muehlenbachs 2003: fig. 14; Walton and Schiffman 2003: fig. 8). Staining of volcanic glass samples from the Costa Rica Rift for example, found that the most concentrated biological material occurred right at the interface between the fresh and altered glass, especially in the tips of tubular structures and that this decreased in concentration towards the centre of fractures (Furnes et al. 1996; Giovannoni et al. 1996). Partially fossilised, mineral encrusted microbial cells have also been observed on the surface of altered glasses using scanning electron microscopy (SEM) and these include filamentous, coccoid, oval and rod shaped forms (e.g., Thorseth et al. 2001). Moreover, these often occur in or near etch marks in the glass that exhibit forms and sizes which resemble the attached microbes suggesting that they may be responsible for pitting (e.g., Thorseth et al. 2003). SEM studies have also revealed delicate hollow and filled filaments attached to microtube walls within glass fragments from the Ontong Java Plateau (e.g., Banerjee and Muehlenbachs 2003: figs. 5-9), along with spherical bodies and thin films interpreted to be desiccated biofilms (e.g., Banerjee and Muehlenbachs 2003: figs. 5-9).

Thin linings less than 1 μm wide of C, N, and P have also been detected by electron probe mapping of modern and ancient bioalteration textures (e.g., Giovannoni et al. 1996; Torsvik et al. 1998; Furnes and Muehlenbachs 2003). Importantly, these do not correlate with the distribution of Ca suggesting that the C is not derived from a carbonate phase and rather, the presence of organic carbon within the microtubes has been confirmed by X-ray absorption spectroscopy (Benzerara et al. 2007). These linings are therefore interpreted to represent the decayed remains of cellular materials that once inhabited the tubular and granular textures (e.g., Torsvik et al. 1998; Banerjee and Muehlenbachs 2003). Marked depletions in Mg, Fe, Ca, and Na along with slight depletions in Al and Mn have also been documented by transmission electron microscopy (TEM) across 0.1-0.3 μm wide zones at the interface of fresh and altered glass from in situ oceanic crust (Alt and Mata 2000). These are accompanied by an order of magnitude increase in the concentration of K which implies the uptake of K from large volumes of seawater, estimated to exceed 100 fracture volumes, and which may also have supplied important nutrients for microbial activity (Alt and Mata 2000).

Carbon isotopic measurements made upon disseminated carbonate found within pillow basalts also give clues as to the potential metabolisms that may be involved in these bioalteration processes. For instance, it is reported that disseminated carbonate preserved in altered pillow basalt rims is ^{13}C-poor, typically between +3.9‰ to –16.4‰ compared to carbonate within the unaltered pillow interiors, which has $\delta^{13}C$ values of +0.7‰ to –6.9‰ and are comparable to mantle values. This much greater range exhibited by the pillow rims is interpreted to reflect the microbial oxidation of organic matter that gives the more negative values and perhaps also the loss of ^{12}C-enriched methane from Archea to give the more positive values (e.g., Furnes et al. 2001b; Furnes et al. 2002b: fig. 9b; Banerjee et al. 2006 and references therein). It has also been found that volcanic glasses from slow spreading ridges such as the mid-Atlantic Ocean have a wider range in $\delta^{13}C$ values from -17‰ to +3‰ compared to those from the faster spreading Costa Rica Rift for example, which show a narrower range from -17‰ to -7‰ (Furnes et al. 2006). This has led to the speculation that at slow spreading ridges the positive $\delta^{13}C$ values may reflect the lithotrophic utilisation of CO_2 by methanogenic bacteria (Furnes et al. 2006).

Establishing the age of bioalteration textures in volcanic glasses is necessary to demonstrate that the euendoliths responsible are not recent contaminants. Relative dating of the textures is possible by investigating the juxtaposition of fracture and pore filling phases. For example, studies of the Hawaii Scientific Drilling Project (HSDP) material have shown that the alteration of the basaltic glass produces well defined zones of incipient alteration, followed by smectitic and palagonitic alteration with depth in the core and moreover, that the formation of tubules occurs relatively early in the sequence of pore lining cements (Walton and Schiffman 2003). Absolute dating of bioalteration textures may also be possible for some ancient examples using ^{206}Pb / ^{238}U radiometric dating of the titanite phases that infill some candidate bioalteration textures (Banerjee et al. 2007). This new approach will be explained below.

Microbiological constraints

A consortium of microorganisms including heterotrophs and chemolitho-autotrophs is thought to be involved in the bioalteration of volcanic glass. It is envisaged that the heterotrophs use organic carbon derived from circulating seawater as a carbon source. The chemolithoautotrophs are thought to use oxidised compounds such as Fe^{3+}, Mn^{4+}, SO_4^{-2} and CO_2 in the glass and / or circulating seawater as electron acceptors along with reduced forms of Fe and Mn found in the glass as electron donors (e.g., Staudigel et al. 2004). In addition the microbial consortia may derive key nutrients especially phosphorus from the glass which is otherwise scarce in oligotrophic sub-seafloor environments. The suggestion that Mn oxidation is a potentially important chemolithoautotrophic metabolism is supported by the isolation of diverse manganese oxidising bacteria from basaltic seamounts where they are argued to enhance the rate of Mn oxidation (e.g., Templeton et al. 2005). The possibility that these microbial consortia may also employ iron oxidation during bioalteration is consistent with the resemblance of bacterial moulds found on volcanic glass fragments to the branched and twisted filaments of the Fe-oxidising bacteria *Gallionella* Ehrenberg, 1836 (e.g., Thorseth et al. 2001, 2003). Moreover, it has recently been discovered that a group of bacteria distantly related to the heterotrophic organisms *Marinobacter* Gauthier, Lafay, Christen, Fernandez, Acquaviva, Bonin and Bertrand, 1992 and *Hyphomonas* (Pongratz, 1957) are also capable of chemolithoautotrophic growth, employing Fe-oxidation at around pH 7 on a range of substrates including basaltic glass (Edwards et al. 2003). It therefore appears likely that the contribution of heterotrophic and chemolithotrophic metabolisms to the bioalteration of volcanic glass may change depending on the prevailing environmental conditions.

Culture independent molecular profiling studies meanwhile have found that basaltic glass is colonised by microorganisms that are distinct from those found in both deep seawater and seafloor sediments. For example, indigenous microbial sequences obtained from samples dredged from the Arctic seafloor were found to be affiliated to eight main phylogenetic groups of bacteria and a single marine Crenarchaeota group (Lysnes et al. 2004). Furthermore, it is reported that autotrophic microbes tend to dominate the early colonising communities and that heterotrophic microbes are more abundant in more altered, basaltic samples (Thorseth et al. 2001). In other words, it appears that a prokaryotic microbial consortia which includes microorganisms that employ Fe and Mn oxidation are plausible candidates for the bioerosion of basaltic glass and that these are associated with a heterotrophic community, the total diversity of which, can at present only be imagined. There are even reports of eukaryotes from within the oceanic crust, with the finding of microbial remains argued to be marine, cryptoendolithic fungi in carbonate filled amygdules from Eocene Pacific seafloor basalts (Schumann et al. 2004).

Efforts to generate bioalteration textures in laboratory experiments using natural inoculums and various glass substrates have generated useful insights, although each with their own limitations. This work was motivated in part by etch pits found in Icelandic hyaloclastites that show ‘growth rings’, which were taken to suggest that they might develop into tubular shaped alteration structures (Thorseth et al. 1992), although no such extended tubular morphologies have yet to be produced in the laboratory. These early studies involved basaltic glass inoculated with microbes taken from the submarine Surtsey volcano that were cultivated in 1% glucose solution at room temperatures for one year and produced etch pits and alteration rinds (Thorseth et al. 1995). Monitoring of these experiments over time suggested that the microbes corrode the volcanic glass first via congruent dissolution then by incongruent dissolution and it was hypothesised that these involved the secretion of organic acids and metal complexing agents by the microbes (Thorseth et al. 1995). The limitation of this work was the use of a nutrient rich media that is not comparable to sub-seafloor conditions. More recent microcosm experiments have been designed to mimic cold, oligotrophic seafloor environments but these have failed to produce enhanced bioalteration rates relative to sterile controls (Einen et al. 2006). Another experimental approach was utilised by Staudigel et al. (1998) who constructed flow through experiments with basaltic glass that was continuously flushed with a natural seawater microbial population and monitored both chemically and isotopically for periods of up to 583 days. These experiments produced twice the mass of authigenic phases compared to the abiotic controls and caused particularly marked Sr exchange. However, again these experiments are not directly analogous to sub-seafloor conditions because surface seawater inoculums were used.

In summary it appears that heterotrophic bacteria along with chemolithoautotrophs which utilise Fe and Mn oxidation are responsible for the bioalteration of volcanic glass. The full diversity of microorganisms involved however, is yet to be fully documented and the conditions under which tubular alteration structures are formed, are yet to be replicated in the laboratory, perhaps because of the long time frames required.

Ichnofossils in volcanic glass

To date the textures produced by the microbial bioerosion of volcanic glass have not been characterised as ichnofossils. This is a significant omission given that an estimated 10 to 20% of the upper oceanic crust comprises volcanic glass (Staudigel and Hart 1983) and therefore represents a volumetrically important potential habitat for euendolithic microorganisms. In addition, ongoing work suggests that their fossil record in meta-volcanic glasses is extensive (Fig. 1) and may include some of the earliest forms on life on Earth (e.g., Furnes et al. 2004; Banerjee et al. 2006, 2007).

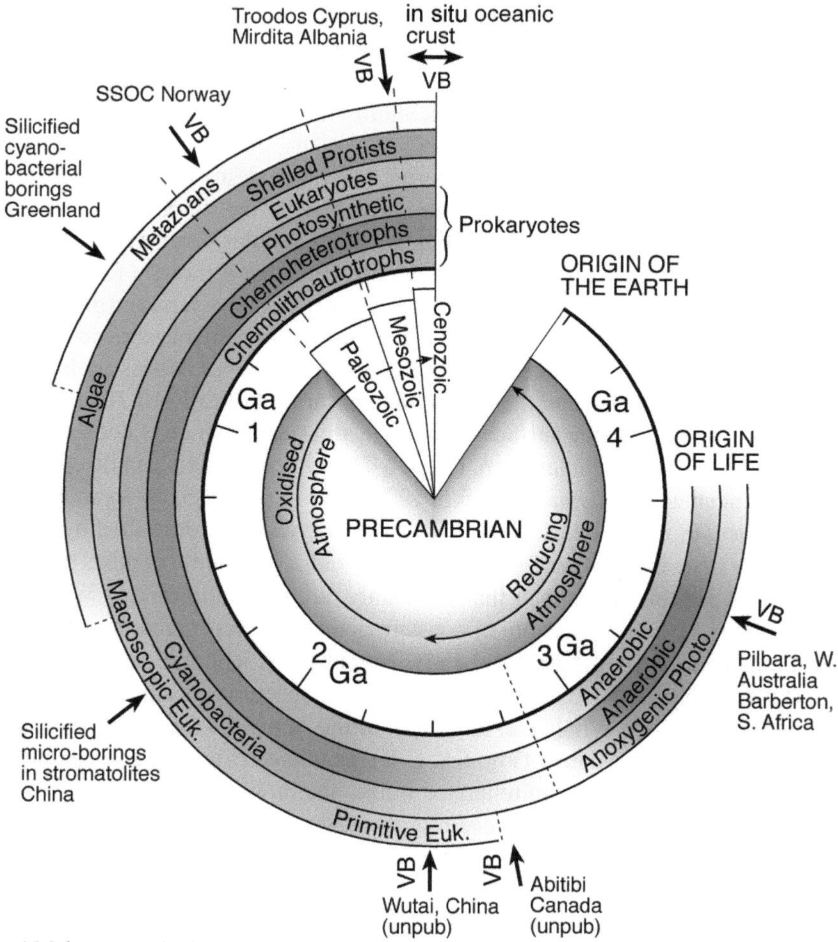

Fig. 1 Major events in the Earth's biogeological 'clock' (modified from des Marais 2000; lighter shading shows periods of uncertainty). At the centre is the geological timescale in billions of years and the surrounding rings show the time ranges of different animal groups and microbial metabolisms. The outer labelled arrows show selected examples of fossilised euendoliths in sedimentary rocks and microbial ichnofossils known to date from volcanic substrates (VB). Many of these volcanic examples lie within intervals of the Precambrian that are of particular evolutionary interest, such as: the earliest evidence for life on earth and the oxygenation of the atmosphere. (SSOC is the Solund Stavfjord ophiolite complex in NW Norway)

For comparison, a recent inventory of marine bioerosion documented 65 ichnogenera, many of which are constructed by microorganisms, but this inventory exclusively dealt with sedimentary and biological substrates (Bromley 2004). A rather separate body of literature that is introduced above, concerns micro-bioerosion traces found in volcanic glass and this has largely been presented by biogeochemists and petrologists. One of the goals of this paper is therefore to try and begin to bridge the gap between these two fields and to begin to apply concepts from the field of ichnology to bioalteration in volcanic glasses. In the Appendix we use an open nomenclature to characterise two ichnofossil forms the 'Granular form' (Fig. 5) and 'Tubular form' (Figs. 6-7) which are found in volcanic glasses and we comment upon the morphological and preservational variations that are seen within each of these. We go on to explain what is currently known about the spatial and temporally variability in the distribution of these ichnofossils and the ichnofabrics that are produced.

A brief review of established microboring ichnotaxa described from calcareous substrates highlights morphological similarities with the 'Granular form' and 'Tubular form' described here. In particular, the 'Granular form' (Fig. 5) can be compared to the ichnotaxon *Planobola* Schmidt, 1992 that is thought to include the initial microborings of various trace makers especially cyanobacteria and green algae, as well as the mature traces of unicellular or globular-multicellular cyanobacteria such as *Cyanosaccus piriformis* Lukas and Golubic, 1981. *Planobola* traces are however, significantly larger in diameter than the 'Granular form', typically 10-30 µm and are sometimes connected to the surface by a single, thick stalk which is not seen in the 'Granular form'. The 'Tubular form' (Figs. 6-7) in volcanic glass can be compared to the ichnogenus *Fascichnus* (Radtke, 1991) that is produced by cyanobacteria and in particular, ichnotaxa such as *Fascichnus dactylus* first described by Radtke (1991). This trace is characterised by a carpet, or radiating bundle of tubes up to 150 µm long and 3-8 µm in diameter that show only rare bifurcations (e.g., Wisshak et al. 2005: fig. 6A). These tubes are of constant diameter except sometimes for slight thickening seen towards their distal ends. Notably, this ichnospecies is produced homeomorphologically and at least 3 different cyanobacteria species are capable of forming this trace (see Wisshak et al. 2005). In the Appendix the 'Tubular form' is subdivided further into simple (Fig. 6A, C-E), annulated (Fig. 6B) and spiral (Fig. 6F-G) morphotypes that are summarised in the line drawing shown in Figure 7A. The *Fascichnus* ichnogenus includes morphologies that are similar to the simple and annulated tubular subtypes (Fig. 6A-E), but we are not aware of ichnotaxa from carbonate substrates that are comparable to the spiralled form (Fig. 6F-G).

It is again emphasised that a consortia of microorganisms is thought to be responsible for producing these ichnofossils in volcanic glasses. It is conceivable that comparable microbial consortia may produce different ichnofossil morphotypes under varying environmental conditions. For instance, consider microbial colonies grown on agar plates that form dendritic shaped colonies when nutrient concentrations are low, as opposed to compact, concentrically zoned colonies when nutrient concentrations are high (e.g., Matsushita et al. 2004). In a similar manner we envisage that the same microorganisms may be capable of producing tubular as opposed to granular ichnofossils in volcanic glasses under varying nutrient or substrate conditions. Conversely, it is also possible that different microorganisms may produce similar trace fossil morphologies and we doubt that there is a one-to-one relationship between ichnofossil morphologies in volcanic glass and the constructing microorganisms. For these reasons in addition to the limitations of the microbiological and experimental constraints explained above, we have chosen to use an open nomenclature to describe the 'Granular form' and 'Tubular form' found in volcanic glass.

How microorganisms bioerode volcanic glass

Microorganisms are believed to cause enhanced, localised dissolution of volcanic glass that causes the progressive development of granular and tubular bioerosion textures (e.g., Thorseth et al. 1992, 1995; Staudigel et al. 1995, 1998). The exact biochemical mechanisms of how they do this are only partially understood however, and are thought to include the secretion of organic acids, production of siderophores and complexing agents. A model for the textural development of such bioerosion traces is given in Figure 2.

In Figure 2, the top line shows abiogenic alteration which results in the production of banded palagonite around glass fragments and along the margins of fractures, with a relatively smooth interface between the fresh and altered glass. This should be contrasted with the granular and tubular ichnofossils shown in the lower lines. These are formed by microorganisms which are carried into fractures in the rock by circulating fluids. These microbial consortia progressively etch the fresh glass, generating more abundant tubes and granular aggregates around fractures, that creates an increasingly ramified alteration front between the fresh and altered glass. In Figure 2 this is schematically shown from left to right across the diagram and illustrated by real examples shown in the back-scatter electron (BSE) images. With time these textures are filled with authigenic phases and the final column shows the chemical signatures that are preserved, namely enrichment in C, N and P along the margins of the bioerosion traces and depletions in Mg, Fe, Ca, and Na in the surrounding modified glass. We stress that this is a schematic diagram and that the distribution of bioalteration textures will differ in fractures of varying geometries; under different fluid flow regimes; in and around vesicles; and as authigenic minerals precipitate modifying the diffusion processes.

During microbially driven glass dissolution the total surface area of fresh glass available progressively increases. Staudigel et al. (2004) calculated that the surface area of fresh glass would increase by factors of 2.4 and 200 during the formation of tubular and granular traces, respectively. In contrast, abiotic glass alteration causes the surface area to progressively decrease and this acts as a negative feedback that inhibits further alteration. Glass dissolution is accompanied by precipitation of fine-grained authigenic minerals such as clays, iron-oxy-hydroxides and zeolites. Principal component analysis of the composition of clay-minerals found in recent

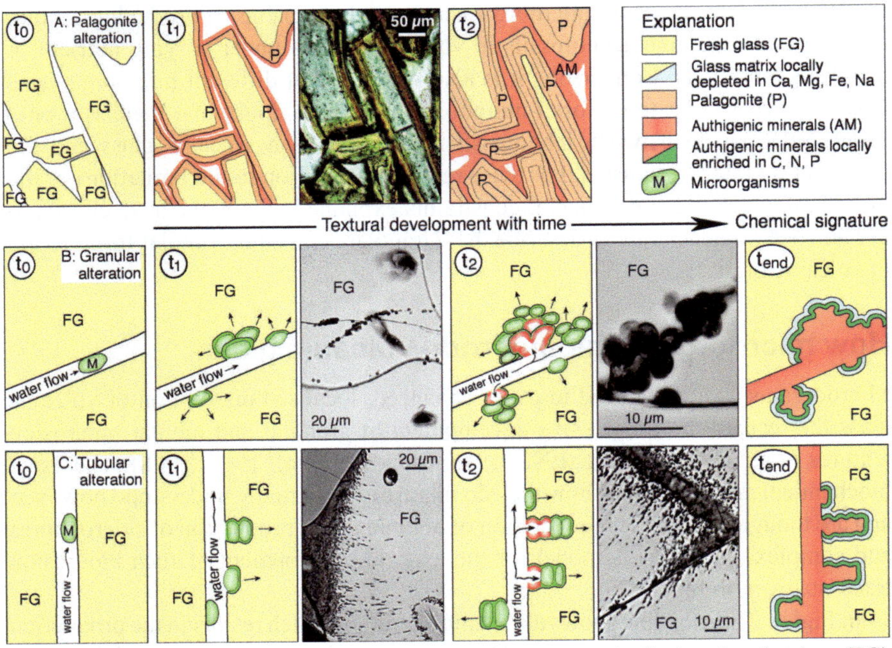

Fig. 2 Diagram showing the generation of alteration textures in fresh volcanic glass (FG) from initial (t_0) to the final time (t_{end}) with secondary electron and thin-section images. The top line shows abiotic palagonite alteration which produces banded rims and authigenic minerals around glass fragments. The middle line shows the development of the 'Granular form' from isolated spheres (t_1) to dense aggregates (t_2). The lower line shows the development of the 'Tubular form' from incipient short tubes (t_1) to longer tubes (t_2). The right column shows the resulting compositional signatures when these structures are infilled by authigenic minerals

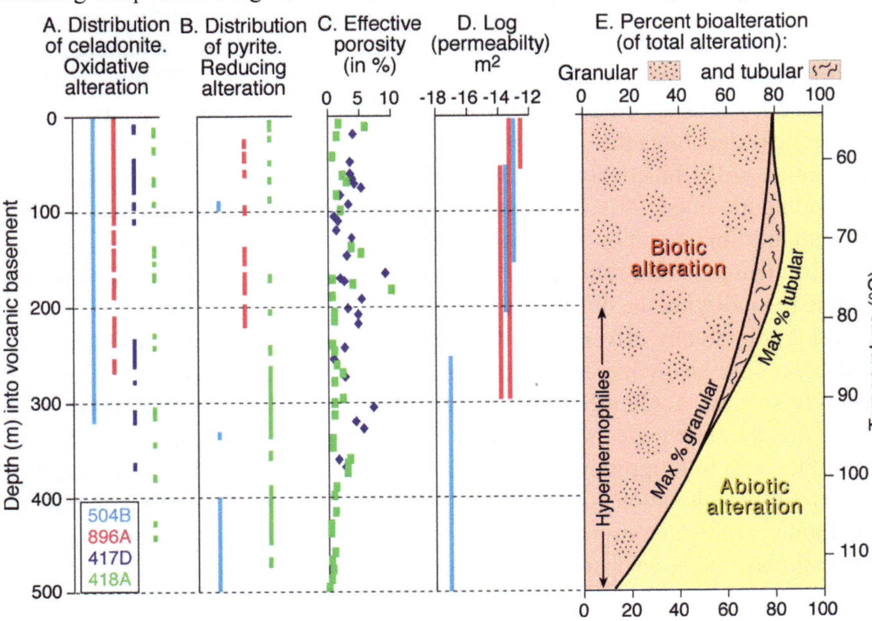

volcanic glasses which contain bioalteration textures have shown that these are compositionally distinct from both clay minerals in abiotically altered zones and those in unaltered zones (Storrie-Lombardi and Fisk 2004). These findings suggest that not only is the dissolution processes microbially mediated but that authigenic mineral precipitation is also biologically influenced. This is supported by studies of hydrothermal altered basaltic glasses in steam vents where it is reported, that microorganisms play a significant role in enhancing authigenic clay precipitation and only a minor role in glass dissolution (Konhauser et al. 2002). In many samples we see that biotic and abiotic alteration processes are not mutually exclusive, with the typical yellow to brown, smooth palagonite zones in the core of fractures accompanied by granular and tubular alteration textures on the outer edges adjacent to the fresh glass (Fig. 5D-E). Bacterial moulds or encrustations are also sometimes found embedded within banded palagonite that lines fractures in pillow basalts (e.g., Thorseth et al. 2001). It is probable therefore, that as long as water continues to flow through cavities in the oceanic crust re-supplying nutrients and organic carbon whilst removing waste products, then the bioalteration processes will continue concomitant with abiotic alteration, providing that the ambient temperature is sufficiently low for life to exist. Bioalteration will cease, only when the fractures and voids become completely sealed by precipitation of authigenic phases or energy and nutrient resources are exhausted.

Lithological and environmental controls on the distribution of ichnofabrics in volcanic glass

Bioalteration textures in volcanic glass are documented from the glassy rims of sub-aqueous volcanic pillow lavas and also volcanic breccias known as hyaloclastites. The majority of reports come from marine environments, with abundant examples described from all of the major oceanic basins, spanning the youngest to the oldest in situ oceanic crust from 0 to 170 million years. These occurrences are shown by white diamond symbols on Figure 4. Non-marine exceptions include etch pits and associated microbes in sub-glacial hyaloclastites from Iceland (Thorseth et al. 1992), putative 'bacteriomorphs' found within fractures in the Columbia River flood basalts (McKinley et al. 2000), and isotopic and microcosm evidence for

Fig. 3 Down hole data to show the controls on the distribution of bioalteration textures with depth in the oceanic crust, from holes 417D, 418A in the Atlantic and 504B and 896A from the Costa Rica Rift in the Pacific, recalculated from Furnes and Staudigel (1999); and Furnes et al. (2001d). **A** Occurrence of celadonite which is an indicator of oxidative alteration. **B** Occurrence of pyrite which is an indicator of reducing conditions. **C** Effective porosity determined by comparing air-dried and water-saturated sample weights. **D** Permeability. **E** Estimates of the relative percentages of granular alteration textures (pink dots); tubular alteration (pink lines); and abiotic alteration (yellow) calculated from thin-section observations and redrawn from Furnes et al. (2001d) and Staudigel et al. (2006). Note these percentage bioalteration estimates are likely to be an underestimate, because ambiguous textures were regarded as abiotic in these calculations and glass hydration processes may also have obscured some textures. It should also be appreciated that most drill holes in the oceanic crust have rather low recoveries which can be less than ~20%

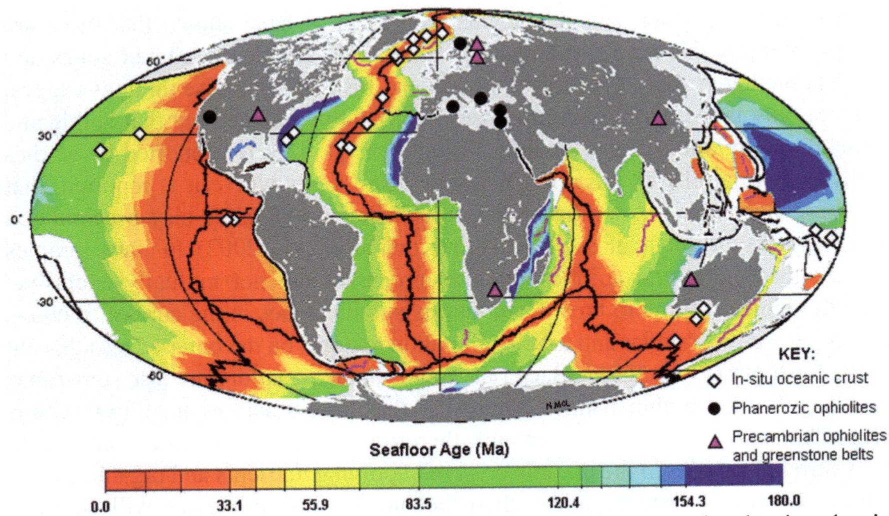

Fig. 4 Map showing the distribution of bioalteration textures documented to date in volcanic glass from in situ oceanic crust. Examples from oceanic basins are shown by white diamonds; from fragments of oceanic crust preserved in Phanerozoic ophiolites by black circles; and from Precambrian ophiolites and greenstone belts by pink triangles. The age of the in situ oceanic crust in millions of years is shown by the colour scale (seafloor spreading map from Müller et al. 1997)

extant, lithoautotrophic microbes in terrestrial aquifers (e.g., Stevens and McKinley 1995). All of the aforementioned glasses are basaltic and therefore rich in iron and manganese which are key to the metabolisms of the constructing organisms.

Systematic studies of ODP and DSDP drill cores have been undertaken to investigate the variation in bioalteration with depth, temperature, permeability and porosity into the oceanic crust (e.g., Furnes et al. 2001d and references therein). The percentage bioalteration was estimated in these studies visually from geological thin sections collected at Sites 417 and 418 of the 110 Ma Western Atlantic, and at Sites 504B and 896A of the 5.9 Ma Costa Rica Rift (see Furnes and Staudigel 1999 and Furnes et al. 2001d). The compilation illustrated in Figure 3 shows that of the two bioalteration morphologies the 'Granular form' is by far the most abundant and can be found at all depths into the volcanic basement where the presence of fresh glass allows bioalteration to be traced, down to ~550 m. At the surface as well as below ~350 m the tubular textures are absent or very rare. In the upper ~350 m of the crust the 'Granular form' is dominant, being most abundant in the upper 200 m and decreasing steadily to become subordinate at temperatures of ~115°C near the currently known upper limits of hyperthermophilic life. The tubular alteration textures meanwhile, constitute only a minor fraction, at most ~20% of the total bioalteration and show a clear maximum at ~120-130 m depth, corresponding to temperatures of ~70°C. The abundance of bioalteration with depth at Sites 504B and 417 / 418 increases with permeability and also the presence of mineral phases such as celadonite which is indicative of relatively oxygenated waters (e.g., Furnes and Staudigel 1999; Furnes et al. 2001d). It is noteworthy that both the 5.9 Ma Costa

Rica Rift and the 110 Ma Western Atlantic oceanic sections show similar maxima in the amount of bioalteration as a percentage of the total alteration, despite their very different ages (see Furnes et al. 2001d: fig 11). This suggests that a substantial portion of the bioalteration happens very early and that the net bioalteration pattern or ichnofabric is established early in the crustal history, i.e., within ~6 Ma. As long as fresh glass is present however, and seawater circulation persists the bioalteration processes are likely to continue.

A range of microbial ichnofabrics are generated by the time-integrated product of microbial bioerosion in the oceanic crust as it cools, subsides, moves away from the ridge flanks and is buried by sediments. The complete oceanic crustal sequence may take 10s to 100s of thousands of years to accumulate depending on the spreading rate, and during which time several episodes of bioerosion and abiotic alteration may occur. For example, the time required to build the ~650 m thick volcanic sequence found at the intermediate-spreading rate Costa Rica Rift is estimated to be between 15,000 and 20,000 years (Pezard et al. 1992). The typical stratigraphy of such a volcanic succession indicates construction by 2-7 volcanic cycles and each with several sub-cycles (see summary in Furnes et al. 2001a, 2003). This implies that during the build-up of the volcanic pile microbial colonisation and bioalteration may have commenced several times, immediately after a new eruption and as this subsequently waned. Systematic investigation of the controls on the development of these ichnofabrics using the wealth of ODP and DSDP drill core that is yet to be studied may enable their future application as a tool for mapping of the deep oceanic biosphere. We must remember however, that the resulting volcanic ichnofabrics also include taphonomic variables. For instance, the differential precipitation of mineral phases within the bioalteration textures from different parts of the oceanic crust may act to enhance or reduce their relative preservation potentials.

Geological record of ichnofabrics in ophiolites

The study of ichnofabrics in volcanic glass can be extended far beyond the age of in situ oceanic crust by investigating ophiolites which represent fragments of former ocean crust (e.g., Furnes and Muehlenbachs 2003). A recent summary table that lists modern to ancient bioalteration textures reported to date is given in Staudigel et al. (2006). Here we focus upon pillow lavas from four ophiolites, ranging in age from Cretaceous to Early Proterozoic that have been investigated for bioalteration textures and are shown by black circles (Phanerozoic) and pink triangles (Precambrian) on Figure 4. These are the 92 Ma Troodos ophiolite complex in Cyprus (e.g., Schmincke and Bednarz 1990); the 160 Ma Mirdita ophiolite complex in Albania (e.g., Dilek et al. 2005); the 443 Ma Solund-Stavfjord ophiolite complex (SSOC) in western Norway (e.g., Furnes et al. 2003); and the 1,953 Ma Jormua ophiolite complex (JOC) in Finland (e.g., Kontinen 1987). In the two youngest of these ophiolites, i.e., the Troodos and Mirdita, which are both of zeolite to lowest greenschist facies metamorphic grade and have experienced only minor to no deformation, there are spectacular textural features of purported microbiological origin (Furnes et al. 2001c). The Troodos in particular includes excellent examples of the tubular shaped ichnofossils that are described in the Appendix and Furnes et al. (2001c). These morphological remains from the Troodos and Mirdita are accompanied by C isotopes shifts that strongly support their biological origin (Furnes and Muehlenbachs 2003: fig 7). In the older ophiolite examples, pillow lavas of the least deformed part of the SSOC, which are of lower greenschist facies metamorphism, possible traces of biogenerated structures have also been recorded in association with marked C isotope shifts (Furnes et al. 2002a). Moreover, even in the strongly recrystallised glassy rims of the lower amphibolite facies JOC the C isotope patterns are consistent with microbial bioalteration and there are areas of high carbon content associated with mineralised features that tentatively resemble bioalteration textures (Furnes et al. 2005). Thus it appears that these carbon isotope patterns which are argued to be a signature of the bioerosion processes may be more resilient to metamorphic degradation than the textural evidence. Nonetheless, the most robust case for bioalteration in ancient meta-volcanic glasses requires all three lines of evidence to be coincident i.e., morphological remains that display localised elemental enrichments and are also associated with carbon isotopic shifts.

The geological record of ophiolites in the early Precambrian is sparse and highly dismembered. Rare exceptions included the 2.5 Ga Dongwanzi ophiolite in the North China Craton (Kusky et al. 2001); the 2.5 Ga Wutai Greenstone belt also of the North China Craton (Wilde et al. 2004); and the recently recognised 3.8 Ga 'Isua Ophiolite Complex' of Greenland (Furnes et al. 2007a), although many of these reports are debated. Notwithstanding, it has been proposed that Archaean greenstone belts may represent important archives in which to seek fragments of the earliest oceanic crust (e.g., de Wit 2004). With this in mind pillow lavas from two of the oldest and best preserved greenstone belts in the world, have been sampled for microbial ichnofossils and will be discussed in the following section.

Candidate ichnofossils in Archaean meta-volcanic glass and the evolution of life

The major events in the Earth's bio-geological 'clock' are summarised in Figure 1 (modified from des Marais 2000). This shows the evolution of the different microbial metabolisms and major biological groups through geological time. To this are added currently recognised examples of euendolithic organisms in volcanic glasses (VB) and sedimentary substrates. Given that pillow lavas are volumetrically significant in many early Archaean terranes this diagram emphasises their potential as an archive for deciphering perhaps some of the earliest evidence for life on earth at ~3.5 Ga.

An endolithic mode of life may have offered many advantages to the early biosphere, with oceanic crustal habitats providing abundant electron donors, principally Fe and Mn found in basaltic rocks; also electron acceptors and carbon sources carried by circulating fluids; as well as protection from the elevated UV irradiation, meteoritic and cometary impacts that early Earth experienced. This hypothesis has been tested by searching for ichnofossils in volcanic glass from both the Palaeo- to Mesoarchaean Barberton Greenstone Belt (BGB) of South Africa and the East Pilbara Terrane of the Pilbara Craton, Western Australia (PWA). Below, we will first briefly introduce the stratigraphy of these two greenstone belts before discussing the micro-textures and candidate ichnofossils that have been found.

The 3,480 to 3,220 Ma old volcano-sedimentary sequence of the BGB comprises 5-6 km of submarine komatiitic and basaltic pillow lavas, with interbedded sheet flows and related intrusions that are interlayered with cherts, banded iron formations and shales (de Wit et al. 1987; de Ronde and de Wit 1994). The search for bioerosion textures has focused on the main part of the Onverwacht Group (including the Theespruit, Komati, Hooggenoeg and Kromberg Formations), which comprises 5-6 km of predominantly basaltic and komatiitic extrusive rocks that include pillow lavas, sheet flows, and minor hyaloclastite breccias, along with intrusive rocks and subordinate metasedimentary rocks (chert and felsic tuffaceous rocks). These rocks are relatively well-preserved, locally little-deformed and have experienced prehnite-pumpellyite to greenschist facies metamorphism (de Wit et al. 1987).

The second Archaean study region is the East Pilbara Terrane of the Pilbara Craton, Western Australia. The Pilbara Supergroup within this terrane contains a ~20 km thick succession of low-grade metamorphic, dominantly volcanic, supracrustal rocks that were deposited in four autochthonous groups from 3,530-3,165 Ma (Van Kranendonk et al. 2007). Pillow basalts and interpillow hyaloclastites were collected from the ~3,490 Ma Dresser Formation and ~3,460 Ma Apex Basalt of the Warrawoona Group, and from the ~3,350 Ma Euro Basalt of the Kelly Group for bioalteration studies. Both of these groups have previously yielded carbonaceous microfossil and stromatolite remains from cherty metasedimentary rocks that are interbedded within the volcanic sequences and have proved contentious to debates concerning the origins of life on Earth (e.g., Brasier et al. 2006; Van Kranendonk 2006).

Mineralised tubular structures have been observed in the formerly glassy rims of low-grade metamorphosed pillow basalts and hyaloclastites from the Hooggenoeg Formation and the lower part of the Kromberg Formation of the BGB and also from the Euro Basalt of the PWA (Furnes et al. 2004; Staudigel et al. 2006; Banerjee et al. 2006, 2007). These are 1-9 µm in width (averages of 4.0 µm for the BGB samples and 2.4 µm for the PWA) and up to 200 µm in length (average ~50 µm). A comparison of the size distribution of the modern and ancient tubular bioalteration textures reveals that these Archaean examples are wider than the modern equivalents, indicating that the mineralisation process may cause some expansion (see Furnes et al. 2007b: fig. 7). The Archaean tubes are infilled with extremely fine-grained titanite and some also contain minor amounts of chlorite and quartz. They extend away from healed fractures and / or the margins of hyaloclastite fragments along which seawater may once have flowed. Some of these tubular structures exhibit segmentation caused by chlorite overgrowths which formed during early stage metamorphism. In addition to their morphological similarities to modern tubular ichnotaxa in volcanic glass, linings enriched in C, N and P localised within the Archaean tubular structures have also been mapped (e.g., X-ray element maps are given in Furnes et al. 2004: fig. 3, for the Hoeggenoeg examples; and in Banerjee et al. 2007: fig. 2, from the Euro Basalt examples). Taken together, the morphologies of these textures and their chemical signatures have been used to argue that these structures represent ancient mineralised trace fossils formed by bioerosion of these formerly glassy volcanic rocks (in detail see Furnes et al. 2004; Banerjee et al. 2006; Staudigel et al. 2006; Banerjee et al. 2007). In other words, these structures represent the oldest candidate ichnofossils in the rock record.

An Archaean age for the Euro Basalt textures has been confirmed by radiometric dating using a new multicollector-ICP-MS technique for measuring ^{206}Pb / ^{238}U ratios within titanite from the 'root zones' at the centre of tubular clusters (Banerjee et al. 2007; and methodology in Simonetti et al. 2006). A minimum age of 2,921 ± 110 Ma was obtained which is ~400 Ma younger than the accepted ~3,350 Ma eruptive age of the Euro Basalt pillows, given by a U-Pb zircon age from an interbedded tuff (Nelson 2005). This titanite date corresponds to the age of regional metamorphism related to the last phase of deformation and widespread granite intrusion, the North Pilbara Orogeny that affected the sample site in the western part of the East Pilbara Terrane at ~2,930 Ma (Van Kranendonk et al. 2002). Moreover, given that chlorite from the meta-volcanic glass matrix overprints the titanite we therefore interpret this age to represent a minimum estimate for titanite formation. This implies a <400 Ma post eruptive period during which the bioalteration textures in the Euro Basalt were mineralised, but does not exclude the possibility that the candidate ichnofossils formed soon after eruption and were mineralised somewhat later. In contrast, the Barberton examples show an overlap in the radiometric ages obtained from the metamorphic chlorites and the igneous minerals in the host pillow lavas and this has been taken to support a syn-eruptive, early Archaean age for the bioalteration textures (Furnes et al. 2004; Banerjee et al. 2006). This means that these candidate ichnofossils from South Africa and Australia predate the oldest, previously known euendolithic microfossils described from ~1.7 billion year old silicified stromatolites in China (Zhang and Golubic 1987).

These candidate Archaean bioerosion structures are only found within apparently narrow horizons of the thick submarine lava sequences that comprise the greenstone belts. When they do occur, however, they are generally present in dense populations. In contrast to bioalteration textures from in situ oceanic crust where granular textures dominate, those recognised from Archaean greenstone belts are dominated by tubular textures. This apparent predominance of the mineralised tubular alteration textures is attributed to the masking of the finer, granular textures by titanite mineralisation and re-crystallisation of the host rock. Conversely, the (early) precipitation of titanite to infill many of the larger tubular textures may have enhanced their preservation, by limiting morphological changes resulting from ongoing re-crystallisation of the host rock.

Summary

Micro-bioerosion of volcanic glass is an under appreciated component of the trace fossil record. Multiple lines of evidence have been collected over the last ~15 years from in situ oceanic crust worldwide to support a biogenic origin for these microbial ichnofossils. Two traces, the 'Granular form' and 'Tubular form' are described in the

Appendix and the 'Tubular form' is further subdivided into simple, annulated and spiralled forms (Figs. 6-7). In this review we explain what is currently known about the consortia of microorganisms that are thought to be involved and present a model for how these microorganisms bioerode volcanic glass (Fig. 2). These ichnofossils in volcanic glass are found at depths of up to ~550 m in the modern oceanic crust with the 'Granular form' being most abundant and the 'Tubular form' comprising less than ~20% of the total bioalteration (Fig. 3). We also review the geological record of such ichnofossils known to date from meta-volcanic glasses in ophiolites and Archaean greenstone belts (Fig. 1). Lastly, we describe how radiometric dating of titanite ($CaTiSiO_4$) mineralised tubular micro-textures in meta-volcanic glass from West Australia has confirmed their antiquity, making these the oldest candidate ichnofossils in the rock record (Banerjee et al. 2007).

Acknowledgements

We acknowledge the help of many collaborators with the work reviewed herein describing the microbial bioerosion of volcanic glass, including: I. Thorseth, T. Torsvik, O. Tumyr, and A. Simonetti. We thank D. McIlroy for suggesting to us that bioalteration textures in volcanic glass should be considered as ichnofossils and to M. Wisshak, I. Glaub and an anonymous reviewer for constructive comments on this manuscript. Financial support for this research was provided by the Norwegian Research Council; the National Sciences and Engineering Research Council of Canada; the US National Science Foundation; the Agouron Institute and the National Research Foundation of South Africa. We thank Fred Daniel of Nkomazi Wilderness for hospitality and the Mpumalanga Parks Board for access during field work in the Barberton Mountain Land of South Africa. The Geological Survey of Western Australia, Arthur Hickman and Cris Stoakes are also thanked for assistance with fieldwork in the Pilbara. Jane Ellingsen kindly helped with the illustrations. This paper is published with the permission of the Executive Director of the Geological Survey of Western Australia.

References

Alt JC, Mata P (2000) On the role of microbes in the alteration of submarine basaltic glass: a TEM study. Earth Planet Sci Lett 181:301-313

Banerjee NR, Muehlenbachs K (2003) Tuff life: Bioalteration in volcaniclastic rocks from Ontong Java Plateau. Geochem Geophys Geosyst 4(4), doi: 10.1029/2002GC000470

Banerjee NR, Furnes H, Muehlenbachs K, Staudigel H, de Wit M (2006) Preservation of microbial biosignatures in 3.5 Ga pillow lavas from the Barberton Greenstone Belt, South Africa. Earth Planet Sci Lett 241:707-72

Banerjee NR, Simonetti A, Furnes H, Muehlenbachs K, Staudigel H, Heaman L, Van Kranendonk MJ (2007) Direct dating of Archean microbial ichnofossils. Geology 35:487-490

Benzerara K, Menguy N, Banerjee NR, Tyliszczak T, Brown Jr GE, Guyit F (2007) Alteration of submarine basaltic glass from the Ontong Java Plateau: A STXM and TEM study. Earth Planet Sci Lett 260:187-200

Brasier MD, McLoughlin N, Green OR, Wacey D (2006) A fresh look at the fossil evidence for early Archaean cellular life. Phil Trans Roy Soc B 361:887-902

Bromley RG (2004) A stratigraphy of marine bioerosion. In: McIlroy D (ed) The application of ichnology to palaeoenvironmental and stratigraphic analysis. Geol Soc London Spec Publ 228:455-479

De Ronde CEJ, de Wit M (1994) Tectonic history of the Barberton greenstone belt, South Africa: 490 million years of Archean crustal evolution. Tectonics 13:983-1005

De Wit M (2004) Archean Greenstone Belts do contain fragments of ophiolites. In: Kusky TM (ed) Precambrian ophiolites and related rocks. Dev Precambrian Geol 313:599-614

De Wit M, Hart RA, Hart RJ (1987) The Jamestown Ophiolite Complex, Barberton mountain belt: a section through 3.5 Ga oceanic crust. J Afric Earth Sci 6:681-730

Des Marais DJ (2000) When did photosynthesis emerge on Earth? Science 289:1703-1705

Dilek Y, Shallo M, Furnes H (2005) Rift-drift, seafloor spreading, and subduction tectonics of Albanian ophiolites. Int Geol Rev 47:147-176

Edwards KJ, Rogers DR, Wirsen CO, McCollom TM (2003) Isolation and characterization of novel psychrophilic, neutrophilic, Fe-oxidising, chemolithoautotrophic α- and γ-Proteobacteria from the Deep Sea. Appl Environ Microbiol 69:2906-2913

Ehrenberg CG (1836) Vorläufige Mitteilungen über das wirkliche Vorkommen fossiler Infusorien und ihre große Verbreitung Poggendorf's. Ann Phys Chem 38:213–227

Einen J, Kruber C, Øvreås L, Thorseth IH, Torsvik T (2006) Microbial colonization and alteration of basaltic glass. Biogeosci Discuss 3:273-307

Fisk MR, Giovannoni SJ, Thorseth IH (1998) The extent of microbial life in the volcanic crust of the ocean basins. Science 281:978-979

Fisk MR, Popa R, Mason OU, Storrie-Lombardie MC, Vicenzi EP (2006) Iron-magnesium silicate bioweathering on Earth (and Mars?). Astrobiology 6:48-68

Furnes H, Muehlenbachs K (2003) Bioalteration recorded in ophiolitic pillow lavas. In: Dilek Y, Robinson PT (eds) Ophiolites in Earth's History. Geol Soc London Spec Publ 218:415-426

Furnes H, Staudigel H (1999) Biological mediation in ocean crust alteration: how deep is the deep biosphere? Earth Planet Sci Lett 166:97-103

Furnes H, Thorseth IH, Tumyr O, Torsvik T, Fisk MR (1996) Microbial activity in the alteration of glass from pillow lavas from Hole 896A. Proc Ocean Drill Program, Sci Result 148:191-206

Furnes H, Hellevang B, Dilek Y (2001a) Cyclic volcanic stratigraphy in a Late Ordovician marginal basin, west Norwegian Caledonides. Bull Volcanol 63:164-178

Furnes H, Muehlenbachs K, Torsvik T, Thorseth IH, Tumyr O (2001b) Microbial fractionation of carbon isotopes in altered basaltic glass from the Atlantic Ocean, Lau Basin and Costa Rica Rift. Chem Geol 173:313-330

Furnes H, Muehlenbachs K, Tumyr O, Torsvik T, Xenophontos C (2001c) Biogenic alteration of volcanic glass from the Troodos ophiolite, Cyprus. J Geol Soc London 158:75-84

Furnes H, Staudigel H, Thorseth IH, Torsvik T, Muehlenbachs K, Tumyr O (2001d) Bioalteration of basaltic glass in the oceanic crust. Geochem Geophys Geosyst 2(8): doi:10.129/2000GC000150

Furnes H, Muehlenbachs K, Torsvik T, Tumyr O, Lang S (2002a) Bio-signatures in metabasaltic glass of a Caledonian ophiolite West Norway. Geol Mag 139:601-608

Furnes H, Thorseth IH, Torsvik T, Muehlenbachs K, Staudigel H, Tumyr O (2002b) Identifying bio-interaction with basaltic glass in oceanic crust and implications for estimating the depth of the oceanic biosphere: A review. In: Smellie JL, Chapman MG (eds) Volcano-ice interactions on Earth and Mars. Geol Soc London Spec Publ 202:407-421

Furnes H, Hellevang H, Hellevang B, Skjerlie KP, Robins B, Dilek Y (2003) Volcanic evolution of oceanic crust in a Late Ordovician back-arc basin: The Solund-Stavfjord Ophiolite Complex, West Norway. Geochem Geophys Geosyst 4(10), doi: 10.1029/2003GC000572

Furnes H, Banerjee NR, Muehlenbachs K, Staudigel H, de Wit M (2004) Early life recorded in Archean pillow lavas. Science 304:578-581

Furnes H, Banerjee NR, Muehlenbachs K, Kontinen A (2005) Preservation of biosignatures in the metaglassy volcanic rocks from the Jormua ophiolite complex, Finland. Precambrian Res 136:125-137

Furnes H, Dilek Y, Muehlenbachs K, Banerjee NR (2006) Tectonic control of bioalteration in modern and ancient oceanic crust as evidenced by carbon isotopes. Island Arc 15:143-155

Furnes H, de Wit M, Staudigel H, Rosing M, Muehlenbachs K (2007a) A vestige of Earth's oldest ophiolite. Science 315:1704-1707

Furnes H, Banerjee NR, Staudigel H, Muehlenbachs K, McLoughlin N, de Wit M, Van Kranendonk M (2007b) Comparing petrographic traces of bioalteration in recent to Mesoarchean pillow lavas: tracing subsurface life in oceanic igneous rocks. Precambrian Res 158:156-176

Gauthier MJ, Lafay B, Christen R, Fernandez L, Acquaviva M, Bonin P, Bertrand JC (1992) *Marinobacter hydrocarbonoclasticus* gen. nov., sp. nov., a new, extremely halotolerant, hydrocarbon-degrading marine bacterium. Int J Syst Bacteriol 42:568–576

Giovannoni SJ, Fisk MR, Mullins TD, Furnes H (1996) Genetic evidence for endolithic microbial life colonizing basaltic glass/seawater interfaces. Proc Ocean Drill Program, Sci Result 148:207-214

Golubic S, Friedmann I, Schneider J (1981) The lithobiontic ecological niche, with special reference to microorganisms. J Sediment Petrol 51:475-478

Jones J, Goodbody QH (1982) The geological significance of endolithic algae in glass. Canad J Earth Sci 19:671-978

Konhauser KO, Schiffman P, Fisher QJ (2002) Microbial mediation of authigenic clays during hydrothermal alteration of basaltic tephras, Kilauea Volcano. Geochem Geophys Geosyst 3(12), doi: 10.1029/2002GC000317

Kontinen A (1987) An early Proterozoic ophiolite – the Jormua mafic-ultramafic complex, northeastern Finland. Precambrian Res 35:313-341

Krumbein WE, Urzi CECA, Gehrman C (1991) Biocorrosion and biodeterioration of antique and medieval glass. Geomicrobiol J 9:139-160

Kusky TM, Li J-H, Tucker RD (2001) The Archean Dongwanzi ophiolite complex, North China craton: 2.505-billion-year-old oceanic crust and mantle. Science 292:1142-1145

Lukas KJ, Golubic S (1981) New endolithic cyanophytes from the North Atlantic Ocean: I. *Cyanosaccus piriformis* gen. et sp. nov. J Phycol 17:224-229

Lysnes K, Thorseth IH, Steinsbu BO, Øvreas L, Torsvik T, Pedersen RB (2004) Microbial community diversity in seafloor basalts from the Arctic spreading ridges. FEMS Microbiol Ecol 50:213-230

Matsushita M, Hiramatsu F, Kobayashi N, Ozawa T, Yamazaki Y, Matsuyama T (2004) Colony formation in bacteria: experiments and modeling. Biofilms 1:305-317

McKinley JP, Stevens TO, Westall F (2000) Microfossils and paleoenvironments in deep subsurface basalt samples. Geomicrobiol J 17:43-54

McLoughlin N, Brasier MD, Wacey D, Green OR, Perry RS (2007) On biogenicity criteria for endolithic microborings on early earth and beyond. Astrobiology 7:10-26

Mellor E (1922) Les lichen vitricole et la déterioration dex vitraux d'église. PhD Thesis, Sorbonne, Paris, 128 pp

Müller RD, Roest WR, Royer JY, Gahagan LM, Sclater JG (1997) Digital isochrons of the world's ocean floor. J Geophys Res B 102:3211-3214

Nelson DR (2005) Geological Survey of West Australia (GSWA) geochronology dataset. In: Compilation of geochronology data, June 2005 update. Western Australia Geological Survey, autorun compact disc, GSWA 178042

Peacock MA (1926) The petrology of Iceland. Part 1. The basic tuffs. Trans Roy Soc Edinburgh 55:53-76

Pezard PA, Anderson RN, Ryan WBF, Becker K, Alt JC, Gente P (1992) Accretion, structure and hydrology of intermediate spreading-rate oceanic crust from drill hole experiments and seafloor observations. Mar Geophys Res 14:93-123

Pongratz E (1957) D'une bactérie pediculé isolé d'un pus de sinus. Schweiz Z Allg Pathol Bakteriol 20:593-608

Radtke G (1991) Die mikroendolithischen Spurenfossilien im Alt-Tertiär West Europas und ihre palökologische Bedeutung. Courier Forschinst Senckenberg 138:1-185

Ross KA, Fisher RV (1986) Biogenic grooving on glass shards. Geology 14:571-573

Schmidt H (1992) Mikrobohrspuren ausgewählter Faziesbereiche der tethyalen und germanischen Trias (Beschreibung, Vergleich und bathymetrische Interpretation). Frankfurter Geowiss Arb A 12:1-228

Schmincke H-U, Bednarz U (1990) Pillow, sheet flow and breccia flow volcanoes and volcano-tectonic hydrothermal cycles in the extrusive series of the northeastern Troodos ophiolite (Cyprus). In: Malpas J, Moores EM, Panayiotou A, Xenophontos C (eds) Ophiolites oceanic crustal analogues. Proc Symp 'Troodos 1987', Geol Surv Dept, Nicosia, Cyprus, pp 185-206

Schumann G, Manz W, Reitner J, Lustrino M (2004) Ancient fungal life in North Pacific Eocene oceanic crust. Geomicrobiol J 21:241-246

Simonetti A, Heaman LM, Chacko T, Banerjee NR (2006) In situ petrographic thin section U-Pb dating of zircon, monazite, and titanite using laser ablation-MC-ICP-MS. Int J Mass Spectromet 253:87-97

Staudigel H, Hart SR (1983) Alteration of basaltic glass: mechanisms and significance for the oceanic crust-seawater budget. Geochim Cosmochim Acta 47:337-350

Staudigel H, Chastain RA, Yayanos A, Bourcier R (1995) Biologically mediated dissolution of glass. Chem Geol 126:119-135

Staudigel H, Yayanos A, Chastain R, Davies G, Verdurmen EAT, Schiffman P, Bourcier R, De Baar H (1998) Biologically mediated dissolution of volcanic glass in seawater. Earth Planet Sci Lett 164:233-244

Staudigel H, Furnes H, Kelley K, Plank T, Muehlenbachs K, Tebo B, Yayanos A (2004) The oceanic crust as a bioreactor. In: Wilcock W, Delong E, Kelley D, Baross J, Cary S (eds) The subseafloor biosphere at mid-ocean ridges. Geophys Monogr, Ser 144, Amer Geophys Union, pp 325-341

Staudigel H, Furnes H, Banerjee NR, Dilek Y, Muehlenbachs K (2006) Microbes and volcanoes: A tale from the oceans, ophiolites and greenstone belts. GSA Today 16(10):4-10

Stevens TO, McKinley JP (1995) Lithoautotrophic microbial ecosystems in deep basalt aquifers. Science 270(5235):450-454

Storrie-Lombardi MC, Fisk MR (2004) Elemental abundance distributions in suboceanic basalt glass: Evidence of biogenic alteration. Geochem Geophys Geosyst 5(10), doi:10.129/2004GC000755

Stroncik N, Schmincke H-U (2001) Evolution of palagonite: Crystallization, chemical changes, and element budget. Geochem Geophys Geosyst 2(7), doi:10.1029/2000GC000102

Templeton AS, Staudigel H, Tebo BM (2005) Diverse Mn(II)-oxidizing bacteria isolated from submarine basalts at Loihi Seamount. Geomicrobiol J 22:127-139

Thorseth IH, Furnes H, Tumyr O (1991) A textural and chemical study of icelandic palagonite of varied composition and its bearing on the mechanism of the glass-palagonite transformation. Geochim Cosmochim Acta 55:731-749

Thorseth IH, Furnes H, Heldal M (1992) The importance of microbiological activity in the alteration of natural basaltic glass. Geochim Cosmochim Acta 56:845-850

Thorseth IH, Furnes H, Tumyr O (1995) Textural and chemical effects of bacterial activity on basaltic glass: an experimental approach. Chem Geol 119:139-160

Thorseth IH, Torsvik T, Torsvik V, Daae FL, Pedersen RB, Keldysh-98 Scientific party (2001) Diversity of life in ocean floor basalts. Earth Planet Sci Lett 194:31-37

Thorseth IH, Pedersen RB, Christie DM (2003) Microbial alteration of 0-30-Ma seafloor and sub-seafloor basaltic glasses from the Australian Antarctic Discordance. Earth Planet Sci Lett 215:237-247

Thorseth IH, Kruber C, Hellevang H, Pedersen RB (2007) Seafloor alteration of basaltic glass: Textures, geochemistry and endolithic microorganisms. Geophys Res Abstr 9:09890

Torsvik T, Furnes H, Muehlenbachs K, Thorseth IH, Tumyr O (1998) Evidence for microbial activity at the glass-alteration interface in oceanic basalts. Earth Planet Sci Lett 162:165-176

Van Kranendonk MJ (2006) Volcanic degassing, hydrothermal circulation and the flourishing of early life on Earth: A review of the evidence from c. 3490-3240 Ma rocks of the Pilbara Supergroup, Pilbara Craton, Western Australia. Earth-Sci Rev 74:197-240

Van Kranendonk MJ, Hickman AH, Smithies RH, Nelson DN, Pike G (2002) Geology and tectonic evolution of the Archaean North Pilbara terrain, Pilbara Craton, Western Australia. Econ Geol 97:695-732

Van Kranendonk MJ, Smithies RH, Hickman AH, Champion DC (2007) Secular tectonic evolution of Archaean continental crust: interplay between horizontal and vertical processes in the formation of the Pilbara Craton, Australia. Terra Nova 19:1-38

Walton AW (2005) Petrography of peridophyllic endolithic microborings from hyaloclastites of Kilauea's Hilina slope: comparison with microborings in HSDP hyaloclastites. Geol Soc Amer Abstr Programmes 37:253

Walton AW, Schiffman P (2003) Alteration of hyaloclastites in the HSDP 2 Phase 1 Drill Core: 1. Description and paragenesis. Geochem Geophys Geosyst 4(5), doi: 10.1029/2002GC000368

Walton AW, Schiffman P, Macperson GL (2005) Alteration of hyaloclastites in the HSDP 2 Phase 1 Drill Core: 2. Mass balance of the conversion of sideromelane to palagonite and chabazite. Geochem Geophys Geosyst 6(9), doi:10.1029/2004GC000903

Wilde SA, Cawood PA, Wang KY, Nemchin A, Zhao GC (2004) Determining Precambrian crustal evolution in China: a case-study from Wutaishan, Shanxi Province, demonstrating the application of precise SHRIMP U-Pb geochronology. In: Malps J, Fletcher CJN, Ali JR, Aichison JC (eds) Aspects of the Tectonic Evolution of China. Geol Soc Spec Publ London 226:5-26

Wisshak M, Gektidis M, Freiwald A (2005) Bioerosion along a bathymetric gradient in a cold temperature setting (Kosterfjord, SW Sweden): an experimental study. Facies 51:93-117

Zhang Z, Goloubic S (1987) Endolithic microfossils (cyanophyta) from early Proterozoic Stromatolites, Hebei China. Acta Micropaleont Sin 4:1-12

Appendix – Systematic ichnology

'Granular form'

Fig. 5

Description: Individual, spherical bodies filled with very fine-grained phyllosilicate phases, zeolites and iron-oxy-hydroxides and perhaps also titanite

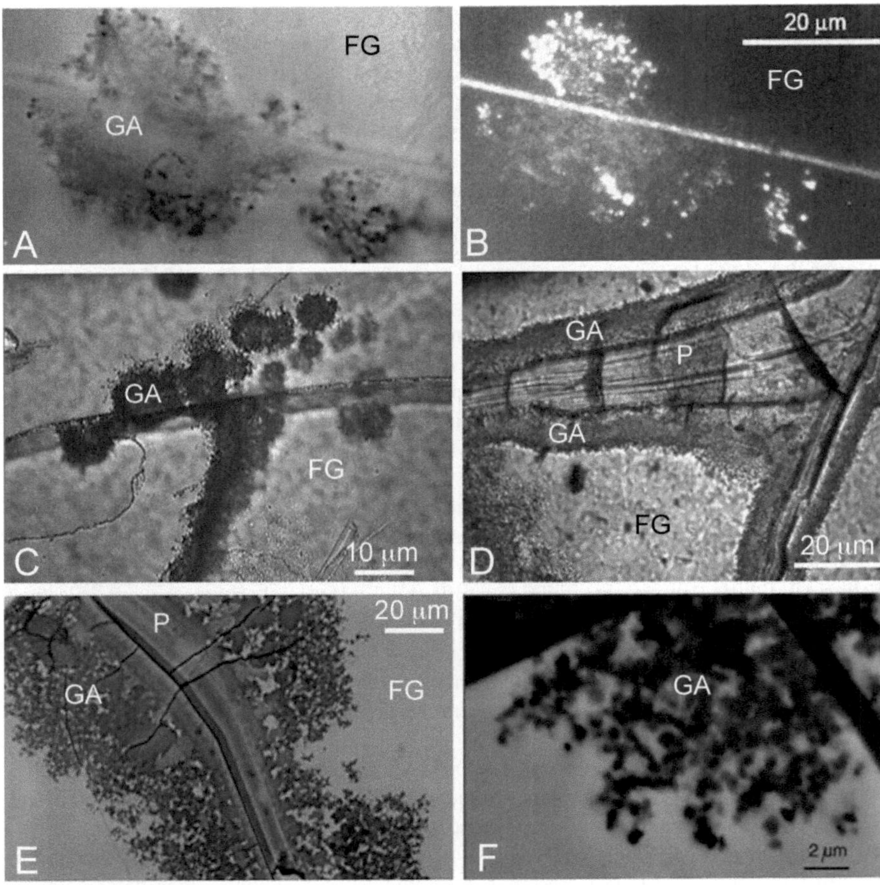

Fig. 5 Images of the 'Granular form' in fresh glass (FG) with banded palagonite (P) and coalesced, granular alteration zones (GA). **A** Transmitted light image of the 'Granular form', sample 148-896A-11R-1, 73-73 cm from the Costa Rica Rift (Furnes et al. 1996). **B** Epiflourescence image of the sample stained with 10 g/mL DAPI, a fluorescent dye that binds to DNA and shows that the biological material is concentrated along the edges of the granular alteration and in the fracture. **C** Granular textures along two intersecting fractures from hole 396B in the mid-Atlantic. **D** Optical image showing a fracture with banded palagonite in the centre and granular alteration on the outer margins with the fresh glass, from sample 418, 56-5, 129-132 on the Bermuda Rise. **E** Back scatter electron image of a fracture filled with banded palagonite and coalesced granules (GA) at the interface with fresh glass, sample 418A, 62-4, 64-7 from the Bermuda Rise. **F** Back scatter electron image of granular alteration at the junction of two fractures, DSDP sample 69-504B, 4-2, 0-20

($CaTiSiO_4$). The diameter of the granules, regardless of age, location, and depth in the oceanic crust varies from around 0.1 µm up to rare examples that are 1.5 µm across, with the most common size around half a micron (Furnes et al. 2007b: fig. 6). In the initial stages of bioerosion the granules occur as isolated spherical bodies along fractures in the glass; as the bioalteration proceeds they become more numerous and coalesce into aggregates, especially at the intersection of fractures (e.g., middle line of Fig. 2).

Distribution: Reported from numerous in situ oceanic crust sites worldwide, these are reviewed in the supplementary table in Staudigel et al. (2006). These include: the north to central Atlantic Ocean, Lau Basin in the South West Pacific and the Costa Rica Rift in the Western Pacific (reviewed in Furnes et al. 2001d) and the Ontong Java Plateau (Banerjee and Muehlenbachs 2003). Examples described from ophiolites include: the 92 Ma Troodos ophiolite from Cyprus (Furnes et al. 2002c) and the 160 Ma Mirdita ophiolite from Albania (Furnes et al. 2007b).

Trace maker: Microorganisms unknown

Remarks: This is by far the most abundant ichnofossil morphotype found in recent volcanic glass. The granular aggregates develop into irregular bands that protrude into the fresh glass, vary in thickness and can exhibit little or no symmetry about the fractures along which they develop. The relatively small size of granular microborings means that they are highly susceptible to re-crystallisation of the glass during metamorphism however, and as a consequence their geological record is less extensive than that of the tubular shaped microborings. Granular textures on the surface of fresh glass have to be carefully distinguished from abiotic, chemical etch pits that can form in the absence of microbes (c.f. Thorseth et al. 2007) and which are thought to display a more uniform size and shape distribution.

'Tubular form'

Figs. 6-7

Description: Tubes which protrude into fresh glass with diameters that range from: ~0.4 µm up to ~6 µm, although most are between 1 and 2 µm, with an average diameter of 1.4 µm (Furnes et al. 2007b: fig. 7). The length of the tubes is highly variable, from a few micrometres up to several hundred micrometres. In recent volcanic glasses the tubes may be hollow, but more usually in the geological record they are (partially) infilled with fine-grained phyllosilicates, zeolites, iron-oxy-hydroxides and sometimes titanite ($CaTiSiO_4$). In most cases the tubes propagated perpendicular to the original fracture plane, although random orientations are also seen. We further sub-divide the 'Tubular form' into the following morphotypes:

Simple, tubular – this is the most abundant 'Tubular form', lacks surface ornamentation and is typically unbranched (e.g., Fig. 6A-B). Rare examples have been found to contain fine sub-micron wide filaments (Fig. 6C) that are interpreted to be of biological origin (Banerjee and Muehlenbachs 2003). More unusually, the simple tubular subtype may exhibit bifurcating branches and build dense dendritic clusters (e.g., Fig. 6D-E).

Annulated, tubular – these show annulations that are continuous around the circumference of the tube and display near equal spacing along the tubes, except at the tips where they may become more closely spaced (e.g., Figs. 6B, 7A).

Spiral, tubular – these are coiled or helical tubes. The diameter of the helix may decrease towards the termination of the tubes and the whorls may become more closely spaced (e.g., Figs. 6F-G, 7A). Occasionally, an outer, spiralled tube is seen to wrap around a central, linear tube with larger diameter (e.g., Fig. 6G). This elaborate morphology strongly supports a biogenic origin for these structures.

Distribution: The simple, tubular subtype is widely reported from in situ oceanic crust, known examples are reviewed in the supplementary table in Staudigel et al. (2006). Examples include: the north to central Atlantic Ocean, Lau Basin in the South West Pacific and the Costa Rica Rift in the Western Pacific (reviewed in Furnes et al. 2001d) and the Ontong Java Plateau (Banerjee and Muehlenbachs 2003). Examples described from ophiolites include: the 92 Ma Troodos ophiolite from Cyprus (Fig. 6F-G; Furnes et al. 2002c); the 160 Ma Mirdita ophiolite from Albania (Fig. 7B; Furnes et al. 2007b); and the 160 Ma Stonyford Ophiolite from California. Examples described from greenstone belts include: the 443 Ma Solund Stavfjord Ophiolite in Norway (Furnes et al. 2001c); the ~2.5 Ga Wutai Greenstone Belt of China; the ~2.7 Ga Abitibi Greenstone Belt of Canada (the ~3.5 Ga Barberton Greenstone Belt of South Africa (Furnes et al. 2004; Banerjee et al. 2006, 2007) and the ~3.5 Ga Pilbara Craton of West Australia (Banerjee et al. 2007).

The annulated, tubular subtype is most clearly seen in volcanic glass fragments from the Ontong Java Plateau (Banerjee and Muehlenbachs 2003).

The spiral, tubuar form is only known to date from the Troodos Ophiolite of Cyprus (Furnes et al. 2001c) and Figure 6F-G.

Trace maker: Microorganisms unknown

Remarks: Sub-parallel tubes are sometimes seen to abruptly change growth direction by up to 180° when they meet another tube or fracture and this has been interpreted to reflect biogenic behaviour, as opposed to abiotic phenomenon, that would be expected to cross-cut (Furnes et al. 2007b: fig. 4C). In glass samples that contain vesicles, tubes may radiate outwards from the vesicle walls into the fresh glass (e.g., Furnes et al. 2007b; Figs. 5A-B). Tubular textures are also often concentrated around varioles that are created by quenching of the volcanic glass (e.g., Furnes et al. 2007b; Fig. 5C), and along concentric fractures formed around phenocrysts such as olivine. This had lead to the suggestion that the microorganisms responsible may be peridophyllic i.e., olivine seeking (Walton 2005). Alternatively, the bioerosion may exploit mechanical weaknesses in the glass such as those provided by fractures.

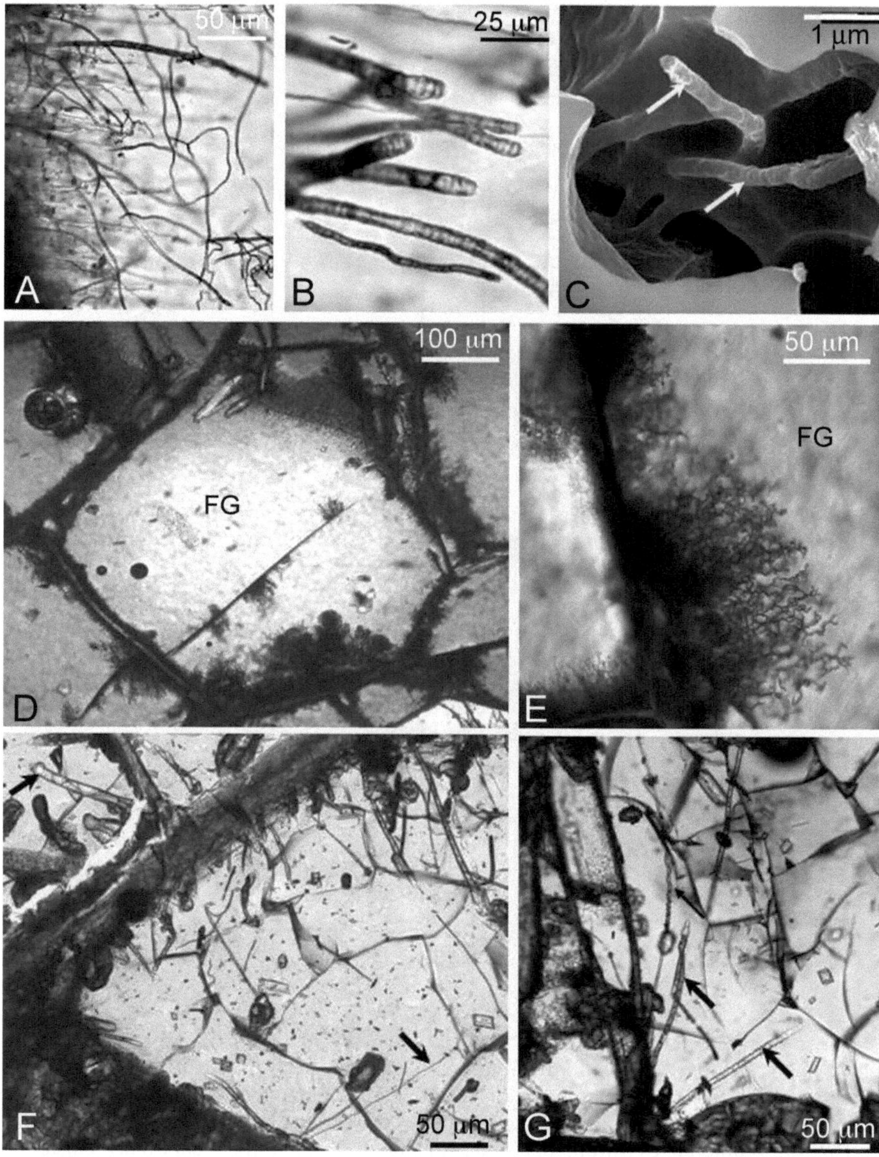

Fig. 6 Images of the 'Tubular form' in fresh volcanic glass (FG). **A** Simple unbranched tubes radiating from the margin of a glass clast on the left. **B** Annulated tubes. **C** Back scatter electron image of sub-micron sized filaments (arrowed) within a hollow tubular cavity. **D** Optical image of branched tubes radiating from fractures in a volcanic glass. **E** Enlarged view of D showing the dendritic branching patterns. **F** Glass from an ophiolite containing an assemblage of 'Tubular form' traces that includes spiral shaped tubes (white arrow) and annulated tubes (black arrow), sample CY-1-30. **G** Enlarged view showing the spiralled tubular subtype, with simple large tubes surrounded by a finer helical tube. Images A-C from sample 1184A-13R-3-4, the Ontong Java Plateau (Banerjee and Muehlenbachs 2003); D-E from hole 396 in the Mid-Atlantic (Furnes et al. 2001d) and F-G from the 92 Ma Troodos Ophiolite, sample CY-1-30 (Furnes et al. 2001c)

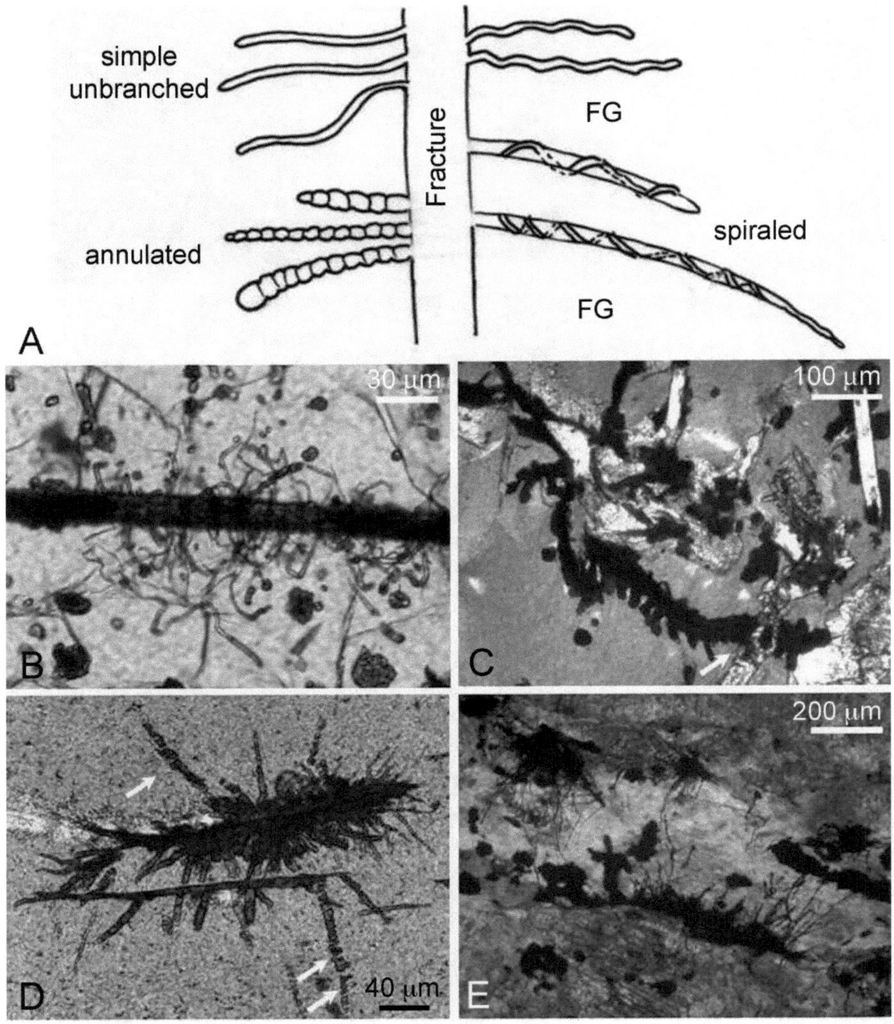

Fig. 7 Images of the 'Tubular form' from ophiolites and greenstone belt samples. **A** Line drawings of the tubular morphological subtypes: simple, annulated and spiraled forms. **B** Hollow tubes radiating away from a fracture in the centre of the image, sample 3-Al-00 from the 162 Ma, zeolite facies Mirdita ophiolite, Albania. **C** Titanite filled, dark black cluster of tubes in a chlorite matrix that are cross-cut by a later stage calcite vein (white arrow), sample 114-S-00 from the greenschist facies metamorphosed ~443 Ma Solund-Stavfjord ophiolite, Norway (Furnes et al. 2002a). **D** A cluster of titanite mineralised tubes that are septate due to chlorite overgrowths (arrowed), sample from the ~3.4 Ga Barberton Greenstone Belt of South Africa (Furnes et al. 2004; Banerjee et al. 2006). **E** Clusters of titanite mineralised tubes in a greenschist facies hyaloclastite, sample 74-PG-04a from the ~3.4 Ga Euro Basalt of Western Australia (Banerjee et al. 2007)

Microbial bioerosion of bone – a review

Miranda M. E. Jans[1]

[1] Vrije Universiteit, Institute for Geo and Bioarchaeology, 1081 HV Amsterdam, the Netherlands, (miranda.jans@falw.vu.nl)

Abstract. Microbial alteration is an important pathway for bone degradation. Organisms involved in bioerosion of bone, mainly bacteria, fungi and cyanobacteria, create different types of alteration. While fungi and cyanobacteria dissolve the bone matrix resulting in branching tunnels, bacteria create microscopical focal destructions with a complex morphology, reorganising the mineral rather than removing it. Different environmental and early post mortem circumstances characterise each type. Bacterial alteration occurs in the early post mortem interval, probably in the first decades after death. A strong link with putrefaction can be observed, indicating that early putrefactive organisms are important contributors to bacterial alteration of bone. Fungal and cyanobacterial alteration occurs when the environment is favourable, i.e., oxygen is present and bone still has sufficient nutrient value. Although microbial alteration causes loss of information in archaeological and palaeontological bone, the study of microbial bioerosion also represents a powerful tool for taphonomic reconstruction.

Keywords. Bioerosion, archaeological bone, bacteria, fungi, cyanobacteria, BSE SEM, histology

Introduction

Bioerosion

Bone microboring organisms include fungi (Wedl 1864; Marchiafava et al. 1974), bacteria (Hackett 1981; Baud and Lacotte 1984; Jackes et al. 2001), and in marine environments, cyanobacteria (Bell et al. 1996; Davis 1997). Bioerosion, or microbial alteration, is found in archaeological bone from all over the world (e.g., Solomon and Hasse 1967: Israel; Hanson and Buikstra 1987: North America; Hackett 1981: Europe, Asia, Australia; Gutierrez 2001: South America; Trueman et al. 2004: Africa). It is a common determinant of bone preservation (Smith et al. 2007). Consequently, bioerosion of bone and teeth has been the focus of several publications over the span of more than 140 years. The first description dates from 1864, when Wedl found tunnels of approximately 8 µm in diameter in recent human teeth exposed to untreated well water as well as in fossil reptile teeth. Marchiafava et al. (1974) succeeded in identifying a bone-boring fungus in an experimental

M. Wisshak, L. Tapanila (eds.), *Current Developments in Bioerosion*. Erlangen Earth Conference Series, doi: 10.1007/978-3-540-77598-0_20, © Springer-Verlag Berlin Heidelberg 2008

setting. In his key paper on microbial alteration of bone – or microscopical focal destruction (MFD) – Hackett (1981) used the term Wedl tunnels for those tunnels caused by fungi. Three other types of MFD were described by Hackett (1981): linear longitudinal, budded and lamellate. These are generally assumed to be produced by bacteria (Child 1995). Backscattered electron scanning electron microscopy (BSE SEM) and mercury intrusion porosimetry (HgIP) revealed the intricate spongiform porosity of bacterial alteration (Jackes et al. 2001; Turner-Walker et al. 2002). Unlike fungal alteration, bacterial alteration has not yet been replicated in an experimental context. As a result, the exact mechanism of bacterial alteration remains elusive (Child 1995). Currently research focuses on how early taphonomy influences bioerosion. Studies in archaeology on how patterns of bioerosion can provide information about the taphonomical history of bone show promising results (Parker Pearson et al. 2005).

Bone

Bone is an important source of information in archaeological and palaeontological research. In these contexts, mineralised tissue (bone, teeth) is generally the only direct physical remain of people and animals of the past. Consequently, bone preservation in its burial environment has been researched extensively (e.g., Hedges et al. 1995; Kars and Kars 2002). The interest in bone degradation has been increased by the discovery that proteins and even DNA can be recovered from ancient mineralised tissue. It has become clear that a good understanding of bone diagenesis is crucial for the success of biomolecular studies (Cooper and Poinar 2000).

Bone tissue has a complex micro-anatomy. The basic building block of bone is the mineralised collagen fibril, which through its organisation, together with bone vascularisation, determines bone microstructure. Bone contains an elaborate interconnected pore system of vascular canals (Haversian canals and Volkmann canals; ranging between 50-250 μm in diameter) and small osteocyte canaliculi which connect the bone cells to each other and the vascular system. In the bone matrix, several highly mineralised discontinuity layers or cementing lines are present, separating anatomical units (e.g., osteons) from the rest of the bone matrix (Francillon-Vieillot et al. 1990).

On the molecular level, bone is composed of both inorganic, mainly hydroxyapatite ($Ca_{10}(PO_4 CO_3)_6(OH)_2$), and organic material (collagen, and non-collagenous proteins like osteocalcin, osteonectin etc.). This combination of organic and inorganic components gives bone the hardness and elasticity required during life and makes it resistant to decay processes after death.

Diagenesis

Bone mineral crystals are held within a framework of collagen and are thus protected from rapid dissolution (Trueman and Martill 2002). Collagen is a stable, insoluble protein in normal conditions. After death slow chemical processes like hydrolysis will degrade collagen. However, this reaction can be sped up in certain environments (e.g., alkaline systems; Collins et al. 1995). If the rate of collagen breakdown is slow enough to allow the mineral to recrystallise, fossilisation of the bone and long term preservation is possible. In corrosive environments or in the case of microbial degradation this 'equilibrium' cannot be reached.

It has often been observed that the preservation of bone is unpredictable based solely upon the soil environment as an indicator; a benign burial environment is by no means a guarantee for good preservation (Mant 1987; Cox and Bell 1999; Nielsen-Marsh et al. 2007). The outcome of a large-scale project characterising bone diagenesis in the European burial environment was surprising, considering the complexity of bone and the variability of possible environmental conditions. Only four general types of bone diagenesis were identified (Smith et al. 2007). Microbial alteration was identified as a common determinant of bone preservation, affecting human bone more than animal bone (Jans et al. 2004). Furthermore, the incidence of bacterial alteration did not relate to the burial environment, but could be related to early post mortem history of the bone (Jans et al. 2004).

Detecting and quantifying microbial alteration

Histology and BSE SEM

Histology (thin section analysis) of bone has a long history of use in research on archaeological bone degradation (e.g., Wedl 1864; Stout 1978; Hackett 1981; Garland 1988; Grupe and Garland 1993; Jans 2005). It has been established as a standard approach in determining bone preservation in archaeological heritage management studies in the Netherlands where it is used to evaluate and monitor the preservation of archaeological bone (Van Heeringen et al. 2004). Using the description by Hackett (1981) it is possible to determine which type of organism altered the bone. Indeed histology and BSE SEM are the only methods that can differentiate between bacterial and fungal alteration. To quantify the extent of microbial alteration in a cross section of bone, Hedges et al. (1995) developed the Oxford Histological Index (OHI), a scale ranging from 5 (pristine bone) to 0 (no original bone left; Millard 2001). To achieve more detailed information on distribution of microbial alteration, a bone section can be divided in three areas (periostal area, middle area and endosteal area), with a separate OHI value assigned to each (Jans et al. 2002). Other types of diagenetic change, such as cracking, inclusion of foreign material or a decrease in natural birefringence, can also be observed and quantified (Jans et al. 2002).

The increased resolution as well as added compositional information using energy dispersive X-ray spectroscopy (EDS), make BSE SEM a very informative method to study bone alteration (Turner-Walker et al. 2004), providing a detailed analysis of the porosity and changes to the mineral in bioeroded bone. Turner-Walker and Syversen (2002) have developed a method for quantifying bacterial alteration using BSE SEM and digital image analysis. This method results in a Bioerosion Index (BI), which offers information on the percentage of original bone matrix destroyed by bacterial alteration. This method is more detailed than the OHI, but has the disadvantages of being more time consuming as well as analysing a relatively small part of the bone.

Mercury porosimetry

Bioerosion causes characteristic changes in the bone porosity which can be measured using mercury intrusion porosimetry (HgIP; Turner-Walker et al. 2002). HgIP allows a rapid and detailed assessment of pore size distribution in bone for systematic quantification of porosity in large sample sets (Turner-Walker et al. 2002). The technique measures interconnected pores by forcing mercury into a bone sample at increasing pressures. The method estimates porosity by assuming cylindrical pores and a contact angle between bio apatite and mercury of 163.1° (Joscheka et al. 2000; see Turner-Walker et al. 2002 for details). In microbially altered bone, HgIP shows an increased volume in pores with diameters in the <0.1 µm region, with the largest increase at approximately 0.6 and 1.2 µm (Turner-Walker et al. 2002), consistent with the average diameter for bacteria (0.5 µm; Hawker and Linton 1979).

Bone boring microorganisms

Bone boring microorganisms are usually no longer present in archaeological bone upon excavation. Consequently, identification of the organism responsible is based on the type of tunnelling present in the bone. Table 1 provides an overview of the different types of bioerosion found in archaeological and palaeontological bone, listing morphological aspects as well as causative organisms. A scheme of the boring types (except for the superficial Hackett tunnels) is given in Figure 1.

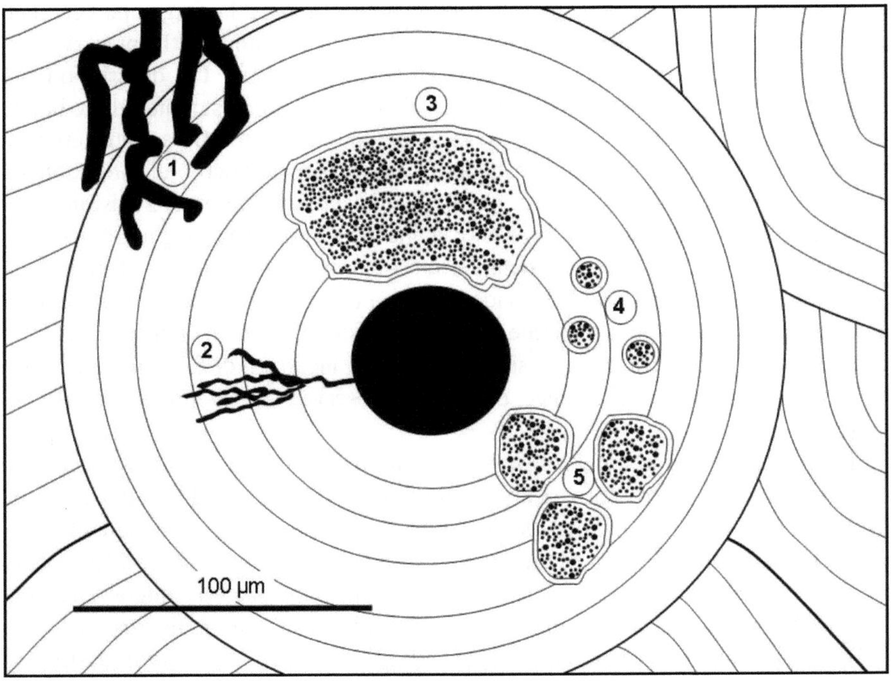

Fig. 1 Schematic representation of a bone osteon with the different types of microbial tunnelling found in transversal sections of bone. 1: Type 1 Wedl tunnelling, 2: Type 2 Wedl tunnelling, 3: Lamellate MFD, 4: Linear longitudinal MFD, 5: Budded MFD

Table 1 Description of tunnelling types, their diameter (* = larger tunnels up to 100-250 μm are occasionally found), presence of a hypermineralised rim, affiliated producing organism, and distribution of tunneling in relation to bone microstructure

Type of tunnelling	Wedl type 1	Wedl type 2	Hackett	Linear Longitudinal	Budded	Lamellate
Diameter	10-15 μm	5 μm	50-250 μm	5-10 μm	30-60* μm	10-60* μm
Infill	Empty	Empty	Empty	Spongiform	Spongiform	Spongiform (lamellate)
Hypermin-eralised rim	-	-	-	+	+	+
Organism	Fungi Cyanobacteria	Fungi? Cyanobacteria?	Fungi? Cyanobacteria?	Bacteria	Bacteria	Bacteria
Distribution	Random	Random	Random (superficial)	Microanatomy	Microanatomy	Microanatomy

Fungi

Fungal alteration of bone was first described in 1864 by Wedl, who observed borings 8 μm in diameter in fossil reptile bone as well as in recent human teeth that were immersed in untreated well water for ten days. A study by Marchiafava et al. (1974) showed that inoculating sterile bone with the fungus *Mucor* Fresenius, 1850 resulted in tunnelling in 25-30 days; confirming that fungi can cause this type of tunnelling. Transmission electron microscopy (TEM) analysis of these specimens revealed that the invading fungal hyphae contained mineral in solution, whilst the tunnels hold no redeposited mineral (Marchiafava et al. 1974). This indicates that the dissolved mineral is transported out of the bone. Hackett (1981) described fungal tunnelling in bone as centrifugal tunnels or Wedl tunnels. These usually start from an exposed bone surface such as the cortex or a fracture surface and penetrate the bone matrix with branching tunnels (Fig. 2). Fungi can, unlike bacteria, cross cement lines, and as a result their distribution in the bone microstructure is random. It is not clear whether fungi dissolve the bone to harvest nutrients or to use the bone as a medium.

Trueman and Martill (2002) further distinguish two types of Wedl tunnelling. Type 1, which is the most common type, consists of simple, randomly branching, three-dimensional networks not influenced by bone micro-anatomy. The tunnel diameter is between 10-15 μm. Type 2 Wedl tunnels are found less commonly and have a smaller diameter (5 μm). They extend from Haversian canals into the bone matrix and show increased branching frequency and form more complex networks (Trueman and Martill 2002).

A third type of tunnelling, Hackett tunnelling, may also be caused by fungi (Davis 1997). This alteration consists of superficial, radiating tunnels, approximately 50-250 μm in diameter.

Fungal structures (hyphae, fruit bodies, spores) are regularly found in archaeological bone (Jans 2005); commonly there is no associated destruction of the bone microstructure, suggesting that several species may use bone as a medium.

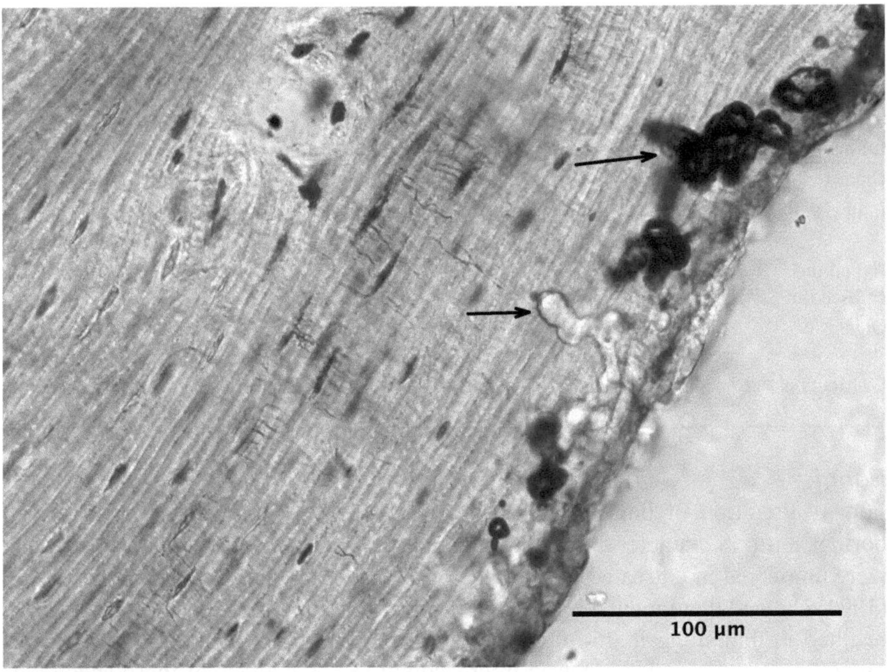

Fig. 2 Micrograph of a bone sample with well preserved microstructure and beginning Wedl tunnelling at the outer edge of the bone (arrow). The bone micro-anatomy has little or no influence on how the tunnels invade the bone

Cyanobacteria

Cyanobacteria are aquatic, photosynthetic bacteria, formerly referred to as blue-green algae. Some species grow in long filaments. Several types of cyanobacteria are euendoliths (organisms that tunnel into a substrate; as opposed to organisms that take advantage of existing pores). Borings in teeth by cyanobacteria, recovered from marine environments, have been described by Bell et al. (1996). The tunnelling caused by cyanobacteria is very similar to that caused by fungi, i.e., of the Wedl type (Davis 1997). The tunnels are approximately 5-7 µm in diameter (in teeth) and do not show redeposited mineral (Bell et al. 1996). As cyanobacteria depend on light for their metabolism, their tunnels are only found on exposed surfaces of the bone, where light can still reach. Indeed, as observed in an experimental set up by Davis (1997), rapid covering of bone by sediment prevents cyanobacterial bioerosion.

Bacteria

Bacteria are thought to produce pores in bone, causing substantial degradation. Hackett (1981) described three types of bacterial boring: linear longitudinal, budded and lamellate, named after their microscopical morphology.

Linear longitudinal tunnels (Fig. 3) are the smallest of the bacterial foci, about 10 µm in diameter and are generally circular. They are surrounded by a hypermineralised rim or 'cuff' that consists of mineral, redeposited after its dissolution to expose the

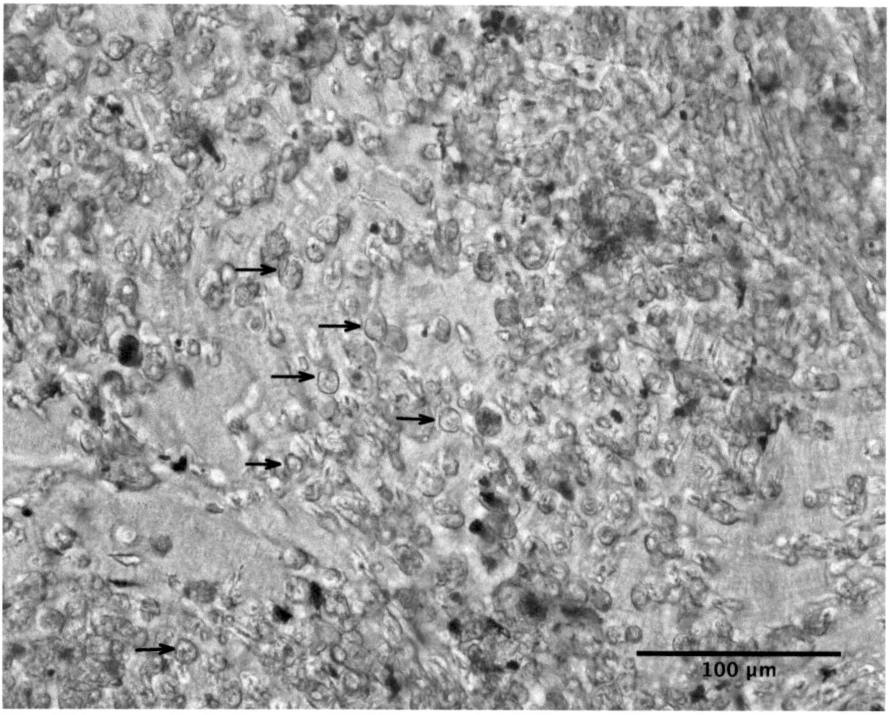

Fig. 3 Micrograph of a bone sample with extensive linear longitudinal tunnelling (arrows). These MFD appear in a transversal section as round foci, about 10 µm in diameter with a thick rim and a stippled infill (the infill is not visible at this magnification)

bone collagen (Hackett 1981). This rim, appearing brighter when imaged with BSE SEM (Fig. 4) in fact contains more mineral than the surrounding unaltered bone and is more dense (Jackes et al. 2001). The infill of the MFD appears stippled or grainy in light microscopy, but shows spongiform porosity in BSE SEM imaging (Fig. 4) that can also be analysed using HgIP (Turner-Walker et al. 2002).

Budded tunnels are larger (30-60 µm in diameter; Fig. 5) and more irregular in shape than linear longitudinal MFD. Similarly to linear longitudinal tunnels, budded tunnels have a hypermineralised rim and their infill appears stippled or grainy in transversal sections in normal light microscopy. In BSE SEM the spongiform porosity is clearly visible (Fig. 4). In a longitudinal section they can be seen stretching along a Haversian canal with regular off-shoots or buds (hence their name; Hackett 1981).

Lamellate MFD are relatively large (10-60 µm in diameter), and have a mineralised rim. What distinguishes this type of MFD from other types is the grainy infill which reflects the lamellar pattern of the bone (Fig. 6). Lamellate foci are less common than budded and linear longitudinal (Jans et al. 2004). Mercury porosimetry profiles of bone containing lamellate tunnels show a slightly smaller spongiform pore diameter (Jans et al. 2004). This difference in pore size has not yet been visually confirmed with BSE SEM.

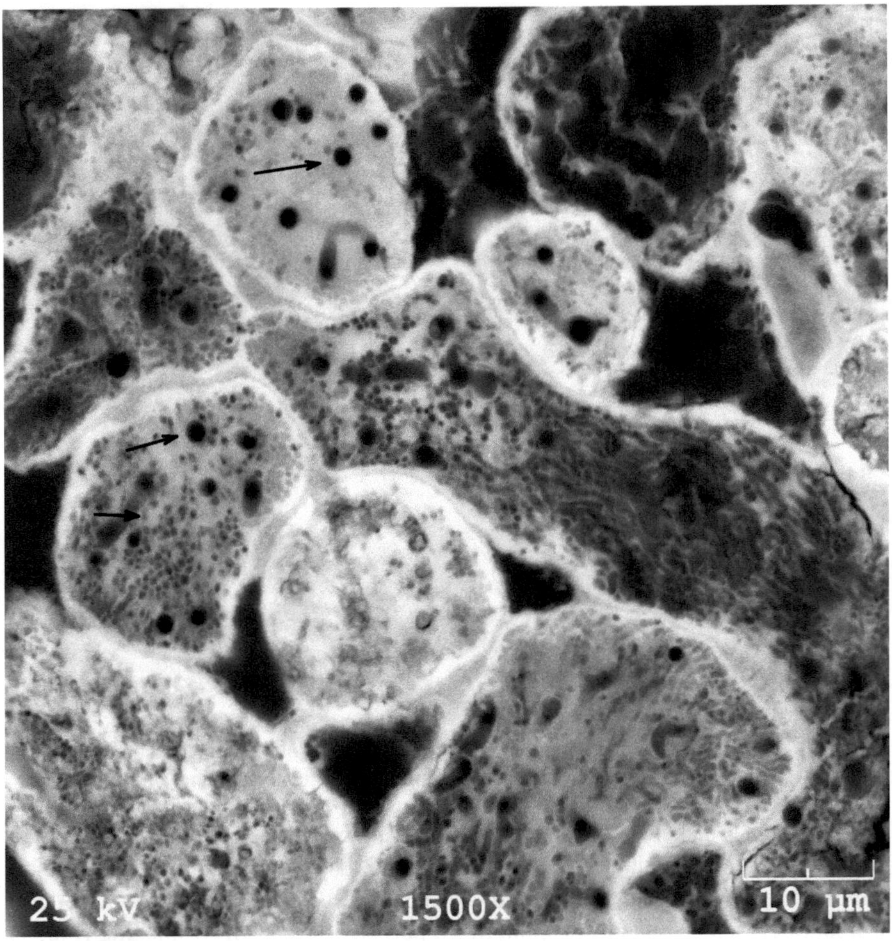

Fig. 4 BSE SEM image of bacterially altered bone showing MFD with sub-micron spongiform porosity (dark circles) surrounded by a hypermineralised rim (arrows)

All three types of bacterial MFD spread out from the blood vessel canals and, when able to proceed uninhibited, 'pack' or fill in the osteon until they reach a cement line (Hackett 1981). Hackett (1981) found no association of the MFD location with the location of osteocyte lacunae. In contrast, Turner-Walker et al. (2002) suggest that early change is typically associated with osteocyte lacunae. Bell (1990) also observed that MFD do not often cross surrounding cement lines, and generally they do not overlap each other or coalesce. This strongly suggests that the osteocyte canaliculi play a role in the expansion of bacteria through the bone. Notwithstanding the focal appearance of MFD in light microscopy, HgIP and to a lesser extent BSE SEM clearly demonstrate that bacterial tunnelling consists of an interconnected pore system, instead of isolated foci (Fig. 7; Turner Walker et al. 2002).

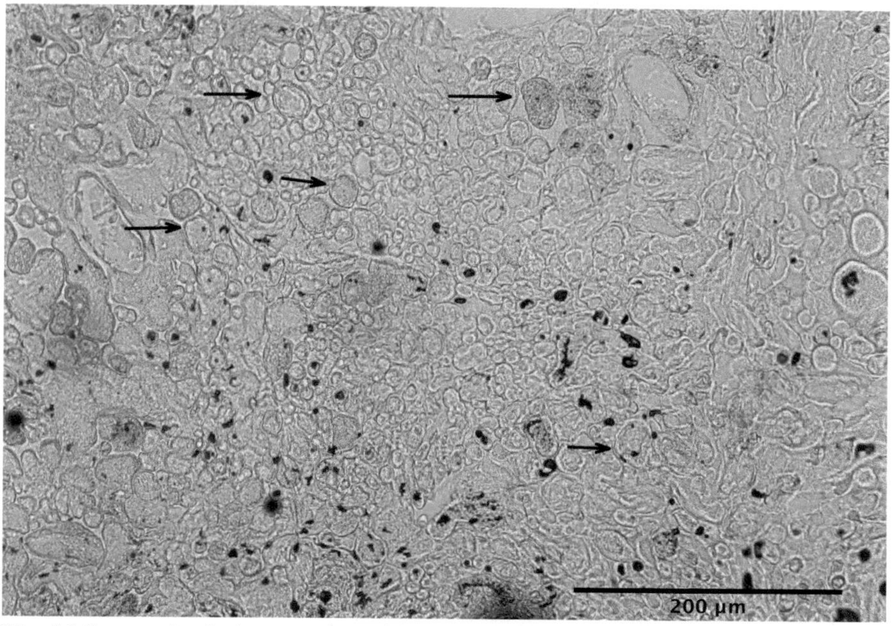

Fig. 5 Micrograph of a bone section with extensive budded tunnelling (arrows), nearly no original bone microstructure is preserved. Some linear longitudinal tunnels are present as well. The rims surrounding the MFD are visible, as is the stippled infill of the MFD

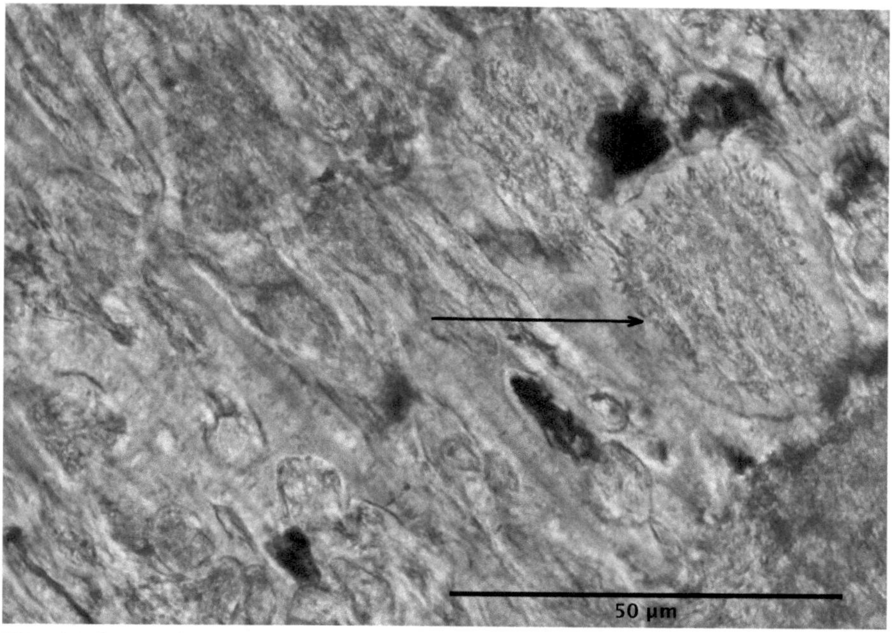

Fig. 6 Micrograph of a bone section with lamellate tunnelling (arrow). Here also the hypermineralised rim is present, but the lamellar pattern of the bone is reflected in the pattern of the MFD infill

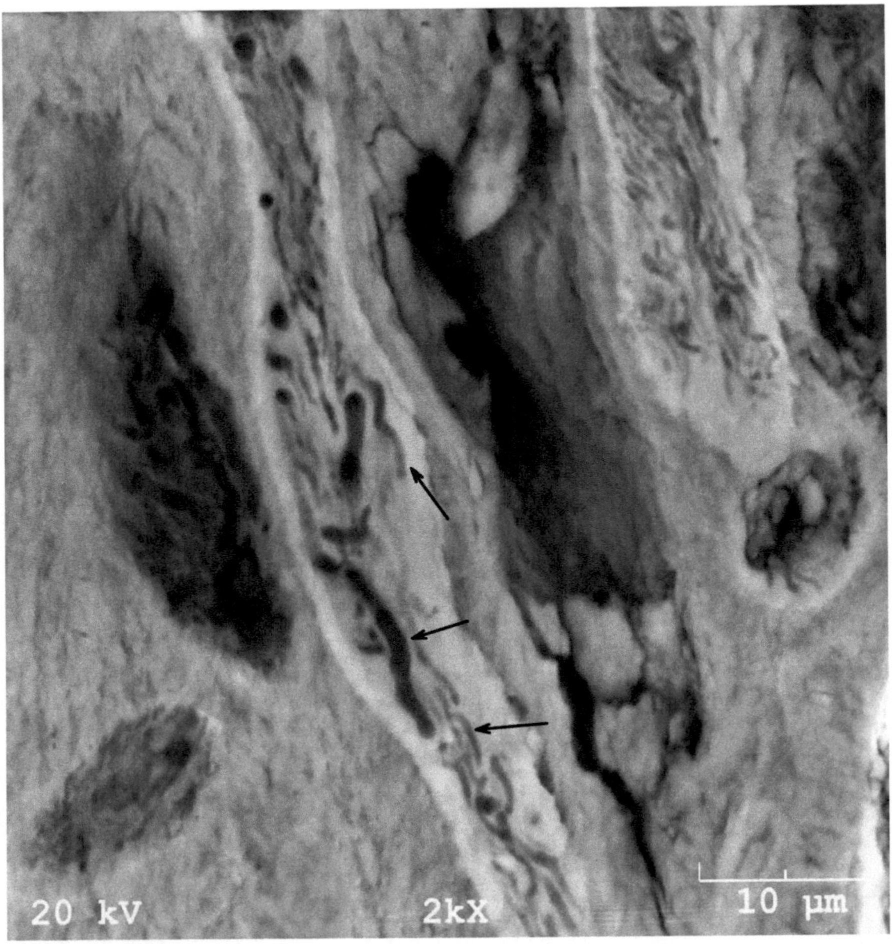

Fig. 7 BSE SEM image of a bone sample showing MFD in a longitudinal section, indicating that the spongiform porosity consists of connected pores (arrows)

Factors influencing alteration

Fungal boring

Although it was the first type of microbial alteration described in calcified tissue, fungal tunnelling is not the most common in archaeological bone (Garland 1988; Trueman and Martill 2002). Jans et al. (2004) found a significant difference in the frequency of fungal tunnelling between animal and human archaeological bone; Wedl tunnelling was more common in animal bone. Trueman and Martill (2002) found that in fossil bone (ranging in age from Pleistocene to Silurian), half of the tunnels found were fungal. Bone decomposing fungi are dependent on environmental factors like the presence of oxygen and a certain level of humidity

(20%; Carlisle et al. 2001); anaerobic fungi have only been found in the rumen of certain herbivores, where they play a role in cellulose digestion (Carlisle et al. 2001). Once the environmental parameters are favourable, fungal attack becomes possible. As mentioned previously, fungi can produce alteration within two to three weeks after exposure (Wedl 1864; Marchiafava et al. 1974; Davis 1997).

This could indicate that fungi are opportunistic in their colonisation of bone, altering bone at points in time when the environmental conditions are suitable (i.e., presence of oxygen and moisture) and probably if the bone is not bacterially altered. Bacterial and fungal alterations rarely occur together in the same bone (Jans et al. 2004). It is unknown if this is a consequence of competition between organisms, a lack of remaining nutrients when the bone has been previously altered or environmental incompatibility.

Cyanobacterial boring

Cyanobacterial alteration has been observed within 11 days after deposition of bone in marine and lacustrine environments (Davis 1997). Cyanobacterial alteration was found in bone two years post mortem by Bell et al. (1996). As with fungal degradation, the environment and the presence of light, determines largely when and to what extent cyanobacterial tunnelling occurs. The state of preservation of the bone will also play a role, i.e., flesh covered remains will not be affected by cyanobacteria (Davis 1997).

Bacterial boring

Bacterial alteration in mineralised tissue has been observed shortly after death. Bell et al. (1996) found evidence of alteration in bone as soon as three months post mortem, though this sample was recovered from a predator scat and may therefore not be representative. More reliable is the identification – in the same study – of bacterial alteration of human bone from a waterlogged muskeg bog at 15 months post mortem (Bell et al. 1996). Yoshino et al. (1991) observed evidence of microbial alteration in bone five years after burial, and Haversian canals in those samples contained bacteria. Hedges et al. (1995) observed that bacterial alteration did not progress after 100-200 years post mortem.

The morphology of bacterial decay is very uniform, independent of (archaeological) date and location, as found by a study by Kars and Kars (2002) analysing archaeological bone in detail, from geographies as diverse as northern Sweden to southern Italy. This uniformity could indicate that bone microstructure dictates the morphology of the MFD in bacterial decay. Alternatively, the bacterial species responsible for decay could be similar in all locations. Bacterial alteration could not be related to burial environment (Nielsen-Marsh et al. 2007). Jans et al. (2004) observed that bone from complete burials (animal and human) was more likely to be affected by bacterial attack than bone, which entered the archaeological record as fragments (e.g., butchered animals). This, as well as the apparent relation of bacterial alteration with bone micro-vascularisation, indicates that bacterial degradation is linked to putrefaction and the very early stages of degradation.

It has been observed that, in soils without a large bacterial population, putrefaction of a body can still be quite extensive (Mant 1987). In fact, endogenous bacterial populations play a large role in the early putrefaction of the body (Janaway 1996). Endogenous commensal bacteria have been shown in vitro to transmigrate from the intestines to the rest of the body within 2 hours (at 37°C; Melvin et al.1984) and 17 hours (Kellerman et al. 1976) after death. Initial spread of bacteria through the body may be slowed by residual immunity, as shown in vivo by Gill et al. (1978), and as a result, bacteria will likely start colonising the body 24-48 hours post mortem. Kainer et al. (2004) hypothesise that the source for *Clostridium* Prazmowski, 1880 infection in patients who received a musculoskeletal allograft, may be haematogenous contamination with transmigrated gut flora. This may have been caused by a prolonged interval between death of the donor and tissue recovery, delays in refrigeration (i.e., refrigeration after 12 hours post mortem) or death from trauma (Kainer et al. 2004). This would suggest that in normal conditions (i.e., no refrigeration or sterilisation) putrefactive bacteria are present in the bone within a day post mortem.

Microbial attack is a pathway for fast collagen degradation involving enzymatic hydrolysis through collagenases (Child 1995). This process has not yet been directly observed, so it is only possible to speculate about the mechanisms at work, by extrapolating from dental caries and microbial degradation of living bone (Child 1995). Collagen is effectively protected from bacterial alteration through its association with bone mineral, as the spaces between the crystallites are too small to allow collagenases to penetrate. Consequently, microorganisms cannot access collagen until the mineral has been dissolved, for example by means of metabolically produced organic acids (Child 1995). Most likely, these metabolites are by-products from the putrefaction of soft tissue. An apparent paradox in this scenario is that at low pH collagenases are permanently inactivated (Jackes et al. 2001). Rapid removal of soft tissue from a complete body by scavengers or environmental processes may slow and finally stop bacterial decay while the slow degradation of soft tissue in a deep coffin burial (Rodriguez and Bass 1985) may allow it to proceed until the whole bone microstructure is destroyed.

Based on the morphology of the sub-micron spongiform pores, Jackes et al. (2001) hypothesise that *Clostridium* spp., and specifically *Clostridium histolyticum* Weinberg and Séguin, 1916 (Bergey, 1923) is a good candidate as both size and known behaviour of this bacterium fit the profile.

Bioerosion as an indicator for early post mortem history

If endogenous putrefactive bacteria are mostly responsible for bacterial alteration of bone, lack of bacterial alteration must indicate exceptional early post mortem circumstances. Examples are the presence of bactericidal substances (like copper, mercury or lead) near or in the body, mummification, or the absence of endogenous bacteria for example in neonates (Janaway 1987; Mant 1987; Yamada et al. 1990; Janaway 1996). Another possibility is dismemberment or butchering shortly after death, preventing endogenous bacteria from invading bone tissue, as intestines and bone are no longer connected through the vascular system (Trueman and Martill 2002; Jans et al. 2004). Most archaeological animal bone fragments will therefore not be affected by this type of putrefaction, which explains the relatively low frequency of bacterial attack found in these bones (Jans et al. 2004).

That the presence or absence of bacterial alteration is useful as an indicator of early post mortem taphonomy was shown by a study by Parker Pearson et al. (2005). In a case of suspected mummification in a Bronze Age burial in Britain, the level of bacterial alteration of the bone was lower than would be expected (Parker Pearson et al. 2005). This was seen as supporting evidence that the body could have been mummified prior to burial. In fact, examples of studies on bone diagenesis on known mummies are minimal but seem to point to a pattern of inhibited or arrested bacterial alteration. Perhaps the most famous example of such a mummy is the Tyrolean Ice Man (aka Ötzi), discovered in a glacial field in the Alps in 1991. Histological analysis of a rib bone of Ötzi showed no bacterial alteration, though gram negative bacteria were observed in the Haversian canal as well as in the digestive tract (Hess et al. 1998) which showed evidence of disintegration and transformation processes due to bacterial decay and autolysis.

In fossilised bone, ranging in age from Silurian to Pleistocene, Trueman and Martill (2002) found that the majority of samples showed little or no microbial alteration. The predominant type of alteration was fungal tunnelling. As this type of bioerosion is determined by environmental factors, its presence can potentially serve as an indicator of (past) burial environments.

Consequences for bone preservation

The majority of archaeological bone is likely to be biologically altered (Jans et al. 2004). Aside from changes in porosity, microbial alteration of bone causes changes in bone mineral (e.g., increases in bone crystallinity), and a decrease in percent bone collagen, ultimately resulting in complete loss (Smith et al. 2007). Biologically altered bone buried in a corrosive environment is more vulnerable to further diagenetic processes, such as leaching. The poorly mineralised and porous infill of the MFD will be easily leached out, leaving behind a honeycomb structure, where only the hypermineralised rims remain (Jans 2005). Though morphological information will generally still be available in the archaeological timeframe, molecular information (DNA) may already be affected as soon as the bone is microbially altered. Gilbert et al. (2005) showed a correlation between bone histological preservation and preservation of DNA as well as contamination of the sample. Moreover, the low frequency of microbial alteration in Pleistocene to Silurian bone suggests that microbially altered bone will generally not survive into the fossil record (Trueman and Martill 2002). This will be the fate of most human bone when inhumed as a complete body.

Conclusions

A better understanding of microbial alteration will benefit diverse areas of research, such as taphonomy, forensic sciences, archaeological heritage management and biomolecular archaeology. Fungal and cyanobacterial alteration are indicative of the (past) burial environment. In contrast, bacterial alteration is influenced by very early post mortem taphonomy. As bacterial degradation in the form of MFD will not occur in situations where bacterial growth has been inhibited (e.g., extreme temperatures, bactericidal chemicals), or where gut bacteria have had no access to the bone (after butchering or dismemberment), it can be used as an indicator of early post mortem history.

The increase in porosity resulting from microbial alteration means that a large portion of archaeological bone is more vulnerable to other diagenetic processes. Indeed, the relative absence of bacterial MFD in palaeontological bone shows that bacterially altered bone is not expected to survive into the fossil record (Trueman and Martill 2002). This information may prove useful in site evaluation and development of archaeological heritage management strategies for in or ex situ preservation. For example, change in the burial environment that would lead to increased oxygen availability should be avoided as this will enable fungal degradation.

Lastly, biological degradation has consequences for biomolecular preservation of mineralised tissue. Specifically, it increases the chances for contamination of samples or causes loss of information in organic structures such as DNA (e.g., Hagelberg et al. 1991; Colson et al. 1997; Gilbert et al. 2005). Knowledge of bioerosion in mineralised tissue will help validate and interpret results as well as contribute to sample selection.

Acknowledgements

This paper was completed when MJ was employed in a Framework 6 Marie Curie Outgoing International Fellowship, 'Bacbone' (project no: FP6-CT2006-022210) at the Central Identification Laboratory, Joint POW / MIA Accounting Command (JPAC-CIL), USA. The use of histological and SEM facilities at the CIL is gratefully acknowledged. Franklin E. Damann (JPAC CIL), Prof. Matthew J. Collins (BioArch, University of York / Institute for Geo and Bioarchaeology, VU University Amsterdam) and Prof. Henk Kars (Institute for Geo and Bioarchaeology, Vrije Universiteit, Amsterdam) are thanked for advice and stimulating discussions. Special thanks go to Toni M. Diegoli, Jessica L. Saunier and Kimberly A. Sturk (Armed Forces DNA Identification Laboratory, USA) for their careful reading of the manuscript. Thanks are also due to two anonymous reviewers who made valuable additions to the manuscript.

References

Baud CA, Lacotte D (1984) Étude au microscope électronique à transmission de la colonisation bacteriènne de l'os mort. C R Acad Sci Paris 298:507-510

Bell LS (1990) Paleopathology and diagenesis; a SEM evaluation of structural changes using backscattered electron imaging. J Archaeol Sci 17:86-102
Bell LS, Skinner SMF, Jones J (1996) The speed of post mortem change to the human skeleton and its taphonomic significance. Forensic Sci Int 82:129-140
Bergey DH, Harrison FC, Breed RS, Hammer BW, Huntoon FM (1923) Bergey's manual of determinative bacteriology. Williams and Wilkins, Baltimore, 442 pp
Carlisle MJ, Watkinson SC, Gooday GW (2001) The fungi. Acad Press, New York, 588 pp
Child AM (1995) Towards an understanding of the microbial decomposition of archaeological bone in the burial environment. Journal of Archaeological Science 22:165-174
Collins MJ, Riley MS, Child AM, Turner-Walker G (1995) A basic mathematical model for the chemical degradation of ancient collagen. J Archaeol Sci 22:175-83
Colson I, Bailey JF, Vercauteren M, Sykes B, Hedges REM (1997) The preservation of ancient DNA and bone diagenesis. Ancient Biomolecul 1:109-117
Cooper A, Poinar HN (2000) Ancient DNA: Do it right or not at all. Science 289:1139
Cox M, Bell LS (1999) Recovery of human skeletal elements from a recent UK murder inquiry: preservational signatures. J Forensic Sci 44:945-950
Davis PG (1997) Bioerosion of bird bones. Int J Osteoarchaeol 7:388-401
Francillon-Vieillot HV, de Buffrenil V, Castanet J, Geraudie J, Meunier FJ, Sire JY, Zylberberg L, de Ricqlès A (1990) Microstructure and mineralization of vertebrate skeletal tissues. In: Carter JG (ed) Skeletal biomineralization: patterns, processes and evolutionary trends. Van Nostrand Reinhold, New York, pp 471-530
Fresenius JBGW (1850) Beiträge zur Mykologie. Brönner, Frankfurt a.M., 111 pp
Garland AN (1988) A histological study of archaeological bone decomposition. In: Boddington A, Garland AN, Janaway RC (eds) Death, decay and reconstruction. Manchester Univ Press, Manchester, pp 109-126
Gilbert MTP, Rudbeck L, Willerslev E, Hansen AJ, Smith C, Penkman K, Prangenberg K, Nielsen-Marsh CM, Jans MME, Arthur P, Lynnerup N, Turner-Walker G, Biddle M, Kjølbye-Biddle B, Collins MJ (2005) Biochemical and physical correlates of DNA contamination in archaeological bones and teeth. J Archaeol Sci 32:785-793
Gill CO, Penney N, Nottingham PM (1978) Tissue sterility in uneviscerated carcasses. Appl Environ Microbiol 36:356-359
Grupe G, Garland AN (1993) Histology of ancient bone: methods and diagnosis. Springer, Berlin, 223 pp
Gutierrez MA (2001) Bone diagenesis and taphonomic history of the Paso Otero 1 bone bed, Pampas of Argentina. J Archaeol Science 28:1277-1290
Hackett CJ (1981) Microscopical focal destruction (tunnels) in exhumed human bones. Med Sci Law 21:243-265
Hagelberg E, Bell LS, Allen T, Boyde A, Jones SJ, Clegg JB (1991) Analysis of ancient bone DNA – techniques and applications. Phil Trans Roy Soc London, Ser B, 333:1268-1399
Hanson DB, Buikstra JE (1987) Histomorphological alteration in buried human bone from the Lower Illinois Valley: Implications for palaeodietary research. J Archaeol Sci 14: 549-63
Hawker LE, Linton AH (1979) Microorganisms. Function, form and environment. Arnold, London, 400 pp
Hedges REM, Millard AR, Pike AWG (1995) Measurements and relationships of diagenetic alteration of bone from three archaeological sites. J Archaeol Sci 22:201-209
Hess MW, Klima G, Pfaller K, Künzel KH, Gaber O (1998) Histological investigations on the Tyrolean Ice Man. Amer J Phys Anthropol 106:521-532

Jackes M, Sherburne R, Lubell D, Barker C, Wayman M (2001) Destruction of microstructure in archaeological bone: a case study from Portugal. Int J Osteoarchaeol 11:415-432

Janaway RC (1987) The preservation of organic materials in association with metal artefacts deposited in inhumation graves. In: Boddington A, Garland AN, Janaway RC (eds) Death, decay and reconstruction. Approaches to archaeology and forensic science. Manchester Univ Press, Manchester, pp 127-148

Janaway RC (1996) The decay of buried human remains and their associated materials. In: Hunter J, Roberts C, Martin A (eds) Studies in crime: an introduction to forensic archaeology. Batsford, London, pp 58-85

Jans MME (2005) Histological characterisation of the degradation of archaeological bone. PhD Thesis, Inst Geo- Bioarchaeol, Vrije Univ, Amsterdam, 163 pp

Jans MME, Kars H, Nielsen-Marsh CM, Smith CI, Nord AG, Arthur P, Earl N (2002) In situ preservation of archaeological bone. A histological study within a multidisciplinary approach. Archaeometry 44:343-352

Jans MME, Nielsen-Marsh CM, Smith CI, Collins MJ, Kars H (2004) The characterisation of microbial attack in archaeological bone. J Archaeol Sci 31:87-95

Joscheka S, Nies B, Krotz R, Gopferich A (2000) Chemical and physicochemical characterisation of porous hydroxyapatite ceramics made of natural bone. Biomaterials 21:1645

Kainer MA, Linden JV, Whaley DN, Holmes HT, Jarvis WR, Jernigan DB, Archibald LK (2004) *Clostridium* infections associated with musculoskeletal allografts. New England J Med 350:2564-2571

Kars EAK, Kars H (2002) The degradation of bone as an indicator for the deterioration of the European archaeological property. Final Rep, Rijksdienst Oudheidkundig Bodemonderzoek, Amersfoort, 279 pp

Kellerman GD, Waterman BG, Scharfenberger LF (1976) Demonstration in vitro of post-mortem bacterial transmigration. Amer J Clinical Pathol 66:911-915

Mant AK (1987) Knowledge from post-war exhumations. In: Boddington A, Garland AN, Janaway RC (eds) Death, decay and reconstruction. Approaches to archaeology and forensic science. Manchester Univ Press, Manchester, pp 65-80

Marchiafava V, Bonucci L, Ascenzi A (1974) Fungal osteoclasia: a model of dead bone resorption. Calcified Tissue Res 14:195-210

Melvin JR, Cronholm LS, Simson LR, Isaacs AM (1984) Bacterial transmigration as an indicator of time of death. J Forensic Sci 29:412-417

Millard AM (2001) The deterioration of bone. In: Pollard AM, Brothwell DR (eds) Handbook of archaeological science. Wiley, New York, pp 633-643

Nielsen-Marsh CM, Smith CI, Jans MME, Nord A, Kars H, Collins MJ (2007) Bone diagenesis in the European Holocene II: taphonomic and environmental considerations. J Archaeol Sci 34:1523-1531

Parker Pearson M, Chamberlain A, Craig O, Marshall P, Mulville J, Smith H, Chenery C, Collins MJ, Cook G, Craig G, Evans J, Hiller J, Montgomery J, Schwenninger J-L, Taylor G, Wess T (2005) Evidence for mummification in Bronze Age Britain. Antiquity 79:529-546

Prazmowski A (1880) Untersuchungen über die Entwicklungsgeschichte und Fermentwirkung einiger Bakterien-Arten. Voigt, Leipzig, 58 pp

Rodriguez WC, Bass WM (1985) Decomposition of buried bodies and methods that may aid in their location. J Forensic Sci 30:836-852

Smith CI, Nielsen-Marsh CM, Jans MME, Collins MJ (2007) Bone diagenesis in the European Holocene I: patterns and mechanisms. J Archaeol Sci 34:1485-1493

Solomon CD, Hasse N (1967) Histological and histochemical observations of decalcified sections of ancient bones from excavations in Israel. Israel J Med Sci 3:747-754

Stout SD (1978) Histological structure and its preservation in ancient bone. *Curr Anthropol* 19:601-604

Trueman CNG, Martill DM (2002) The long-term survival of bone: the role of bioerosion. Archaeometry 44:371-382

Trueman CNG, Behrensmeyer AK, Tuross N, Weiner S (2004) Mineralogical and compositional changes in bones exposed on soil surfaces in Amboseli National Park, Kenya: diagenetic mechanisms and the role of sediment pore fluids. J Archaeol Sci 31:721-739

Turner-Walker G, Syversen U (2002) Quantifying histological changes in archaeological bones using BSE-SEM image analysis. Archaeometry 44:461-468

Turner-Walker G, Nielsen-Marsh CM, Syversen U, Kars H, Collins MJ (2002) Sub-micron spongiform porosity is the major ultrastructural alteration occurring in archaeological bone. Int J Osteoarchaeol 12:407-414

Van Heeringen RM, Mauro G, Smit A (2004) A pilot study on the monitoring of the physical quality of three archaeological sites on the Unesco Monument of Schokland, province of Flevoland, the Netherlands. Nederl Archeol Rapp 26, Rijksdienst Oudheidkundig Bodemonderzoek, Amersfoort, 133 pp

Wedl C (1864) Ueber einen im Zahnbein und Knochen keimenden Pilz. Akad Wiss Wien, math-natw Kl (I) 50:171-193

Weinberg M, Séguin P (1916) Contribution à l'étiologie de la gangrène gazeuse. C R Acad Sci Paris 163:449-451

Yamada TK, Kudou T, Takahashi-Iwanaga H (1990) Some 320-year-old soft tissue preserved by the presence of mercury. J Archaeol Sci 17:383-392

Yoshino M, Kimijima T, Miyasaka S, Sato H, Seta S (1991) Microscopical study on estimation of time since death in skeletal remains. Forensic Sci Int 49:143-158

Xylic substrates at the fossilisation barrier: oak trunks (*Quercus* sp.) in the Holocene sediments of the Labe River, Czech Republic

Radek Mikuláš[1]

[1] Institute of Geology, Academy of Sciences of the Czech Republic, 165 00 Praha 6, Czech Republic, (mikulas@gli.cas.cz)

Abstract. Sediments of the Holocene floodplain of the Labe River (central Bohemia) provided an accumulation of oak trunks (*Quercus* sp.) bearing relatively rich and diverse assemblages of borings. Among the recognised morphotypes of borings, four can be attributed to insect feeding, one resulted probably from an enzymatic fungal activity, and the last one is probably a mammal 'scratch'. The borings record three phases of activity: (1) on living trees, (2) on dead trees before their burial by sediment, and (3) on exhumed trunks (i.e., during the last several years). Generally, the wood mass comes mostly from live, 'healthy' floodplain forests, which shows that these were affected by extremely large floods during certain intervals in the Holocene. The borings found support the idea of erecting terrestrial wood ichnofacies, but the Holocene material itself is not suitable for this purpose.

Keywords. Xylic substrate, Holocene, trace fossils, insects, fungi, fluvial settings, Czech Republic

Introduction

In ichnology, xylic substrates typically rank among substrates that can be exploited by organisms. Wood and similar substrates are considered so specific that trace fossils made in wood are given ichnogeneric names different from morphologically identical traces made in softgrounds, firmgrounds and rocky substrates (Bertling et al. 2006). Within the ichnological theoretical background, wood as the substrate for the life activity of fauna was studied in detail especially in the marine realm, where wood transported by rivers or floodwater to the sea is often colonised (pholadid and teredinid bivalves; cf. Kříž and Mikuláš 2006). Diversity of the ichnologic record in marine-submerged wood is quite low; however, only two species, *Teredolites clavatus* Leymerie, 1842 and *T. longissimus* Kelly and Bromley, 1984 are generally recognised and accepted. The spectrum of forms and possible tracemakers is considerably larger in modern terrestrial settings. Among the producers recognised, fungi (Genise 2004), insects (e.g., Walker 1938; Scott 1992; Genise 1995), birds (Buchholz 1986; Elbroch and Marks 2002) and mammals (e.g., Elbroch 2003) have

M. Wisshak, L. Tapanila (eds.), *Current Developments in Bioerosion*. Erlangen Earth Conference Series, doi: 10.1007/978-3-540-77598-0_21,

been reported. However, only a small portion of the wealth of recent wood borings has already been reported in the fossil record, and if so, most of them have not received formal ichnotaxonomic names. In comparison to the subaerial settings, borings in wood submerged to fresh water are a rare phenomenon, though one example (Thenius 1979) has been described from the fossil record; this study has even an ichnotaxonomic part.

The lack of interest in ichnotaxonomy in this case is probably also the reason of low interest in other aspects of wood substrates, namely taphonomic and ichnofacies-related. The present study is, in this context, related mostly to taphonomy; it demonstrates the accumulation of wood mass in fluvial sediments, the manner of its preservation in the Holocene geologic record, and the degree of attack by boring organisms. The study cannot have the ichnotaxonomical impact because (according to the generally accepted definition of the term 'fossil') the Holocene material is considered 'modern' and therefore not allowed as the type material for ichnotaxa (cf. Bertling et al. 2006). Aspects of woodground ichnofacies are briefly discussed.

Human-made changes of the floodplain pose a considerable restriction for the use of actualistic observations in the floodplain. Plant and adjacent animal assemblages, hydrologic conditions, and substrates available for erosion during flood events, were strongly modified during the past several hundred years. Therefore, the described geologic record, though young, is the record of a true geologic history. This brings yet another welcome aspect to the geological studies of the Labe floodplain: the possibility to compare the Holocene and present processes in the floodplain. Because of the human change in the floodplain environment, numerous economic activities in the flood area, and possible climatic change, any forecast of future geologic development of the Labe floodplain is a welcome contribution to the discussion on the future development of the area.

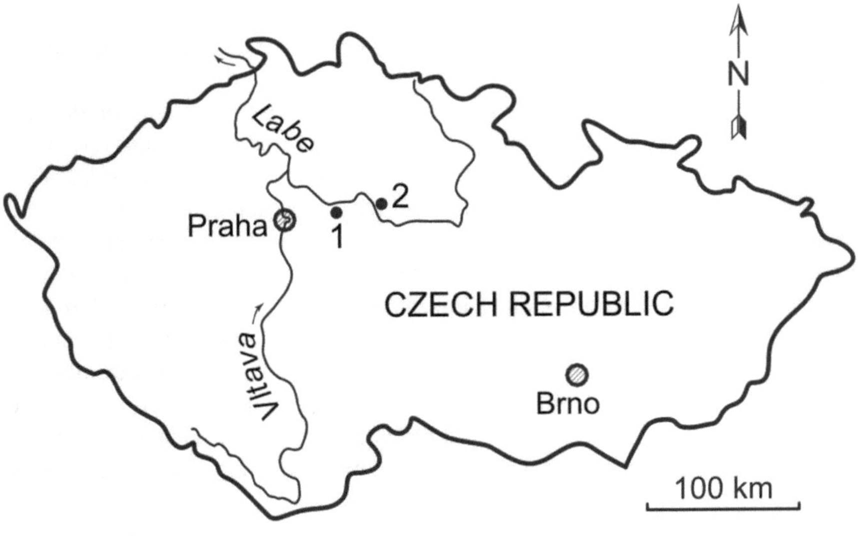

Fig. 1 Sketch map with localities: 1 = Čelákovice; 2 = Libice nad Cidlinou

Location and geologic settings

The Labe River (named Elbe in German and English literature) is the largest Czech river, having its source in the Krkonoše Mountains at the altitude of 1,384 metres above sea level. After leaving the Krkonoše Mountains, the Labe runs across their foothills towards Hradec Králové and turns west across the Labe Lowland. Near the town of Mělník, the Vltava River flows into Labe, thus doubling its average discharge to 248 m^3 per second. After passing two mountain ranges (volcanic České středohoří Mts. and sandstone tectonic blocks of the Bohemian-Saxonian Switzerland), the Labe leaves the area of the Bohemian Massif; its total length on the Bohemian Massif is ca. 400 km (Fig. 1).

Recent summary of the geology of the floodplain deposits of the Labe was presented by Růžičková and Zeman (1994a). The middle reach of Labe (in Czech context usually considered the area between Kolín and Mělník, cf. Růžičková and Zeman 1994a) shows a distinct Holocene floodplain, several kilometres to tens of kilometres wide. It can be divided into three levels: upper floodplain terrace (4.0 to 4.5 metres above the present river level), lower floodplain terrace (2-3 m above the river), and recent floodplain (0-2 m above the river). These levels represent erosional steps due to hydrological changes. The overall thickness of the Holocene sedimentary fill in the middle reach of the Labe River is ca. 10 metres. Sedimentation of the upper floodplain terrace began at ca. 9,500 years BP; the lower terrace shows the range of ages from 7,700 BP and 3,700 BP. The recent floodplain dates to the 16th century and corresponds to the onset of a cold climatic fluctuation in central Europe called the 'Little Ice Age' (cf. Ložek 1980; Růžičková and Zeman 1994a).

Fluvial sediments of all the three levels consist dominantly of sand and gravel (Růžičková and Zeman 1994a; Fig. 4A). These sediments contain a surprising amount of well-preserved tree trunks – the estimate of their number still buried in the sediments is on the order of thousands (Růžičková and Zeman 1994b). More than 90% of them are oaks (*Quercus* cf. *robur* Linné, 1758); the remaining ones have been determined as species of *Acer* Linné, 1758 and *Ulmus* sp. Linné, 1758 (Růžičková and Zeman 1994b).

Čelákovice

The locality of 'Čelákovice', for the purpose of this paper, is limited to abandoned sandpits lying ca. 1500 m to the NW of the town centre. It lies at the middle reach of the Labe River in the sense of Růžičková and Zeman (1994a, b). The floodplain east of the locality is not strongly disturbed by rail and road constructions, sand quarrying etc.; therefore, its pre-modern state is well traceable (Fig. 2). The floodplain width ranges from 1.5 to 3.0 km. A schematic profile of the floodplain in the above mentioned part of the floodplain is shown in Figure 3. The sedimentary record comprises Pleistocene fluvial sediments, coarse clastics of the upper and middle floodplain terrace (lying approx. 3 m above the present water level to 2 m below the level) were the subject of exploitation.

Fig. 2 Schematic map of the Labe floodplain east of Čelákovice with the location of the Čelákovice locality (modified from Růžičková and Zeman 1994a)

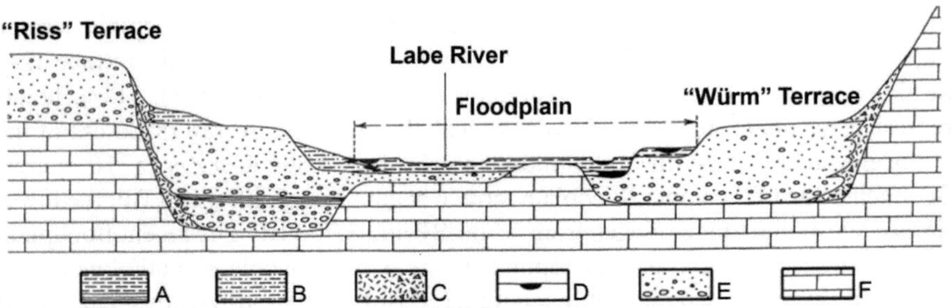

Fig. 3 Schematic section of the Labe floodplain east of Čelákovice (modified from Růžičková and Zeman 1994a). A = argillaceous fluvial sediments, B = air-borne sands, C = slope loams and debris, D = humic horizons, E = fluvial sands with gravel lenses and layers, F = Late Cretaceous marine sediments, chiefly marlstones

During the sandpit operation (ca. 1996-2002), dozens of huge tree trunks (commonly 5-12 m long) were excavated from sand. All of them were determined as *Quercus* Linné, 1758 following Balabán (1955). The trunks were put on three stacks at the pit margin (Fig. 4C). Most of them are still there; few other trunks were discarded elsewhere. Position of the trunks was not recorded; therefore, the wood layer can be only indirectly paralleled with the layer described by Petrbok (1937). The disadvantage of unclear location is, however, compensated by the large amount of material for study and the uniqueness of its exposure.

During fieldwork in 2005, 28 trunks were studied by a thorough surface observation and by cuts produced with a hammer or a hatchet. Three trunks yielded larger amounts of borings attributable, considering the way of preservation, to a pre-depositional suite; all other trunks were strongly to sporadically bored with uniform, narrow passages. The passages sometimes show a fresh xylic dust, and few provided live tracemaker-larvae attributable to *Hadrobregmus* Thomson, 1859 (Insecta, Coleoptera, Anobiidae). Therefore, all of them were made probably after the exhumation of the trunks.

Fig. 4 A Sediments of the lower Labe floodplain terrace at Mělník – Cítov sandpits. **B** Remnants of floodplain forests at the middle reach of the Labe at the village of Jiřice nad Labem. **C** Čelákovice locality – a stack of exhumed oak trunks at the margin of an abandoned sandpit

Libice – a museum sample

A unique sample of wood of *Quercus* sp. with a boring and a tracemaker preserved in situ – insect larva 7 cm long – was conveyed to the National Museum in Prague in the early 1990s. The sandpit at Libice nad Cidlinou (ca. 35 km E of Čelákovice)

was recorded as the sampling site. The sandpit, now abandoned and overgrown, lies 1 km SE of the centre of the village. A detailed section of the sandpit was not published and the finding does not bear any details on its original position. Considering the elevation of the sandpit above the river level, the site is most likely the lower floodplain terrace as defined by Růžičková and Zeman (1994a, b).

Previous study of wood from the Labe sediments

Although tree trunks are a conspicuous feature of the Holocene sediments of the Labe, the history of their research is short. It started by a popular article by Petrbok (1937) who mentioned finds of carbonised trunks in the floodplain of the Labe near Čelákovice and Káraný. Petrbok found a layer with numerous carbonised trunks (almost exclusively oak trunks), oak branches, leaves and acorns in a cut bank at Káraný and at the construction site of a flood gate at Čelákovice. The mentioned depth of the finds below the floodplain level (i.e., 2-2.5 m) corresponds well to the presumed situation at the present locality of Čelákovice (sandpits).

Further literature sources on this topic come nearly exclusively from the volume edited by Růžičková and Zeman (1994c). Růžičková and Zeman (1994b) presented a table of selected, well-documented findings of trunks, giving their size, locations and approximate geologic settings, from several distinct stratigraphic levels. Leuschner and Kyncl (1994) completed a tree-ring analysis of sixteen trunks; four of them were dated against the Göttingen master chronology for oaks from the alluvial gravels in Germany. Radiocarbon datings of selected samples of trunks made by Šilar et al. (1994) yielded ages from 9,386 ± 239 years BP to the present. Coalification processes of the trunks were studied by Sýkorová and Čermák (1994). Finally, Krs et al. (1994) reported on remanent magnetism of the wood. No ichnologic features, e.g., wood borings, have so far been reported.

Modern Labe River

Human impact on the environment of the Labe floodplain in the study area is a very long-lasting one. First, the Neolithic period brought a deforestation of fertile lowlands (Dreslerová 1994) which changed the power and effect of floods; the floods became the most intensive process responsible for sedimentation and erosion in the floodplain until the 19th century, where direct human impact became dominant. During the 'Medieval landscape change' (13th century AD), the deforestation progressed, reaching also a considerable degree in highlands (cf. Sádlo et al. 2005).

During the 19th century and the first half of the 20th century, the river was regulated for navigation of larger ships; therefore, the river channel was straightened and shortened up to one-third of its original length. Floodgates were made at intervals of 10-20 km, and the new channel was lined by embankments. At present, forests cover only ca. 10% of the floodplain area. In addition, the composition of the forests was changed: poplar, alder, and pine were set out and grow at the expense of oaks (Fig. 4B).

Bioerosive traces – description and interpretation

Fungal pits
Fig. 5A-B

Material: A single cluster of more than 60 specimens on the sample of wood of *Quercus* sp.; Čelákovice locality.

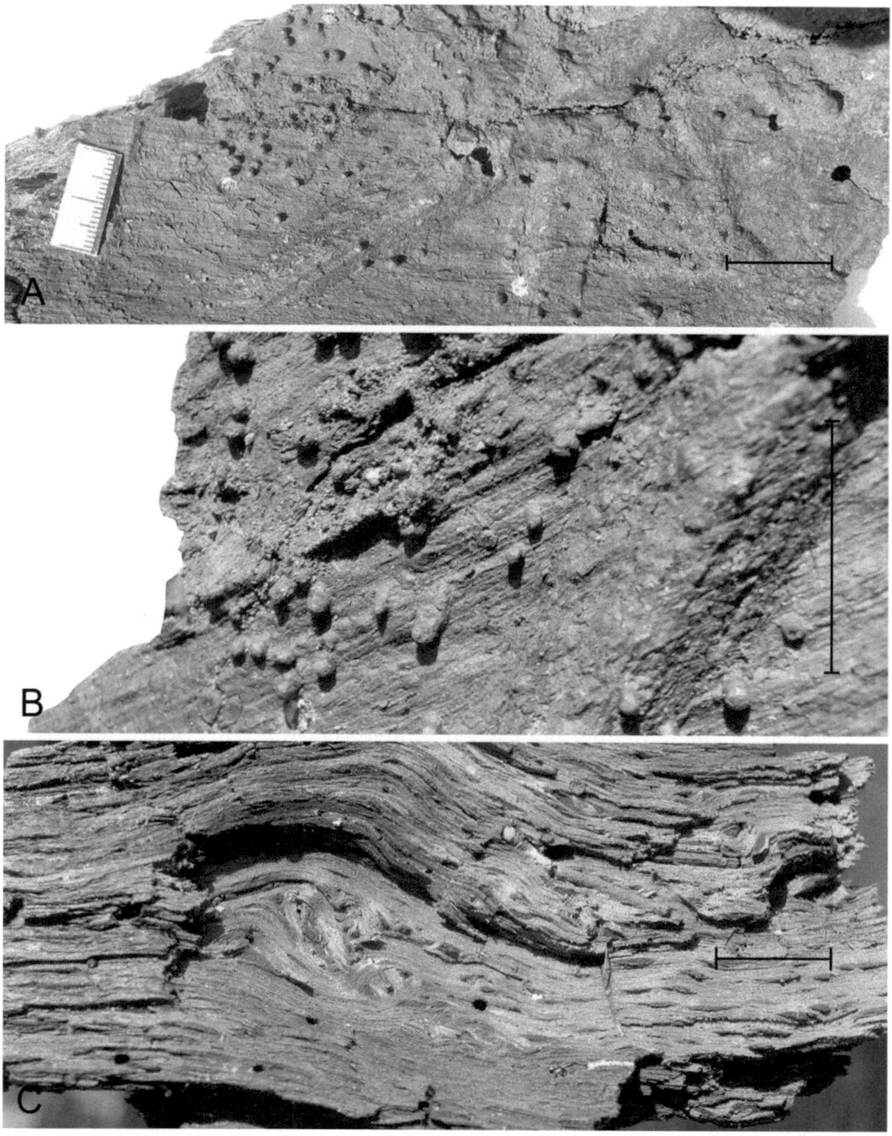

Fig. 5 A Fungal pits and mammal (?human) scratches. **B** Detail of A. **C** Insect borings Form C. Holocene, wood of *Quercus* sp., Čelákovice. Scale bears: 2 cm

Description: Minute, smooth, regular, deep, hemispherical to cylindrical pits, preserved as casts (i.e., as tubercles). The negative preservation originated as the bark was detached, the bare wood was attacked by the tracemaker; subsequently, the exposed wood was healed, i.e., covered with younger wood. After the exhumation of the trunk, the older, attacked wood was crumbly, and the 'wood cast' appeared as the older substrate crumbled away. This results in a structure analogous to convex hyporeliefs typical for rhythmical clastic rocks originated. Pits / tubercles are 0.9 to 2.5 mm wide; their depth is almost constant, around 1.5 mm. The smallest pits / tubercles, therefore, show a higher depth / diameter ratio (ca. 1.6) than the largest ones (ca. 0.7). Therefore, larger pits are close to a hemisphere, while the small ones are cylindrical with a hemispherical termination.

Remarks: Wood borings of the above-described shape cannot be attributed to insects. The insect borings are usually composed of tunnels inside the wood and / or subsurface chambers. Considering the data published by Genise (2004), the structures originated most likely by the activity of wood-destroying fungi. Using the accessible literature data on recent wood-destroying fungi in central Europe (Černý 1989), the tracemaker cannot be determined in more detail, and the assumption on fungi as the tracemaker cannot be fully confirmed. According to the present central European references, cavities / pits made by enzymatic activity of fungi are smaller, less regular and more closely spaced (M. Tomšovský pers. comm.; cf. also Černý 1989). However, fungal pits may be as large as or larger than those described herein, and very regular too. Regarding spacing, even when attacks produce commonly clustered marks, space patterns may also result from taphonomic processes and wood weathering (i.e., a selective preservation of the deeper ones; J. Genise pers. comm.).

Mammal (?human) scratches

Fig. 5A

Material: A single finding on a wood sample (*Quercus* sp.) from Čelákovice, preserved on the same surface as the above described fungal pits.

Description: The find is preserved in the same manner as the fungal pits, i.e., the healing of the unstuck bark produced a convex cast. The original concave structure was star-like, with six arms. It could appear by three to five scratches: in one case, two opposite arms evidently represent one scratch; the remaining ones could appear as solitary. The overall size of the trace is ca. 120 × 80 mm, the width of arms is max. 18 mm. Terminations of three arms were outside the collected sample; two arms fade away gradually; the remaining one is terminated abruptly. The depth of the scratches / height of their casts is 0-6 mm. Surface of the arms is sulcate; orientation and size of the ornament correspond to the presumed course of fibres in the presently crumbled substrate. One of the arms is broadly 'V'-shaped in cross-section; sections of the others are 'U'-shaped. Surface of the arms bears several (six at minimum) specimens of fungal pits as described above; these, however, occur on the surrounding bedding plane with higher density.

Remarks: Morphology of the trace and taphonomic aspects show that the trace originated by scratching into wood of live oak at a place with stripped off bark. It is possible that the scratching itself led to the bark removal. The star-like shape does not resemble scratching traces made by claws (cf. Mikuláš 2001) or teeth (cf. Mikuláš et al. 2006). Therefore, human origin of the scratches is the possible explanation (but other possibilities of the origin of the scratches cannot be ruled out). This conclusion is not inconsistent with the geologic record, as the origin of the whole wood accumulation falls into the period of relatively dense population of the floodplain (cf. Dreslerová 1994).

Insect borings Form A

Fig. 6C

Material: Two finds, one of which bearing the tracemaker preserved in situ (Libice nad Cidlinou locality), the second one is empty and comes from Čelákovice.

Description: A pouch-like hollow, elliptical in cross-section, length of the specimen from the Libice is 75 mm, maximum width 18 mm. The specimen from Čelákovice is 17 mm long and up to 4 mm wide. Axes of the borings are oblique to nearly perpendicular to wood fibres, reaching deeply to the trunks of *Quercus* sp. Both specimens were sliced during the cutting of the trunks, and only one half was found. The preserved parts do not show the orientation of the 'pouch'. Boring walls are nearly smooth, with only indistinct wood xenoglyph; hollows are empty, without remains of frass.

Remarks: The mummy of the tracemaker preserved in situ is an excellent guideline for the determination of the origin of the borings. Thanks to the mummified tracemaker, the structure was collected and housed in the museum. The tracemaker was determined as the larva of *Cerambyx cerdo* Linné, 1758 (Coleoptera, Cerambycidae) (D. Král pers. comm.), see also Polášek and Špaček (2006).

Insect borings Form B

Material: Three intensively bored wood samples from the trunk of *Quercus* sp. from Čelákovice.

Description: A network, an incomplete network to a mere branched system of cylindrical tunnels showing a nearly constant diameter of 5 mm. Most of the tunnel systems (ca. 90%) lie on cylindrical surfaces of growth rings; the remaining 10% of the tunnels are oblique to these surfaces.

Remarks: Similar tunnels are presently made by beetles of the family Cerambycidae (J. Prokop and D. Král pers. comm.; Polášek and Špaček 2006). Considering the lack of morphologic features and the absence of finds of preserved tracemakers, also other beetle families can be taken into account. Less probably, Hymenoptera might form these tunnel systems.

Fig. 6 A Empty or inhabited passages made by / with finds of live *Hadrobregmus* spp. Recent, exhumed Holocene oak trunks, Čelákovice. **B** Insect borings Form C. Holocene, wood of *Quercus* sp., Čelákovice. **C** Insect borings Form A with a larva of *Cerambyx cerdo* Linné, 1758. Holocene, wood of *Quercus* sp., Libice nad Cidlinou. Scale bars: 2 cm

Insect borings Form C
Figs. 5C, 6B

Material: A dozen of collected wood fragments with the below-described boring, collected from two trunks at Čelákovice.

Description: Narrow, smooth, cylindrical, in most parts of their course nearly straight but then abruptly bending tunnels. Their straight parts are parallel to subparallel to wood fibres. Tunnel diameter is constant, also 2 mm; 10-30% of the attached wood is consumed. The passages are filled with firm dark frass. Length of the tunnels may exceed 10 cm.

Remarks: Various insect groups, most probably beetles, can be considered as tracemakers (J. Prokop and D. Král pers. comm.; Polášek and Špaček 2006).

Empty or inhabited passages by / with *Hadrobregmus* spp.
Fig. 6A

Material: Several thousand observations of tunnel openings on the trunk surfaces. A dozen samples of wood collected and cut for the study of the whole systems at the locality of Čelákovice.

Description: Cylindrical tunnels, 0.8 to 1.2 mm in diameter. Close to their openings, they are perpendicular to nearly perpendicular to the wood surface, and reach the maximum depth of several centimetres. Deeply in the substrate, they are sparsely branching. Passages parallel to wood fibres run from the main tunnel (or, more accurately, 'shaft' sensu Bromley 1996). The second-order passages are blind, typically 1 to 3 cm long. Wood can be intensively attacked by this type of boring. Typical distance of the openings on wood surfaces is 1 cm; sparse to solitary openings are also common.

Remarks: Some of the tunnel finds showed a fresh wood powder at openings, others provided live larvae of *Hadrobregmus* sp. (cf. Polášek and Špaček 2006). It is, therefore, evident that at least a certain part of the borings originated after their exhumation, i.e., during the last several years; the spacing obviously corresponds to the duration of the substrate colonisation.

Bioerosive traces – succession

The fact that the wood attack was multiphase has already been mentioned at remarks of individual morphotypes of borings. The individual phases are specified below:

1. ***living trees***. Traces preserved as 'wood casts' on surfaces exposed by sloughing the bark were produced undoubtedly in lived trees. Such traces can often be

observed nowadays, as well as their overgrowing by new wood. Most often, bite marks by roe deer, wood decaying fungi, and especially human-made scratches are present. Similar tracemakers are presumed for the two above-described Holocene forms, i.e., fungal pits and mammal (?human) scratches. The boring activity on living trees was probably not very intensive (as it is not at present), and only one convincing example was found. However, these situations can be discerned only by probing cuts in wood; therefore, many findings may be overlooked.

2. ***dead trees***. In the present floodplain forests, intensely bored wood is found only on dead, commonly fallen trunks, or on necrotised parts of alive trees. It can be presumed that the situation in the Holocene floodplain forests was analogous, and probably even the tracemakers were the same as nowadays (Coleoptera, Cerambycidae and Lucanidae; e.g., *Lucanus cervus* Linné, 1758). Subtle details in the preservation of Insect borings Form A and Form B (e.g., thin surface film) point to their Holocene age, when floodplain forests favourable for the occurrence of large beetles prevailed in the study area. In contrast, the present stock-piles of exhumed wood are related instead to suburban settings.

3. ***exhumed xylic substrate***. *Hadrobregmus* spp. occurs not only in woodlands, but also in urban and suburban settings. It is likely that semifossil wood exhumed outside woodlands is attacked by 'wood-destroying' insects well known as pests of building constructions. Spontaneous exhumation of wood on undercut-slope banks (cf. Petrbok 1937) would probably have a different effect; first, the surroundings would probably be afforested; second, the wood would probably be periodically moistened. The moisture would probably repel most insects; fungal decay could prevail instead. Such a situation, unfortunately, cannot be documented nowadays.

Potential fresh-water borings – candidates of which are may-fly larvae, well known from freshwater firmgrounds – or traces described by Thenius (1979) were not found. In fluvial settings, where the wood transport is connected with flood events and where preserved wood is commonly completely buried during the flood events, any attack of borers to submerged wood is unlikely.

Discussion and conclusions

The described material shows the following limitations: (1) most borings cannot be attributed to their tracemakers, as it can be done for modern substrates, still exploited by organisms. Significance of data is comparable to that of the fossil record. (2) Nevertheless, the material cannot be considered fossil; conventionally, only pre-Holocene findings are regarded as fossil. Therefore, formal ichnotaxonomic assignment is excluded. And (3), despite these limitations, the described structures represent a rare example of relatively rich and diverse collection of wood borings from non-marine sediments in which wood has already undergone some steps of the taphonomic process, which can be described and analysed. Most of the present report refers to a single morphotype of a boring from one or several wood fragments (e.g., Mikuláš and Dvořák 2000). Exceptions to this are papers by Walker (1938) and Genise (1995) that described associations of traces in fossil woods for single forests. Rather than the paucity in the fossil record, a lack of knowledge during the fieldwork is a real reason of that situation.

The described geologic record bearing a large amount of well-preserved wood mass, which was subaerially exposed for a long time before its burial, stimulates thoughts on its ichnofacies classification. The vast majority of trace fossil assemblages, both marine and terrestrial, can be placed in any of the recurring ichnofacies, as presently defined on the basis of substrate, composition and dynamics of the surrounding settings (cf. Buatois et al. 1998). The above cited paper, however, shows that 'wood-substrate' subaerial ichnofacies have not been defined yet. If established in the future, it will be probably a welcome contribution to the theoretical integrity of the ichnofacies concept as well as a tool for solving practical problems in the geologic record. To define such ichnofacies, the following circumstances should be accomplished: (1) analogous situations should be described on pre-Holocene material; (2) respective ichnotaxa should be named formally; (3) the associations should be found recurrently – otherwise they cannot requisite the status of archetypal ichnofacies; and (4) the substrates should be laterally persistent, which also pose a limitation for woodground associations of trace fossils.

If a subaerial 'wood ichnofacies' can be defined in the future, it will be necessary as well to resolve the question on how to treat associations of trace fossils on leaves within the ichnofacies model. There are many records of traces in leaves for many localities and geologic ages (recurrence) and some ichnogenera have been defined in the last years and others are under study (J. Genise pers. comm.). Certain types of forests are a mosaic of leaves and wood substrates, some of them are not (dead trees, conifer forests), and, finally, various plant growths with leaves have very often no wood. As a result, there will be certain overlap of the ichnofacies in terms of their occurrence (or an unusually small mosaic of ichnofacies) for vegetation types.

Subaerial wood borings can be divided into three groups: (1) those made in living trees (in the studied case, climax oak forests); (2) those made in dead trees (dead wood in climax oak forests); and (3) those made in exhumed wood (restricted environments outside oak forests). This succession and recognition of pertinence of individual borings to its groups can be useful for understanding the geologic record. At the Čelákovice locality, most trunks show no bioerosion; these were probably uprooted during flood events. Uprooting may correspond to large floods or periods with frequent floods (a question for dendrochronology and a comparison of data with the existing climatic models, which is the suggestion for further research).

Acknowledgements

Thanks are due to the Grant Agency of the AS CR, Praha, for the grant support (project No. A300130505). Tomáš Matoušek (Praha) kindly called the attention of the author to the Čelákovice locality. Josef Brožek (Institute of Geology, AS CR, Praha) assisted during the fieldwork. Vojtěch Turek (National Museum, Praha) provided valuable help during the study of the museum sample. Jakub Prokop, David Král, and Michal Tomšovský (Praha) contributed with comments to the recent wood bioerosion and tentative tracemakers. Luis Buatois (Saskatchewan) and Jorge Genise (Buenos Aires) are acknowledged for constructive reviews of the paper.

References

Balabán K (1955) Nauka o dřevě. Státní Zemědělské Nakladatelství, Praha, 216 pp

Bertling M, Braddy S, Bromley RG, Demathieu GD, Mikuláš R, Nielsen JK, Nielsen KSS, Rindsberg A, Schlirf M, Uchman A (2006) Names for trace fossils: a uniform approach. Lethaia 39:265-286

Bromley RG (1996) Trace fossils: biology, taphonomy and applications. Chapman and Hall, London, 361 pp

Buatois LA, Mángano MG, Genise JF, Taylor TN (1998) The ichnologic record of the continental invertebrate invasion: evolutionary trends in environmental expansion, ecospace, utilization, and behavioral complexity. Palaios 13:217-240

Buchholz H (1986) Die Höhle eines Spechtvogels aus dem Eozän von Arizona, USA (Aves, Piciformes). Verh Natw Ver Hamburg, New Ser 28:5-25

Černý A (1989) Parazitické dřevokazné houby. Státní Zemědělské Nakladatelství, Praha, 160 pp

Dreslerová D (1994) Archaeology and the Labe River flood plain: recent discoveries. In: Růžičková E, Zeman A (eds) Holocene flood plain of the Labe River. Geol Inst Acad Sci, Praha, pp 84-88

Elbroch M (2003) Mammal Tracks & Sign. A guide to North American Species. Stackpole Books, Mechanicsburg, 792 pp

Elbroch M, Marks E (2002) Bird Tracks & Sign. A guide to North American Species. Stackpole Books, Mechanicsburg, 464 pp

Genise JF (1995) Upper Cretaceous trace fossils in permineralized plant remains from Patagonian Argentina. Ichnos 3:287-299

Genise JF (2004) Fungus trace in wood: a rare bioerosional item. In: Buatois LA, Mángano MG (eds) Ichnia 2004, First International Congress on Ichnology, April 19-23, 2004. Mus Paleont Egidio Feruglio, Trelew, Patagonia, Argentina, Abstr, p 37

Kelly SRA, Bromley RG (1984) Ichnological nomenclature of clavate borings. Palaeontology 27:793-807

Kříž J, Mikuláš R (2006) Bivalve wood borings of the ichnogenus *Teredolites* Leymerie from the Bohemian Cretaceous Basin (Upper Cretaceous, Czech Republic). Ichnos 13: 159-174

Krs M, Krsová M, Pruner P, Čápová J, Parés JM (1994) Magnetism of subfossil and fresh wood: initial reports. In: Růžičková E, Zeman A (eds) Holocene flood plain of the Labe River. Geol Inst Acad Sci, Praha, pp 51-65

Leuschner HH, Kyncl J (1994) Dendrochronologische Untersuchungen an subfossilen Eichen aus Labe-Schotten. In: Růžičková E, Zeman A (eds) Holocene flood plain of the Labe River. Geol Inst Acad Sci, Praha, pp 35-38

Leymerie A (1842) Suite de mémoire sur le terrain Crétacé du département de l'Aube. Mém Soc Géol France 5:1-34

Linné C (Linnaeus C) (1758) Systema naturae per regna tria naturae, secundum classes, ordines, genera, species, cum characteribus, differentiis, synonymis, locis. Tomus I. Editio decima, reformata. Laurentii Salvii, Holmiae, 824 pp

Ložek V (1980) Vývoj přírody středních Čech v nejmladší geologické minulosti. Stud Československé Akad Věd 1:9-43

Mikuláš R (2001) Modern and fossil traces in terrestrial lithic substrates. Ichnos 8:177-184

Mikuláš R, Dvořák Z (2000) Hmyzí chodbičky v xylickém materiálu z terciéru Severočeské hnědouhelné pánve (Insect borings in fossil xylic tissues from the Tertiary of the North Bohemian Brown Coal Basin). Zpr Geol Výzk Roce 1999, pp 64-67

Mikuláš R, Kadlecová E, Fejfar O, Dvořák Z (2006) Three new ichnogenera of biting and gnawing traces on reptilian and mammalian bones: a case study from the Miocene of the Czech Republic. Ichnos 3:113-127

Petrbok J (1937) Zkamenělý prales pod Labem. Národní Politika, September 7, Praha, p 4

Polášek J, Špaček T (2006) Dřevokazné houby, dřevokazný hmyz. Mendelova Lesnická Univ, Brno, 122 pp

Růžičková E, Zeman A (1994a) Holocene fluvial sediments of the Labe River. In: Růžičková E, Zeman A (eds) Holocene flood plain of the Labe River. Geol Inst Acad Sci, Praha, pp 3-25

Růžičková E, Zeman A (1994b) Trunks in Holocene fluvial sediments of the Labe River. In: Růžičková E, Zeman A (eds) Holocene flood plain of the Labe River. Geol Inst Acad Sci, Praha, pp 31-34

Růžičková E, Zeman A (1994c) Holocene flood plain of the Labe River. Geol Inst Acad Sci, Praha, 116 pp

Sádlo J, Pokorný P, Hájek P, Dreslerová D, Cílek V (2005) Krajina a revoluce. Významné přelomy ve vývoji kulturní krajiny. Malá Skála, Praha, 247 pp

Scott AC (1992) Trace fossils of plant-arthropod interactions. In: Maples CG, West RR (eds) Trace Fossils. Short Courses Paleont 5, Palaeont Soc Univ Tennessee, Knoxville, pp 197-223

Šilar J, Jílek P, Melková J (1994) Radiocarbon dating of samples of wood. In: Růžičková E, Zeman A (eds) Holocene flood plain of the Labe River. Geol Inst Acad Sci, Praha, pp 39-42

Sýkorová I, Čermák I (1994) Microscopic and chemical investigation of woods from fluvial sediments of the Labe River. In: Růžičková E, Zeman A (eds) Holocene flood plain of the Labe River. Geol Inst Acad Sci, Praha, pp 43-50

Thenius E (1979) Lebensspuren von Ephemeropteren-Larven aus dem Jung-Tertiär des Wiener Beckens. Ann Nathist Mus Wien 82:177-188

Thomson CG (1859) Skandinaviens Coleoptera, synoptiskt bearbetade. Vol 1, Berlingska Boktryckeriet, Lund, 290 pp

Walker MV (1938) Evidence of Triassic insects in the petrified forest national monument, Arizona. Proc US Natl Mus 85:137-141

Trace fossil assemblages on Miocene rocky shores of southern Iberia

Ana Santos[1], Eduardo Mayoral[1], Carlos M. da Silva[2], Mário Cachão[2], Rosa Domènech[3], Jordi Martinell[3]

[1] Dept. de Geodinâmica y Paleontologia, Universidad de Huelva, 21071 Huelva, Spain, (asantos@dgyp.uhu.es)
[2] Dept. e Centro de Geologia, Universidade de Lisboa, 1749-016 Lisboa, Portugal
[3] Dept. d'Estratigrafia, Paleontologia i Geociències Marines, Universitat de Barcelona, 08028 Barcelona, Spain

Abstract. The use of rocky palaeoshore bioerosion analysis in the study of palaeontological and geological questions is beginning to bear fruit. Five southern Iberian Neogene rocky shores have been analysed and their bioerosion structures have been identified. The observed ichnodiversity is rather low; eleven ichnospecies were identified. These include bioerosion structures produced by polychaete annelids (*Caulostrepsis*, *Maeandropolydora*), clionaid sponges (*Entobia*), echinoids (*Circolites*), and endolithic bivalves (*Gastrochaenolites*). The different ichnoassemblages present in Miocene rocky shores in both Portuguese and Spanish sectors correspond to the *Entobia* ichnofacies. Comparison with the northeastern counterparts of these shores has also been carried out. The study of southern Iberian Miocene rocky shores made it possible to correlate them with the regional tectonic evolution and the main Neogene transgressive events affecting the region.

Keywords. Bioerosion, endolithic communities, rocky shores, transgression, Neogene, Iberian Peninsula

Introduction

As recently as two decades ago, ancient rocky shores were still considered geological rarities (Boucot 1981). Since then, growing scientific interest in these palaeoenvironments has generated a wealth of publications (e.g., Palmer 1982; Brett and Brookfield 1983; Wilson 1985, 1987; Johnson 1988a, b, 1992; Pirazzoli et al. 1994; Brett 1998; Bertling 1999; Johnson and Baarli 1999; Ekdale and Bromley 2001; Benner et al. 2004; Johnson 2006; Plag 2006). This proliferation clearly demonstrates that fossil rocky shores are more common than previously believed.

Because of their particular location, bordering marine and coastal environments, rocky shores react to any modification of physical or chemical parameters. So, they may be successfully used to estimate ancient sedimentation and erosion rates,

M. Wisshak, L. Tapanila (eds.), *Current Developments in Bioerosion*. Erlangen Earth Conference Series, doi: 10.1007/978-3-540-77598-0_22,

oceanic geochemistry, and eustatic fluctuations in sea level. Moreover, rocky coasts are special environments, with exceptional conditions for colonisation by boring and encrusting organisms due to their reduced or null sedimentation rates.

The study of bioerosion structures produced in palaeoshores is invaluable for the identification and quantification of the relative oscillations of the sea level, past shoreline positions, the erosion / sedimentation rates associated with them, movements and reliefs of tectonic origin, depositional and / or stratigraphic hiatuses and the intensity of the physical disturbance of coastal environments. In addition, from a strictly palaeontological point of view, the analysis of rocky palaeoshore biota makes it possible to investigate ecological parameters such as spatial distribution, competitive overgrowth interactions and community succession (Brett 1988). Lastly, it is paramount for the study of the evolution of marine communities in this type of environment (Johnson 1992) since their remains are often preserved in situ, attached to the hard surface itself (Brett 1988; Johnson 1992; Taylor and Wilson 2003). Examples of ancient marine hard substrate communities have been described from the Lower Cambrian (Kobluk et al. 1978; Kobluk and James 1979; Kobluk 1981a, b) to the Neogene (e.g., Radwanski 1970; Bromley and D'Alessandro 1987; Watkins 1990; Bromley and Asgaard 1993; Martinell and Domènech 1995; Gibert et al. 1998; Domènech et al. 2001).

Bioerosion studies on different kinds of hard substrates in the Spanish Cenozoic are numerous, especially those dealing with the northeast sector of the Mediterranean coast of Iberia (Martinell 1982a, b; Martinell and Domènech 1995; Gibert et al. 1998; Domènech et al. 2001, concerning rocky shores; Perry 1996, concerning bioerosion on reefs; Siggerud et al. 2000, regarding bioerosion in conglomerates). Up to now only two studies have been published on rocky shores in the southern part of Spain (Doyle et al. 1998; Mayoral et al. 1998). In these papers, the authors reported preliminary results on evidence of bioerosion structures in conglomerates, providing new information on *Gastrochaenolites-Entobia* association on Neogene palaeoshores of the Iberian Peninsula.

In Portugal, the study of bioeroded rockgrounds is still in its inception. The oldest reference to Portuguese Neogene trace fossils in lithified substrates goes back to Choffat (1903-1904) who recognised the presence of clavate borings made by Middle Pliocene bivalves on Lower Jurassic dolomitic pebbles in Senhora da Vitória, near Nazaré (Central-West Portugal). In the nineteen eighties, Pais (1982) and Antunes (1984) briefly commented on the presence of bivalve borings on contact surfaces between Jurassic and Neogene formations in southern Portugal (Algarve). However, Silva et al. (1995, 1999) presented the first evidence of Neogene bioerosion structures clearly indicating the occurrence of ancient rocky shores in several places in Portugal. Domènech et al. (1999) studied the bioeroded megasurface at Oura (Albufeira), one of the longest bioeroded surfaces recognised in the geological record, which correlates with a transgressive surface that shows the existence of an important intra-Miocene stratigraphic discontinuity, with vast implications for the understanding of the palaeogeographic and tectonic (Betic) evolution of the Algarve Basin.

The aim of the present paper is to characterise the occurrence of Miocene bioeroded rocky shores in southern Iberia, as well as the ichnoassemblages related to them, and thus, to obtain quantitative data that will enable us to compare them with other marine hard substrate communities.

Geographical setting and methodology

The study area covers the southern Portuguese coast (southern coast of the Algarve region, Western Betic foreland), and the southern coast of Spain, namely the north of the Sierra Tejeda (Granada, central sector of the Betic Range).

The present study was based on the following localities where there is evidence of bioerosion structures indicative of ancient rocky shores: Sagres (Vila do Bispo), Arrifão (Albufeira), Oura (Albufeira) and Cacela (Tavira), in Portugal, and La Resinera (Sierra Tejeda, Granada), in Spain (Fig. 1).

The surfaces studied and the bioerosive structures affecting them are Middle to Late Miocene in age. However, the affected substrates span different ages (Palaeozoic to Cenozoic) and a wide range of lithologies (limestones, sandstones and marbles). Information on these rockgrounds is synthesised in Figure 1 and Table 1.

To perform this study, several well-preserved 20 x 20 cm square areas of substrate were randomly selected, and their bioerosion structures identified and quantified. After identification, the composition of the ichnocenoses was identified. Remains of epilithozoans colonising some of these hard substrates were also recorded. The bioerosive structures were counted, measured and their spatial relationships recorded whenever possible (Arrifão and Oura in Portugal and La Resinera in Spain). Barranco das Mós and Cacela exposures (both in Portugal) turned out to be too small to apply the same methodology, thus observations were made on the entire surfaces.

Measurements taken are for the diameter of *Gastrochaenolites* Leymerie, 1842. This would be simply a tool to characterise and compare the different outcrops, as physical Miocene marine erosion affected all the traces.

Description of the locations of the Miocene rocky shores

Barranco das Mós (Sagres, SW Algarve, Portugal)

The Barranco das Mós outcrop is located three km East of Sagres, in the southwestern tip of the Algarve, and about seven km inland from the present shoreline (Fig. 1).

The bioeroded substrate is visible both in surface and cross-sectional exposures (Fig. 2A-B). The surface corresponds to an unconformity in the contact between Jurassic limestones (Oxfordian-Kimmeridgian) and Miocene bioclastic limestones (Langhian-Serravalian), and it is contemporaneous to the beginning of the sedimentation of the Lagos-Portimão Formation (mainly Langhian-Serravalian), which represents the lower fossiliferous carbonate unit of the Neogene sequence of the Algarve region (Cachão and Silva 2006).

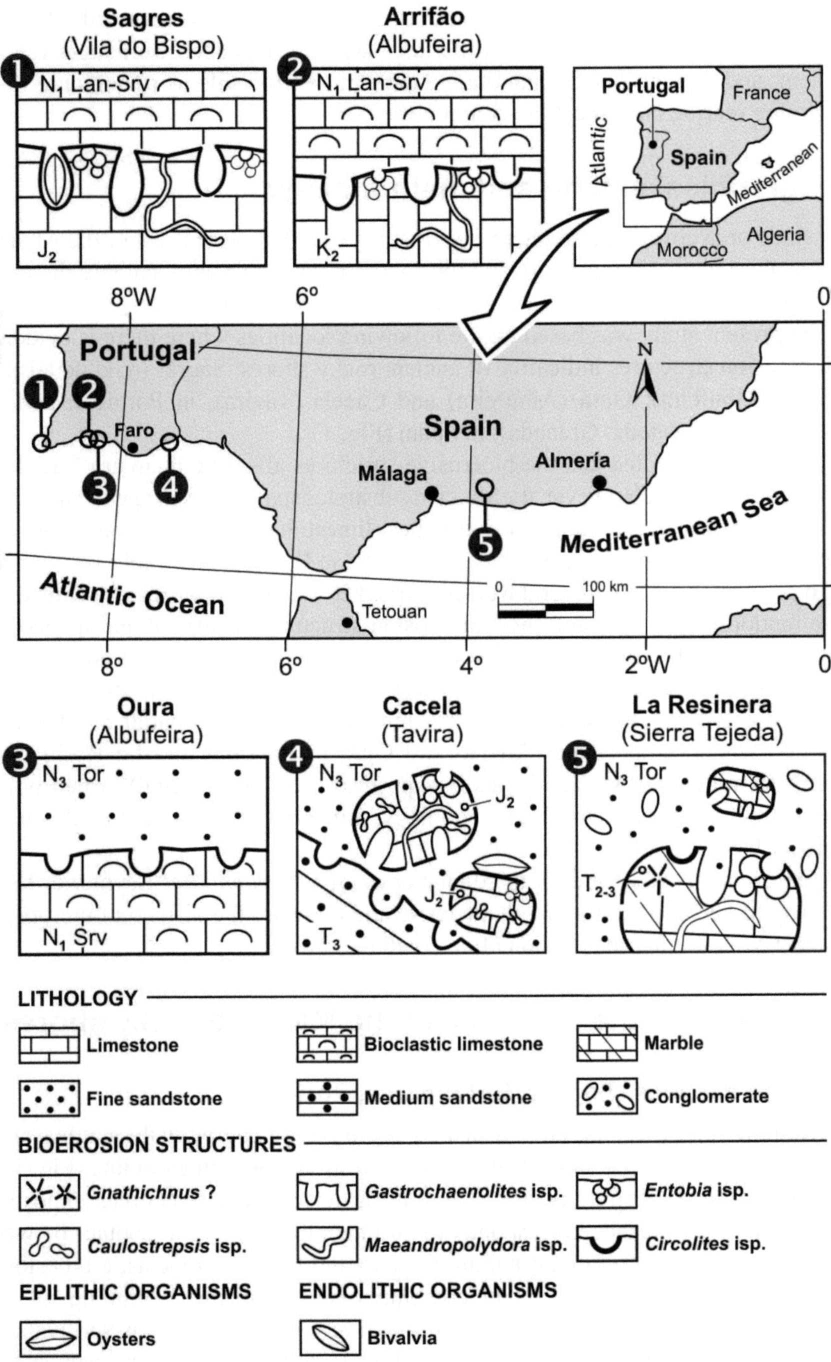

Fig. 1 Geographical setting of the Portuguese and Spanish outcrops featuring bioerosion structures associated with the rocky palaeoshores studied (see Table 1 for abbreviations)

Table 1 Information on Portuguese and Spanish Miocene sites displaying bioerosion structures in rocky palaeoshores. Abbreviations and symbols used in the table: Cs – Cross section exposure; S = Surface exposure; B = Boulders; * = Age symbols and stage abbreviations after Harland et al. 1989: N1 = (Lower Neogene); N3 = (Upper Neogene); Lan-Srv = (Langhian-Serravalian); Tor = (Tortonian); J2 = (Middle Jurassic); K2 = (Upper Cretaceous), T2-3 = (Middle and Upper Triassic); T3 = (Upper Triassic)

Outcrops / Substrate	Portugal				Spain
	Sagres N_1 Lan-Srv	Arrifão N_1 Lan-Srv	Oura N_3 Tor	Cacela N_3 Tor	La Resinera N_3 Tor
Age*	J2	K2	N1 Srv	T3 + J2	T2-3
Lithology	limestone	limestone	limestone	(T3) sandstone (J2) limestone	marble
Exposure type	Cs; S	S	S	(T3) S (J2) B	B
Surface preservation	regular	regular	regular	(T3) very poor (J2) regular	good
Ichnotaxa					
Gastrochaenolites isp.	•	•	•	• (T3 + J2)	•
G. torpedo Kelly and Bromley, 1984			•		
G. turbinatus Kelly and Bromley, 1984			•		•
G. lapidicus Kelly and Bromley, 1984					•
G. ornatus Kelly and Bromley, 1984					•
Entobia isp.	•	•		• (J2)	•
E. geometrica Bromley and D'Alessandro, 1984					•
E. ovula Bromley and D'Alessandro, 1984					•
Caulostrepsis isp.				• (J2)	
Maeandropolydora sulcans Voigt, 1965	•	•		• (J2)	•
Circolites isp.					•
Gnathichnus ?					•
Epilithobionts					
Oysters				•	

Neogene bioerosion structures at Barranco das Mós have been previously reported by several authors (Pais 1982; Antunes 1984; Silva et al. 1999), who described this contact surface as a hardground with abundant bivalve borings. The recognised trace fossil assemblage is basically composed of the bivalve boring *Gastrochaenolites* isp. Average size of the observed sections is 1-1.5 cm. The areas between the *Gastrochaenolites* are occupied by occasional xenomorphic *Entobia* Bronn, 1837 and *Maeandropolydora sulcans* Voigt, 1965. *Gastrochaenolites* appear perpendicular to the surface and are filled with the overlying carbonate sediments (Fig. 2C). Some of them contain evidence of their trace maker, presumably *Petricola* Lamarck, 1801 bivalves (Fig. 2D). No remains of epilithobionts are observed.

Arrifão (Albufeira, Central Algarve, Portugal)

This locality is 5 km southwest of Albufeira, near Praia do Arrifão (Fig. 1). This occurrence corresponds to a sub-horizontal karstified surface with a total area of 600 m^2, developed unconformably on the contact between Cretaceous limestones, below, and Miocene bioclastic limestones (Langhian-Serravalian), above (Fig. 2E). As with the bioeroded surface of Sagres, the Arrifão rocky shore developed prior to the deposition of the Lagos-Portimão Formation, which directly onlaps the folded and tilted Cretaceous units (Cachão 1995).

The bioeroded surface is located on the present-day shoreline, affected by tides, and as a result the exposure is highly weathered and not extensively exposed. However, the distal portions of bivalve borings (*Gastrochaenolites* isp.) are easily identified. These borings are oriented perpendicularly to the bored surface, and show diameters ranging between 1 and 1.5 cm. Counting of specimens yielded an amount of 955 per square metre. Usually, the borings are filled with a cemented bioclastic fine sandstone (Fig. 2F). The annelid trace *Maeandropolydora sulcans* and the clionaid sponge borings *Entobia* isp. are also present, although, apparently, with a lesser development. So far, no remains of epilithozoans have been observed in Arrifão.

Oura (Albufeira, Central Algarve, Portugal)

The Oura site is located on the sea cliff to the west of the sandy beach of the locality with the same name and about 3 km east of Albufeira, on central Algarve (Fig. 1).

The bioeroded surface of Oura corresponds to a sub-horizontal transgressive surface cutting through a previous emerged and karstified carbonate relief. This

Fig. 2 A Plan view of abundant *Gastrochaenolites* isp. in cross-section at Sagres (Portugal). **B** Angular unconformity between Jurassic limestones (J2) and the carbonate 'Lagos-Portimão Formation' (Langh-Srv) at Sagres (Portugal). Bioerosion structures occur below the planar surface that bounds both rock units. **C** Detail of the surface in 2B with *Gastrochaenolites* isp. fill of the overlying sediments. **D** Internal mould of a bivalve (*Petricola* ?) preserved inside a *Gastrochaenolites* on the Sagres hardground, showing surface contact. **E** Angular unconformity between Cretaceous limestones (K2) and the fossiliferous 'Lagos-Portimão Formation' (Langh-Srv) at Arrifão (Portugal). **F** Plan view of abundant *Gastrochaenolites* isp. in cross-section filled with cemented sandstone on the Arrifão hardground (Portugal). **G** Surface of contact (arrows) between the carbonate (Langh-Srv; below) and the siliciclastic (Tortonian; above) sequences at Oura (Portugal)

A
Langh-Srv
J2
B
C
D
Langh-Srv
K2
E
F
G

surface was identified as an intra-Miocene (Early Tortonian) marine abrasion platform by Domènech et al. (1999). In fact, the bioeroded surface of Oura corresponds to an unconformity that separates the two main lithostratigraphic units in the central Algarve region: the lower fossiliferous cool carbonate units (Brachert et al. 2003), represented by the Lagos-Portimão Formation (Langhian-Serravalian in age) and the upper siliciclastic unit, represented by the Cacela Formation (Late Tortonian) (Cachão and Silva 2000). This surface can be followed continuously over 1.5 km, along the present-day Oura sea cliffs and, at present, is the largest fossil bioeroded surface in the literature (Domènech et al. 1999; Fig. 2G).

The marine abrasive platform developed itself on a very fossiliferous carbonate unit containing mainly rhodophyceans, oysters and echinoids. At the top of the unit, bioturbation traces (*Thalassinoides* Ehrenberg, 1944) are abundant. Significantly, *Thalassinoides* chimneys are truncated and refilled by a slightly different sediment. These characteristics show that a significant erosive episode affected them prior to the bioerosion episode (Domènech et al. 1999). Two ichnospecies of *Gastrochaenolites* have been differentiated there on the basis of the shape of the distal part of the boring: *G. torpedo* Kelly and Bromley, 1984 (with an acutely parabolic base) and *G. turbinatus* Kelly and Bromley, 1984 (with a short, rounded base).

Field counts, (including both ichnospecies) point to a potential maximum concentration of 300 bivalve borings per square metre in some areas, although presenting a very irregularly clustered distribution. The bivalve borings are perpendicular to the surface and truncated, with a maximum diameter of 2.5 cm and a maximum depth of 5 cm. The upper part of the borings has been destroyed by the marine erosion itself and by subsequent bioerosion, so it can be assumed that several cm of rock were eroded after their production (Fig. 3A). More delicate bioerosion structures such as *Entobia*, *Caulostrepsis* Clarke, 1908 or *Maeandropolydora* are absent, as are epilithozoans.

Cacela (Tavira, SE Algarve, Portugal)

The outcrop is located 0.5 km northeast of Cacela Velha, in the southwestern sector of Algarve, about 30 km from the Spanish border (Fig. 1). The bioeroded substrate corresponds to a scantily exposed and poorly preserved bored surface developed unconformably on Triassic ferruginous sandstones overlapped by the Miocene siliciclastic sediments of the Cacela Formation (Late Tortonian). On this surface, large Jurassic limestone boulders with signs of bioerosion are also observed.

Fig. 3 A Plan view of *Gastrochaenolites* isp. in cross-section; Oura (Portugal). **B** Bioeroded boulder of Jurassic limestone from Cacela (Portugal) showing *Gastrochaenolites* isp., *Entobia* isp. and *Maeandropolydora sulcans* Voigt, 1965. **C** Cross-section (detail) of *Gastrochaenolites lapidicus* Kelly and Bromley, 1984 on a marble boulder from La Resinera (Spain). **D** Intensively bioeroded boulder of Triassic marble with *Gastrochaenolites* isp. at La Resinera (Spain); scale bar: 10 cm. **E** Dwelling structures of regular echinoids (*Circolites* isp.) on a boulder from La Resinera (Spain). **F** *Entobia ovula* Bromley and D'Alessandro, 1984, and *Maeandropolydora sulcans* Voigt, 1965, on the upper surface of bored pebble from La Resinera (Spain); scale bar: 10 cm. **G** *Gastrochaenolites* isp. and *Entobia* isp. on a boulder from La Resinera (Spain); scale bar: 10 cm. **H** Large *Maeandropolydora sulcans* Voigt, 1965 on a marble boulder from La Resinera (Spain); scale bar: 10 cm

A
B
2 cm
C
D
E
F
G
H

Probably due to poor preservation – or because the nature of the substrate – only *Gastrochaenolites* borings were observed affecting the Triassic basal materials, with a diameter between 1 and 2 cm. However, the carbonatic Jurassic boulders show a more diversified ichnoassemblage with *Gastrochaenolites* and other ichnotaxa such as *Entobia, Caulostrepsis* and *Maeandropolydora.* Oysters on the boulders are the only epilithozoans present (Fig. 3B).

La Resinera (Granada, SW Spain)

The La Resinera site is located 50 km southwest of Granada (Fig. 1). The outcrop studied corresponds to the contact between the first marine sediments of the Neogene Granada Basin and the underlying Middle to Late Triassic marbles of the Alpujarride Complex. The Sierra Tejeda, a small mountain range with a NW-SE orientation, located in the central sector of the Betic Range (southern Spain), is made up of rocks of the Alpujarride Complex. The Neogene deposits are conglomerates, sands and bioclastic calcarenites, from the Middle to Upper Miocene, with a maximum thickness of about 30 m. Among the bioclastic calcarenites (Tortonian), decimetre- to metre-sized Triassic marble cobbles and boulders with signs of bioerosive structures are observed. These boulders are extensively bored, and in some of them at least 80% of their exposed surface is bioeroded. The rock-boring assemblage is dominated by *Gastrochaenolites* isp., *G. lapidicus* Kelly and Bromley, 1984, *G. turbinatus* (Fig. 3C) and *G. ornatus* Kelly and Bromley, 1984, accompanied by xenomorphic *Entobia* isp., *E. geometrica* Bromley and D'Alessandro, 1984, *E. ovula* Bromley and D'Alessandro, 1984, *Maeandropolydora sulcans*, *Circolites* isp. Mikulás, 1992, and possible but very badly preserved *Gnathichnus* Bromley, 1975 (the scratches are less than 5 mm long). All *Gastrochaenolites* bivalve borings are perpendicular to the surface and truncated, and none of them contain evidence of their trace-producing organism. In almost all cases, the upper part of the *Gastrochaenolites* borings has been destroyed (Fig. 3D). Counting of specimens in the boulders yielded an amount of 195 per square metre. Diameters of the borings are between 0.7 and 4.3 cm, with an average of 1.9 cm. The total number of boulders considered was 15, with a total number of 595 *Gastrochaenolites* in them. Moreover, 429 other *Gastrochaenolites* were counted outside the square areas, with an average diameter of 1.29 cm.

Also present in the assemblage are concave, subcircular pits 3.5 to 5.2 cm in diameter and 1 to 2.5 cm in depth, which were interpreted as having been dwelling structures of regular echinoids (*Circolites* isp.; Fig. 3E). Borings are common on the upper or lateral surfaces of the boulders. Fewer boulders are bored on all sides. Epilithozoans are not present in the boulders. The areas around the bivalve and echinoid borings are occupied by *Entobia*, while other areas are without trace fossils (Fig. 3F-G). Large and well developed *Maeandropolydora sulcans* (20 to 25 cm long and 0.5 to 0.8 cm wide) are also present but rare (Fig. 3H).

Results

In spite of the poor preservation of some of the bioeroded rocky surfaces (such as Arrifão and Cacela), which made specific identification of some of the bioerosion structures more difficult, the systematic study of these Miocene rocky shores supplies a considerable amount of new data.

The morphological analysis of the bioerosion structures preserved, which represent in situ rocky shore palaeocommunities, identified twelve ichnospecies belonging to six ichnogenera (Table 1). These include structures produced by polychaete annelids (*Caulostrepsis* isp., *Maeandropolydora sulcans*), clionaid sponges (*Entobia* isp., *E. geometrica*, *E. ovula*), echinoids (*Circolites* isp., *Gnathichnus*?), and endolithic bivalves (*Gastrochaenolites torpedo*, *G. turbinatus*, *G. lapidicus*, *G. ornatus*, and *Gastrochaenolites* isp). No detailed taxonomic descriptions of trace forms are presented here since the authors' descriptions are regarded as adequate for their recognition. In general, these bioerosive structures correspond to the boring activity of endolithic organisms and, possibly, to the scraping activity of epilithic mobile organisms. From an ethological point of view, only dwelling structures (Domichnia) are present: *Entobia*, *Caulostrepsis*, *Maeandropolydora*, *Gastrochaenolites* and *Circolites*. Possible *Gnathichnus* from La Resinera would be the only gnawing structures (Pascichnia).

Ichnodiversity, size of the traces (mainly the recurrent *Gastrochaenolites*) and boring density vary among the different localities. The nature of substrates is also quite varied, both in lithological composition and age: Middle to Late Triassic marbles in La Resinera, Triassic sandstones and Jurassic limestone boulders in Cacela, Jurassic limestones in Sagres, Cretaceous limestones in Arrifão and Miocene limestones in Oura. In terms of ichnodiversity, cliff surfaces are less diverse than boulders. *Gastrochaenolites* is the more prominent and abundant trace on the rocky shore surfaces, whereas the boulders are characterised by a higher diversity, with a minimum of four ichnotaxa identified at Cacela and a maximum of 10 at La Resinera. At all sites, *Gastrochaenolites* appears perpendicularly to the rocky surface. Moreover, *Gastrochaenolites* is rarely preserved completely, and if so, only in boulders and cobbles. *Gastrochaenolites* specimens have usually been eroded to the base, producing the circular cross-sections seen at all the localities. The same occurs with *Entobia*, which has been eroded in most areas to expose chambers and canals.

Except in the cobbles and boulders of La Resinera and Cacela, *Gastrochaenolites* is the only or almost the only trace on the eroded substrates. Diameter sizes of this ichnotaxon in the surfaces vary between 1 and 2.5 cm. In the boulders, they range from small ones at Cacela (1-2 cm) to the bigger ones at La Resinera (0.7-4.3 cm). All of these are sections based on erosive processes, and these numbers are just for comparison.

Regarding the density of bioerosion, counts using square areas method yields different values for the outcrops: the density of *Gastrochaenolites* varies from 195 specimens per square metre on the La Resinera boulders to 955 on the Arrifão rocky surface.

It was possible to distinguish two ichnoassemblages which characterise these marine rocky exposures: (1) an ichnoassemblage dominated by *Gastrochaenolites-Entobia-Circolites* (G-E-Ct), present only in La Resinera, composed of *Gastrochaenolites* isp., *G. turbinatus*, *G. lapidicus*, *G. ornatus*, *Maeandropolydora sulcans*, *Entobia geometrica*, *E. ovula*, *Entobia* isp., *Circolites* isp. and doubtful *Gnathichnus;* and (2) an ichnoassemblage dominated by *Gastrochaenolites-Entobia* (G-E), present in the remaining outcrops and made up of *Gastrochaenolites* isp, *G. torpedo*, *G. turbinatus*, *Entobia* isp. and *Maeandropolydora sulcans*. However, the dominant ichnotaxon in both assemblages is *Gastrochaenolites*.

The G-E-Ct trace fossil assemblage in the boulders of La Resinera exhibits the highest level of ichnodiversity and the best preservation of all sites studied. This way, it allows to establish a hypothesis about the substrate colonisation sequence, whilst the G-E assemblage frequently presents only one ichnotaxon (*Gastrochaenolites*) tending to occupy the whole available surface. Sometimes, this bioerosion structure is associated with *Entobia* and with the *Caulostrepsis-Maeandropolydora* suite.

Epilithozoans are not well represented in the rocky shore exposures studied. Among the investigated sites, only Cacela presents remains of encrusters (oysters), fixed to the pebbles and / or boulders, and never directly to the cliffs.

Discussion

With the data obtained it is possible to formulate several considerations. The geologic significance of the fossil rocky shore occurrences described in the areas studied is related to the existence of at least two transgressive events that took place during the Miocene (Langhian-Serravalian and Tortonian) and affected the Algarve and Granada basins. Although the outcrops analysed correspond to distinct geographical locations, substrate ages, lithologies, kinds of exposures and bioeroded surface preservation, they present common characteristics, namely those concerning palaeoenvironmental interpretations.

The palaeoenvironmental approach

The environments represented (ancient rocky shores) are the same for all the sites studied in both sectors, as the abundance of *Gastrochaenolites* and *Entobia* on the investigated ichnoassemblages is interpreted as corresponding to very shallow marine environments with a low or null rate of sedimentation (Bromley and Asgaard 1993; Gibert et al. 1998). The locally dense concentrations of *Gastrochaenolites* (195 to 955 specimens per square metre) and its vertical position in the rockground reconfirm this environmental assignation. *Gastrochaenolites* displays a preference for very shallow marine environments, usually appearing in waters only a few metres deep (Bromley 1994). Deeper occurrences are possible, but with significantly lower density. The La Resinera site is the only bored surface studied which can be related to a sea-level highstand. Today, in the Mediterranean Sea, regular echinoids responsible for *Circolites* mainly inhabit depths of 0 to 2 m (Martinell 1981). In accordance with that, the presence of the trace fossil *Circolites* in La Resinera records the highest shoreline position.

Despite the varied lithology and age of the bored surfaces, it does not appear that their nature had a significant effect on Miocene colonisation, as the same main ichnotaxa appear in all the outcrops.

The scarcity of epilithobiont remains in these Miocene rocky exposures may be due to a set of environmental variables which could affect colonisation and preservation. Limited time of exposure at the sediment / water interface could be responsible for the pattern observed (Brett and Brookfield 1984). In the outcrops studied, however, the lack of encrusters attached to the surfaces is clearly related to the strong erosion associated with planar hard surfaces, as is suggested by the shallowness of *Gastrochaenolites* truncated borings. Boulders and pebbles represent the highest-energy settings, at least during their formation (Brett and Brookfield 1984), but they are the only surfaces where scarce, low-diversity encrusters are found. Colonisation could take place after their detachment from the cliff and the sedimentation has protected the traces from subsequent erosion. For that reason they were not affected by the physical erosive processes identified on the cliff surfaces.

The boulders from La Resinera appear to have been more or less stable during colonisation, since only a few of them show borings on the entire surface. Seawater energy levels probably were particularly important in the development of this association, but they were not sufficient to overturn the huge boulders in some cases. According to Wilson (1987), the frequency of overturned blocks is inversely correlated with their size. This is confirmed by the large boulders present at La Resinera and their bioerosive evidence.

The borings identified represent typical rocky shore communities of bivalves, sponges, worms and echinoids. Only for the Resinera outcrop could a hypothesis about the colonisation sequence be established. This sequence starts with the pioneer echinoids (producing *Circolites* + *Gnathichnus*?), clionaid sponges (producing *Entobia*) and the lithophagous bivalves (producing *Gastrochaenolites*). At this stage, an erosion event of the blocks' surfaces was initiated and probably followed by a new *Gastrochaenolites-Entobia* production. *Maeandropolydora* borings were registered in the last phases, and a subsequent erosion phase could have taken place a short time before their final burial.

The two ichnoassemblages present in both sectors correspond to the *Entobia* ichnofacies (Bromley and Asgaard 1993); they have close parallels in other widespread assemblages in carbonate rocky shores since the Jurassic (Gibert et al. 1998).

Tectonic and stratigraphic implications

The major difference between Portuguese and Spanish sites concerns the Betic Range tectonic deformation which affected both sectors in different ways. The Spanish sector of Sierra Tejeda has been affected more intensely than the Algarve sector in Portugal. During the Miocene, in the Sierra Tejeda sector, high-angle faults uplifted the Sierra and an extensive marble formation was exposed, generating slopes and cliffs of tectonic origin. Their erosion produced huge blocks, which fell and were deposited at the base, being submerged and more or less reworked in shallow, high-energy conditions.

On the Algarve platform, a foreland basin of the Betic orogen, all the bioeroded Miocene unconformities are of a much lower angle when compared to the tectonically highly disturbed Spanish unconformities in the inner sector of the Betic orogen.

Since it is closer to the Betic orogen, the Miocene bioeroded unconformity surface at La Resinera and Cacela represents a much larger temporal hiatus (between the Triassic and the Tortonian) than the area to the West. At Central Algarve (Oura, Arrifão), the Late Tortonian unconformity overlays Late Serravalian limestones with only a 3 Ma hiatus (Cachão and Silva 2000). Further West (Sagres), the Late Tortonian unconformity is not preserved, indicating a stronger uplift of the Miocene units due to subsequent tectonic activity. Instead, the other Miocene (Langhian-Serravalian) unconformity surface overlays Jurassic and Cretaceous but not Triassic units. This indicates an uplift in this region prior to the deposition of the Lagos-Portimão Formation that is not as strong as the overall uplift at Cacela and La Resinera.

The bioerosion data made it possible to identify and correlate two major Miocene unconformities along the Portuguese and Spanish southern borders westward from the Betic chain. The occurrence of two unconformities on the Portuguese (westernmost) side and a larger hiatus for the upper unconformity on the Spanish and eastern Algarve sector both reflect a direct response to the Betic tectonic activity.

From a stratigraphic point of view, all the above rockgrounds show hard substrate marine flooding (transgressive) surfaces. The density of borings and the lateral extent of bored rocky surfaces indicate that during a certain time window marine platforms were extensively abraded. Coastal communities of infaunal bioeroders and epifaunal encrusters intensively colonised sediment-free rock surfaces in both Portuguese and Spanish study sectors.

Comparison with Miocene rocky shores of northern Iberia

Domènech et al. (2001) studied six Middle Miocene rockgrounds of the Camp de Tarragona area, northern Iberia. They are interpreted as erosional transgressive surfaces where an important bioerosion activity took place. In contrast to the more heterogeneous rocky shores of southern Iberia, almost all of them developed on the same kind of substrate, the dolomitic limestones of the Jurassic, and constitute horizontal to sub-horizontal discordant surfaces.

Ichnotaxa identified in northern Iberia are listed in Table 2. Similarities with southern Iberia are evident, although some peculiarities may be noted. The *Gastrochaenolites-Entobia* association characterises both areas, but *Entobia* is more conspicuous in the majority of northern outcrops, both on the surfaces and in the cobbles / boulders. In contrast, it has not been possible to identify these traces in terms of ichnospecies due to preservation, whereas at least two ichnospecies are known in the South. Regarding *Gastrochaenolites*, only two ichnospecies have been reported in the northern outcrops as compared to four in the south. *Phrixichnus* Bromley and Asgaard, 1993, a boring also produced by lithophagous bivalves, is only present in the Camp de Tarragona rockgrounds, as is *Trypanites rectus* Mägdefrau, 1932. On the other hand, *Caulostrepsis* isp. and *Circolites* isp. are only found in the South.

Table 2 Information on northeast Spanish Neogene sites displaying bioerosion structures on rocky palaeoshores (according to Domènech et al. 2001). Abbreviations and symbols used: S = Surface exposure; B/C = Boulders/Cobbles; * = Age symbols and stage abbreviations after Harland et al. 1989: N1 = (Lower Neogene); Srv = (Serravalian); J = (Jurassic)

Outcrops / Substrate	NE Iberia					
	Fortí de la Reina	L' Arrabassade	Punta Savinosa	Cala Romana	Sant Pere de Sescelades	Vespella
Age*	J	J	J	J	J	N1 Srv
Lithology	dolomite	dolomite	dolomite	dolomite	dolomite	limestone calcarenites
Exposure type	S	S, B/C	S, B/C	S, B/C	S, B/C	S
Surface preservation	good	good	good	regular	good	regular
Ichnotaxa						
Gastrochaenolites isp.			•			
G. torpedo Kelly and Bromley, 1984	•	•		•		
G. lapidicus Kelly and Bromley, 1984	•	•		•	•	•
Phrixichnus isp.	•		•	•		
Entobia isp.	•	•	•	•	•	•
Maeandropolydora sulcans Voigt, 1965	•					
Maeandropolydora decipiens Voigt, 1965					•	
Trypanites rectus Mägdefrau, 1932					•	
Epilithobionts						
Bryozoans		•				
Oysters		•				

It is worth noting that in both areas an intra-Miocene erosive platform has been identified: Oura in Portugal and Vespella in Camp de Tarragona. In both cases, truncated *Gastrochaenolites* appear on a sub-horizontal surface. The differences concern the presence of *Entobia*, which appears only at Vespella, and *Gastrochaenolites* density: 300 specimens per square metre at Oura and up to 700 at Vespella, although in both cases the distribution is patchy. Moreover, the ichnospecies represented are different: *G. torpedo* and *G. turbinatus* at Oura and *G. lapidicus* at Vespella. Also, boring diameters slightly differ, as they are between a few mm to 4 cm at Vespella while the maximum diameter at Oura is 2.5 cm. The maximum density of *Gastrochaenolites* as been found at Fortí de la Reina (Camp de Tarragona), with 1,500 specimens per square metre. In spite of these differences, the ichnoassociations represented in both areas clearly correspond to the *Entobia* ichnofacies defined by Bromley and Asgaard (1993). Their identification in locations in southern Iberia corroborates the validity of this ichnofacies.

On the whole, these small variations would not imply significant environmental differences. The geographical areas concerned are extensive, especially regarding the South.

Conclusions

During the Miocene the southern part of Iberia was affected by at least two marine transgressions during the Langhian-Serravalian and Tortonian. These isostatic events are reflected in the conspicuous bioerosive structures produced by marine organisms. They belong to a marine hard substrate community related to sedimentary omission or starvation, usually accompanied by fast inland migration of the coastline during transgressions. It was possible to identify eleven ichnotaxa: *Caulostrepsis* isp., *Maeandropolydora sulcans*, *Entobia* isp., *E. geometrica*, *E. ovula*, *Circolites* isp., *Gastrochaenolites* isp, *G. torpedo*, *G. turbinatus*, *G. lapidicus* and *G. ornatus*. The presence of a twelfth ichnotaxon (*Gnathichnus*) is yet to be confirmed.

These borings represent an in situ rocky substrate community of bivalves, sponges, worms and echinoids. The bivalve boring *Gastrochaenolites* is dominant in all outcrops studied. The high energy marine environment could be the reason for this low level of ichnodiversity and the minimal presence of epilithozoans remains.

These fossil assemblages can be related to the *Entobia* ichnofacies of Bromley and Asgaard (1993), which is recurrent worldwide on Neogene rocky shores (Gibert et al. 1998). In this sense, comparison with their counterparts in northeastern Iberia Miocene sites (Camp de Tarragona area) reveals both slight similarities and differences concerning the ichnological composition and the relative abundance of traces. *Entobia* shows a greater presence in the Camp de Tarragona outcrops, and the density of *Gastrochaenolites* is locally higher there than in the South.

The major difference between Portuguese and Spanish southern Miocene sites lies in the geometry of the rockgrounds and the tectonic phases associated with them. The Portuguese sector farthest away from the Betic orogen is characterised by planar surfaces, while in the Spanish sector, the stronger Betic tectonics produced highly heterometric deposits developing bioeroded surfaces on cataclastic boulders. Another important aspect is related to the temporal hiatus of the later unconformity surface, which decreases westward.

Acknowledgements

Financial support was given by the Fundação para a Ciência e a Tecnologia (Portuguese government) in the form of a post-doctoral fellowship (Praxis / BPD / 20562) given to A. Santos and co-financed by EU (POCI 2010). Financial support was also given by the Junta de Andalucía (Spanish government) to the Research Group RNM316 (Tectonics and Palaeontology) and by the Spanish Ministry of Education and Culture (Project BTE2003-01356). The paper is also a contribution to the Centro de Investigação Marinha e Ambiental (Universidade do Algarve, Portugal), and to the Grup de Recerca Paleobiologia del Neogen Mediterrani (Universitat de Barcelona, Spain).

The authors are grateful to Jordi M. de Gibert, of the University of Barcelona, for his valuable comments during the preparation of this paper, and also to the referees Markes E. Johnson (Williams College, Massachusetts) and an anonymous reviewer for their useful comments.

References

Antunes MT (1984) Orla Algarvia. In: Oliveira JT (ed) Carta Geológica de Portugal, 1/200000. Serv Geol Portugal, Lisboa, Not Explic 7:60-64

Benner JS, Ekdale AA, Gibert JM de (2004) Macroborings (*Gastrochaenolites*) in Lower Ordovician hardgrounds of Utah: sedimentologic, paleoecologic, and evolutionary implications. Palaios 19:543-550

Bertling M (1999) Taphonomy of trace fossils at omission surfaces (Middle Triassic, East Germany). Palaeogeogr Palaeoclimatol Palaeoecol 149:27-40

Boucot J (1981) Principles of benthic marine palaeoecology. Acad Press, New York, 463 pp

Brachert T, Forst M, Pais J, Legoinha P, Reijmer J (2003) Lowstand carbonates, highstand sandstones? Sediment Geol 155:1-12

Brett CE (1988) Paleoecology and evolution of marine hard substrate communities: An overview. Palaios 3:374-378

Brett CE (1998) Sequence stratigraphy, paleoecology, and evolution: Biotic clues and responses to sea-level fluctuations. Palaios 13:241-262

Brett CE, Brookfield ME (1984) Morphology, faunas and genesis of Ordovician hardgrounds from Southern Ontario, Canada. Palaeogeogr Palaeoclimatol Palaeoecol 46:233-290

Bromley RG (1975) Comparative analysis of fossil and recent echinoid bioerosion. Palaeontology 18:725-739

Bromley RG (1994) The palaeoecology of bioerosion. In: Donovan SK (ed) The palaeobiology of trace fossils. Wiley and Sons, Chichester, pp 134-154

Bromley RG, Asgaard U (1993) Endolithic community replacement on a Pliocene rocky coast. Ichnos 2:93-116

Bromley RG, D'Alessandro A (1984) The ichnogenus *Entobia* from the Miocene, Pliocene and Pleistocene of southern Italy. Riv Ital Paleont Stratigr 90:227-296

Bromley RG, D'Alessandro A (1987) Bioerosion of the Plio-Pleistocene transgression of Southern Italy. Riv Ital Paleont Stratigr 93:379-442

Bronn HG (1837) Lethaea Geognostica oder Abbildungen und Beschreibungen der für die Gebirgs-Formationen bezeichnendsten Versteinerungen. Schweizerbart, Stuttgart, 1:1-544 and plates

Cachão M (1995) Utilização de nanofósseis calcários em biostratigrafia, paleoceanografia e paleoecologia. Unpubl PhD Thesis, Univ Lisbon, Lisbon, 356 pp

Cachão M, Silva CM da (2000) The three main marine depositional cycles of the Neogene of Portugal. Ciênc Terra 14:303-312

Cachão M, Silva CM da (2006) A Bacia do Algarve: estratigrafia, paleogeografia e tectónica. In: Dias R, Araújo A, Terrinha P, Kullberg JC (eds) Geologia de Portugal no Contexto da Ibéria. Univ Évora, Évora, pp 245-317

Choffat P (1903-1904) L'Infralias et le Sinémurien du Portugal. Comunic Comiss Serv Geol Portugal 5:49-114

Clarke JM (1908) The beginnings of dependent life. New York State Mus Bull 121:146-196

Domènech R, Silva CM da, Cachão M, Martinell J (1999) Una megasuperficie bioerosionada en Oura (Albufeira): Implicaciones para la evolución sedimentaria del Mioceno de Algarve (S de Portugal). Temas Geol Mineros ITGE 26:226-230

Domènech R, Gibert JM de, Martinell J (2001) Ichnological features of a marine transgression: Middle Miocene rocky shores of Tarragona, Spain. Géobios 34:99-107

Doyle P, Bennett M, Cocks F (1998) Borings in a boulder substrate from the Miocene of southern Spain. Ichnos 5:277-286

Ehrenberg K (1944) Ergänzende Bemerkungen zu den seinerzeit aus dem Miozän von Burgschleinitz beschriebenen Gangkernen und Bauten dekapoder Krebse. Paläont Z 23:354-359

Ekdale AA, Bromley RG (2001) Bioerosional innovation for living in carbonate hardgrounds in the Early Ordovician of Sweden. Lethaia 34:1-12

Gibert JM de, Martinell J, Domènech R (1998) *Entobia* ichnofacies in fossil rocky shores, Lower Pliocene, Northwestern Mediterranean. Palaios 13:476-487

Harland WB, Armstrong RL, Craig LE, Smith AG, Smith DG (1989) A geologic time scale. Cambridge Univ Press, Cambridge, 279 pp

Johnson ME (1988a) Why are ancient rocky shores so uncommon? J Geol 96:469-480

Johnson ME (1988b) Hunting for ancient rocky shores. J Geol Educ 36:147-154

Johnson ME (1992) Studies on ancient rocky shores: A brief history and annotated bibliography. J Coast Res 8:797-812

Johnson ME (2006) Uniformitarianism as a guide to rocky-shore ecosystems in the geological record. Canad J Earth Sci 43:1119-1147

Johnson ME, Baarli BG (1999) Diversification of rocky-shore biotas through geologic time. Géobios 32:257-273

Kelly SRA, Bromley RG (1984) Ichnological nomenclature of clavate borings. Palaeontology 27:793-807

Kobluk DR (1981a) The record of early cavity-dwelling (coelobiontic) organisms in the Paleozoic. Canad J Earth Sci 18:181-190

Kobluk DR (1981b) Earliest cavity-dwelling organisms (coelobiontis), Lower Cambrian Poleta Formation, Nevada. Canad J Earth Sci 18:669-679

Kobluk DR, James NP (1979) Cavity-dwelling organisms in Lower Cambrian patch reefs from southern Labrador. Lethaia 12:193-218

Kobluk DR, James NP, Pemberton SG (1978) Initial diversification of macroboring ichnofossils and exploitation of macroboring niche in the Lower Paleozoic. Paleobiology 4:163-170

Lamarck JB (1801) Système des animaux sans vertèbres. Déterville, Paris, 432 pp

Leymerie A (1842) Suite de mémoire sur le terrain Crétacé du département de l'Aube. Mém Soc Géol France 5:1-34

Mägdefrau K (1932) Über einige Bohrgänge aus dem unteren Muschelkalk von Jena. Paläont Z 14:150-160

Martinell J (1981) Actividad erosiva the Paracentrotus lividus (Lmk.) (Echinodermata, Echinoidea) en el litoral gerundense. Oecol aquat 5:219-225

Martinell J (1982a) Echinoid bioerosion from the Pliocene of NE Spain. Geobios 15:249-253

Martinell J (1982b) Borings produced by presumed Pliocene brachiopods from l'Empordà (Catalunia, Spain). Bull Inst Catalan Hist Nat, Sec Geol 48 (3):91-97

Martinell J, Domènech R (1995) Bioerosive structures on the Pliocene rocky shores of Catalonia (Spain). Rev Españ Paleont 10:37-44

Mayoral E, Chaves FM, Muniz F, Gibert JM de, Domènech R, Martinell J (1998) New evidences of fossil rocky shores in the upper Miocene of the Iberian Peninsula (Sierra de Tejeda, Betic Range, SE Spain). Comunic 2nd Int Bioerosion Workshop, Fort Pierce, Florida, pp 44-46

Mikulás R (1992) Early Cretaceous borings from Stramberk (Czechoslovakia). Čas Mineral Geol 37: 297-312

Pais J (1982) O Miocénico do litoral Sul Português. Ensaio de Síntese. Unpubl PhD Thesis, Univ Lisbon, Lisbon, 47 pp

Palmer TJ (1982) Cambrian to Cretaceous changes in hardground communities. Lethaia 15:309-323

Perry CT (1996) Distribution and abundance of macroborers in an Upper Miocene reef system, Mallorca, Spain, implications for reef development and framework destruction. Palaios 11:40-56

Pirazzoli PA, Stiros SC, Laborel J, Laborel-Deguen F, Arnold M, Papageorgious S, Morhange C (1994) Late-Holocene shoreline changes related to palaeoseismic events in the Ionian Islands, Greece. Holocene 4:397-405

Plag H-P (2006) Recent relative sea-level trends: an attempt to quantify the forcing factors. Phil Trans Roy Soc, A 364:821-844

Radwanski A (1970) Dependence of rock-borers and burrowers on the environmental conditions within the Tortonian littoral zone of southern Poland. In: Crimes TP, Harper JC (eds) Trace Fossils. Seel House Press, Liverpool, pp 371-390

Siggerud EIH, Steel RJ, Pollard JE (2000) Bored pebbles and ravinement surface clusters in a transgressive systems tract, Sant Llorenç del Munt fan-delta complex, SE Ebro Basin, Spain. Sediment Geol 138:161-177

Silva CM da, Cachão M, Martinell J, Domènech R (1995) Estruturas bioerosivas como indicadores de paleolitorais rochosos. O exemplo do Miocénico da Foz da Fonte (Sesimbra, Portugal). Dados preliminares. Mem Mus Lab Mineral Geol Univ Porto 4:133-137

Silva CM da, Cachão M, Martinell J, Domènech R (1999) Bioerosional evidence of rocky palaeoshores in the Neogene of Portugal: environmental and stratigraphical significance. Bull Geol Soc Denmark 45:156-160

Taylor PD, Wilson MA (2003) Palaeoecology and evolution of marine hard substrate communities. Earth-Sci Rev 62:1-103

Voigt E (1965) Über parasitische Polychaeten in Kreide-Austern sowie einige andere in Muschelschalen bohrende Würmer. Paläont Z 39:193-211

Watkins R (1990) Paleoecology of a Pliocene rocky shoreline, Salton Trough Region, California. Palaios 5:167-175

Wilson MA (1985) Disturbance and ecological succession in an Upper Ordovician cobble dwelling hardground fauna. Science 228:575-577

Wilson MA (1987) Ecological dynamics on pebbles, cobbles and boulders. Palaios 2:594-599

Role of bioerosion in taphonomy: effect of predatory drillholes on preservation of mollusc shells

Patricia H. Kelley[1]

[1] Department of Geography and Geology, University of North Carolina Wilmington, Wilmington, NC 28403-5944, USA, (kelleyp@uncw.edu)

Abstract. Although bioerosion is recognized as causing significant destruction of hard substrates, few studies have assessed loss of hard substrates by taphonomic bias against shells with predatory borings (the trace fossil *Oichnus*). Results of point-load compression experiments suggest preferential loss of *Oichnus*-bearing shells; however, studies of drilling predation often assume lack of taphonomic bias against drilled shells. This project tested the hypothesis that taphonomic processes preferentially destroyed shells with predatory borings for the Miocene Choptank Formation of Maryland and the Plio-Pleistocene Caloosahatchee Formation of Florida. If bias against drilled shells occurred, drilled shells should be more pristine than undrilled shells because they do not survive exposure, and shells should break through drillholes.

Drilled and undrilled shells from bulk samples of the Drumcliff and Boston Cliffs members of the Choptank Formation were analyzed for taphonomic condition, and shell fragments were examined for breakage through drillholes. Taphonomic condition was also determined for five species of Caloosahatchee bivalves that exhibited predatory bioerosion: *Dosinia elegans, Chione latilirata, Chione elevata, Macrocallista nimbosa*, and *Stewartia anodonta.*

Most measures of taphonomic condition showed no statistically significant difference between drilled and undrilled shells for both formations. In the Choptank, condition of shell sculpture (Boston Cliffs Member) suggested taphonomic bias against drilled shells. However, distinctness of the pallial line (Drumcliff Member) supported the opposite conclusion. In addition, numerous shell fragments contained intact drillholes; breaks passed through drillholes in only 17% of Choptank *Oichnus*-bearing fragments, suggesting that drillholes did not weaken shells significantly. In the Caloosahatchee, taphonomic condition did not vary significantly between drilled and undrilled shells for most of the 24 cases studied. One case, nonpredatory bioerosion in *Chione elevata,* supported the hypothesis that drillholes cause taphonomic loss of shells. In contrast, visibility of the pallial line and muscle scars and condition of sculpture in *Stewartia anodonta* indicate that specimens with *Oichnus* survived more taphonomic damage than undrilled specimens. The predominance of shells with breaks that did not pass through drillholes in the Choptank Formation, and the lack of difference between drilled and undrilled shells

M. Wisshak, L. Tapanila (eds.), *Current Developments in Bioerosion*. Erlangen Earth Conference Series, doi: 10.1007/978-3-540-77598-0_23,

for most taphonomic measures in both formations, indicates that drillholes did not weaken shells significantly.

Keywords. *Oichnus*, drilling, predation, taphonomic bias, Choptank Formation, Caloosahatchee Formation

Introduction

Although the concept of bioerosion is frequently associated with the production of domiciles in hard substrates (rock, carbonate hardgrounds, shells or wood), bioerosion may also result from the process of feeding. For instance, chitons and gastropods may produce bioerosion during the process of grazing (e.g., the trace fossil *Radulichnus* Voigt, 1977) and gnawing traces (*Gnathichnus* Bromley, 1975) are produced by echinoids (Bromley 1994; Taylor and Wilson 2003). Perhaps the most widely recognized result of bioerosion produced during feeding is the trace fossil *Oichnus* Bromley, 1981, usually produced by predatory molluscs in the shells of their prey.

Three ichnospecies of *Oichnus* were named by Bromley (1981, 1993). *Oichnus ovalis* Bromley, 1993 traces are small, rhomboidal to oval, tapering holes typical of drilling by octopods. Two ichnospecies are usually formed by gastropod predation: *O. simplex* Bromley, 1981 (cylindrical to subcylindrical drillholes that are typically produced by muricid gastropods and more rarely by octopods; Bromley 1981, 1993); and *O. paraboloides* Bromley, 1981 (beveled holes that are parabolic in cross section). Naticid gastropods are primarily responsible for *Oichnus paraboloides*, although the trace may also be produced by some nassariid, marginellid, nudibranch, and muricid gastropods (Ponder and Taylor 1992; Morton and Chan 1997; Harper et al. 1998; Carriker and Gruber 1999; Herbert and Dietl 2002). Additionally, Nielsen and Nielsen (2001) named three new ichnospecies discovered on foraminiferans: *O. coronatus* (which resembles *O. simplex* but with an etched halo), *O. asperus* (elongate-oval, non-tapering holes), and *O. gradatus* (with a stepped profile in cross section). Donovan and Jagt (2002) erected the ichnospecies *O. excavatus* for nonpenetrative cavern-like borings, typically with a prominent central boss; these most likely represent domiciles, however.

The process of predatory bioerosion has been studied in detail for naticid and muricid gastropods (Ziegelmeier 1954; Carriker 1981). Muricids typically hunt and drill prey epifaunally, whereas naticids usually drill infaunal prey within the substrate, although exceptions are known (e.g., Guerrero and Reyment 1988; Savazzi and Reyment 1989; Grey 2001; Dietl 2003a). Drilling involves both mechanical rasping by the radula and chemical secretions from the accessory boring organ, which may include hydrochloric acid, enzymes, and chelating agents (Carriker and Gruber 1999).

Bioerosion by grazers and borers can result in significant destruction of hard substrates (Taylor and Wilson 2003); Kiene and Hutchings (1994) reported loss by bioerosion of up to 2-3 $kg/m^2/yr$ in some environments on the Great Barrier Reef, Australia. Although less often considered, predatory bioerosion may also lead to loss of hard substrates by preferential destruction of bored shells.

Roy et al. (1994) tested for taphonomic bias against drilled shells by subjecting valves of the mactrid bivalve *Mulinia lateralis* (Say, 1822) to point-load compression until breakage occurred. Breaking load was compared for undrilled shells and those bearing *Oichnus paraboloides* over a range of shell sizes and for two environments (shells were collected from beach and tidal channel environments). Drilled valves were significantly weaker. Roy et al. (1994: 416) concluded that, at least for *'small, thin-shelled, umbonally drilled bivalves under local compressive loads,'* stress concentrations due to the presence of a drillhole should lead to taphonomic bias against drilled shells.

Hagstrom (1996) determined breaking strengths of various sized shells of the venerid bivalves *Macrocallista nimbosa* (Solander, 1786) and *Chione elevata* (Say, 1822), referred to as *C. cancellata* (Linnaeus, 1767) in her work; however, Roopnarine and Vermeij (2000) have indicated that *C. cancellata* is a Caribbean species and that the Florida species is *elevata*. Hagstrom (1996) conducted point-loading experiments similar to those of Roy et al. (1994) and determined that boreholes located near the umbo served as stress concentrators and that the mean strength of undrilled shells was significantly greater than that for drilled shells for both species. However, boreholes that were not located near the umbo of the shell did not serve as local stress concentrators and did not affect the strength of the valve.

Zuschin and Stanton (2001) subjected valves of the arcid bivalve *Anadara ovalis* (Bruguière, 1789) to point-load compression and found that valves bearing *Oichnus paraboloides* were significantly weaker than undrilled valves. However, a set of experiments that more realistically simulated the stresses placed on shells in nature yielded different results. Drilled and undrilled beach-collected *Mulinia lateralis* were embedded in sediment either as isolated shells or in 1-cm thick shell layers and subjected to compaction. Isolated *Mulinia lateralis* valves, with or without drillholes, did not suffer breakage during sediment compaction. When shell layers (in which shells were in contact with one another) were compacted, breakage did occur, but drilled valves suffered less breakage (19%) than undrilled valves (32%). In addition, only 10% of the fragments of drilled shells showed breakage through the drillhole. Zuschin and Stanton (2001) concluded that shell breakage during compaction is due to point contacts between shells rather than to the presence of a drillhole.

Kaplan and Baumiller (2000) tested experimentally for taphonomic bias against bored brachiopod shells. They manufactured equal-mass casts of valves of the Ordovician brachiopod *Onniella meeki* (Miller, 1875) and subjected them to point-load compression. Results were similar to that of Roy et al. (1994) in that undrilled valves required greater loads to fracture, although the fractures in *Onniella* passed through the sulcus rather than through the borings. However, under contact-compaction experiments, in which bored and unbored valve casts were placed together in a piston and compressed, unbored and bored valve loads were similar (though there was some evidence that, when breaks occurred, they preferentially passed through drillholes in drilled valves).

Experimental studies thus provide mixed results concerning the potential taphonomic bias against shells bearing predatory bioerosion traces. Studies that examine such traces in the fossil record would benefit from detection of taphonomic bias when present. A few studies have suggested ways of determining the degree of taphonomic bias against bored specimens that might have affected a fossil assemblage (see summary by Kowalewski 2002). Roy et al. (1994) recommended assessing the degree of fragmentation within an assemblage and in particular whether fragments bear intact drillholes or 'partial holes' (in which breakage passed through the drillhole). Nebelsick and Kowalewski (1999) proposed a test involving comparison of taphonomic condition of drilled and undrilled specimens. They hypothesized that, if drillholes do not have a taphonomic effect, then drilled and undrilled specimens should not differ in their degree of taphonomic alteration. On the other hand, if drillholes weaken shells, drilled shells should not be as resistant to taphonomic processes of mechanical, chemical, and biological alteration. Drillholes should thus occur preferentially in more pristine specimens, because drilled specimens should not survive long enough to accumulate taphonomic damage. (Though not explicitly stated, a requirement of this method is that drilled and undrilled shells should not differ in their potential for exposure to taphonomic processes due to differences in life mode or whether mortality occurred within the substrate or epifaunally.)

Nebelsick and Kowalewski (1999) applied these criteria to Recent echinoid material from the Red Sea, Egypt, and found no relationship between degree of pre-burial taphonomic alteration and presence of a drillhole. Similarly, Kowalewski and Hoffmeister (2004) found no statistically significant relationship between drilling frequencies and shell taphonomy for material from Holocene beach ridges of Baja California and Miocene fossil assemblages of Europe. In contrast to the work by Roy et al. (1994), these results suggest that predatory bioerosion does not lead to preferential destruction of shell material.

Relatively few studies have applied these criteria specifically to fossil assemblages to test for taphonomic bias against drilled specimens. In this paper I present the results of two case studies from the US Coastal Plain in which the criteria of Roy et al. (1994) and Nebelsick and Kowalewski (1999) are used to test whether predatory bioerosion caused taphonomic loss of shells. The first case study focuses on the middle Miocene Choptank Formation of the Chesapeake Group of Maryland. Predation on Choptank Formation molluscs has figured prominently in studies of the history and dynamics of drilling gastropod predator-prey systems (e.g., Dudley and Vermeij 1978; Kelley 1988, 1989,1991; Hagadorn and Boyajian 1997; Kowalewski et al. 1998; Kelley et al. 2001; Kelley and Hansen 2001, 2003, 2006; Grey et al. 2006). Thus knowledge of whether the fossil record is biased against drilled shells has implications concerning the validity of much previous work. I also apply Nebelsick and Kowalewski's (1999) criteria to a mollusc assemblage from the Plio-Pleistocene Caloosahatchee Formation of Florida to determine the generality of the results from the Miocene. Caloosahatchee samples have also been used in previous work on drilling predation (e.g., Anderson 1992; Roopnarine and Beussink 1999; Beatty 2003; Dietl 2003b).

Materials and methods

Choptank Formation study

The Chesapeake Group formations were divided into a series of numbered 'zones' by Shattuck (1904) based on lithology and in particular the occurrence of shell beds. The 'zones' of the middle Miocene Choptank Formation were subsequently designated as formal members by Gernant (1970). Two shell beds are exposed at Matoaka Cottages, near the town of St. Leonard on the western shore of Chesapeake Bay in Calvert County, Maryland. The Drumcliff Member (Shattuck's 'zone' 17) is a brownish muddy fine sand with ~3 meters of densely packed shell material. The Boston Cliffs Member ('zone' 19) consists of a dense accumulation of molluscs within a light orange-brown, muddy fine sand. At Matoaka the Boston Cliffs is less than a meter thick, but it is much thicker at the type locality (~ 5 meters; Gernant 1970). Both members were deposited under open marine conditions, during sea-level rise, mostly between storm and fair-weather wave bases (Kidwell 1989).

Teams of student researchers in an invertebrate paleontology course at University of North Carolina Wilmington participated in the study (Kelley et al. 2006). We collected six bulk samples from the Drumcliff Member and five bulk samples from the Boston Cliffs Member of the Choptank Formation at Matoaka Cottages. Samples were wet sieved through 2-mm screens and all whole pelecypod valves and gastropod specimens were picked and identified to the species level using Vokes (1957). In addition, all shell fragments bearing drillholes were picked and identified to species level to the extent possible. All fragments were examined by at least two students and by me, to avoid overlooking any 'partial drillholes' on the edges of fragments (i.e., holes through which breakage had occurred).

For whole specimens we recorded the presence of predatory drillholes and identified them as either *Oichnus paraboloides* or *O. simplex*. We also tabulated the occurrence of incomplete drillholes (those that do not completely penetrate the shell, and which are usually interpreted as representing unsuccessful drilling; Kelley et al. 2001). Taphonomic condition of all whole specimens was determined as follows. We coded each of the following characteristics as either present (1) or absent (0): nonpredatory bioerosion; encrustation; edge rounding; edge chipping. The condition of shell sculpture was coded from 0 (worn) to 2 (pristine). Visibility of the muscle scars and pallial line was coded from 0 (not visible) to 2 (distinct). Luster was coded from 0 (chalky) to 2 (pristine). Finally, for shell fragments with drillholes, we determined if breakage passed through drillholes or if fragments possessed intact drillholes, as recommended by Roy et al. (1994).

For each taphonomic character, I used a Mann-Whitney U test to compare the condition of drilled and undrilled specimens across all taxa within each member of the Choptank; values with $p < 0.05$ were considered statistically significant. Specimens with incomplete drillholes were rare in both the Choptank and Caloosahatchee samples and were treated as undrilled shells; *Oichnus simplex* was rare and was not treated separately from *O. paraboloides*. We also determined the percent of drillhole-containing fragments in which breakage passed through the drillhole. If gastropod boreholes do not affect a shell's strength and preservation potential, we hypothesized that: 1) taphonomic condition of drilled and undrilled shells will not differ; and 2) breaks will not pass through drillholes in fragmented shells.

Caloosahatchee Formation study

The Caloosahatchee Formation was initially studied by Heilprin in 1886 from exposures along the Caloosahatchee River, Florida. Although Heilprin considered the formation to be Pliocene in age, based on the ratio of extinct to extant molluscs present, some subsequent workers have assigned it to the Pleistocene (Perkins 1968; Allmon 1993). Rasmussen (1997) and Missimer (2001b) considered it to be Plio-Pleistocene in age. The formation includes well-indurated fossiliferous limestone, soft marly limestone, and densely fossiliferous quartz sands. Deposition occurred within a range of environments, from shallow shelf to brackish bay or lagoon, during several sea-level rises and regressions (Rasmussen 1997; Missimer 2001a). The materials employed in this study came from a bulk sample collected by Squires and Heaslip in 1955 from a spoil bank along Highway 80, 3.5 km west of La Belle, in Hendry County, Florida. The bulk sample was housed at the American Museum of Natural History (AMNH) and loaned to the University of North Carolina Wilmington in 2006 for curation and use in student research projects.

As part of a graduate paleoecology course at the University of North Carolina Wilmington (Lewis et al. 2006; McCoy et al. 2006), the bulk sample was separated into species and specimens were identified to species level based on the following references: Olsson and Harbison 1953; Olsson and Petit 1964; Olsson 1993; Petuch 1991, 1994. Although the bulk sample yielded 7126 bivalve specimens, overall drilling frequency was low (3.6%; McCoy et al. 2006), and many species were not drilled. Therefore, rather than examining the taphonomic condition of all specimens in the sample, I focused on several species that exhibited drilling: the venerid bivalves *Dosinia elegans* (Conrad, 1843), *Chione (Lirophora) latilirata* (Conrad, 1841), *Chione elevata,* and *Macrocallista nimbosa*, and the lucinid *Stewartia anodonta* (Say, 1824). Two of the species, *Chione elevata* and *Macrocallista nimbosa*, had been subjected previously to point-load compression experiments by Hagstrom (1996). The five species studied also vary in susceptibility to taphonomic damage; for instance, *Macrocallista* is thin shelled relative to shell length, and *Chione* is thick shelled. They differ in shell curvature as well; Hagstrom (1996) postulated that the flatter *Macrocallista* should be less subject to breakage than the more hemispherical *Chione.* All taxa are infaunal, and potential for taphonomic alteration should have been similar between drilled and undrilled shells.

For the five species selected for examination, the same taphonomic characteristics were used as in the Choptank study, with the exception of the smooth-shelled *Macrocallista nimbosa*, for which the characteristic of condition of ornamentation did not apply. I coded the taphonomic condition of all specimens of each species except *Chione elevata,* which was represented by >1600 valves. For *Chione elevata,* I coded the condition of all 55 drilled valves but I only coded an equal-sized, randomly selected subsample of undrilled valves (55 specimens were drawn from the pool of undrilled valves without inspecting them, so that the sample is random with respect to shell size and taphonomic condition).

Stewartia anodonta and *Chione elevata* included enough drilled specimens to test statistically for taphonomic differences between drilled and undrilled shells using the Mann-Whitney U test. Because only three drilled specimens were present for each of the other Caloosahatchee species, data for those three species were combined for statistical comparison of drilled and undrilled shells. As for the Choptank Formation, U tests compared drilled and undrilled shells with respect to each taphonomic characteristic.

Because the Caloosahatchee sample included relatively few fragments of the species of interest, the criterion of Roy et al. (1994) could not be applied. A detailed record of the collection and sieving (if any) methods employed by Squires and Heaslip was not available, and it is unclear if any fragments had been discarded prior to depositing at the AMNH.

Results

Choptank Formation

The 11 bulk samples from the Choptank Formation yielded a total of 643 whole specimens, including 299 pelecypod valves and 112 gastropods from the Drumcliff Member and 221 pelecypod valves and 11 gastropod specimens from the Boston Cliffs Member. Abundances and drilling frequencies of all species are shown in Table 1. Samples were dominated by the bivalves *Dallarca, Astarte*, and *Bicorbula*; *Turritella plebeia* was the most common gastropod. Almost all specimens (98%) in the sample belonged to infaunal taxa, which would have been preyed upon infaunally by their primary predator, naticid gastropods. Thus individuals both lived infaunally and were drilled infaunally, yielding comparable potentials for taphonomic alteration of drilled and undrilled specimens.

The results of the Mann-Whitney U tests comparing taphonomic condition of drilled vs. undrilled shells are given in Table 2 for the Drumcliff Member and in Table 3 for the Boston Cliffs Member. In the Drumcliff Member, the Mann-Whitney test revealed no statistically significant differences between drilled and undrilled shells for muscle scar visibility, preservation of sculpture, retention of luster, and presence of edge rounding, edge chipping, encrustation and predatory boring. As shown in Figure 1, the pallial line was more distinct for undrilled shells than for drilled shells ($p < 0.05$). In the Boston Cliffs Member, only one taphonomic characteristic showed a statistically significant difference between bored and unbored shells. Shell sculpture was better preserved on drilled shells than on undrilled shells (Fig. 2; $p < 0.01$).

Most fragments (83%) that contained *Oichnus* did not break through the drillhole. In the Drumcliff Member, only 4 (10%) of 41 drillhole-bearing fragments had breaks that passed through predatory drillholes. Eleven of 45 drillhole-bearing fragments (24%) in the Boston Cliffs Member were broken through the drillhole. Drilling frequencies would have changed as follows if fragments with partial drillholes had been included in the calculation. In the Drumcliff Member, *Dallarca elnia* drilling

Table 1. Abundance and drilling data for molluscs from bulk samples of the Drumcliff and Boston Cliffs members of the Choptank Formation. N, number of specimens; CD, number of specimens with at least one complete drillhole; DF, drilling frequency. DF calculated as (number of drilled specimens) / (total number of specimens) for gastropods and as (number of drilled valves) / (½ total number of valves) for bivalves

	Species	N	CD	DF
Drumcliff Member	*Astarte thisphila* Glenn, 1904	44	7	0.32
	Bicorbula idonea (Conrad, 1833)	21	0	0.00
	Cyclocardia sp. Conrad, 1867	1	0	0.00
	Chesapecten nefrens (Ward and Blackwelder, 1975)	1	0	0.00
	Corbula inaequalis Say, 1824	26	4	0.31
	Dallarca elnia (Glenn, 1904)	198	21	0.21
	Marvacrassatella turgidula (Conrad, 1843)	4	1	0.50
	Glossus fraterna (Say, 1824)	4	0	0.00
	Total bivalves	**299**	**33**	**0.22**
	Turritella plebeia Say, 1824	111	9	0.08
	Turritella variabilis d'Orbigny, 1852	1	1	1.00
	Total gastropods	**112**	**10**	**0.09**
Boston Cliffs Member	*Anomia aculeata* Gmelin, 1791	1	0	0.00
	Astarte obruta Conrad, 1834	61	5	0.16
	Bicorbula idonea (Conrad, 1833)	18	1	0.11
	Chesacardium laqueatum (Conrad, 1830)	1	0	0.00
	Chesapecten nefrens (Ward and Blackwelder, 1975)	15	1	0.13
	Corbula inaequalis Say, 1824	4	2	1.00
	Dallarca elevata (Conrad, 1840)	114	8	0.14
	Dosinia acetabulum (Conrad, 1832)	3	0	0.00
	Marvacrassatella marylandica (Conrad, 1832)	2	0	0.00
	Mercenaria cuneata Conrad, 1869	2	0	0.00
	Total bivalves	**221**	**17**	**0.15**
	Turritella plebeia Say, 1824	8	5	0.63
	Turritella subvariabilis d'Orbigny, 1852	3	1	0.33
	Total gastropods	**11**	**6**	**0.55**

Table 2. Results of Mann-Whitney U test comparing taphonomic condition of undrilled ('sample A') vs. drilled ('sample B') specimens in the Drumcliff Member of the Choptank Formation. n_A and n_B, sample size for undrilled and drilled samples respectively; U_A, value of U based on sample A. Asterisk (*) indicates taphonomic characteristic for which difference between drilled and undrilled specimens is significant at $p < 0.05$

	Rounding	Chipping	Encrustation	Nonpred. Boring	Pallial line*	Muscle scar	Sculpture	Luster
n_A – undrilled	371	371	371	371	272	272	370	371
n_B – drilled	43	43	43	43	31	31	43	43
mean rank A	209.4	206.6	205.5	207.2	155.5	151.7	204.6	208.3
mean rank B	191.5	215.4	224.8	210.1	121.3	154.5	227.8	200.8
U_A	7289	8316.5	8718.5	8089.5	3264	4294	8851.5	7690.5
z	0.920	-0.460	-1	-0.15	2.06	-0.17	-1.21	0.38
p (one-tailed)	0.179	0.323	0.1587	0.4404	0.0197	0.4325	0.1131	0.352

Table 3. Results of Mann-Whitney U test comparing taphonomic condition of undrilled ('sample A') vs. drilled ('sample B') specimens in the Boston Cliffs Member of the Choptank Formation. n_A and n_B, sample size for undrilled and drilled samples respectively; U_A, value of U based on sample A. Asterisk (*) indicates taphonomic characteristic for which difference between drilled and undrilled specimens is significant at $p < 0.05$

	Rounding	Chipping	Encrusta- tion	Nonpred. Boring	Pallial line	Muscle scar	Sculp- ture*	Luster
n_A – undrilled	212	212	212	212	206	206	203	212
n_B – drilled	23	23	23	23	17	17	22	23
mean rank A	117.1	117.2	118.2	118.4	112.6	110.4	109.1	118.9
mean rank B	125.9	125.4	116.5	114.5	104.2	131.7	149.3	110.1
U_A	2620	2608.5	2403.5	2357.5	1619	2086.5	3032.5	2256.5
z	-0.590	-0.550	0.11	0.26	0.51	-1.31	-2.76	0.58
p (one-tailed)	0.278	0.291	0.4562	0.3974	0.305	0.0951	0.0029	0.281

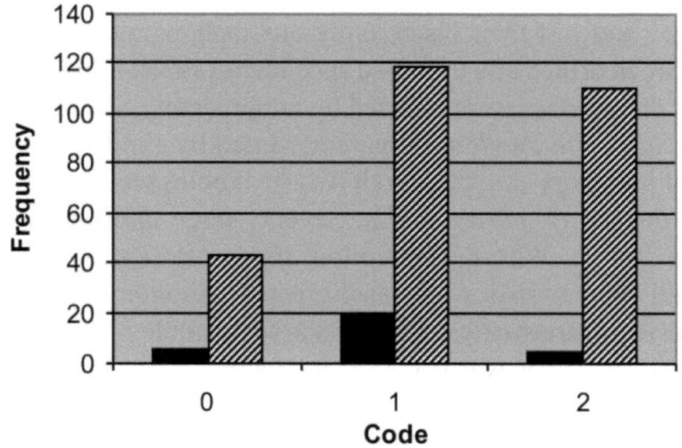

Fig. 1 Visibility of pallial line for drilled (shown in black) and undrilled (diagonal lines) specimens in the Drumcliff Member of the Choptank Formation. Visibility of pallial line coded from 0 = not visible to 2 = distinct

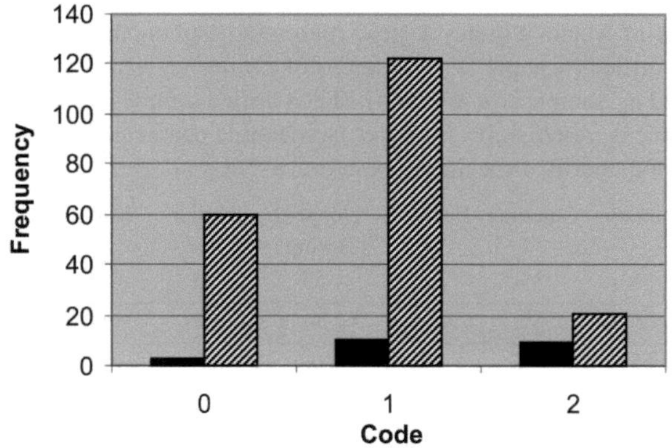

Fig. 2 Condition of sculpture for drilled (shown in black) and undrilled (diagonal lines) specimens in the Boston Cliffs Member of the Choptank Formation. Condition of sculpture coded from 0 = worn to 2 = pristine

frequencies would have increased from 0.21 to 0.22; *Turritella plebeia*, from 0.08 to 0.10; the bivalve fauna as a whole, from 0.22 to 0.23; and the gastropod fauna as a whole, from 0.09 to 0.11. In the Boston Cliffs Member, some partial drillholes occurred in taxa for which too few specimens were present to calculate drilling frequencies reliably (*Corbula inaequalis*, *Marvacrassatella marylandica*). For the more common taxa, drilling frequencies would have changed as follows with inclusion of specimens with partial drillholes: *Dallarca elevata*, from 0.14 to 0.16; *Turritella plebeia*, from 0.63 to 0.67; bivalve fauna as a whole, from 0.15 to 0.23; and gastropod fauna as a whole, from 0.55 to 0.58. The differences in drilling frequency calculated with and without inclusion of partial drillholes are not statistically significant, based on chi-squared tests, for either member.

Caloosahatchee Formation

Taphonomic alteration was common for specimens of *Chione elevata*, regardless of whether they were drilled. Most of the 110 specimens examined had rounded or chipped edges and lost their luster. Only one taphonomic characteristic differed statistically between drilled and undrilled specimens (Table 4; Fig. 3): more undrilled specimens than drilled ones were affected by nonpredatory bioerosion ($p < 0.05$).

The lucinid *Stewartia anodonta* was represented by eight valves with drillholes thought to be of predatory origin, though two such holes were identified as *Oichnus simplex* rather than as *O. paraboloides*. Seventy-three undrilled valves were also included in the sample. With the exception of encrustation, which was not found on any valve, all characteristics indicated greater taphonomic alteration for drilled valves than undrilled valves (Figs. 4-7). As shown in Table 5, the difference between drilled and undrilled valves was statistically significant for visibility of pallial line ($p < 0.05$) and muscle scar ($p < 0.05$) and distinctness of sculpture ($p < 0.01$).

Drillholes were uncommon in the remaining three Caloosahatchee species studied. A combined analysis of all three species yielded no statistically significant differences in taphonomic condition between drilled and undrilled shells for any of the eight characteristics examined (Table 6).

Table 4. Results of Mann-Whitney U test comparing taphonomic condition of undrilled ('sample A') vs. drilled ('sample B') specimens of *Chione elevata* from the Caloosahatchee Formation. n_A and n_B, sample size for undrilled and drilled samples respectively; U_A, value of U based on sample A. Asterisk (*) indicates taphonomic characteristic for which difference between drilled and undrilled specimens is significant at $p < 0.05$

	Rounding	Chipping	Encrustation	Nonpred. Boring*	Pallial line	Muscle scar	Sculpture	Luster
n_A – undrilled	55	55	55	55	55	55	55	55
n_B – drilled	55	55	55	54	51	52	55	55
mean rank A	55	56.5	56	60.6	58	55.8	52.7	56.5
mean rank B	56	54.5	55	49.3	48.6	52.1	58.3	54.5
U_A	1540	1457.5	1485	1177.5	1155	1330	1664	1457
z	-0.16	0.33	0.16	1.86	1.56	0.62	-0.9	0.33
p (one-tailed)	0.4364	0.3707	0.4364	0.0314	0.0594	0.2676	0.1841	0.3707

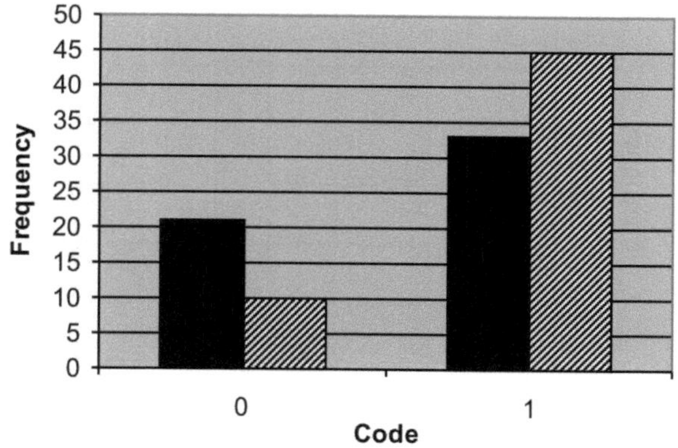

Fig. 3 Occurrence of nonpredatory bioerosion on drilled (shown in black) and undrilled (diagonal lines) specimens of *Chione elevata* in the Caloosahatchee Formation. Nonpredatory bioerosion coded as 0 = absent and 1 = present

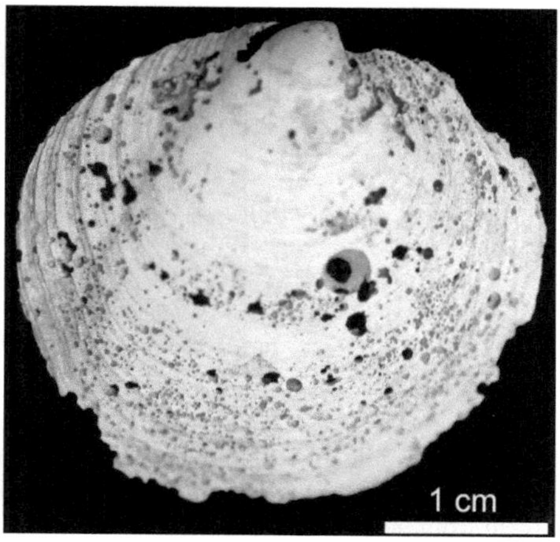

Fig. 4 Drilled specimen of *Stewartia anodonta* from the Caloosahatchee Formation exhibiting edge chipping and rounding and extensive nonpredatory bioerosion

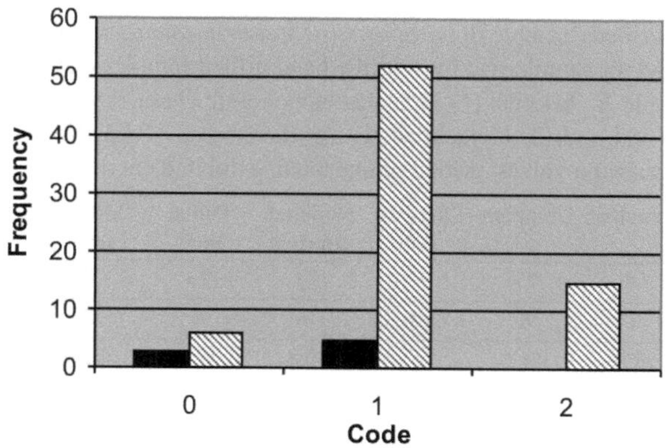

Fig. 5 Visibility of pallial line for drilled (shown in black) and undrilled (diagonal lines) specimens of *Stewartia anodonta* in the Caloosahatchee Formation. Visibility of pallial line coded from 0 = not visible to 2 = distinct

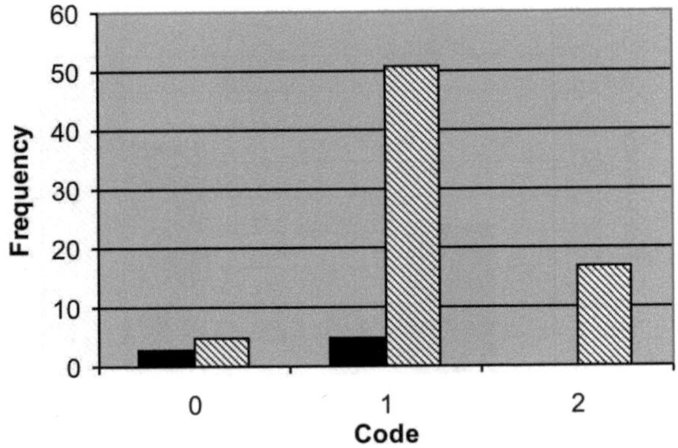

Fig. 6 Visibility of muscle scars for drilled (shown in black) and undrilled (diagonal lines) specimens of *Stewartia anodonta* in the Caloosahatchee Formation. Visibility of muscle scars coded from 0 = not visible to 2 = distinct

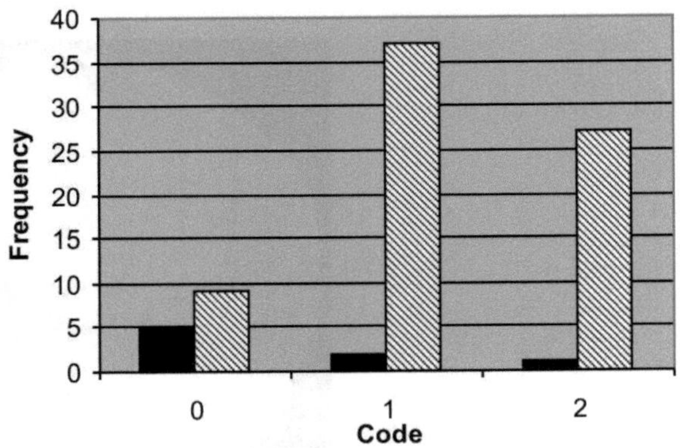

Fig. 7 Condition of sculpture for drilled (shown in black) and undrilled (diagonal lines) specimens of *Stewartia anodonta* in the Caloosahatchee Formation. Condition of sculpture coded from 0 = worn to 2 = pristine

Table 5. Results of Mann-Whitney U test comparing taphonomic condition of undrilled ('sample A') vs. drilled ('sample B') specimens of *Stewartia anodonta* from the Caloosahatchee Formation. n_A and n_B, sample size for undrilled and drilled samples respectively; U_A, value of U based on sample A. Asterisk (*) indicates taphonomic characteristic for which difference between drilled and undrilled specimens is significant at $p < 0.05$. No results are given for encrustation because no valves, drilled or undrilled, exhibited encrustation

	Rounding	Chipping	Encrustation	Nonpred. Boring	Pallial line*	Muscle scar*	Sculpture*	Luster
n_A - undrilled	68	73	73	73	73	73	73	73
n_B - drilled	8	8	8	8	8	8	8	8
mean rank A	37.9	40.2	-	40.4	42.7	42.8	43.1	41.8
mean rank B	43.3	48.5	-	46.3	25.6	24.5	21.6	34
U_A	310	352	-	334.5	169	160	137	236
z	-0.63	-0.94	-	-0.66	1.94	2.08	2.45	0.88
p (one-tailed)	0.2643	0.1736	-	0.2546	0.0262	0.0188	0.0071	0.1894

Table 6. Results of Mann-Whitney U test comparing taphonomic condition of undrilled ('sample A') vs. drilled ('sample B') specimens of *Dosinia elegans*, *Macrocallista nimbosa*, and *Chione latilirata* (combined) from the Caloosahatchee Formation. n_A and n_B, sample size for undrilled and drilled samples respectively; U_A, value of U based on sample A

	Rounding	Chipping	Encrustation	Nonpred. boring	Pallial line	Muscle scar	Sculpture	Luster
n_A - undrilled	76	76	70	72	72	72	47	74
n_B - drilled	9	9	9	9	9	9	6	9
mean rank A	43.3	42.9	39.7	41.2	41.1	40.7	27.1	42.6
mean rank B	40.8	43.9	42.3	39.5	39.9	43.2	26.3	37.4
U_A	322.5	350	335.5	310.5	314.5	344	137	292
z	0.27	-0.11	-0.31	0.2	0.14	-0.29	0.1	0.59
p (one-tailed)	0.3936	0.4562	0.3783	0.4207	0.4443	0.3859	0.4602	0.2776

Discussion and conclusions

Nearly all taphonomic criteria examined showed no significant difference between shells with *Oichnus* and those lacking predatory bioerosion for both the Miocene Choptank Formation and Plio-Pleistocene Caloosahatchee Formation. In the Boston Cliffs Member of the Choptank Formation, shell sculpture was better preserved on drilled shells than on undrilled shells, suggesting that drilled shells experienced less exposure on the seafloor than did undrilled shells. Distinctness of the pallial line was greater for undrilled shells than drilled shells in the Drumcliff Member, however; this result is opposite expectations if drillholes weaken shells. For all the other taphonomic characteristics examined (14 cases), no significant difference was found between drilled and undrilled shells. The preponderance of characteristics that do not vary significantly between drilled and undrilled shells suggests that predatory bioerosion did not significantly bias the Choptank samples.

In addition, for the shell fragments that contained drillholes, only 17% of breaks passed through drillholes (10% in the Drumcliff Member and 24% in the Boston Cliffs Member). Thus most fragmentation in these bulk samples did not occur as a result of breakage through drillholes. This lack of breakage through drillholes also indicates that predatory bioerosion did not significantly affect estimates of drilling frequencies in the Choptank. This conclusion is confirmed by examining how drilling frequencies would have changed had fragments with partial drillholes been included in the calculations. For most taxa exhibiting breakage through drillholes, and for the bivalve and gastropod faunas as a whole, drilling frequencies would have increased by ~1-2% in most cases. None of the differences between drilling frequency calculated with and without inclusion of partial drillholes was statistically significant.

For the Caloosahatchee species examined, I found only four examples in which drilled specimens differed statistically in taphonomic condition from undrilled specimens (out of 24 cases). One example supports the hypothesis that drillholes cause taphonomic loss of shells: more undrilled specimens than drilled specimens of *Chione elevata* had nonpredatory bioerosion. In contrast, three cases indicate that specimens with *Oichnus* suffered more taphonomic damage than did undrilled specimens. Although Caloosahatchee *Stewartia anodonta* in general showed a high degree of taphonomic alteration (Lewis et al. 2006), drilled specimens of *Stewartia* were especially subject to alteration of the pallial line, muscle scars, and sculpture. The Caloosahatchee results support the conclusion that taphonomic bias against drilled shells was not significant for all five species examined, regardless of shell morphology.

The results of this study are contrary to experimental work that indicates that drilled shells are significantly weaker than undrilled shells. Point-load compression experiments conducted by Roy et al. (1994), Hagstrom (1996), Kaplan and Baumiller (2000), and Zuschin and Stanton (2001) resulted in undrilled shells requiring greater loads to break than drilled shells for a variety of taxa. These compression experiments examined taxa similar to those included in the present study; for instance, two of the Caloosahatchee species examined were those used by Hagstrom. In addition, Hagstrom (1996) calculated that flow-induced forces may exceed the strength of drilled shells in nature, at least in wave-swept environments. Why then was there so little evidence of taphonomic bias against drilled shells in the Choptank and Caloosahatchee formations?

One explanation is that the point-load compression experiments may not be realistic simulations of the conditions shells normally experience during their incorporation into the fossil record. Shells may experience forces that would cause breakage either while exposed on the sea floor or after burial due to sediment compaction. If burial is rapid, or if infaunal shells are never exposed on the sea floor, sediment compaction would be the primary agent causing preferential breakage of shells. The results of compaction experiments (Kaplan and Baumiller 2000; Zuschin and Stanton 2001) contrast with those of the point-load compression experiments, which showed reduced strength of drilled shells. Breakage of shells enclosed in sediment was not influenced by the presence of a drillhole but instead was affected by the degree of contact between individual shells (Zuschin and Stanton 2001). Leighton (personal communication) has found similar results in compaction experiments with brachiopods; number of shells compacted was more significant in determining breakage than the presence of a drillhole.

Shells exposed on the sea floor experience forces of impact, drag, lift, and acceleration reaction (Hagstrom 1996). Hagstrom calculated that only impact forces would be strong enough to break drilled shells, even at extreme water velocities. She also calculated that high-energy, wave-swept environments should experience impact forces great enough to break drilled shells. The fact that taphonomic bias is not apparent in the Choptank and Caloosahatchee may indicate either that shells were rapidly buried and thus sequestered from impact forces or that the environments of deposition were relatively calm.

The Choptank shell beds have been interpreted as being condensed units formed primarily at depths between storm and fair-weather wave base (Kidwell 1989). Lewis et al. (2006) concluded that the Caloosahatchee was a parautochthonous deposit in which the amount of transportation outside the habitat was limited, but that shells were alternately exposed and buried in a process of repeated reworking. This conclusion is consistent with that of Rasmussen (1997), who found mixing of species representing a variety of habitats (brackish bay or lagoon, shallow euryhaline bay or lagoon, and shallow shelf) in the Plio-Pleistocene shell beds of southern Florida. Rasmussen inferred time averaging as the cause of such mixing; Allmon (1993) arrived at a similar conclusion for the Pliocene Pinecrest beds of Florida. In addition, the presence of some taphonomic alteration on many Choptank and Caloosahatchee shells indicates that shells were not immediately and / or permanently buried. Thus Choptank and Caloosahatchee shells were probably exposed to taphonomic alteration on the sea floor but not to forces that preferentially destroyed drilled shells.

An alternative explanation for the apparent lack of taphonomic bias in these samples is that predatory bioerosion does weaken shells significantly, but that most drilled shells were so severely fragmented that they did not survive to be observed in the present studies. Hagstrom (1996) found a high degree of shell fragmentation in high-energy environments, which would support this hypothesis. However, this explanation does not seem adequate for the discrepancy between the experimental compression studies and the results reported here for several reasons: the Choptank and Caloosahatchee environments were lower in energy than the high-energy environments examined by Hagstrom (1996); many Choptank and Caloosahatchee shells with drillholes had undergone taphonomic alteration; and the Choptank study directly accounted for the possibility of fragmentation of drilled shells by specifically examining fragments for predatory drillholes and assessing whether breakage occurred through drillholes. Assessment of breakage through drillholes could potentially be affected by two factors: extreme abrasion of fragments might have obliterated the evidence for partial drillholes, and partial drillholes are more difficult to recognize than intact drillholes. However, Choptank fragments did not exhibit the excessive wear required to obliterate partial drillholes, and all fragments were examined by at least three observers in order to ensure that partial drillholes were not overlooked.

Both Hagstrom (1996) and Roy et al. (1994) only included valves with umbonal drillholes in their studies of drilled valve strength; Hagstrom stated that drillholes that are not located near the umbo do not weaken shells because they do not act as local stress concentrators. Lack of taphonomic bias against drilled shells in the present study therefore could be related to non-umbonal placement of drillholes. This explanation may apply to some species (e.g., *Astarte*, which is usually drilled at the center of the valve, or corbulids, in which drillhole placement is wide ranging; Kelley 1988). However, within the Choptank, almost half of the drilled specimens belonged to *Dallarca*, a species displaying a high degree of stereotypy in placement of the drillhole at the umbo (Kelley 1988). Fragments of *Dallarca* that contained drillholes were in two cases broken through holes, but six *Dallarca* fragments contained intact drillholes. *Marvacrassatella* is also preferentially drilled at the umbo (Kelley 1988) and in this study four fragments were broken through the drillhole, but 10 fragments contained intact drillholes. It is possible that preferential breakage through umbonal holes could be an artifact of Hagstrom's methodology, assuming point-load force was applied close to the umbo, and that hole position is not important under natural taphonomic circumstances (Leighton, personal communication). However, even if Hagstrom (1996) is correct and only umbonal drillholes concentrate stress, the species most susceptible to breakage in this study, based on drillhole site, were not substantially affected by taphonomic bias.

Our results are similar to those reported by Kowalewski and Hoffmeister (2004). They tested whether drilling frequencies were independent of the taphonomic condition of shells for Holocene beach ridges of Baja California and Miocene mollusc assemblages of Europe. Based on a lack of statistical relationship between occurrence of drillholes and taphonomic condition of shells, Kowalewski and Hoffmeister concluded that drilling did not affect pre-burial survival of shells. These results are also consistent with those obtained by Nebelsick and Kowalewski (1999) for Recent Red Sea echinoids.

Results are also consistent with data reported by Vermeij (1982), who compared gastropod shell breakage produced by the crab *Calappa* with post mortem breakage. Vermeij considered breakage post mortem if it occurred in a drilled shell. Six Recent shallow-water soft-bottom samples from the tropical Pacific had at least 10 shells with drillholes, and of those, only one sample showed greater than 10% breakage of drilled shells. Three of six samples exhibited no post mortem breakage of drilled shells.

Despite the greater susceptibility to breakage of drilled shells compared to undrilled shells in point-load experiments, bias against specimens with predatory bioerosion was not significant for the fossil shell beds examined in this study. Nevertheless, studies of predatory bioerosion would benefit from testing for such bias by applying the criteria of Roy et al. (1994) and Nebelsick and Kowalewski (1999). In addition, drilling data will be most reliable for samples from lower-energy environments in which taphonomic damage due to impact forces is less likely. Results of this study indicate that time-averaged deposits need not suffer taphonomic bias against drilled shells; such deposits will continue to yield valuable information on the geologic history of predatory bioerosion.

Acknowledgments

I thank B. Hussaini, Senior Scientific Assistant in the Division of Paleontology at the American Museum of Natural History, for loan of the bulk sample of the Caloosahatchee Formation used in this study. The following students in the Fall 2005 Invertebrate Paleontology course at UNCW assisted in collecting taphonomic data for the Choptank Formation: Brianne Catlin, Kyle Godwin, Carolyn Gomes, Robert Johnson, Elizabeth Mitchell, Kelly Robertson, and Christopher Whitehead. The following students in the Spring 2006 Paleoecology course at UNCW curated the Caloosahatchee sample, identified species, and conducted preliminary paleoecological analysis: René Lewis, Patricia Mason, and Michelle McCoy. Timothy Kelley assisted with illustrations. Lindsey Leighton and Michael Savarese provided extremely helpful reviews of the manuscript.

References

Allmon WD (1993) Age, environment, and mode of deposition of the densely fossiliferous Pinecrest Sand (Pliocene of Florida): implications for the role of biological productivity in shell bed formation. Palaios 8:183-201

Anderson LC (1992) Naticid gastropod predation on corbulid bivalves: effects of physical factors, morphological features, and statistical artifacts. Palaios 7:602-620

Beatty WL (2003) The role of spinose ornament in predator deterrence and epibiont colonization: the bivalve *Arcinella*, Pinecrest (Pliocene) of Florida. PhD Thesis, Univ Pittsburgh, 90 pp

Bromley RG (1975) Comparative analysis of fossil and recent echinoid bioerosion. Palaeontology 18:725-739

Bromley RG (1981) Concepts in ichnotaxonomy illustrated by small round holes in shells. Acta Geol Hisp 16:55-64

Bromley RG (1993) Predation habits of octopus past and present and a new ichnospecies, *Oichnus ovalis*. Bull Geol Soc Denmark 40:167-173

Bromley RG (1994) The palaeoecology of bioerosion. In: Donovan SK (ed) The palaeobiology of trace fossils. Johns Hopkins Univ Press, Baltimore, pp 134-154

Bruguière JG (1789) Encyclopédie méthodique. Histoire naturelle des vers. Tome premier. Panckoucke, Paris, 757 pp

Carriker MR (1981) Shell penetration and feeding by naticacean and muricacean predatory gastropods: a synthesis. Malacologica 20:403-422

Carriker MR, Gruber GL (1999) Uniqueness of the gastropod accessory boring organ (ABO): comparative biology, an update. J Shellfish Res 18:579-595

Conrad TA (1830) On the geology and organic remains of a part of the Peninsula of Maryland. J Acad Nat Sci Philadelphia 6(2):205-230

Conrad TA (1832) Fossil shells of the Tertiary Formations of North America. Vol I. Judah Dobson, Philadelphia, 20 pp

Conrad TA (1833) On some new fossil and Recent shells of the United States. Amer J Sci Arts 23(17):339-346

Conrad TA (1834) Descriptions of new Tertiary fossils from the Southern States. J Acad Nat Sci Philadelphia 7(1):130-157

Conrad TA (1840) Fossils of the Medial Tertiary of the United States, No. 2. Judah Dobson Philadelphia, pp 33-56

Conrad TA (1841) Twenty-six new species of fossil shells, Medial Tertiary deposits of Calvert Cliffs, Maryland. Proc Acad Nat Sci Philadelphia 1(3):28-33

Conrad TA (1843) Descriptions of a new genus, and of twenty-nine new Miocene, and one new Eocene fossil shells of the United States. Proc Acad Nat Sci Philadelphia 1(3):305-311

Conrad TA (1867) Notes of fossil shells and descriptions of new species. Amer J Conchol 3(2):188-190

Conrad TA (1869) Descriptions of and reference to Miocene shells of the Atlantic Slope and descriptions of two new supposed Cretaceous species. Amer J Conchol 4(4):278-279

Dietl GP (2003a) Traces of naticid predation on the gryphaeid oyster *Pycnodonte dissimilaris*: epifaunal drilling of prey in the Paleocene. Hist Biol 16:13-19

Dietl GP (2003b) Coevolution of a marine gastropod predator and its dangerous bivalve prey. Biol J Linn Soc 80:409-436

Donovan SK, Jagt JWM (2002) *Oichnus* Bromley borings in the irregular echinoid *Hemipneustes* Agassiz from the Type Maastrichtian (Upper Cretaceous, The Netherlands and Belgium). Ichnos 9:67-74

Dudley EC, Vermeij GJ (1978) Predation in time and space: drilling in the gastropod *Turritella.* Paleobiology 4:436-441

Gernant RE (1970) Paleoecology of the Choptank Formation (Miocene) of Maryland and Virginia. Maryland Geol Surv, Rep Invest 12, 90 pp

Glenn LC (1904) Mollusca, Pelecypoda, in systematic paleontology, Miocene. Maryland Geol Surv, Miocene Vol, pp 274-401

Gmelin JF (1791) Caroli a Linné, systema naturae per regna tria naturae. Editio decima tertia. Beer, Lipsiae, Vol 1, Pars 6, pp 3021-3910

Grey M (2001) Predator-prey relationships of naticid gastropods and their bivalve prey. MSc Thesis, Univ Guelph, 56 pp

Grey M, Boulding EG, Brookfield ME (2006) Estimating multivariate selection gradients in the fossil record: a naticid gastropod case study. Paleobiology 32:100-108

Guerrero S, Reyment RA (1988) Predation and feeding in the naticid gastropod *Naticarius intricatoides* (Hidalgo). Palaeogeogr Palaeoclimatol Palaeoecol 68:49-52

Hagadorn JW, Boyajian GE (1997) Subtle changes in mature predator-prey systems: an example from Neogene *Turritella* (Gastropoda). Palaios 12:372-379

Hagstrom KM (1996) Effects of compaction and wave-induced forces on the preservation and macroevolutionary perception of naticid predator-prey interactions. MSc Thesis, Indiana Univ, 63 pp

Harper EM, Forsythe GTW, Palmer T (1998) Taphonomy and the Mesozoic marine revolution: preservation state masks the importance of boring predators. Palaios 13:352-360

Herbert GS, Dietl GP (2002) Tests of the escalation hypothesis: the role of multiple predators. Geol Soc Amer Abstr Prog 34(6):538-539

Kaplan P, Baumiller TK (2000) Taphonomic inferences on boring habit in the Richmondian *Onniella meeki* epibole. Palaios 15:499-510

Kelley PH (1988) Predation by Miocene gastropods of the Chesapeake Group: stereotyped and predictable. Palaios 3:436-448

Kelley PH (1989) Evolutionary trends within bivalve prey of Chesapeake Group naticid gastropods. Hist Biol 2:139-156

Kelley PH (1991) The effect of predation intensity on rate of evolution of five Miocene bivalves. Hist Biol 5:65-78

Kelley PH, Hansen TA (2001) The role of ecological interactions in the evolution of naticid gastropods and their molluscan prey. In: Allmon W, Bottjer D (eds) Evolutionary Paleoecology. Columbia Univ Press, New York, pp 149-170

Kelley PH, Hansen TA (2003) The fossil record of drilling predation on bivalves and gastropods. In: Kelley PH, Kowalewski M, Hansen TA (eds) Predator-prey interactions in the fossil record. Kluwer Acad / Plenum Press, New York, pp 113-139

Kelley PH, Hansen TA (2006) Comparisons of class- and lower taxon-level patterns in naticid gastropod predation, Cretaceous to Pleistocene of the U.S. Coastal Plain. Palaeogeogr Palaeoclimatol Palaeoecol 236:302-320

Kelley PH, Hansen TA, Graham SE, Huntoon AG (2001) Temporal patterns in the efficiency of naticid gastropod predators during the Cretaceous and Cenozoic of the United States Coastal Plain. Palaeogeogr Palaeoclimatol Palaeoecol 166:165-176

Kelley PH, Catlin B, Godwin K, Gomes C, Johnson R, Mitchell E, Robertson K, Whitehead C (2006) Effect of predatory drillholes on taphonomy of mollusc shells from the Miocene Choptank Formation of Maryland. Geol Soc Amer Abstr Prog 38(3):17

Kidwell SM (1989) Stratigraphic condensation of marine transgressive records: origin of major shell deposits in the Miocene of Maryland. J Geol 97:1-24

Kiene WE, Hutchings PA (1994) Long-term bioerosion of experimental coral substrates from Lizard Island, Great Barrier Reef. Proc 7th Int Coral Reef Symp 1:397- 403

Kowalewski M (2002) The fossil record of predation: An overview of analytical methods. In: Kowalewski M, Kelley PH (eds) The fossil record of predation. Paleont Soc Pap 8:3-42
Kowalewski M, Hoffmeister A (2004) Taphonomy and time-averaging of drill holes: assessing the quality of the fossil record left by hunting gastropods. Geol Soc Amer Abstr Prog 36(5):478
Kowalewski M, Dulai A, Fürsich FT (1998) A fossil record full of holes: The Phanerozoic history of drilling predation. Geology 26:1091-1094
Lewis RA, Kelley PH, McCoy ML, Mason PH (2006) Taphonomy of a molluscan assemblage from the Caloosahatchee Formation (Plio-Pleistocene) type area, La Belle, Florida. Geol Soc Amer Abstr Prog 38(7):63
Linnaeus C (1767) Systema naturae per regna tria naturae, secundum classes, ordines, genera, species, cum characteribus, differentiis, synonymis, locis. Tomus I, pars 2. Editio duodecima, reformata. Laurentii Salvii, Holmiae, pp 533-1327
McCoy ML, Kelley PH, Hansen TA, Lewis RA, Mason PH (2006) Drilling predation in the Caloosahatchee Formation (Plio-Pleistocene of Florida): Test of a latitudinal predation gradient. Geol Soc Amer Abstr Prog 38(7):66
Miller SA (1875) Monograph of the class Brachiopoda of the Cincinnati group. Cincinnati Quart J Sci 1875:6-62
Missimer TM (2001a) Late Neogene geology of northwestern Lee County, Florida. In: Missmer TM, Scott TM (eds) Geology and hydrology of Lee County, Florida. Florida Geol Surv Spec Publ 49:21-34
Missimer TM (2001b) Late Paleogene and Neogene chronostratigraphy of Lee County, Florida. In: Missmer TM, Scott TM (eds) Geology and hydrology of Lee County, Florida. Florida Geol Surv Spec Publ 49:67-90
Morton B, Chan K (1997) The first report of shell boring predation by a member of the Nassariidae (Gastropoda). J Mollusc Stud 63:476-478
Nebelsick JH, Kowalewski M (1999) Drilling predation on Recent clypeasteroid echinoids from the Red Sea. Palaios 14:127-144
Nielsen KSS, Nielsen JK (2001) Bioerosion in Pliocene to late Holocene tests of benthic and planktonic foraminiferans, with a revision of the ichnogenera *Oichnus* and *Tremichnus*. Ichnos 8:99-116
Olsson AA (1993) Papers on Neogene mollusks. Paleont Res Instit Spec Publ 19, 163 pp
Olsson AA, Harbison (1953) Pliocene Mollusca of southern Florida with specific reference to those from North Saint Petersburg. Acad Nat Sci Philadelphia Monog 8:1-304
Olsson AA, Petit RE (1964) Some Neogene Mollusca from Florida and the Carolinas. Bull Amer Paleont 47(217):509-575
Orbigny A de (1852) Prodrome de paléontologie stratigraphique universelle des animaux mollusques et rayonnés faisant suite au cours elémentaire de paléontologie et de géologie stratigraphiques. Masson, Paris, vol 3, 196 pp and 190 pp table alphabétique
Perkins RD (1968) Second annual field trip of the Miami Geological Society. Miami Geol Soc Publ, 110 pp
Petuch EJ (1991) New gastropods from the Plio-Pleistocene of southwestern Florida and the Everglades Basin. WH Dall Paleont Res Cent Spec Pub 1, 59 pp
Petuch EJ (1994) Atlas of Florida fossil shells (Pliocene and Pleistocene marine gastropods). Spectrum Press, Chicago, 394 pp
Ponder WF, Taylor JD (1992) Predatory shell drilling by two species of *Austroginella* (Gastropoda: Marginellidae). J Zool London 228:317-328
Rasmussen DL (1997) Biostratigraphic analysis of southern Florida's Plio-Pleistocene shell beds. MSc Thesis, Old Dominion Univ, 154 pp

Roopnarine PD, Beussink A (1999) Extinction and naticid predation of the bivalve *Chione* Von Mühlfeld in the late Neogene of Florida. Palaeont Electr 2(1), 24 pp [http://palaeo-electronica.org/1999_1/bivalve/issue1_99.htm]

Roopnarine PD, Vermeij GJ (2000) One species becomes two: The case of *Chione cancellata*, the resurrected *C. elevata*, and a phylogenetic analysis of *Chione*. J Mollusc Stud 66:517-534

Roy K, Miller DK, LaBarbera M (1994) Taphonomic bias in analyses of drilling predation: effects of gastropod drill holes on bivalve shell strength. Palaios 9:413-421

Savazzi E, Reyment RA (1989) Subaerial hunting behaviour in *Natica gualteriana* (naticid gastropod). Palaeogeogr Palaeoclimatol Palaeoecol 74:355-364

Say T (1822) An account of some of the marine shells of the United States. J Acad Nat Sci Philadelphia 2(2):221-248

Say T (1824) An account of some of the fossil shells of Maryland. J Acad Nat Sci Philadelphia 4(1):124-155

Shattuck GB (1904) Geological and paleontological relations, with a review of earlier investigations. In: Clark WB (ed) Miocene. Maryland Geol Surv, pp 33-94

Solander D (1786) A catalogue of the Portland Museum, lately the property of the Duchess Dowager of Portland, deceased: Which will be sold by auction by Mr. Skinner and Co. On Monday the 24th of April, 1786, and the thirty-seven following days (...) at her late dwelling-house, in Privy-Garden, Whitehall, by order of the Acting Executrix. Skinner, London, 194 pp

Taylor PD, Wilson MA (2003) Palaeoecology and evolution of marine hard substrate communities. Earth-Sci Rev 62:1-103

Vermeij GJ (1982) Gastropod shell form, breakage, and repair in relation to predation by the crab *Calappa*. Malacologia 23:1-12

Vokes HE (1957) Miocene fossils of Maryland. Maryland Geol Surv Bull 20, 85 pp

Voigt E (1977) On grazing traces produced by the radula of fossil and Recent gastropods and chitons. Geol J Spec Issue 9:335-346

Ward LW, Blackwelder BW (1975) *Chesapecten*, a new genus of Pectinidae (Mollusca: Bivalvia) from the Miocene and Pliocene of eastern North America. US Geol Surv Prof Pap 861, 24 pp

Ziegelmeier E (1954) Beobachtungen über den Nahrungserwerb bei der Naticide *Lunatia nitida* Donovan (Gastropoda Prosobranchia). Helgoländer Wiss Meeresunters 5:1-33

Zuschin M, Stanton RJ (2001) Experimental measurement of shell strength and its taphonomic interpretation. Palaios 16:161-170

V

Bibliography

An online bibliography of bioerosion references

Mark A. Wilson[1]

[1] The College of Wooster, Department of Geology, Wooster, OH 44691, USA, (mwilson@wooster.edu)

Abstract. A bibliography of bioerosion-related books and papers has been available on the Internet for several years. The list now includes almost 2200 references. This bibliography is freely available to the public and frequently edited, so it grows in size, coverage, and accuracy. The bibliography shows the diversity of interests in the bioerosion community. References range from the mostly geological to the botanical; from articles concerned with macrobioerosion by bivalves, worms, sponges, and barnacles to numerous papers on microbioerosion by bacteria and fungi. The online nature of this resource means it will grow with the needs of the bioerosion research community.

Keywords. Bioerosion, bibliography, online, microbioerosion, macrobioerosion

Introduction

Since 1990 I have maintained a bibliography of references related to bioerosion. It began as a list of articles which included carbonate hardgrounds and their faunas, both encrusting and boring, which was later published in Wilson and Palmer (1992). It was annually updated and circulated within the small community of hardground researchers as an unpublished booklet. In 2000, Paul Taylor and I began work on a review of all marine hard substrate communities, both ancient and modern. Together we accumulated hundreds of references which covered sclerobionts (organisms which live on hard substrates). Upon the publication of this paper (Taylor and Wilson 2003), I separated the bioerosion-related articles from the hardground papers and placed both lists on the Internet at this website:

http://www.wooster.edu/geology/bioerosion/bioerosion.html

The bioerosion bibliography is online as a pdf:

http://www.wooster.edu/geology/bioerosion/bioerosionbiblio.pdf

It is also online as a Microsoft Word file:

http://www.wooster.edu/geology/bioerosion/bioerosionbiblio.doc

M. Wisshak, L. Tapanila (eds.), *Current Developments in Bioerosion*. Erlangen Earth Conference Series, doi: 10.1007/978-3-540-77598-0_24,

As of the time of this writing, the bibliography has 2198 references, and it has been growing by about a dozen new papers a month. This bibliography has been downloaded an average of 130 times each month over the past three years. I also send free copies of the bibliography on a compact disk to all who request it.

Construction of the bioerosion bibliography

This bibliography is a product of the growing community of paleontologists and biologists interested in bioerosion. Since it is an online resource which is frequently updated, anyone can send the compiler additions and corrections to the reference list. Dozens of people have done so, with the most prominent contributors listed in the acknowledgments for the bibliography and in this paper. In a sense, then, this bibliography is like a 'Wikipedia' for bioerosion specialists where all participants can see the full information resource and help develop its content.

The bibliography is updated about once a month, with the date of the newest version always posted in the header. New articles are added from Internet searches, professional databases (GeoRef and Biological Abstracts especially), and contributions from users. To be included they must make a significant contribution to our knowledge of marine bioerosion processes, the distribution in space and time of bioerosion, or the systematics of bioeroding organisms. Often one or more of the above topics will be the primary purpose of the paper, but commonly the bioerosion portion of the article will be relatively small yet important to bioerosion experts. The selected articles also must be accessible to the community. Unpublished manuscripts and dissertations are not included, nor are abstracts or sections in locally-produced guidebooks. Exceptions are made, though, if there are no other sources covering the particular topic.

Accuracy of the citations is a concern. I have seen my own mistakes in the reference list passed into published articles. These errors are usually small but are surprisingly common. With the increasing speed and efficiency of Internet searches, we can quickly compile a list of how a reference has been cited in dozens of different sources. When we do this we often see an accumulation of scribal errors, some of which make the original reference impossible to find. I work on reducing errors in this bibliography by collecting actual paper reprints when I can, checking questionable citations with the original authors where possible, and continually asking bioerosion experts for corrections. The most common error I have observed among bioerosion citations is a misspelling of the journal's title, especially if it is not in English and has been listed with more than one abbreviation. If I could I would require all journals to list the references in their articles by the full name of the journal. This would greatly limit this pervasive bibliographic error. I have reproduced the full journal titles in the bibliography and encourage future contributors to do the same.

I have attempted to make this bibliography 'comprehensive', but of course it can never be. Every topic covered by these articles can be probed in more detail, revealing additional secondary and tertiary references. This is especially true with the biological articles. I have endeavored to include the most important modern bioerosion references; each one, though, opens doors into further exploration of systematics, evolution and biochemistry of the bioeroders. A bibliography of ten thousand or more references would be so large as to be unwieldy, and so there is a limit to how many literary rabbit holes we can safely go down.

Review of other bioerosion-related bibliographies

The most comprehensive bibliography of an important aspect of bioerosion is that by Clapp and Kenk (1963) on marine borers. It is over 1100 pages long and thoroughly annotated. It was produced for the United States Navy and as such is directed toward practical concerns such as the maintenance of wooden wharf pilings, limestone coastal defenses, and the hulls of wooden boats. There is considerable overlap with the online bioerosion bibliography, but many of the papers in Clapp and Kenk (1963) are not included because they are engineering reports on the stability of the substrates rather than the process of boring itself. Radtke et al. (1997) produced an extensive bibliography of bioerosion, integrating macrobioerosion and microbioerosion. This is still a valuable bibliography, but its contents are now almost entirely subsumed in this online version. The best general trace fossil bibliography is the references cited list in Bromley (1996). While this book contains many citations of papers on borings as trace fossils (all of which are in this online bibliography), it was not intended to include a thorough examination of bioerosion. The most recent extensive bioerosion-related bibliographies are the long references cited sections in Taylor and Wilson (2003) and Bromley (2004). Almost all of these references are now in the online bioerosion bibliography.

The bibliography as a reflection of the bioerosion community

I began this bibliography as part of my own research program in bioerosion and other aspects of sclerobiology. It was thus centered on macroborings and the fossil record. I relied upon the geological literature for most of the references. Later my colleagues impressed upon me both the importance of modern bioerosion and the growing number of papers covering microborings. With the help of several German bioerosion experts, I have added hundreds of articles on microbioerosion to the list. They are now more than one-third of the bioerosion references. The result is an unusual mix of journal types in the list from the strictly geological (such as, well, Geology) to the botanical (The Canadian Journal of Botany is an example) and zoological (Zoologica Scripta comes to mind). This shows that the study of bioerosion in its entirety is highly interdisciplinary, more so than would be expected from an apparently narrow focus on a biological process.

Statistics gathered from the bioerosion bibliography

As of May 2007 there are 2198 references in the bibliography. The earliest is a 1733 book by Gottfried Sellius who showed the molluscan affinities of the 'shipworm'

Teredo, which at the time was devastating the wooden dikes holding back the sea in Holland. The distribution of citations through time is predictable. There are four references from the 18th century, 55 from the 19th, 1827 from the 20th, and thus far 302 from the 21st century. The number of references through the 20th century by decades is shown in Figure 1. Except for a dip in the war years of the 1940s, there has been a steadily increasing number of references through the century. About half of all the bioerosion papers in the bibliography come from the 1980s and 1990s.

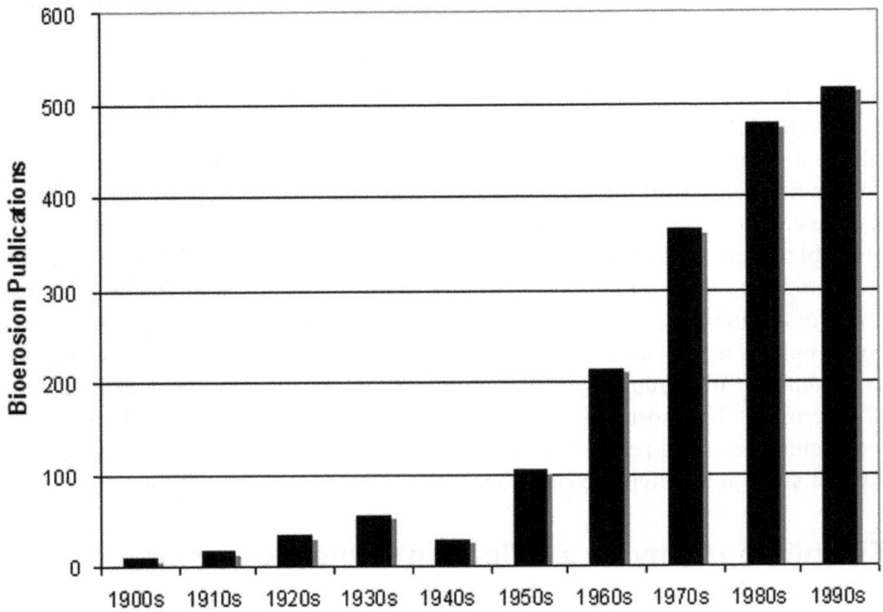

Fig. 1 Bioerosion-related papers and books published during each decade of the 20th century

The most prolific author on the list is Stjepko (Steve) Golubic of Boston University. He has 81 articles in the bibliography, most of them on microbioerosion. The second most common author in the bibliography is Richard Bromley of the University of Copenhagen who has 58 articles here primarily covering macrobioerosion. The most appropriately named author must be J.E. Barnacle of Australia's Commonwealth Scientific and Industrial Research Organisation (CSIRO)!

Dozens of journals host these bioerosion papers. The most common journals in the bibliography are the Journal of Paleontology (67 articles), Palaios (60), Lethaia (53), Palaeogeography Palaeoclimatology Palaeoecology (52), Ichnos (39), and Palaeontology (38). These journals are all from the geological community. This appears to show that the geologists in our community tend to publish in relatively few journals and the biologists in many more.

The advantages of an online bibliography

The primary reason of an online bibliography is valuable is that it can be continually corrected and updated. The bioerosion bibliography is prominent on the web within our discipline – type 'bioerosion' into the Google or Yahoo search engines and it is the second link to appear, preceded only by the Wikipedia page on bioerosion (which has a link to the bibliography). Authors are not shy about telling me when I have made an error with one of their citations, and some students have taken it as a challenge to find bioerosion references elsewhere which are not on the list. The Microsoft Word document version of the bibliography can be quickly downloaded and the references can be electronically searched for words in the article titles, the journals, or the dates. These references can then be cut out and pasted into other documents.

The disadvantages of an online bibliography

The Internet does not yet have the permanence, or at least the illusion of permanence, we have with books and journals. The words placed on paper will be outdated soon, and they cannot be easily corrected, but at least history shows they have a chance of surviving indefinitely. A power failure at The College of Wooster will shut down the server, making the bibliography temporarily inaccessible. A cyber-attack on our server could erase the bibliography from its memory. I keep numerous back-up copies on other computers and on compact disks for these eventualities, but I may not be around to reload the bibliography. Right now I would have to count on a colleague uploading a copy of the documents on another server if I could not.

The future of the online bioerosion bibliography

To increase the longevity of the online bioerosion bibliography, we will soon have at least one mirror website to serve as a real-time back-up. To increase the usefulness of the bibliography as a database of bioerosion research, I hope to soon convert the text into data fields for a bibliographic program such as EndNote. I also hope to add keywords to the references so that we can search on text-strings not necessarily in the article titles. With the continual editing and help of my colleagues, the bioerosion bibliography will also become more accurate and comprehensive with time.

Acknowledgements

Karl Kleemann, Leif Tapanila, Leslie Eliuk, Tim Palmer, Paul Taylor, Max Wisshak, Gudrun Radtke and Elizabeth Heise provided significant contributions to this bibliography; many other paleontologists and biologists sent me numerous additions and corrections. Jeff Bowen and Amy Wilson did considerable organization of the references and their associated reprints. Grateful acknowledgement is made to the Donors of the American Chemical Society Petroleum Research Fund for support of this work over the past five years.

References

Bromley RG (1996) Trace fossils: biology, taphonomy and applications. Chapman and Hall, London, 361 pp

Bromley RG (2004) A stratigraphy of marine bioerosion. In: McIlroy D (ed) The application of ichnology to palaeoenvironmental and stratigraphic analysis. Geol Soc London Spec Publ 228:455-481

Clapp WF, Kenk R (1963) Marine borers: An annotated bibliography. Office Naval Res, Dept Navy, Washington DC, 1136 pp

Radtke G, Hofmann K, Golubic S (1997) A bibliographic overview of micro- and macroscopic bioerosion. Courier Forschinst Senckenberg 201:307-340

Sellius G (1733) Historia naturalis teredinis seu xylophagi marini, tubulo-conchoidis speciatim Belgici cum tabulis ad vivum delineatis. Bessling, Utrecht, 353 pp

Taylor PD, Wilson MA (2003) Palaeoecology and evolution of marine hard substrate communities. Earth-Sci Rev 62:1-103

Wilson MA, Palmer TJ (1992) Hardgrounds and hardground faunas. Univ Wales, Aberystwyth, Inst Earth Stud Publ 9:1-131

Index

A

C

H

I

M

P

Q

R

U

V

W

X

Y

Z